T0298673

Water and Wastewater Engineering Technology

Water and Wastewater Engineering Technology presents the basic concepts and applications of water and wastewater engineering technology. It is primarily designed for students pursuing programs in civil, water resources, and environmental engineering, and presents the fundamentals of water and wastewater technology, hydraulics, chemistry, and biology. The book examines the urban water cycle in two main categories, water treatment and distribution, and wastewater collection and treatment. The material lays the foundation for typical one-semester courses in water engineering and also serves as a valuable resource to professionals operating and managing water and wastewater treatment plants. The chapters in this book are standalone, offering the flexibility to choose combinations of topics to suit the requirements of a given course or professional application.

Features:

- Contains example problems and diagrams throughout to illustrate and clarify important topics.
- Problems both in SI and USC system of units.
- The procedure of unit cancellation followed in all solutions to the problems.
- Design applications and operation of water and wastewater system emphasized.
- Includes numerous practice problems with answers, and discussion questions in each chapter cover a range of engineering interventions to help conserve water resources and preserve water quality.

Water and Wastewater Engineering Technology

Subhash Verma

CRC Press
Taylor & Francis Group
Boca Raton London New York

CRC Press is an imprint of the
Taylor & Francis Group, an **informa** business

Designed cover image: Shutterstock

First edition published 2024
by CRC Press
6000 Broken Sound Parkway NW, Suite 300, Boca Raton, FL 33487-2742

and by CRC Press
4 Park Square, Milton Park, Abingdon, Oxon, OX14 4RN

CRC Press is an imprint of Taylor & Francis Group, LLC

Library of Congress Cataloging-in-Publication Data

Names: Verma, Subhash (Professor), author.
Title: Water and wastewater engineering technology / Subhash Verma.
Description: First edition. | Boca Raton : CRC Press, 2024. | Includes index.
Identifiers: LCCN 2022060669 (print) | LCCN 2022060670 (ebook) | ISBN 9781032390055 (hardback)
| ISBN 9781032390062 (paperback) | ISBN 9781003347941 (ebook)
Subjects: LCSH: Water-supply engineering. | Sewage disposal. |
Environmental engineering.
Classification: LCC TD345 .V47 2024 (print) | LCC TD345 (ebook) | DDC 628.1--dc23/eng/20230111
LC record available at https://lccn.loc.gov/2022060669
LC ebook record available at https://lccn.loc.gov/2022060670

ISBN: 978-1-032-39005-5 (hbk)
ISBN: 978-1-032-39006-2 (pbk)
ISBN: 978-1-003-34794-1 (ebk)

DOI: 10.1201/9781003347941

Typeset in Times
by Deanta Global Publishing Services, Chennai, India

Contents

List of Figures xxv
List of Tables xxix
About the Author xxxiii

SECTION I Basic Sciences

Chapter 1   Introduction 3

1.1   Historical Perspective 3
    1.1.1   Water Supply 3
    1.1.2   History of Sanitary Engineering 4
1.2   Hydrologic Cycle 4
1.3   Urban Water Cycle 6
1.4   Essentials of a Water Supply System 7
1.5   Need for Wastewater Treatment System 8
1.6   Global Issue 9
1.7   Role of the Engineer 9

Chapter 2   Standards of Measurement 11

2.1   Systems of Units 11
2.2   Measures of Dimension 11
2.3   Dimensions and Units 12
2.4   Derived Units 12
    2.4.1   Force 13
    2.4.2   Mass and Weight 13
    2.4.3   Pressure 14
    2.4.4   Energy and Power 14
2.5   Symbols and Suffixes 14
2.6   Significant Figures 15
2.7   Numerical Precision 16
    2.7.1   Absolute Precision 16
    2.7.2   Relative Precision 16
2.8   Conversions 17
    2.8.1   Temperature Conversions 17
    2.8.2   Steps for Unit Cancellation 20

Chapter 3   Basic Hydraulics 23

       3.1   Flow Velocity...23
       3.2   Continuity Equation 24
       3.3   Energy and Head 25
             3.3.1   Flow Energy 25
             3.3.2   Kinetic Energy 25
             3.3.3   Gravitational Potential Energy 26
             3.3.4   Hydraulic Head...26
             3.3.5   Total Head 27
       3.4   Bernoulli's Equation..28
             3.4.1   Limitations of Bernoulli's Equation...........28
             3.4.2   Static Flow Conditions 29
       3.5   General Energy Equation 30
       3.6   Power 31
       3.7   Flow Equations..33
             3.7.1   Darcy-Weisbach Flow Equation 33
             3.7.2   Hazen-Williams Flow Equation 36
             3.7.3   Manning's Flow Equation 38

Chapter 4   Basic Chemistry 43

       4.1   States of Matter 43
             4.1.1   Structure of Atom.......................................43
             4.1.2   Periodic Table..44
       4.2   Compounds...45
       4.3   Acids and Bases...47
             4.3.1   Neutralization...47
             4.3.2   The pH Scale and Alkalinity.......................47
       4.4   Solutions 49
       4.5   Expressing Concentrations 50
             4.5.1   Mass per Unit Volume, $C_{m/V}$ 50
             4.5.2   Mass per Unit Mass, $C_{m/m}$ 50
       4.6   Stoichiometry 51
       4.7   Chemical Feeding...52
       4.8   Common Chemicals 55
             4.8.1   Activated Carbon.......................................55
             4.8.2   Alum (Aluminium Sulphate).......................56
             4.8.3   Chlorine (CI) 56
             4.8.4   Lime and Soda Ash 57

Chapter 5   Microbial Water Quality...61

       5.1   Basics of Microbiology...61
             5.1.1   Bacteria..61

     5.1.2   Algae 62

     5.1.3   Fungi 62

     5.1.4   Protozoa 63

     5.1.5   Viruses 63

5.2   Microbiological Contaminants 63

5.3   Microbiological Tests 64

     5.3.1   Indicator Organisms 64

     5.3.2   Membrane Filtration Method 65

     5.3.3   Multiple-Tube Fermentation Method 65

     5.3.4   Sample Collection for Microbiological Testing 66

5.4   Biochemical Oxygen Demand 67

5.5   Nitrogen (N) 69

5.6   Solids 70

5.7   Hazardous Contaminants 70

5.8   Sampling 70

     5.8.1   Grab Samples 71

     5.8.2   Composite Samples 71

SECTION II   Water Treatment

Chapter 6   **Sources of Water Supply** 77

6.1   Surface Water 77

     6.1.1   Lakes and Ponds 77

     6.1.2   Rivers and Streams 77

     6.1.3   Artificial Reservoirs 78

     6.1.4   Seawater 78

     6.1.5   Wastewater Reclamation 78

     6.1.6   Stored Rainwater 79

     6.1.7   Yield Assessment 79

6.2   Intake Works 79

     6.2.1   Reservoir Intakes 79

     6.2.2   Twin Tower River intake 80

     6.2.3   Single Well Type River Intake 80

     6.2.4   Lake Intake 81

6.3   Water Transmission 81

6.4   Groundwater 81

     6.4.1   Water Wells 81

     6.4.2   Springs 82

     6.4.3   Infiltration Galleries 82

     6.4.4   Collector Wells 82

6.5   Well Types 83

6.6   Well Hydraulics 84

6.6.1   Steady Flow to an Artesian Well 84
6.6.2   Unconfined Well Equation 87
6.6.3   Modified Non-Equilibrium Equation 89
6.7   Failure of Wells and Remediation 90
6.8   Sanitary Protection 90
6.9   Well Abandonment 90
6.10 Water Quantity 91
6.11 Water Quality 91
6.12 Groundwater under the Influence (GUDI) 92
6.13 Choice of Source of Water Supply 93

Chapter 7   Water Demand and Water Quality 97
7.1   Design Period 97
7.2   Forecasting Population 97
       7.2.1   Arithmetical Increase 98
       7.2.2   Geometrical Increase Method 98
       7.2.3   Incremental Increase Method 99
7.3   Estimating Water Demand 100
       7.3.1   Domestic 100
       7.3.2   Industrial and Commercial 101
       7.3.3   Public Use 101
       7.3.4   Firefighting 101
7.4   Total Demand 104
7.5   Factors Affecting per Capita Demand 104
7.6   Variation in Demand 106
       7.6.1   Seasonal Variation 106
       7.6.2   Daily Variation 106
       7.6.3   Hourly Variation 107
7.7   Water Quality Standards 108
7.8   Water Quality Parameters 109
7.9   Physical Parameters 109
       7.9.1   Turbidity 109
       7.9.2   Color 110
       7.9.3   Temperature 110
       7.9.4   Taste and Odor 110
       7.9.5   Solids 111
       7.9.6   Water Density 111
       7.9.7   Viscosity 112
7.10 Chemical Parameters 113
       7.10.1 Hydrogen Ion Concentration (pH) 113
       7.10.2 Alkalinity 113
       7.10.3 Hardness 113
       7.10.4 Iron and Manganese 114
       7.10.5 Fluorides 114

7.10.6 Nitrogen 115
7.10.7 Dissolved Gases 115

Chapter 8 Coagulation and Flocculation 119

8.1 Source of Supply and Treatment 119
8.2 Preliminary Treatment 120
8.3 Conventional Treatment 121
8.4 Coagulation 121
8.5 Coagulating Chemicals 122
8.5.1 Primary coagulants 122
8.5.2 Coagulant Aids 123
8.6 Chemistry of Coagulation 124
8.6.1 Chemical Reactions 124
8.6.2 Alum Floc (Sludge) 125
8.7 Flocculation Phenomenon 127
8.7.1 Mixers 127
8.7.2 Flocculation Tanks 128
8.7.3 Factors Affecting Flocculation 130
8.8 Jar Testing 131
8.9 Operational Control Tests 132
8.9.1 Acidity Tests 132
8.9.2 Turbidity Tests 132
8.9.3 Filterability Tests 133
8.9.4 Zeta Potential 133
8.9.5 Streaming Current Monitors 133
8.9.6 Particle Counters 133

Chapter 9 Sedimentation 137

9.1 Theory of Sedimentation 137
9.1.1 Plain Sedimentation 137
9.1.2 Discrete Settling 138
9.2 Sedimentation Aided with Coagulation 141
9.3 Sedimentation Basins and Tanks 141
9.3.1 Rectangular Basins 141
9.3.2 Circular and Square Basins 141
9.3.3 Tube Settlers 142
9.3.4 Solids Contact Units 142
9.3.5 Pulsator Clarifier 143
9.3.6 Ballasted Flocculation 143
9.4 Design Parameters 143
9.4.1 Detention Time 143
9.4.2 Surface Overflow Rate 144
9.4.3 Overflow Rate and Removal Efficiency 144

9.4.4   Effective Water Depth 144
9.4.5   Mean Flow Velocity 145
9.4.6   Weir Loading Rate 146
9.5   Factors Affecting Operation of Sedimentation 146
9.6   Volume of Sludge 147
9.7   Sludge Disposal 149

Chapter 10   Filtration 153

10.1   Filtration Mechanisms 153
10.2   Types of Filters 153
10.2.1   Slow Sand Filters (SSFs) 153
10.2.2   Rapid Gravity Filters (RGFs) 154
10.2.3   High-Rate Filters 155
10.2.4   Pressure Filters 155
10.3   Components of a Gravity Filter 155
10.3.1   Filter Box 156
10.3.2   Filter Media 156
10.3.3   Underdrain System 157
10.3.4   Surface Wash System 158
10.3.5   Wash-Water Troughs 158
10.3.6   Control Equipment 158
10.4   Filtration Operation 158
10.4.1   Filtering 158
10.4.2   Declining Rate Control 159
10.4.3   Split Flow Control 159
10.4.4   Backwashing 159
10.4.5   Backwash Operation 160
10.4.6   Filtering to Waste 161
10.4.7   Backwashing Key Points 161
10.5   Design and Performance Parameters 161
10.5.1   Filtration Rate 161
10.5.2   Unit Filter Run Volume 163
10.5.3   Flow Rate and Volume of Water Filtered 164
10.5.4   Backwash Rate 164
10.6   Operating Problems 165
10.7   Optimum Filter Operation 166

Chapter 11   Disinfection 169

11.1   Definition 169
11.1.1   Primary Disinfection 169
11.1.2   Secondary Disinfection 169
11.2   Disinfection Methods 170

	11.2.1	Removal Processes	170
	11.2.2	Inactivation Processes	170
11.3	Chlorine Compounds		172
	11.3.1	Gas Chlorination	172
	11.3.2	Chlorine Safety	173
11.4	Hypochlorination		173
	11.4.1	Calcium Hypochlorite	173
	11.4.2	Sodium Hypochlorite	173
	11.4.3	Chlorine Dioxide Disinfection	174
11.5	Chemistry of Chlorination		174
11.6	Chlorine Practices		178
	11.6.1	Chloramination	178
	11.6.2	Breakpoint Chlorination	179
	11.6.3	Superchlorination	179
	11.6.4	Dechlorination	179
11.7	Points of Chlorination		179
	11.7.1	Pre-Chlorination	179
	11.7.2	Post-Chlorination	180
	11.7.3	Re-Chlorination	180
11.8	Formation of Trihalomethanes		180
11.9	Factors Affecting Chlorine Dosage		180
11.10	Gas Chlorination Equipment		184
11.11	Chlorine Feed Control		185
	11.11.1	Manual Control	186
	11.11.2	Automatic Proportional Control	186
	11.11.3	Automatic Residual Control	186
	11.11.4	Hypochlorination Facilities	186
	11.11.5	Hypochlorinators	186

Chapter 12   Water Softening 191

12.1	Types of Hardness		191
	12.1.1	Carbonate Hardness	191
	12.1.2	Non-Carbonate Hardness	191
12.2	Softening Methods		193
	12.2.1	Lime-Soda Ash Softening	193
	12.2.2	Chemical Dosages	193
12.3	Types of Lime-Soda Ash Processes		195
	12.3.1	Selective Calcium Removal	195
	12.3.2	Excess Lime Treatment	196
	12.3.3	Split Treatment	197
12.4	Ion-Exchange Softening		198
	12.4.1	Removal Capacity	198
	12.4.2	Water Treatment Capacity	198

Chapter 13   Miscellaneous Methods I...205

    13.1   Fluoridation 205
        13.1.1   Fluoride Chemicals 205
        13.1.2   Fluoridation Systems.....................................206
    13.2   Defluoridation..208
        13.2.1   Calcium Phosphate 208
        13.2.2   Tri-Calcium Phosphate..................................209
        13.2.3   Ion Exchange 209
        13.2.4   Lime 209
        13.2.5   Aluminum Compounds 209
        13.2.6   Activated Carbon...209
    13.3   Iron and Manganese Control 209
    13.4   Control Methods..210
        13.4.1   Phosphate Treatment 210
        13.4.2   Feed System 211
    13.5   Removal Methods..211
        13.5.1   Oxidation by Aeration....................................211
        13.5.2   Oxidation with Chlorine.................................211
        13.5.3   Oxidation with Permanganate 212
        13.5.4   Ion Exchange with Zeolites............................212
    13.6   Arsenic Removal 213
    13.7   Nitrate Removal...214

Chapter 14   Miscellaneous Methods II 217

    14.1   Taste and Odor Control 217
        14.1.1   Organics in Raw Water 217
        14.1.2   Chemical Dosing..218
    14.2   Taste and Odor Removal 218
        14.2.1   Oxidation...218
        14.2.2   Aeration 218
        14.2.3   Chemical Oxidation 219
        14.2.4   Adsorption...220
        14.2.5   Forms of Activated Carbon 220
    14.3   Membrane Filtration..221
        14.3.1   Microfiltration and Ultrafiltration 221
        14.3.2   Nanofiltration and Reverse Osmosis 222
    14.4   Desalination..222
        14.4.1   Membrane Technology....................................222
        14.4.2   Distillation of Seawater 223
    14.5   Water Stabilization 223
        14.5.1   Classifying Water Stability 223
        14.5.2   Chemistry of Corrosion..................................223
        14.5.3   Stability Index 224
        14.5.4   Corrosion Control...226

SECTION III   Water Distribution

Chapter 15   Water Distribution 231

    15.1 System Components 231
    15.2 Methods of Water Distribution 232
          15.2.1 Gravitational System 232
          15.2.2 Pumping System 232
          15.2.3 Combined Gravity and Pumping System 233
    15.3 Equalizing Demand 233
          15.3.1 Equalizing Storage Capacity 233
          15.3.2 Other Purposes of Storage 235
          15.3.3 Types of Storage 235
    15.4 Pipeline Layout 237
          15.4.1 Dead End Systems 237
          15.4.2 Gridiron System 238
          15.4.3 Ring System 238
          15.4.4 Radial system 239
    15.5 Pipe Material 239
          15.5.1 Plastic Pipes 239
          15.5.2 Cast Iron Pipes 240
          15.5.3 Ductile Iron Pipes 240
          15.5.4 Steel Pipes 240
          15.5.5 Cement Concrete and RCC Pipes 240
          15.5.6 Asbestos Cement Pipes 241
    15.6 Pipe Joints 241
          15.6.1 Flanged Joint 241
          15.6.2 Socket and Spigot Joint 241
          15.6.3 Flexible Joint 242
          15.6.4 Mechanical Joint 243
          15.6.5 Expansion Joint 243
          15.6.6 Simplex Joint 243
    15.7 Pipelaying and Testing 244
          15.7.1 Anchoring of Pipes 244
          15.7.2 Backfilling with Earth 245
          15.7.3 Testing of Pipes 245
          15.7.4 Flow Velocity 245
    15.8 Valves 246
          15.8.1 Gate Valve 246
          15.8.2 Globe Valve 246
          15.8.3 Air and Vacuum Relief Valves 246
          15.8.4 Rotary Valves 247
          15.8.5 Special Function Valves 247
          15.8.6 Exercising of Valves 248

15.9   Cross Contamination 248
        15.9.1   Back Pressure 248
        15.9.2   Back Siphoning 249
        15.9.3   Backflow Prevention...249
15.10   Hydrants 250
15.11   Service Connections 250
15.12   Water Meters 250
15.13   Thrust Control 251
15.14   Dual Water Systems 251

Chapter 16   Pipeline Systems..253

16.1   Series and Parallel Systems 253
16.2   Equivalent Pipe...253
16.3   System Classification..255
        16.3.1   Class I Systems..255
        16.3.2   Class II Systems 256
        16.3.3   Class III Systems...260
16.4   Complex Pipe Networks 260
16.5   Hardy Cross Method 261
16.6   Computer Applications..263

Chapter 17   Pumps and Pumping 267

17.1   Positive Displacement Pumps...267
17.2   Velocity Pumps..267
        17.2.1   Types of centrifugal pumps....................................267
        17.2.2   Positive Displacement Pump Characteristics...........268
        17.2.3   Performance Curves of Centrifugal Pumps.............268
17.3   System Head 272
17.4   Affinity Laws...274
17.5   Specific Speed 275
17.6   Homologous Pumps...275
17.7   Multiple Pumps 276
17.8   Cavitation..277
        17.8.1   Net Positive Suction Head 278
        17.8.2   Permissible Suction Lift 279
17.9   Operation and Maintenance 279

Chapter 18   Water Distribution Operation 285

18.1   Head Losses in Water Main 285
        18.1.1   Flow Capacity...285
        18.1.2   Pipe Roughness, Coefficient C..............................286
18.2   Free Flow Velocity and Discharge, Q 287

18.3   Hydrant Testing 287
18.4   Water Quality 290
    18.4.1   Monitoring.. 290
    18.4.2   Secondary Disinfection............................... 290
    18.4.3   Flushing and Cleaning of Water Mains 290
    18.4.4   Repairs and Breaks 291
    18.4.5   Field Disinfection 291

SECTION IV   Wastewater Collection

Chapter 19   Wastewater Collection System.. 297

19.1   Sewer Mains 297
    19.1.1   Combined Sewers....................................... 297
    19.1.2   Storm Sewers 299
19.2   Infiltration & Inflow 299
19.3   Wastewater Flows 300
19.4   Sewer Mains 301
    19.4.1   Pipe Size 301
    19.4.2   Sewer Grade 301
    19.4.3   Pipe Flow Velocity and Capacity 302
    19.4.4   Gravity Sewer Mains.................................... 303
    19.4.5   Force Mains... 304
19.5   Operation and Maintenance 304
    19.5.1   Detecting and Repairing an Obstruction 304
    19.5.2   Crown Corrosion 305
    19.5.3   Repairing Broken Sections............................. 305
    19.5.4   Building Services 306
    19.5.5   Force Main Maintenance 306
19.6   Inspection 306
    19.6.1   Smoke Test 306
    19.6.2   Dye Test.. 307
    19.6.3   Closed Circuit Television 307
19.7   Inverted Siphon.. 307
    19.7.1   Design of Inverted Siphon............................. 308
19.8   Manholes 309
    19.8.1   Ordinary Manhole....................................... 309
    19.8.2   Constructional Details................................. 309
    19.8.3   Drop Manholes and Dead-End Manholes 310
    19.8.4   Manhole Safety 311
    19.8.5   Sewer Ventilation 311
    19.7.4   Manhole Inspection and Maintenance 311
19.9   Sampling and Flow Measurement... 312

            19.9.1   Flow Measurement in Sewers 312
            19.9.2   Sample Collection 312
      19.10 Wastewater Pumping 313
            19.10.1 Wet Well Lift Stations 313
            19.10.2 Dry Well Lift Stations 314
      19.11 Wastewater Flow Pumps 314
            19.11.1 Wet Wells 315
            19.11.2 Screens 315
            19.11.3 Electrical and Controls 315
      19.12 Lift Station Maintenance 315
            19.12.1 Screening Baskets and Bar Screens 315
            19.12.2 Wet Well Floor Maintenance 316
            19.12.3 Sump Pump Operation and Maintenance 317
      19.13 Pump Operating Sequence 317
            19.13.1 Level Setting... 317
            19.13.2 Pumping Rate in Lift Stations..................... 318

Chapter 20 Design of Sewers 327
      20.1 Open Channel Flow... 327
            20.1.1   Flow Classification 327
            20.1.2   Hydraulic Slope 327
      20.2 Manning's Flow Equation 328
            20.2.1   Hydraulic Radius 329
            20.2.2   Uniform Flow Problems............................. 329
            20.2.3   Circular Pipes Flowing Full 329
      20.3 Efficient Conveyance Section 331
      20.4 Maximum and Minimum Flow Velocities 331
            20.4.1   Minimum Flow Velocity 331
            20.4.2   Maximum Velocity or Non-Scouring Velocity 332
      20.5 Partial Full Pipes 333
      20.6 Storm Drainage 335
      20.7 Rational Method 336
            20.7.1   Runoff Coefficient..................................... 336
            20.7.2   Time of Concentration............................... 336
            20.7.3   Rainfall Intensity...................................... 338
            20.7.4   Areal Weighing of Runoff Coefficients 339
            20.7.5   Limitations of Rational Method.................. 339
            20.7.6   Urban Catchments 341

Chapter 21 Construction of Sewers.. 347
      21.1 Materials for Sewers... 347
            21.1.1   Vitrified Clay Pipe (VCP) 347
            21.1.2   Plastic Pipe 347

21.1.3   Fiberglass Polymer Pipe (FRPP)............................... 347
21.1.4   Concrete Pipe (CP).. 348
21.1.5   Asbestos Cement Pipe (ACP)................................. 348
21.1.6   Brick Masonry... 348
21.1.7   Cast-Iron Pipe.. 349
21.1.8   Steel Pipe.. 349
21.1.9   Cast-in-Place Reinforced Concrete 349
21.2   Layout and Installation 349
21.2.1   Setting Out 349
21.2.2   Alignment and Gradient.................................... 350
21.2.3   Excavation of Trenches 350
21.2.4   Bedding 351
21.2.5   Laying... 352
21.2.6   Lasers 353
21.2.7   Jointing 354
21.3   Testing 356
21.3.1   Water Test.. 356
21.3.2   Air Testing 356
21.3.3   Ball Test... 357
21.3.4   Mirror Test 357
21.3.5   Smoke Test 357
21.3.6   Back Filling 357
21.4   Structural Requirements... 358
21.5   Loading Conditions 358
21.6   Dead Loads.. 358
21.7   Field Supporting Strength 362
21.7.1   Load-Carrying Capacity 362
21.7.2   Load Factor... 362

SECTION V   Wastewater Treatment

Chapter 22   Natural Purification 369

22.1   BOD Reaction.. 369
22.2   Natural Process... 371
22.2.1   Zone of Degradation 371
22.2.2   Zone of Active Decomposition............................. 372
22.2.3   Zone of Recovery 372
22.2.4   Clear Water Zone 372
22.3   Oxygen Sag Curve.. 372
22.4   Dilution into Sea... 376
22.5   Disposal by Land Treatment 376
22.6   Comparison of Disposal Methods.................................. 377

Chapter 23   Characteristics of Wastewater..381

    23.1   Treatment Facility..381
    23.2   Domestic Wastewater 382
    23.3   Physical Characteristics..382
    23.4   Chemical Characteristics...383
        23.4.1   Solids...383
        23.4.2   Dissolved Gases...384
        23.4.3   Alkalinity and pH.......................................384
        23.4.4   Biochemical Oxygen Demand.......................384
        23.4.5   Chemical Oxygen Demand 384
        23.4.6   Nutrients 385
        23.4.7   Toxins 385
    23.5   Biological Characteristics..385
    23.6   Percentage Removal 387
    23.7   Industrial Wastewater 388
        23.7.1   Equivalent Population 388
        23.7.2   Composite Concentration 389
    23.8   Infiltration and Inflow..390
    23.9   Municipal Wastewater 390
        23.9.1   Hydraulic and Organic Loading....................390
        23.9.2   Main Points 391
    23.10 Evaluation of Wastewater 391
        23.10.1 Automatic Compositing................................392
        23.10.2 Manual Compositing 392
        23.10.3 Sample Locations 393

Chapter 24   Primary Treatment 399

    24.1   Preliminary Treatment 399
        24.1.1   Screens 399
        24.1.2   Coarse Screens 400
        24.1.3   Fine Screens 400
        24.1.4   Mechanically Cleaned Screens 400
        24.1.5   Volume of Screenings.................................400
        24.1.6   Disposal of Screenings................................400
        24.1.7   Flow through Screens..................................401
    24.2   Comminution of Sewage 401
    24.3   Flow Measurement 401
        24.3.1   Parshall Flume..402
        24.3.2   Palmer-Bowlus Flume 402
        24.3.3   Weirs...402
    24.4   Grit-Removal Units..402
        24.4.1   Settling Velocity 403
        24.4.2   Grit Channels 403

           24.4.3 Aerated Grit Chamber...403

           24.4.4 Detritus Tank..405

           24.4.5 Cyclone Separators...405

           24.4.6 Grit Disposal 405

     24.5 Pre-aeration 405

     24.6 Process Calculations..406

     24.7 Primary Clarification..407

     24.8 Circular Clarifier 407

     24.9 Rectangular Clarifiers 408

     24.10 Scum Removal...409

     24.11 Factors Affecting Settling 409

           24.11.1 Temperature...410

           24.11.2 Short-Circuiting...410

           24.11.3 Settling Characteristics of Solids 410

           24.11.4 Detention Time..410

           24.11.5 Surface Settling or Overflow Rate 410

           24.11.6 Weir Loading...411

     24.12 Secondary Clarifier 414

     24.13 Sludge Handling 415

Chapter 25   Activated Sludge Process 421

     25.1 Biological Treatment 421

           25.1.1 Suspended Growth Systems 421

           25.1.2 Fixed Growth Systems 421

     25.2 Principles of the Activated Sludge Process 422

           25.2.1 Transfer..422

           25.2.2 Conversion...422

           25.2.3 Flocculation...422

     25.3 Components of ASP 423

           25.3.1 Aeration Tanks 423

           25.3.2 Final Settling Tanks 424

           25.3.3 Sludge Recirculation and Wasting 424

     25.4 Factors Affecting ASP...425

     25.5 Process Loading Parameters 425

           25.5.1 Aeration Period..425

           25.5.2 Volumetric BOD Loading 425

           25.5.3 Food to Microorganism (F/M) Ratio 426

           25.5.4 Sludge Age 427

           25.5.5 Substrate Utilization Rate 429

     25.6 Final Clarification..430

           25.6.1 Hydraulic Loading 430

           25.6.2 Solids Loading 430

           25.6.3 Sludge Settlement..431

           25.6.4 Return Rate and SVI 432

25.6.5 Return Ratio and Sludge Thickness 432
25.6.6 State Point Analysis 433
25.7 Variations of ASP 435
25.7.1 Conventional Aeration 435
25.7.2 Contact Stabilization 436
25.7.3 Extended Aeration 437
25.7.4 Oxidation Ditch 437
25.7.5 High-Rate Aeration 438
25.7.6 High Purity Oxygen System 438
25.8 Oxygen Transfer 439
25.8.1 Mass Transfer Equation 440
25.8.2 Specific Uptake Rate (SUR) 440
25.8.3 Oxygen Transfer Efficiency 440
25.9 Operating Problems 441
25.9.1 Aeration Tank Appearance 441
25.9.2 Secondary Clarifier Appearance 442

Chapter 26 Stabilization Ponds 447

26.1 Facultative Ponds 447
26.2 Loading Parameters 448
26.2.1 BOD Removal 449
26.2.2 Winter Storage 450
26.3 Algae 451
26.4 Berms 451
26.5 Daily Monitoring 452
26.5.1 Visual Monitoring 452
26.5.2 Water Color 452
26.5.3 Water Level 452
26.6 Operational Problems 453
26.6.1 Scum Control 453
26.6.2 Odor Control 453
26.7 Lagoon Maintenance 454
26.7.1 Lagoon Weeds 454
26.7.2 Berm Erosion 454
26.7.3 Mosquitos 454
26.7.4 Daphnia 454

Chapter 27 Attached Growth Systems 459

27.1 Trickling Filters 459
27.2 Main Components of the Trickling Filter 460
27.2.1 Filter Media 460
27.2.2 Underdrains 460
27.2.3 Wastewater Distribution 461

27.2.4 Loading on Filters 461
27.2.5 Recirculation 462
27.2.6 Staging.. 463
27.3 BOD Removal Efficiency 463
27.4 Operating Problems... 464
27.4.1 Ponding... 464
27.4.2 Fly Nuisance.. 464
27.4.3 Odor Nuisance... 465
27.5 Secondary Clarification 465
27.6 Rotating Biological Contactors... 467
27.6.1 Staging... 468
27.6.2 Operation... 468
27.7 Process Control Parameters.. 468
27.7.1 Soluble BOD.. 470
27.7.2 Organic Loading 470
27.8 Operation of RBC System 472

Chapter 28   Anaerobic Systems... 477

28.1 Septic Tanks 477
28.2 Design Considerations 477
28.2.1 Capacity.. 477
28.2.2 Free Board.. 477
28.2.3 Inlet and Outlet... 478
28.2.4 Detention time 478
28.2.5 Shape of the Tank... 478
28.2.6 Disposal of the Tank Effluent...................................... 480
28.3 Soil Absorption System 480
28.3.1 Percolation Test 480
28.3.2 Absorption Field 481
28.4 Soak Pit... 481
28.5 Biological Filters 481
28.6 Upflow Filters 482
28.7 Upflow Anaerobic Sludge Blanket 484
28.7.1 Zones and Components 485
28.7.2 Design Approach 486

Chapter 29   Bio-Solids... 491

29.1 Primary Sludge... 491
29.2 Secondary Sludge 491
29.3 Processing of Sludges 493
29.4 Sludge Thickening 493
29.4.1 Gravity Thickener 494
29.4.2 Concentration Factor 494

29.4.3 Floatation Thickener 494
29.4.4 Gravity Belt Thickener..495
29.4.5 Centrifuge Thickening 495
29.5 Mass Volume Relationship...496
29.6 Sludge Stabilization..498
29.7 Sludge Digestion...498
29.7.1 Anaerobic Sludge Digestion.....................................498
29.7.2 Aerobic Sludge Digestion...499
29.7.3 Anaerobic Digester Capacity....................................499
29.7.4 Two-Stage Digestion 501
29.7.5 Volatile Solids Reduction in Digestion 502
29.7.6 Gas Composition 503
29.7.7 Digester Solid Mass Balance....................................503
29.8 Dewatering of Sludges..503
29.8.1 Sludge Drying Beds 504
29.8.2 Mechanical Methods of Dewatering Sludge 505
29.8.3 Sludge Conditioning...507
29.9 Disposal of Sludge..507
29.9.1 Incineration 508
29.9.2 Sanitary Land Fill 508
29.9.3 Disposal in Water or Sea..508
29.9.4 Sludge Composting 509

Chapter 30   Advanced Wastewater Treatment..513

30.1 Suspended Solids Removal...513
30.1.1 Microscreening..514
30.1.2 Ultrafiltration..514
30.1.3 Granular Media Filtration 515
30.2 Control of Nutrients..516
30.3 Phosphorus Removal 516
30.3.1 Biological Phosphorous Removal.............................516
30.3.2 Chemical Phosphorus Removal 517
30.3.3 Biological–Chemical Phosphate Removal 518
30.4 Nitrogen Removal..518
30.4.1 Biological Nitrification–Denitrification 518
30.4.2 Three-Stage Nitrification–Denitrification................520
30.5 Treatment Methods for the Removal of Toxins.....................521
30.5.1 Carbon Adsorption 521
30.5.2 Chemical Oxidation 522
30.6 Wastewater Disinfection...522
30.7 Improved Treatment Technologies 522
30.7.1 Sequencing Batch Reactor (SBR)............................522
30.7.2 Membrane Bioreactor Process (MRP) 523
30.7.3 Ballasted Floc Reactor (BFR) 523

            30.7.4 Biological Aerated Filters (BAFs)...............................523
            30.7.5 Integrated Fixed-Film Activated Sludge (IFAS)......523
    30.8 Water Recycle and Reuse 523
            30.8.1 Water Conservation 524
            30.8.2 Reuse of Processed Wastewater 524
    30.9 Water Quality and Reuse...524
            30.9.1 Urban Landscape...525
            30.9.2 Reclaimed Wastewater 525
    30.10 Industrial Wastewater Treatment..525
    30.11 Industrial Wastewater Discharges 526
    30.12 Industrial Wastewater Treatment..527
            30.12.1 Removal of Chromium...527
            30.12.2 Removal of Phenol 528
            30.12.3 Removal of Mercury 529
    30.13 Common Effluent Treatment Plants 529

Appendices...**533**
Index..**537**

List of Figures

Figure 1.1 Hydrologic cycle 5

Figure 1.2 Urban water cycle 6

Figure 1.3 Elements of a water supply system 7

Figure 1.4 Wastewater collection and treatment 8

Figure 3.1 Continuity of flow 24

Figure 3.2 Hydraulic grade line 26

Figure 3.3 Bernoulli's principle 28

Figure 3.4 Bernoulli's Equation (Ex. Prob. 3.5) 30

Figure 3.5 General energy equation 31

Figure 3.6 Pumping system (Ex. Prob. 3.6) 32

Figure 4.1 Making a solution of desired strength 52

Figure 4.2 Liquid chemical feeding 54

Figure 6.1 Reservoir intake 79

Figure 6.2 River intake 80

Figure 6.3 Types of water wells 83

Figure 6.4 Pumping an artesian well 84

Figure 6.5 Pumping a water table well 88

Figure 7.1 Hourly variation (diurnal variation) 108

Figure 8.1 Process schematic of a filtration plant 120

Figure 8.2 Coagulation and flocculation mixers 127

Figure 8.3 Jar testing apparatus 131

Figure 9.1 Discrete settling 138

Figure 9.2 Plan and sectional view of a rectangular basin 139

Figure 9.3 A solids contact unit 142

Figure 9.4 Detention time and overflow rate 144

Figure 10.1 Slow sand filter 154

Figure 10.2 Components of a gravity filter 156

Figure 10.3 Head loss in a filter 159

Figure 10.4 Backwashing of a gravity filter 160

Figure 11.1 Breakpoint chlorination curve 175

Figure 11.2 Breakpoint chlorination (Ex. Prob. 11.3) 178

Figure 11.3 Chlorine feed system 185

Figure 12.1 Excess lime softening treatment 196

Figure 12.2 Split treatment softening plant 197

Figure 15.1 Mass curve (Ex. Prob. 15.1) 235

Figure 15.2 Grid iron versus tree distribution 237

Figure 15.3 Ring versus radial distribution 238

Figure 15.4 Flanged joint and socket joint 242

Figure 15.5 Flexible joint and dresser coupling 242

Figure 15.6 Expansion joint and simplex joint 243

Figure 16.1 Simplified water distribution system 259

Figure 16.2 Pipe network (Ex. Prob. 16.10) 262

Figure 16.3 Water supply to a pressurized tank 264

Figure 16.4 Simple water distribution system 265

Figure 16.5 Two connected reservoirs 266

Figure 17.1 Centrifugal pump performance curves 269

Figure 17.2 A simple distribution pumping system 271

Figure 17.3 Pump head and system head 273

Figure 17.4 Pump operating point (Ex. Prob. 17.4) 273

Figure 17.5 Two identical pumps in series and parallel 277

Figure 19.1 Sewage collection system 298

Figure 19.2 Building sewer connection 298

Figure 19.3 Dry weather wastewater flow hydrograph 300

Figure 19.4 Inverted siphon 308

Figure 19.5 A drop manhole 310

Figure 19.6 Dry well lift station 314

Figure 20.1 Open channel flow energy 328

Figure 20.2 Standard chart for proportionate elements 333

Figure 20.3 Rational method 336

Figure 20.4 Urban draining system (Ex. Prob. 20.11) 341

Figure 20.5 Storm drainage system 345

Figure 21.1 Laying of sewer 352

Figure 21.2 Bell and spigot joint 354

Figure 21.3 Water test 356

Figure 21.4 Load on pipe (trench conditions) 359

Figure 21.5 Sewer line reaches 365

Figure 22.1 Zones of natural process 372

Figure 22.2 Oxygen sag curve 373

Figure 23.1 Flow schematic of a tertiary plant 382

Figure 23.2 Flow and BOD pollutograph 391

Figure 23.3 Standard sampling locations 393

Figure 24.1 Flow through a grit channel 406

Figure 24.2 Circular settling tank 408

Figure 24.3 Flow through a rectangular clarifier 411

Figure 25.1 A conventional activated sludge plant 422

Figure 25.2 Solids entering and exiting 428

Figure 25.3 Settleometer test 431

Figure 25.4 State point analysis 434

Figure 25.5 Tapered aeration ASP process 435

Figure 25.6 Step aeration 436

Figure 25.7 Contact stabilization 436

Figure 25.8 Schematic of an oxidation ditch 438

Figure 26.1 Facultative pond biology 448

Figure 27.1 A typical trickling filter plant 460

Figure 27.2 Recirculation and staging in TF filter plants 463

Figure 27.3 Rotating biological contactors 467

Figure 28.1 Section of a typical septic tank 478

Figure 28.2 A soak pit 482

Figure 28.3 A single-chamber upflow filter 483

Figure 28.4 A typical UASB reactor 485

Figure 29.1 Primary and secondary sludge flow streams 492

Figure 29.2 Sludge process scheme 494

Figure 29.3 Schematic of a floatation thickener 495

Figure 29.4 Schematic of a two-tank anaerobic sludge digester 499

Figure 29.5 Sludge digestion and storage capacity 500

Figure 30.1 Ultrafiltration and biodegradation combined 515

Figure 30.2 Configurations for wastewater filtration 516

Figure 30.3 Alternative points of chemical addition 518

Figure 30.4 Three-stage nitrification–denitrification 521

Figure 30.5 Flow schemes for industrial wastewaters 528

Figure 30.6 CETP flow diagram 529

List of Tables

Table 2.1   Base units of mass length and time 12

Table 2.2   Derived units 13

Table 2.3   Suffixes in SI units 14

Table 3.1   Absolute roughness 34

Table 3.2   Hazen-Williams coefficient, C 37

Table 4.1   Atomic mass of common elements 44

Table 4.2   Common chemicals in water industry 55

Table 5.1   Diseases and causative organisms 63

Table 6.1   Surface water versus groundwater supplies 91

Table 7.1   Design periods of various components 98

Table 7.2   Worksheet for population forecast 99

Table 7.3   Residential water consumption in USA 101

Table 7.4   Residential water consumption in India 101

Table 7.5   Water consumption in USA 102

Table 7.6   Water consumption in India 102

Table 7.7   Suspended and dissolved impurities in water 112

Table 7.8   Maximum permissible limit (MPL) of chemicals ... 115

Table 8.1   Main treatment processes 120

Table 8.2   Common coagulants and doses 121

Table 8.3   Recommended doses of coagulant aids 123

Table 10.1   Comparison of various types of filters 155

Table 10.2   Filter media characteristics 157

Table 12.1   Classification of hardness 192

Table 12.2   Milliequivalent table (Ex. Prob. 12.1) 192

Table 12.3   Lime and soda ash requirement 194

Table 13.1   Common fluoride compounds 206

Table 14.1   Comparing PAC with GAC 220

Table 14.2 Values of constant A...224

Table 14.3 Values of constant B...225

Table 14.4 Comparison of various indices..225

Table 14.5 Proper use of pH adjustment chemicals..226

Table 15.1 Excel worksheet (constant supply)..234

Table 15.2 Storage capacity problem..252

Table 16.1 Excel worksheet (Ex. Prob. 16.10)..262

Table 17.1 Operating point of a pump...274

Table 17.2 Pumps in series and parallel..277

Table 19.1 Storm sewers versus sanitary sewers...299

Table 19.2 Gravity sewer pipe grading and size...302

Table 20.1 Partial flow and equivalent hydraulic elements....................................334

Table 20.2 Runoff coefficients for rural areas...337

Table 20.3 Typical C values (urban areas)..338

Table 20.4 Table of computations..342

Table 21.1 Factor Kμ for various fill materials..360

Table 21.2 Strength of vinyl chloride pipe of various sizes....................................362

Table 21.3 Load factor for different classes of bedding...363

Table 22.1 Saturation dissolved oxygen...371

Table 22.2 Recommended loading rate..377

Table 23.1 General characteristics of flow streams..383

Table 23.2 Nutrient removal..385

Table 23.3 Table of computations..393

Table 23.4 Flow hydrograph (Practice Problem 14)...396

Table 24.1 Main components of a circular clarifier...408

Table 24.2 Components of a rectangular clarifier..409

Table 24.3 Design parameters for settling tank..411

Table 25.1 Comparison of plant types...437

Table 25.2 Uptake rates and air requirement..439

Table 26.1 Color of algae (visual monitoring)...452

Table 27.1   Design features for trickling filters..462

Table 27.2   Operating conditions and slime color..472

Table 28.1   Design Parameters of UASB Reactor 487

Table 29.1   Aerobic versus anaerobic digestion 500

Table 29.2   Ceiling concentration of heavy metals..507

Table 30.1   Residual constituents in treated wastewater.................................... 514

Table 30.2   Industrial waste characteristics... 526

Table 30.3   Selected industrial wastewaters 526

About the Author

Before leaving for Canada in 1978, Subhash was an assistant professor in the Department of Soil and Water Engineering of Punjab Agricultural University Ludhiana, India. After completing his Master of Engineering in Water resources Engineering from University of Guelph, he stayed with the school of engineering and researched watershed modeling. In 1982, Subhash joined Sault College of Applied Arts and Technology to lead a new program in Water Resources Engineering Technology. In addition to teaching, Subhash developed training programs in many areas of water and wastewater engineering technology. These training programs were delivered to Ministry of Environment personnel in the Province of Ontario. During his stay at Sault College, Subhash developed course manuals to supplement courses in hydraulics, hydrology, water supply, and wastewater engineering technology. Starting in 2007, Subhash was appointed Chair of Sault Ste. Source Protection Committee by the Minister of Environment. After his retirement, Subhash spent time developing and teaching online courses in water and wastewater technology on Ontario Learn (www.ontariolearn.com), a provincial initiative. Subhash has been the lead author of two books: *Water Supply Engineering* and *Environmental Engineering: Fundamentals and Applications.* He is currently working on books on water resources engineering, applied hydraulics, engineering hydrology, and groundwater and wells.

Section I

Basic Sciences

1 Introduction

Water is essential for life and no life is possible without water. All living things: humans, plants, and animals require water. Almost 70% of human body mass is due to water present in body tissues. Essential for life, water is used for many other purposes – to drink, for personal needs, power generation, agricultural irrigation, in industry, and for transportation and recreation. In the USA, fresh water is used for the following: about 36% is for agriculture, 11% for the public water supply, 8% for power generation, and 5% for industrial purposes. In India, about 70% of the total water available is used for agriculture, 20%–22% by industry, and only 8% is employed for domestic use.

The use of water is increasing rapidly with the increase in world population and other factors including urbanization and lifestyle. In the urban setting, the twin services of water supply and sewage disposal are essential. Water used in homes and commercial entities is largely collected in the sewer system, then treated at a water pollution control plant before being returned to surface water. In underdeveloped countries, most places have little or no wastewater treatment. The increase in water use and the lack of proper treatment have exhausted the natural purification capacity of many lakes and rivers. This is a serious problem when the same water downstream is the source of water supply for the neighboring community. In industrialized countries, toxic and persistent chemicals discharged in wastewater are not treatable by conventional methods and endanger surface water supplies. The planning, designing, financing, and operation of water and water systems are complex undertakings, and they require high degrees of skill and judgment.

1.1 HISTORICAL PERSPECTIVE

Historically, many civilizations developed around water bodies that could support agriculture, transportation, and supply for domestic use. The earliest recorded known example of water treatment is found in Sanskrit medical writing and an inscription on a wall in Egypt. Sanskrit writings in India dating from 2000 B.C.E. describe water purification by boiling water in copper vessels, exposing it to sunlight, filtering it through charcoal followed by cooling in earthen pots. There is a mention of digging wells in Rig Veda as far back as 4000 B.C.E.

1.1.1 WATER SUPPLY

The first engineering report on water supply and treatment was written in C.E. 98 by S.J. Frontinus, the water commissioner of Rome. He produced several books on the water supply in Rome. In the seventeenth century, the English philosopher Francis Bacon wrote of experiments in purifying water by filtration, boiling, distillation, and

DOI: 10.1201/9781003347941-2

clarification by coagulation. The first known illustrated description of sand filters was published in 1685 by Luc Porzio, an Italian physician. A comprehensive article on the water supply of Venice was published in 1863 in *The Practical Mechanic's Journal*.

In 1856, Henry Darcy came up with a theory (Darcy's Law) for groundwater flow and the design of filters, and patented water filters in France and England. Filtered water was piped for the first time to homes in Glasgow, Scotland, in 1807. Around 1890, rapid sand filters were developed, and coagulation was later introduced to improve their performance. After this, the techniques of clarification and filter were improved and modified. This was followed by the process of chlorination for disinfection in 1905. With the technical and scientific developments that concluded in the twentieth century and extending into the twenty-first century, water engineers are now in a position to meet the quantity and quality challenges in water supply.

1.1.2 History of Sanitary Engineering

The collection of storm water and drainage dates from ancient times. The systematic collection, treatment, and disposal of wastewater followed in the late 1800s and early 1900s. The development of germ theory by Koch and Pasteur in the latter half of the nineteenth century marked the beginning of a new era in sanitation. Edwin Chadwick (1800–1890), an administrator in the UK, introduced the idea of sanitation. Chadwick called for street and house cleaning by means of a supply of water and improved sewage collection. He specifically determined to obtain the services of and advice from civil engineers rather than physicians. Chadwick proposed the supply of drinking water to every house and the collection of wastewater in a network of pipes. This was the beginning of sanitary or wastewater engineering. It resulted in an improvement in the general health and hygiene of the public. Legislation was subsequently introduced in the UK that prohibited the discharge of untreated sewage into streams and rivers.

A breakthrough in wastewater technology came when in 1914 Edward Arden and William T. Lockett discovered that when sewage was kept aerated in settling tank for some days, it resulted in a significant removal of the biochemical oxygen demand (BOD) of the effluent. This led to the development of what is the activated sludge process. Today, activated sludge is the most widely used wastewater treatment process around the world. Similar developments took place in USA and Canada. At the beginning of the twentieth century, there were public drinking water supplies but sewage was collected in sewer pipes and discharged into rivers. However, there were recurring epidemics of typhoid and other diseases. It was not until the second half of the twentieth century that engineered wastewater treatment were built and sewage was treated before discharge.

1.2 HYDROLOGIC CYCLE

The quantity of water on the Earth, in all its forms, is always the same, although it is continuously in motion, with no start or end points. This water in nature is in a

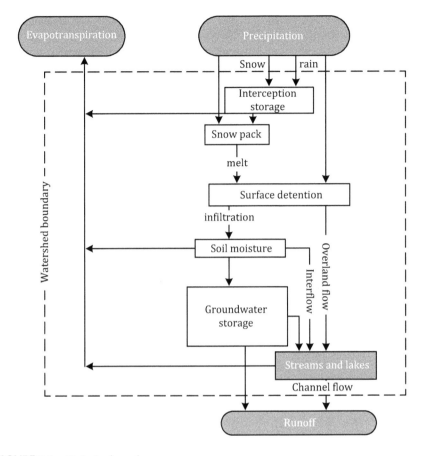

FIGURE 1.1 Hydrologic cycle

constant state of motion, as depicted in the hydrologic cycle shown in **Figure 1.1**. The water cycle is driven by energy from the sun and gravitational forces. The major components of the hydrologic cycle are the evaporation of water from the sea or land, precipitation as rain or snow over the oceans and land, and returns from the land to the sea via streams and rivers. With slight variations, the water content of the sea remains constant.

Water in nature is most pure in its state as vapor. However, as it condenses on surfaces or nuclei, it may acquire impurities, for example acid rain as it dissolves oxides of sulfur and nitrogen emitted from industries and automobile exhausts. Additional impurities are added as it travels the remaining part of the hydrologic cycle as well as due to the impurities contributed by the anthropogenic activities such as domestic or industrial wastes and agricultural chemicals. Urbanization results in urban runoff that is loaded with pollutants and in many cases poorly or untreated wastewater discharges that seriously degrade the quality of receiving water bodies. Since urbanization in the world is increasing, so are problems related to water pollution.

In developed countries, municipal and industrial wastewaters are treated before discharge. However, that is not the case in developing and underdeveloped countries, where still large amounts of wastewater find their way directly into rivers and streams, which may be the source of the water supply for communities downstream.

About 75% of the Earth's surface is covered with water and because of this the Earth is also called the "Water Planet." Of all the water available, 97% is saline and only 3% is fresh water. However, less than 1% of the fresh water is easily available as rivers, lakes, streams, and groundwater.

1.3 URBAN WATER CYCLE

As urban development expands, so does the need for municipal water systems. Within urban areas, water also goes through a cycle to serve the needs of municipal establishments. This is known as the urban water cycle. The main components of the urban water cycle are illustrated in **Figure 1.2**. These include: source of water supply, water treatment, water distribution, wastewater collection, and wastewater treatment. Due to the ever-increasing demand for fresh water, reclamation, and the recycling of

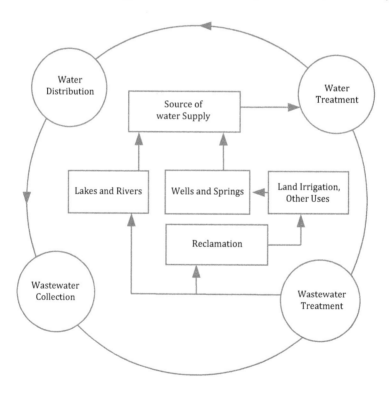

FIGURE 1.2 Urban water cycle

water are gaining popularity. In addition, new technologies to treat water are being developed, with more emphasis on source water protection.

1.4 ESSENTIALS OF A WATER SUPPLY SYSTEM

Figure 1.3 shows the components of a typical water supply system. The primary requirement of a water supply system is the selection of a reliable and safe source of water. After selection of the source of water, it is essential to design and construct a water intake structure to collect and transmit water to the treatment plant. At the treatment plant, water is treated either using chemicals or by simple physical processes to improve its quality to meet consumer requirements. The treated water is stored in clear water reservoirs from where it is distributed to consumers. Also, depending on the elevation of the supply area, water may be pumped into the system, which requires the use of suitable pumping equipment. The reliability of the water supply system depends on the permanency of the source which provides water in adequate quantities and required quality. It should be able to supply the required quantities and quality of water to consumers. Robust design and construction and maintenance of treatment and distribution system are essential to meet these requirements.

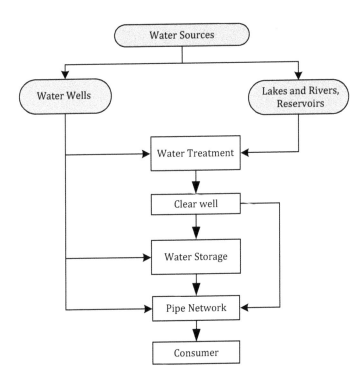

FIGURE 1.3 Elements of a water supply system

1.5 NEED FOR WASTEWATER TREATMENT SYSTEM

Sanitation can be perceived as the conditions and processes relating to people's health, especially the systems that supply water and deal with human waste. Such a task would logically cover other matters, such as solid waste, industrial and other special wastes, and storm water drainage. However, the most potent of these pollutants is sewage. When untreated sewage accumulates and is allowed to become septic, the decomposition of its organic matter leads to nuisance conditions including the production of malodorous gases. In addition, untreated sewage contains numerous pathogens that dwell in the human intestinal tract. Sewage also contains nutrients, which can stimulate the growth of aquatic plants and may contain toxic compounds or compounds that are potentially mutagenic or carcinogenic. For these reasons, the immediate and nuisance-free removal of sewage from its sources of generation, followed by treatment, reuse, or disposal into the environment in an eco-friendly manner, is necessary to protect public health and the environment. A typical wastewater collection and treatment system is shown in **Figure 1.4**.

Water conservation practices will play a key role in our goal to meet tomorrow's water demands. Wastewater system engineers, in addition to the design and operation

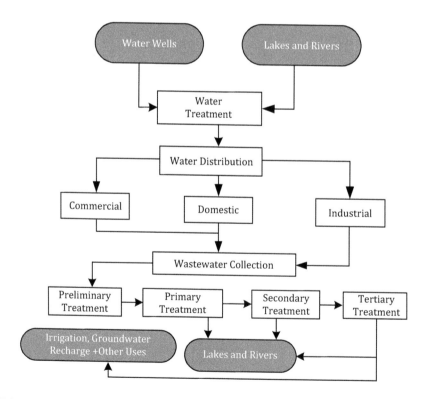

FIGURE 1.4   Wastewater collection and treatment

of wastewater collection and treatment, must be aware of the impact of treated wastewater discharges on the quality of water in the receiving water body.

1.6   GLOBAL ISSUE

The enormous demands being placed on water supply and wastewater facilities have necessitated the development of broader concepts, management scenarios, and improvements and innovations in environmental technology. While the standards of water quality have become more stringent, raw water quality has become poorer due to increases in demand and anthropogenic activities that limit the capacity of natural purification. Toxic and hazardous materials have entered the water system, which have heightened the concerns over environmental issues. As the world population sees an alarming growth, the environment becomes a critical factor. Climate change has added another dimension to environmental issues and problems. Climate change may increase or may reduce rainfall and snowfall, which will impact water supply. Current water droughts have raised concerns about the sustainability of future water supply. Political and social unrest throughout the world has made water supply an urgent security issue. Consequently, the use, management, and control of water resources must be sustainable.

1.7   ROLE OF THE ENGINEER

A water system engineer should understand the origin and movement of water in nature, the different sources of water, its quality and quantity, the processes of the treatment of water to improve its quality, and the operation and maintenance of the systems for collection, conveyance, treatment, and distribution of water.

Practicing wastewater engineers are involved in the conception, planning, evaluation, design, construction, and operation and maintenance of the systems that are needed to meet wastewater management objectives. Knowledge of the methods used for the determination of wastewater flow rates and their characteristics is essential to an understanding of all aspects of wastewater engineering. The engineer must also study the subjects of source control, collection, transmission, and pumping, if truly integrated wastewater systems are to be designed. The issues of treatment, disposal of effluent, and the possible reuse of wastewater are also equally important. In this, old ideas are being re-evaluated and new concepts are being formulated.

Until recently, the role of water and wastewater system engineers and operators was limited to the treatment of water and the distribution, collection, and disposal after treatment. However, in today's world, the role of people working in the water industry has gone beyond that. Modern water and wastewater treatment facilities are now planned and designed with the key objective of environmental sustainability. This includes reducing the carbon footprint as well as the building site footprint. This is achieved by minimizing any fossil fuel power requirements and by using compact and efficient treatment units, incorporating green building design using Leadership in Energy and Environmental Design (LEED) principles. This may include energy saving and tapping solar energy by installing solar panels on rooftops.

Questions for Discussion

1. Write short notes on source water protection and water conservation.
2. Briefly describe the history of water and wastewater systems.
3. With the help of a flow chart, explain the various components of a water supply system?
4. World population is on the rise, but the quantity of fresh water is on decline. How are tomorrow's water engineers going to meet this challenge?
5. Search the internet to find out countries experiencing water shortages in the twenty-first century.
6. Draw a flow chart of the urban water cycle, starting with the source of water supply and the discharge of wastewater effluent.
7. Environmental engineers and scientists have reported that global warming may be a real problem. What are the causes of global warming and how might it impact the water supply?
8. In what sense is the job of future environmental engineers going to be more challenging?
9. Give an account of the need for wastewater treatment.
10. Name the water quality parameters of concern from the public health point of view.
11. List the major microbiological contaminants and the diseases caused by effluent.
12. What role do nutrients play in the biological treatment of wastewater?
13. What basic knowledge is required for decision-making in wastewater treatment technology?
14. List the roles of the engineer in wastewater management.
15. When and how was the discovery of activated sludge treatment made in the UK?
16. Search the internet and find the effluent standards for the USA and India.
17. Explain why secondary treatment has become the norm in developed countries.
18. "Dilution is not a solution." Comment on this.
19. Compare unit operations and unit processes and list common combinations of unit operations and processes used in the processing of wastewater.

2 Standards of Measurement

In engineering technology, one has to deal with numerous types of quantities and their measurements. It may be the flow rate, liters of a chemical, size of a pipeline, or power required by the pumps. All numerical engineering measurements should be dimensioned with proper units. To express a quantity correctly, an appropriate unit must accompany it. In the field of engineering, one needs to have a thorough understanding of the units of measurement to communicate effectively. In engineering calculations, most mistakes are made due to lack of understanding of units and conversion of units. In North America, there is two systems of units currently used – one referring to length in meters and the other in units of feet and inches. History is witnessing that errors made in conversions have led to some serious accidents, including the loss of a spacecraft and a jet running out of fuel. With that thought in mind, a brief review of this topic is presented here.

2.1 SYSTEMS OF UNITS

There are two common systems of units, US customary (USC) and the Système international (SI) often referred to as the metric system. Except for the USA and a few other countries, much of the world has adopted the metric system. Even in the USA, federally funded projects are required to use SI. Even in countries like Canada and India, which went metric a long time ago, many seasoned engineers as well as the general public still use USC. In reality, both systems of units are prevalent in the workplace. Hence, it is important for a professional to master both the USC and SI systems of units. The conversion of units from English to metric and vice versa are a fact of life. This can be tricky sometimes. For this reason, this book uses both types of units. Conventional units are shown in parentheses after the metric units to help in thinking metric. Before we attempt to learn topics related to water, it is fundamental to have a sound knowledge of systems of units in use and their conversions. In engineering practice, it is recommended to use only one system of units on drawings and specifications.

2.2 MEASURES OF DIMENSION

There are three basic dimensions: length, mass, and time. The base units for each dimension is shown in **Table 2.1**. Any system of units of measurement must have standards or units to express mass, length, and time. Whereas the meter is the basic

DOI: 10.1201/9781003347941-3

TABLE 2.1

Base units of mass length and time

Base Unit	SI	USC
mass	kg	slug**
length	m	ft
time	s	S

** In fact, the unit of force (lb) is a base unit and slug is a derived unit in USC.

unit of measure in SI, the corresponding unit of linear measure is foot in USC. To the surprise of some, the unit of mass in the USC system is slug and not pound. It is very important to make the distinction between pound mass and pound weight. Weight is force due to gravity. Hence units of weight are the same as that of force. In addition to these base units, the SI unit for electric current is ampere (A) and the unit for luminous intensity is candela (cd). The base unit of time, the second, is the same in both the systems.

2.3   DIMENSIONS AND UNITS

When expressing units of a quantity, the following rules will apply:

1. The numerical magnitude of a physical quantity has no meaning without an accompanying unit of measurement, e.g., 25 m, km, yd.
2. Each physical quantity has a fundamental dimensional character.
3. The dimensional character of a physical quantity is expressed as some combination of length, mass, time, or other.
4. The dimensions of quantities must be subjected to the same mathematical operations as the quantities themselves.
5. Mathematical operations involved in manipulating dimensions are multiplication and division.
6. Addition and subtraction can take place only with terms having the same dimensions or units.

2.4   DERIVED UNITS

All other quantities, for example, area, velocity, and weight, can be expressed in terms of base units. For example, velocity – usually called speed in everyday life – is how much distance is traveled in a given amount of time. This relation dictates that units of speed must be that of length over time units. In terms of base units, units of velocity are m/s or ft/s. Based on the same logic, units of area of a surface are length units squared. Following the same logic, units of volume of capacity must be length units cubed, like m^3 or ft^3.

2.4.1 FORCE

In scientific terms, weight is force due to gravity, According to Newton's second law of motion, units of weight in SI will be that of mass multiplied by units of acceleration. In honor of Sir Isaac Newton, unit of force is called newton (N). As shown in **Table 2.2**, one newton force in terms of base units is kg·m/s².

$$F = m \times a = kg.m / s^2 = N \left(newton \right)$$

One newton force is defined as the force that will cause a mass of 1 kg to accelerate at the rate of 1 m/s². In USC, lb is a unit of force and unit of mass is slug, which is a derived unit. A mass of 1 slug will accelerate at the rate of 1 ft/s² when acted upon by a force of 1 lb. One-pound force is roughly about 4.5 N.

2.4.2 MASS AND WEIGHT

Mass and weight are different physical quantities. Since weight is force due to gravity, units of weight are the same as of force, N or lb.

$$W = m \times g = 9.81 \text{ kg} \bullet m / s^2 \approx 10 \text{ N}$$

Since acceleration due to gravity on earth is 9.81 N/kg (m/s²), a mass of 1.0 kg will have a weight of 9.81 N – roughly about 10 N (newton). In other words, weight of about 100 g (one apple) has a weight of 1 N. One newton force is, relatively, a smaller unit of force. A larger unit, kN (kilo newton or 1000 N), is more commonly used to express weight or force. Suffixes including kilo are discussed in more detail later. In USC, lb is used for both units of mass and weight or force. The unit of slug is rarely used.

- A mass of 1 lb mass (lb_m) will have weight of 1 lb force (lb_f).
- Mass of an object cannot be zero, but weight can be zero (g = 0 in space).
- Since weight depends on both mass and acceleration due to gravity, weight of the same object will vary from one planet to the other.
- Acceleration due to gravity at moon surface is about 1/6th that of earth. Weight of the same mass will be reduced to 1/6th that on earth.

TABLE 2.2
Derived units

Quantity	Derived Unit		Symbol	
	SI	USC	SI	USC
Force, weight	newton	pound	N = kg·m/s²	lb = slug·ft/s²
Pressure	pascal	pound per square in	Pa = N/m²	psi = lb/in²
Work/energy	joule	foot pound	J = N·m	lb·ft or ft·lb
Power	watt	horsepower	W = J/s	hp = 550 ft·lb/s

2.4.3 PRESSURE

The term pressure refers to hydrostatic pressure in the water and wastewater industry. The basic definition of pressure is force per unit area. Hence the unit of pressure
in SI is N/m^2 or pascal (Pa) and in English system it is lb/in^2 (psi). Typical pressure
in water main is 40 psi or 300 kPa. A pressure of 30 psi roughly equates to 200 kPa
or 6.89 kPa/psi when more precision is required.

2.4.4 ENERGY AND POWER

Based on the unit of force and length, a unit of energy can be derived. Because work
or energy is force multiplied by distance, the unit of energy is force times distance.
The unit of energy in SI is the joule (J). One joule of energy is spent to move an
object through a distance of one meter by applying a force of one newton (J = N.m).
The corresponding unit of energy in USC is ft.lb. Power is the rate of doing work,
and the SI unit of power is J/s or W (watt). In USC, the most common unit of power
is horsepower (hp), which is defined as working at the rate of 550 lb.ft/s. One horsepower is equal to 746 W or 0.746 kW.

2.5 SYMBOLS AND SUFFIXES

The actual size of physical quantities varies over a wide range. In the metric system,
suffixes are added to the basic quantities to express small and large values. The list
of suffixes is shown in **Table 2.3**. One advantage of SI is that small and large quantities can be expressed by using suffixes. Furthermore, these suffixes are based on the

TABLE 2.3
Suffixes in SI units

Suffix	Symbol	10x	Factor
tera	T	10^{12}	1 000 000 000 000
giga	G	10^{9}	1 000 000 000
mega	M	10^{6}	1 000 000
kilo	k	10^{3}	1 000
hecto	h	10^{2}	100
deka	da	10^{1}	10
deci	d	10^{-1}	0.1
centi	c	10^{-2}	0.01
milli	m	10^{-3}	0.001
micro	μ	10^{-6}	0.000 001
nano	n	10^{-9}	0.000 000 001
pico	p	10^{-12}	0.000 000 000 001
femto	f	10^{-15}	0.000 000 000 000 001
atto	a	10^{-18}	0.000 000 000 000 000 001

decimal system, meaning, therefore, in multiples of tens. For example, in the USC system, small units of length are inches and feet, and larger units are miles. Where there are twelve inches in one foot, there are 5280 ft per mile. Compared to this, the unit of length in the metric system remains the meter and you can make it smaller or larger by attaching a suffix, such as mm (millimeter) and km (kilometer). For example, a distance of 1575 m can be expressed as:

$$1575 \ m = 157.5 \ dam = 15.75 \ hm = 1.575 \ km$$

Some rules should be followed when using symbols, units, and suffixes in the SI system. The following is a list of the rules and conventions regarding the use of symbols and suffixes in SI system.

Key points
- Unit symbols are not followed by a period (mm and not mm.).
- Unit symbols are always written in the singular form (mm and not mms).
- Numerical values associated with a symbol should be separated from the symbol by a space (5.5 N and not 5.5N).
- A space is used to separate large numbers in groups of three starting from the decimal point in either direction (6 545 370.0 and not 6,545,370.0).
- Value less than one decimal point is preceded by a zero (0.038 and not .038).
- In designating the product of units, use a centered dot (N.m and not Nm).
- For quotients, use a solidus (/) or a negative exponent (N/m^2 or $N{\cdot}m^{-2}$).
- The suffix becomes part of the symbol (mm and not m m).
- Compounded suffixes should not be used (such as GmL).
- The suffix is combined with the unit to form a new unit. For example, the meaning of km^2 is $(km)^2$ and not $k(m^2)$.
- Where possible, avoid the use of suffixes in the denominator of a compound unit; kg, being one of the base units, is an exception to this rule.

Indicating the result, select a suffix such that the numerical value will fall between 0.1 and 1000. This rule may be disregarded when using the scientific notation or using the same multiple for all the items. Standard usage in the SI system calls for only those suffixes varying in steps of 10^3 (100 mm, not 10 cm)

2.6   SIGNIFICANT FIGURES

Engineers and scientists use significant figures to identify the precision of a measurement. The number of significant figures allows us to describe the level of confidence we have in our measurements. Use the following guidelines to determine which digits are significant in a given measurement.

1. When the number contains a decimal, the leftmost digit that is not zero is considered significant. All other digits within the number are considered significant. For example, the numbers 3.28, 0.328, 0.0328 all contain three significant figures (SFs) but 0.012 and 0.0012 both contain only two.

2. Terminal zeros placed to the right of a decimal point are assumed to be significant. Thus, 1.230 and 1.200 both have four SFs.
3. If the number does not contain a decimal, the rightmost non-zero digit is considered significant, as are all other digits in the number. In a number like 100, it is not certain whether it represent 1 or 3 SFs. To avoid this uncertainty, it is recommended that scientific notation be used. For example, 1×10^2 has 1 SFs and 1.00×10^2 has 3 SFs.
4. Internal zeros are all significant.

2.7 NUMERICAL PRECISION

The precision of a measurement is determined by the type of instrument or gauge used to make that measurement. For example, if you are measuring length with a ruler that has markings to a mm, you can report your reading to 1/10th of a mm, though the last digit in your reading is approximate. Let us say the reading is 11.15 cm. If you write it 11.150, then it would convey the wrong message. So, the position of the last digit in your measurement indirectly indicates the precision with which reading was taken.

2.7.1 ABSOLUTE PRECISION

The finite precision of measured quantities can be discussed in both absolute and relative terms. **Absolute precision** is expressed in terms to the nearest x that is the position of the last **significant digit**. For example, 21 cm to the nearest cm, 21.0 cm to the nearest 0.1 cm (mm). If a given quantity is 200 m, the precision is not clear. Is it to the nearest m, 10 m, or 100 m? One way of avoiding this ambiguity is to use scientific notation and express the quantity as 2.0×10^2 or 2.00×10^2, as appropriate.

2.7.2 RELATIVE PRECISION

The number of **significant figures** (SFs) indicates the accuracy of a number. This number is equal to the leftmost non-zero digit extending to the right to include all digits. For example, 21, 21.0, 21.00 represent 2, 3, and 4 significant figures respectively, and 2×10^2, 2.0×10^2 represent 1 and 2 significant figures. The following rules should be applied when assigning significant figures to the final answer in a calculation.

1. For addition or subtraction, the answer should contain the same number of decimal places as the measurement that contained the least significant figures. For example, $1.23 + 1.0 = 2.2$ and not 2.23.
2. When a calculation involves multiplication and division, the result should contain as many significant figures as the measurement data containing the least significant figures. Based on this rule, the product of $1.23 \times 2.0 = 2.46 = 2.5$. The final answer contains 2 significant figures and the last significant figure is rounded up since the next figure is 6. This is called the rounding-off rule. The last significant figure retained in the result should be increased by one if the next deleted figure is 5 or more than 5.

2.8 CONVERSIONS

Unit conversions are fact of life. In engineering practice, you will need to perform conversions in two ways:

1. Within the same system of units, for example speed in m/s expressed as km/h.
2. Between systems, from imperial to metric. or vice versa.

A simple straightforward procedure called unit cancellation will ensure proper conversions in any kind of conversion or calculation. The basic principle is that multiplying by unity does not affect the value in a calculation. For example, if a given value of time is given in seconds, we can express it in minutes or hours, as shown below.

$$\frac{min}{60\ s} = \frac{h}{60\ min} = \frac{1440\ min}{d} = 1 = unity$$

When doing conversions, the first thing is to know the conversion factor between the desired and to existing units of measurement. For example, if you need change m into ft, the conversion factor can be read from tables as follows:

$$1\ m = 3.28\ ft \quad or \quad 1\ ft = 0.3048\ m$$

First of all, you need to know only one of the factors – that is either ft to m or vice versa. The next step is to write it as a vertical or stacked fraction.

$$\frac{3.28\ ft}{m} = \frac{m}{3.28\ ft} = \frac{0.3048\ m}{ft} = 1 = unity$$

The trick is to go on multiplying by the conversion factors written as a vertical fraction in progression so that you are cancelling out the unwanted units and getting the desired unit. This will happen when you place the conversion factor (60 min/h or h/60 min) so as to cancel out the unwanted unit. The conversion factor must be put in a fashion so that the unwanted unit appears in the opposite place – numerator versus denominator or vice versa. This procedure, when properly executed, will work for any equation or formula. If you are converting m to ft, then the appropriate vertical fraction desired will be ft over m.

2.8.1 TEMPERATURE CONVERSIONS

Temperature is an important parameter that any water system professional must be familiar with. The two common scales to define temperature are the Celsius scale and the Fahrenheit scale. One problem with temperature conversion is that it is not as straightforward as in the case of, for example, feet and meters. It involves a formula which makes the conversion a bit cumbersome.

$$\frac{C}{100} = \frac{(F-32)}{180}$$

In the scientific world, the absolute scale of temperature is degrees kelvin (K). Adding 273 to a temperature in degrees Celsius converts temperatures to degrees kelvin. The symbol for degrees kelvin is K.

Example Problem 2.1

How much volume of water in m³ does 1 in of water over an area of 1 acre represent?

Solution

$$acre.in = \frac{4840\ yd^2}{acre} \times \frac{(3\ ft)^2}{yd^2} \times \frac{ft}{12\ in} \times \frac{m^3}{(3.28\ ft)^3}$$

$$= 102.8 = \underline{100} = 1 \times 10^2\ m^3$$

You might have noticed that the final answer is shown as 100 (one significant figures) and not 102.8 (four significant figures). This decision is based on the precision of the original data, which in this case was accurate to one significant figures. Our answer cannot be more accurate than the original data or information. Since 100 can be mistaken to represent only one significant figure, it is better to write it in scientific notation.

The unit cancellation method is a very simple and foolproof procedure, but some practice is required. Not only does it prevent any chance of making a mistake and eliminate any guessing (e.g., should the factor appear at the top or the bottom), but the other good news is that the same procedure works whether you need to make one conversion or ten conversions. The trick is to go on writing the factors as vertical fractions until you get the desired units. The more you practice it more you will feel comfortable with its use. Make this a habit. This will go a long way in getting the correct answer and getting rid of "math scare". The following examples demonstrate this procedure.

Example Problem 2.2

The overflow rate in the clarifier represents the upward velocity of water in the device. A typical value is 24 m³/m².d /d say. Let us express it as mm/s and US customary units of gal/ft².d.

Solution

$$V_o = \frac{24\ m}{d} \times \frac{1000\ mm}{m} \times \frac{d}{24\ h} \times \frac{h}{60\ min} \times \frac{min}{60\ s} = 0.277 = \underline{0.28\ mm/s}$$

$$= \frac{24\ m}{d} \times \frac{3.28\ ft}{m} \times \frac{7.48\ gal}{ft^3} = 573.08 = \underline{570\ gal/ft^2.s}$$

After calculating the final answer, always ask yourself the question: Does it make sense? If by chance we put the conversion factor upside down, you would not be able to cancel out the unwanted units. This is a hint to go back and make the correction. Sometimes there is a temptation to go straight to the calculator and start multiplying or dividing by the conversion factor. Never do this. Before performing any calculations, make sure that you have got rid of unwanted units and that the result is in the desired units. It will save you lot of frustration and time.

Example Problem 2.3

In Canada, gasoline is sold in liters and in the USA it is sold in US gallons. In addition, the currencies are the Canadian dollar and the US dollar. An American visitor in Canada went to a gas station and wanted to know the Canadian gas price per gallon in US dollars. Although the currency rates vary day to day, let us assume that the Canadian dollar is the equivalent of 78 US cents. To help the visitor, express the gas price of Canadian $1.73/L to American $/gal.

Solution

$$Gas\ price = \frac{\$1.73}{L} \times \frac{3.78\,L}{gal} \times \frac{0.78\ \$\ American}{\$\ Canadian} = 5.100 = \underline{\$5.10\,/\,gal}$$

Note that the imperial (UK) gallon is larger than the US gallon (imperial gallon = 4.54 L). In this book, a gallon refers to a US gallon unless otherwise indicated.

Example Problem 2.4

In the biological treatment of wastewater, organic loading is indicated in units of mass rate of biochemical oxygen demand (BOD) per unit volume. For conventional activated sludge treatment, the BOD loading rate is typically 30 lb/1000 ft^3.d. Express this in g/m^3.d.

Solution

$$BODL = \frac{30\ lb}{1000\ ft^3.d} \times \frac{454\ g}{lb} \times \frac{(3.28\ ft)^3}{m^3} \times = 480.6 = \underline{480\ g\,/\,m^3.d}$$

Example Problem 2.5

In the treatment of water, the last treatment is usually disinfection by adding chlorine. A chlorinator is set to feed 16 kg/d of chlorine to treat a flow of 4.0 MGD (million gallons per day). What is the chlorine dosage applied?

Given:

$Q = 4.0 \times 10^6$ gal/d $M = 16$ kg/d $C = ?$ Formula $M = Q \times C$

Solution:

Rearrange the formula to find the unknown:

$$C = \frac{M}{Q} = \frac{16\ kg}{d} \times \frac{d}{4.0 \times 10^6\ gal} \times \frac{gal}{3.78\ L} \times \frac{1000\ L}{m^3} \times \frac{1000\ g}{kg}$$

$$= 1.05 = \underline{1.1\ g/m^3\ (mg/L)}$$

Example Problem 2.6

Calculate the water power required for delivering water at the rate of 5.5 L/s against a pressure of 450 kPa. Determine the power added to water?

Given:

$$Q = 5.5\ \text{L/s} \qquad p_a = 450\ \text{kPa} \qquad P_a = ?$$

Solution:

$$P_a = Q \times p_a = \frac{5.5\ L}{s} \times 450\ kPa \times \frac{kJ}{kPa.m^3} \times \frac{kW.s}{kJ} \times \frac{m^3}{1000\ L}$$

$$= 2.48 = \underline{2.5\ kW}$$

2.8.2 STEPS FOR UNIT CANCELLATION

Step I

After solving the equation for the unknown and substitute values, including the units.

Decide on the proper units for the result. Usually, the units for the results are specified in the statement of the problem.

Step II

Cancel units that appear in both the numerator and denominator of the expression.

Step III

Use the conversion factor to eliminate unwanted units and obtain the proper units as decided in Step II. To cancel out, the conversion factor should appear opposite the unwanted unit. Make sure the units are given the same exponents as the variable in the equation.

Step IV

In Step IV, if you fail to get the desired unit, it may be due to a mistake made in the preceding steps. Based on the principle of dimension analysis, if you did the right operations, the units after cancellation should correspond to the quantity on the lefthand side of the equation or the unknown. If it is not so, increase your concentration by a notch or two and find your mistake. Getting the right units for the unknown is a good indication that you are on the right track.

Step V

Perform the calculation and check the answer. Weigh your answer and, if satisfied, round up to the appropriate number of significant places. If there is uncertainty about significant figures in the original data, do not produce your final answer in more than three significant figures.

Discussion Questions

1. List the base units in the SI and USC systems.
2. Describe the procedure of unit cancellation.
3. Compare absolute precision with relative precision and describe the rule for rounding off.
4. Power by definition is the rate of doing work. Based on this, derive the expression for power in terms of: pumping rate, Q, unit weight of liquid γ and pumping head h.

$$P = Q \times \gamma \times h$$

6. Manning's flow equation in SI is

$$v = \frac{1}{n} \times R_h^{2/3} \times \sqrt{S}$$

Where v is in m/s and hydraulic radius R_h is in m and slope is in decimal fraction. Modify this formula for USC units, v is in ft/s and R_h is in ft.
7. Define the unit of horsepower. Estimate how much power an average adult can produce.
8. From the definition of energy and power derive the units of power and express a unit of kW.h in kJ.
9. A valid equation must be dimensionally correct. Comment.
10. Find the history of the SI system of units and list those countries who still have not fully adopted SI.
11. List the advantages of the SI system of units.
12. Differentiate between lb mass and lb force. Derive the unit of slug.

Practice Problems

1. Hydraulic loading on a filter is expressed as 3.0 L/s.m². Express it as m/h (11 m/h).
2. For practice problem 1, express it as gpm/ft² (4.4 gpm/ft²).
3. Typical BOD loading on a facultative lagoon is 5 g/m².d. Express it as lb/1000 ft².d. (1 lb/1000 ft².d).
4. Organic loading on a stabilization pond is 50 lb/acre.d. Express this loading as g/m².d (5.6 g/m².d).
5. Show that flow expressed as acre.in/h is about the same as ft³/s.
6. Overflow rate is 24 m³/m².d. Convert it to ft/min. (5.5×10^{-2} ft/min.).

7. Determine the conversion factor to convert m³/m².d to gal/ft².d. ($\times$ 24.5).
8. Show that concentration of 1 mg/L is the same thing as 1kg/ML.
9. The formula for power added by the pump Pa = Q $\times$ p_a, where Q is the pumping rate in m³/s and p_a is pressure added in kPa. Show that plugging the values of pressure and pumping rate in the respective units you get units of power in kW.
10. The formula for power added by the pump in terms of head added h_a is Pa = Q $\times$ γ $\times$ h_a/C. The value of constant C depends on the units. If Q is gpm, γ = 62.4 lb/ft³, h_a in ft, and answer is desired in horsepower (hp = 550 lb·ft/s)), what is the value of C? (3960).
11. Settling velocity of a discrete particle is estimated to be 0.75 mm/s. Express this value as m³/m².d. (65 m³/m².d).
12. Express filter backwash rate of 13 L/s.m² as gal/ft².min (19 gal/ft².min).
13. Kinematic viscosity of water at 15°C is 1.0 $\times$ 10⁻⁶ m²/s. Express it in USC (3.5 $\times$ 10⁻⁵ ft²/s).
14. General formula of flow over a rectangular weir in consistent units is:

$$Q = \frac{2}{3} \times C_d \times L \times \sqrt{2gH^3} = KH^{1.5}$$

Find the value of K in USC units, assuming C_d = 0.62 (3.33).
15. Head loss due to friction per unit length in a pipe of diameter, D carrying a liquid at flow Q, can be found using the Darcy-Weisbach equation (shown below). Where f is the friction factor (dimensionless) and g is acceleration due to gravity. Modify this equation in terms of Q is gpm D in inches, assuming g = 32.2 ft/s² (C = 32).

$$\frac{h_f}{L} = \frac{f}{1.23\,g} \times \frac{Q^2}{D^5} = \frac{f}{C} \times \frac{Q^2}{D^5}$$

3 Basic Hydraulics

Hydraulics comes from the word hydro, which means water, and deals with the study of water and other liquids in states of rest or motion. A good understanding of some basic concepts and key terms is essential for anyone dealing with the study of water systems.

3.1 FLOW VELOCITY

Flow velocity indicates the average rate at which water moves. Thus, velocity of flow dictates if the solids in water will be carried or settle down. For example, in clarifier tanks, it is important to maintain very low velocity to allow the solids to settle under the forces of gravity.

Key points

I. At a given section, point velocity will be maximum in the middle of the pipe and minimum close the surface of the pipe. Thus, when we refer to flow velocity, we usually mean average flow velocity.
II. Average or mean flow velocity can be defined as flow rate per unit flow area. In mathematical form this can be defined as follows.

Flow rate and flow velocity

$$v = \frac{Q}{A} \quad or \quad Q = v \times A$$

Knowing the flow velocity, flow carried by a conduit can be found. This relationship can also be used to select the size of a pipe to carry a required flow without exceeding the maximum flow velocity. Flow velocity in water mains is maintained in the range of 0.5–1.5 m/s to keep the head loss within acceptable limits. Sewer lines are graded to achieve minimum flow velocity of 0.6 m/s to prevent the settling of solids.

Example Problem 3.1

Calculate the flow-carrying capacity of a 300 mm diameter water main at a flow velocity of 1.5 m/s.

Solution

$$Q = A \times v = \frac{\pi (0.30 \, m)^2}{4} \times \frac{1.5 \, m}{s} \times \frac{1000 \, L}{m^3} = 105 = 110 \, L/s$$

Flow velocity exceeding 3.0 m/s will result in high head losses hence hydraulic grade line becomes steeper.

3.2 CONTINUITY EQUATION

Depending on the conduit section, flow velocity may vary from one section to another; however, the mass of the liquid in a fluid system remains the same. Incoming fluid must equal the outgoing fluid as mass flow rate. The relationship based on this concept is called the **continuity equation**. In other words, for a given flow rate, flow velocity would change inversely proportional to flow area, as shown in **Figure 3.1**.

$$\rho_1 \times A_1 \times v_1 = \rho_2 \times A_2 \times v_2$$

This equation is popularly known as the **continuity equation**. It is valid for both gases and liquids. For incompressible fluids, $\rho_1 = \rho_2$. To a greater extent, this is true for liquids. So, the continuity equation applied to liquids becomes:

Continuity Equation

$$\boxed{A_1 \times v_1 = A_2 \times v_2}$$

Based on the continuity of flow, flow velocity is inversely proportional to the flow area. Since flow area is proportional to the square of the diameter, flow velocity is inversely proportional to the square of the pipe diameter. Reducing the pipe diameter by a factor of two will cause the flow velocity to increase by a factor of four.

Example Problem 3.2

The velocity of flow in a 30 cm diameter pipe is 1.60 m/s. If it discharges through a nozzle 7.5 cm in diameter, find the jet velocity.

Given:

Parameter	Pipe = 1	Nozzle = 2
D, cm	30	7.5
v, m/s	1.6	?

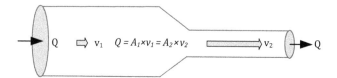

FIGURE 3.1   Continuity of flow

Solution:

$$V_2 = V_1 \times \frac{D_1^2}{D_2^2} = \frac{1.6\ m}{s} \times \left(\frac{30\ cm}{7.5\ cm}\right)^2 = 25.6 = \underline{26\ m/s}$$

3.3 ENERGY AND HEAD

The term **head** is defined as fluid energy expressed per unit weight of the fluid and has dimensions of length. The dimension of length is another reason why this is the preferred term to express fluid energy.

Head as expression of energy

$$Head = \frac{Energy}{Weight} = \frac{J}{N} \times \frac{N \cdot m}{J} = m \quad or \quad \frac{ft.lb}{lb} = ft$$

3.3.1 FLOW ENERGY

Flow energy is the fluid energy due to pressure. The head term corresponding to flow energy is **pressure head**, which represents the height of the liquid to which liquid will rise due to pressure. Mathematically, it is equal to pressure divided by the weight density, γ.

Pressure head

$$h_p = \frac{E_{flow}}{w} = \frac{p \times V}{w} = \frac{p}{\gamma}$$

Weight density of water

$$\gamma_w = \frac{9.81\ kN}{m^3} = 9.81\ kPa/m = \frac{62.4\ lb}{ft^3} = 0.433\ psi/ft$$

3.3.2 KINETIC ENERGY

Kinetic energy refers to energy due to motion. The term corresponding to kinetic energy is called **kinetic** or **velocity head** or **velocity pressure**. Since this energy is due to motion, this term is also called **dynamic pressure**. The formula for velocity head is derived as follows:

Velocity head

$$h_v = \frac{E_{kinetic}}{w} = \frac{1}{2} \times \frac{m \times v^2}{m \times g} = \frac{v^2}{2g}$$

As pressure head can be measured with a piezometer, velocity or **dynamic pressure head** can be measured with the help of a **pitot tube**. A **static pitot tube** reads the sum of pressure head and velocity head at a given point in the flow stream.

3.3.3 GRAVITATIONAL POTENTIAL ENERGY

Gravitational potential energy is energy possessed by the fluid due to its position. The head term corresponding to gravitational potential energy is called elevation head or gravity head and is simply equal to elevation.

Elevation head

$$h_e = \frac{E_{gravity}}{w} = \frac{m \times g \times z}{m \times g} = z$$

3.3.4 HYDRAULIC HEAD

For fluids at rest, flow velocity is zero. Hence, when dealing with static fluids, velocity head is not applicable. Thus, for static conditions, the fluid can have flow energy or potential energy. Sum of pressure head and elevation head is called **hydraulic head** or **grade**. Hydraulic head is also called **piezometric head**.

Hydraulic head

$$h = h_p + h_e = \frac{p}{\gamma} + z$$

In a fluid system, the line indicating hydraulic head at different points is known as the **hydraulic grade line** (HGL) and the slope of this line, the **hydraulic gradient** (**Figure 3.2**). In static flow conditions (valve in closed position), the hydraulic grade line will be horizontal, indicating the hydraulic head remains the same. However, as the valve is opened to introduce the flow, the hydraulic head will decrease in the direction of flow to overcome losses due to friction. Hence HGL will always slope in the direction of flow. The slope of HGL is called the hydraulic gradient or **friction slope**. A steep HGL will indicate high losses.

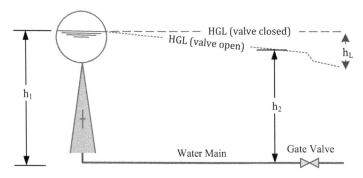

FIGURE 3.2 Hydraulic grade line

3.3.5 TOTAL HEAD

Total head, H, refers to the total energy possessed by the fluid at a given point in the fluid flow system and it is the sum of all the three components.

Total head

$$H = h_p + h_v + h_e = \frac{p}{\gamma} + \frac{v^2}{2g} + z$$

A line representing total head is called the **energy grade line** (EGL). In a given system, the EGL will be above the HGL by an amount equal to velocity head. Under static flow conditions, EGL and HGL will coincide.

Example Problem 3.3

Water is flowing at the rate of 2.0 L/s in a 50 mm diameter pipe. The water pressure in the pipe at a point 5.0 m above the datum is 300 kPa. Calculate the hydraulic head and total head.

Given:

$$Q = 2.0 \text{ L/s} \quad D = 50 \text{ mm} = 0.050 \text{ m} \quad p = 300 \text{ kPa} \quad Z = 5.0 \text{ m}$$

Solution:

$$v = \frac{Q}{A} = \frac{2.0 \, L}{s} \times \frac{m^3}{1000 \, L} \times \frac{4}{\pi (0.05 \, m)^2} = 1.02 = 1.0 \, m/s$$

$$h_v = \frac{v^2}{2g} = \frac{(1.02 \, m)^2}{s^2} \times \frac{s^2}{2 \times 9.81 \, m} = 0.0530 = 0.053 \, m$$

$$h_p = \frac{p}{\gamma} = 300 \, kPa \times \frac{m}{9.81 \, kPa} = 30.58 = 31 \, m$$

$$h = h_p + h_e = 30.58 \, m + 5.0 \, m = 35.58 = \underline{35.6 \, m}$$

$$H = h_p + h_v + h_e = 30.58 \, m + 0.053 \, m + 5.0 \, m$$

$$= 35.63 = \underline{35.6 \, m}$$

Velocity head term is negligibly small in this case. Hence both hydraulic head and energy head have the same value.

3.4   BERNOULLI'S EQUATION

The principle of the conservation of energy states that energy can neither be created nor destroyed. Energy can change its form but the sum total of energy in a given fluid system will remain the same. Potential energy due to elevation in the reservoir transforms into pressure and kinetic energy when water flows in the main line. Consider the pipe flow system shown in **Figure 3.3.** Fluid moves from a larger section 1 to a narrow section 2. Applying the principle of energy conservation, energy at section 1 must be equal to energy at section 2.

Bernoulli's Equation

$$\frac{p_1}{\gamma} + Z_1 + \frac{v_1^2}{2g} = \frac{p_2}{\gamma} + Z_2 + \frac{v_2^2}{2g}$$

3.4.1   LIMITATIONS OF BERNOULLI'S EQUATION

While applying **Bernoulli's equation** to solve fluid flow problems, the following limitations of this equation should be kept in mind:

   i. It is valid only for incompressible fluids (γ = const.). This may restrict the use of this equation to analyzing gas flow problems.
   ii. No hydraulic machines like pumps or turbines can be between the two sections of interest because these would add or remove energy from the fluid.

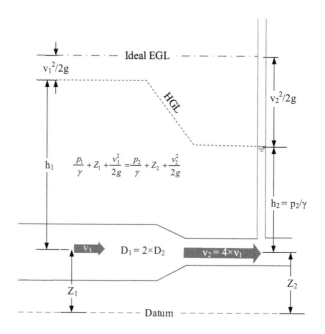

FIGURE 3.3   Bernoulli's principle

iii. No fluid energy can be lost due to friction. This is only true for ideal fluids and very small lengths. Sections having large lengths of pipe or accessories like valves and bends will have significant head losses between two points.

iv. No change in internal energy can occur and there is no heat transfer.

3.4.2 STATIC FLOW CONDITIONS

Under static flow conditions, water is not moving, hence velocity terms are dropped.

$$\frac{p_1}{\gamma} + Z_1 = \frac{p_2}{\gamma} + Z_2 \quad or \quad \left(p_2 - p_1\right) = -\gamma\left(Z_2 - Z_1\right)$$

- For static flow conditions, gain in elevation causes drop in pressure and vice versa.
- Hydraulic head is the same at all points; HGL is horizontal.

Example Problem 3.4

A pressure gauge attached to the bottom of an open water tank reads 55 kPa. How deep is the water in the tank?

Given:
Let us say, open water surface as point 1 and pressure gauge point 2 (reference point or datum)

$Z_1 = ?$ (depth) $Z_2 = 0$ $p_1 = 0$ $p_2 = 55$ kPa $\gamma = 9.81$ kPa/m

Solution:

$$\left(p_2 - p_1\right) = -\gamma\left(Z_2 - Z_1\right) \quad reduces\ to$$

$$Z_1 = \frac{p_2}{\gamma} = 55\ kPa \times \frac{m}{9.81\ kPa} = 5.60 = \underline{5.6\ m}$$

Example Problem 3.5

In **Figure 3.4**, water is flowing from section 1 to section 2. At section 1, which is 25 mm in diameter, the gauge pressure is 450 kPa and the velocity of flow is 3.0 m/s. Section 2 is 4.5 m above section 1 and the pipe diameter at section 2 is twice that of the pipe diameter at section 1. What is the water pressure at section 2?

Given:

Quantity	Section 1	Section 2
p, kPa	450	?
v, m/s	3.0	?
$(Z_2 - Z_1)$, m	4.5	
D, mm	25	50

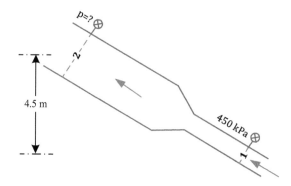

FIGURE 3.4   Bernoulli's Equation (Ex. Prob. 3.5)

Solution:

$$V_2 = V_1 \times \frac{A_1}{A_1} = V_1 \times \left(\frac{D_1}{D}\right)^2 = \frac{3.0\,m}{s} \times 0.5^2 = 0.75\,m/s$$

Doubling the diameter reduces flow velocity to one-fourth. Reduction in flow velocity causes gain in hydrostatic pressure.

$$P_2 = P_1 + \frac{\gamma}{2g}\left(v_1^2 - v_2^2\right) + \gamma(Z_1 - Z_2)$$

$$P_2 = P_1 + \gamma\left[\frac{\left(v_1^2 - v_2^2\right)}{2g} + (Z_1 - Z_2)\right]$$

$$= 450\,kPa + \frac{9.81\,kPa}{m}\left(\frac{\left(3^2 - 0.75^2\right)}{2}\frac{m^2}{s^2} \times \frac{s^2}{9.81\,m} - 4.5\,m\right)$$

$$= 450\,kPa + 4.22\,kPa - 44.15\,kPa = 410.07 = \underline{410\,kPa}$$

3.5   GENERAL ENERGY EQUATION

Bernoulli's equation is modified to include terms for mechanical devices like pumps (h_a) and energy losses (h_l) due to friction (**Figure 3.5**). Accounting for energy additions and subtractions, the general energy equation looks like this:

General energy equation (pump)

$$\boxed{\frac{P_1}{\gamma} + Z_1 + \frac{v_1^2}{2g} + h_a = \frac{P_2}{\gamma} + Z_2 + \frac{v_2^2}{2g} + h_l}$$

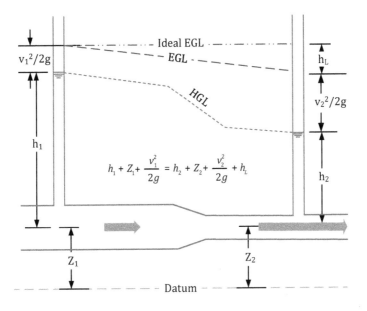

FIGURE 3.5   General energy equation

h_a = head added to the fluid by the pump or **total dynamic head (TDH)**
h_l = head losses due to friction and minor losses = $h_f + h_m$
h_f = head losses due to friction in pipes, major losses
h_m = minor head losses due to changes in flow path and turbulence

Note that not all the terms will be applicable to each and every situation. This is the key when applying the energy equation to solve practical engineering problems. The trick is to select the two sections of interest such that there is only one unknown quantity. For example, if liquid is open to atmosphere at a given point, pressure term can be dropped. Since flow direction is assumed from 1 to 2, the head removed term should either be added on the right-hand side or it should accompany a negative sign if kept on the left side of the equation.

3.6   POWER

Power is rate of doing work or rate at which energy is being spent. Units of power are energy (J) divided by time units (s) or J/s. The power of 1 J/s is also called 1 watt, named for James Watt, who invented the steam engine and introduced the term horsepower. Power added to water – also called water horsepower or **water power** – can be determined knowing the water pumping rate Q and the head added, h_a – also called the **total dynamic head**. Since head is the energy per unit weight, power added to the water can be calculated by multiplying h_a by the **weight flow rate** of water W.

Power added (water power)

$$P_a = W \times h_a = Q \times \gamma \times h_a = Q \times p_a$$

Where p_a is the pressure added to water by the pump. Pressure added is the difference in pressure gauge readings of the pump inlet and outlet.

Example Problem 3.6

A water pumping system is shown in **Figure 3.6**. Calculate the total head on the pump when pumping at the rate of 8.5 L/s. The pumping level in the well (1) is 95 m below the water level in the elevated tank (2). The head losses are estimated to be 3.5 m. Also find the power requirement, assuming the pumping system is 63% efficient.

Given:

$Z_2 - Z_1 = 95$ m $h_l = 3.5$ m $Q = 8.5$ L/s $h_a = ?$ $P = ?$

Solution:

Writing the energy equation between two water levels, and noting that p_1, v_1, p_2, $v_2 = 0$

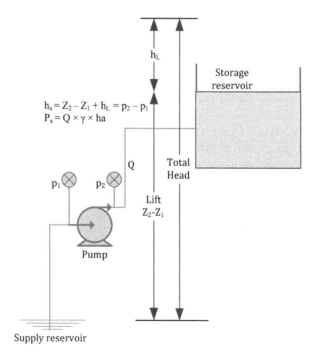

FIGURE 3.6   Pumping system (Ex. Prob. 3.6)

$$h_a = h_l + Z_2 - Z_1 = 3.5\,m + 95\,m = 98.5 = \underline{99\,m}$$

$$P_a = Q \times \gamma \times h_a = \frac{0.0085\,m^3}{s} \times \frac{9.81\,kN}{m^3} \times 98.5\,m \times \frac{kW.s}{kN.m} = 8.25 = 8.3\,kW$$

$$P_l = \frac{P_a}{E_P} = \frac{8.25\,kW}{0.63} = 13.1 = 13\,kW\,(18\,hp)$$

3.7 FLOW EQUATIONS

In the general energy equation, one of the terms is **head loss**. When applying the general energy equation, if the unknown is other than head loss, h_l, then head losses must be estimated. A flow equation relates pipe characteristics, including diameter, roughness, length to the flow rate, and head loss due to friction, h_f. The three equations commonly in use are the Darcy Weisbach, Hazen Williams, and Manning's equations. In addition to head losses due to friction, losses occur due to sudden changes in flow velocity as in the case of bends and valves. Such losses are called **minor losses**. Head losses are estimated using tables or the **equivalent pipe** concept.

Equivalent length, L_e, is the length of straight pipe that will give the same head loss due to friction as the elbow or valve in question. This way, the same formula can be used, by adding equivalent length to the actual length of straight pipe.

3.7.1 DARCY-WEISBACH FLOW EQUATION

The Darcy-Weisbach flow equation is theoretical and can be used for any consistent units. Friction factor f in this equation depends on the **Reynolds number** and **relative roughness**. This equation yields more precise results and is applicable to all kinds of fluids and flow conditions.

Darcy-Weisbach Equation (Consistent Units)

$$h_f = \frac{f}{1.23\,g} \times L \times \frac{Q^2}{D^5}, \quad Q = \sqrt{\frac{1.23\,g}{f} \times \frac{h_f}{L} \times D^5}, \quad D = \sqrt[5]{\frac{f}{12.3\,g} \times \frac{1}{S_f} \times Q^2}$$

Where D, L are the diameter and length of the pipeline and Q is the flow rate. Term h_f represents head loss due to friction and friction factor, f, is a dimensionless number. It is based on two dimensionless parameters: relative roughness and the Reynolds number.

$$\text{Reynolds number, } N_R = \frac{velocity \times diameter}{kinematic\ viscosity} = \frac{vD}{v}$$

$$\text{Relative roughness, } RR = \frac{diameter}{asolute\ roughness} = \frac{D}{\varepsilon}$$

TABLE 3.1

Absolute roughness

Pipe Material	Roughness ε, μm
glass, plastic	smooth
copper, brass	1.5
commercial steel	46
cast iron, 1 uncoated	240
cast iron, 2 coated	120
concrete	1200
riveted steel	1800

$$\text{Friction slope, } S_f = \frac{\text{head loss}}{\text{length of the conduit}} = \frac{h_f}{L}$$

In turbulent flow conditions, the friction factor is dependent on the Reynolds number and the relative roughness of the pipe. Absolute roughness of a pipe is measured in terms of the average thickness of irregularities on the surface of the pipe and is denoted by the symbol ε (epsilon). To make it a dimensionless parameter it is expressed as relative roughness, which is the ratio of the diameter of the pipe to the absolute roughness. Typical values of absolute roughness for commonly used pipe materials are shown in **Table 3.1**.

Moody Diagram

Most of the usable data available for evaluating friction factor has been derived from experimental data. This data presented in graphical form is called a Moody diagram and is widely used. The diagram shows the friction factor f plotted against N_R with a series of parametric curves, each representing a specific value of D/ε. Some authors use the inverse relationship that of ε/D for representing relative roughness. Of course, in the modern world of calculators and computers, not many of us are used to reading graphs as it is more cumbersome and is not suitable for computer use. An explicit relationship like the one shown below is gaining popularity. You can call this the digital form of the Moody diagram.

Friction factor (turbulent flow)

$$f = 0.0055 + 0.0055 \times \sqrt[3]{\frac{20000}{D/\varepsilon} + \frac{1000000}{N_R}}$$

Example Problem 3.7

What lead loss can be expected when a 8 in water main (f = 0.02) is carrying a flow of 500 gpm? The length of water main in question is 840 ft and for losses due to fittings assume equivalent of 50 ft.

Given:

D = 8 in = 0.667 ft   Q = 450 gpm   L+ L$_e$ = 840 ft + 50 ft = 890 ft   h$_l$ = ?

Solution:

$$Q = \frac{500 \, gal}{min} \times \frac{min}{60 \, s} \times \frac{ft^3}{7.48 \, gal} = 1.114 = 1.1 ft^3 / s$$

$$h_l = \frac{f(L+L_e)}{1.23 \, g} \times \frac{Q^2}{D^5} = \frac{0.02}{1.23} \times \frac{s^2}{32.2 \, ft} \times 890 \, ft \times \left(\frac{1.114 \, ft^3}{s}\right)^2 \times \frac{1}{(0.667 \, ft)^5}$$

$$= 4.211 = 4.2 \, ft$$

Example Problem 3.8

A 450 mm diameter water main is an old concrete (ε = 1.2 mm) pressure pipe. Calculate the pressure drop and friction slope between two hydrants 300 m apart when the pipe is carrying 500 L/s of water at 10°C. Assume that hydrants 1 and 2 are at the same elevation.

Given:

D = 450 mm   D$_1$ = D$_2$ ∴v$_1$ = v$_2$,   Z$_1$ = Z$_2$,   v = 1.3 × 10^{-6} m²/s (10°C)   Δp = ?

Solution:

$$h_l = \frac{v_1^2}{2g} - \frac{v_2^2}{2g} + h_a + Z_1 - Z_2 + \frac{p_1}{\gamma} - \frac{p_2}{\gamma} = \frac{p_1}{\gamma} - \frac{p_2}{\gamma} \quad or \quad p_1 - p_2 = \gamma h_l$$

$$v = \frac{Q}{A} = \frac{4Q}{\pi D^2} = \frac{4}{\pi} \times \frac{500 \, L}{s} \times \frac{m^3}{1000 \, L} \times \left(\frac{1}{0.45 \, m}\right)^2 = 3.14 \, m/s$$

$$N_R = \frac{vD}{v} = \frac{3.14 \, m}{s} \times 0.45 \, m \times \frac{s}{1.3 \times 10^{-6} m^2} = 1.1 \times 10^6$$

$$\frac{D}{\varepsilon} = \frac{0.45 \, m}{1.2 \times 10^{-3} m} = 375$$

$$f = 0.0055 + 0.0055 \times \sqrt[3]{\frac{20000}{375} + \frac{1.0 \times 10^6}{1.1 \times 10^6}} = 0.0263$$

$$h_f = \frac{fLv^2}{2gD} = 0.0263 \times \frac{300 \, m}{0.450 \, m} \times \left(\frac{3.14 \, m}{s}\right)^2 \times \frac{s^2}{19.62 \, m} = 8.71 m$$

$$p_1 - p_2 = \gamma h_l = \frac{9.81\,kPa}{m} \times 8.71\,m = 85.45 = 85\,kPa$$

$$S_f = \frac{8.71\,m}{300\,m} \times 100\% = 2.90 = 2.9\%$$

3.7.2 HAZEN-WILLIAMS FLOW EQUATION

The Darcy-Weisbach equation when applied to solve flow capacity, Q, and similar problems is cumbersome and here the Hazen-Williams equation becomes handier. Therefore, empirical flow equations, which provide direct solutions, are commonly used. This equation relates the flow carrying capacity (Q) with the size (D) of pipe, slope of the hydraulic gradient (S_f), and a coefficient of friction C which depends on the roughness of the pipe. However, it is important to note that the value of C is based on the judgment of the designer. Results based on this equation would be less accurate compared to Darcy-Weisbach flow equation.

Hazen-Williams flow equation

$$Q = 0.278C \times D^{2.63} \times S_f^{0.54} \quad or \quad h_f = 10.7 \times \left(\frac{Q}{C}\right)^{1.85} \times \frac{L}{D^{4.87}} - SI$$

$$Q = 0.281C \times D^{2.63} \times S_f^{0.54} \quad or \quad h_f = 10.4 \times \left(\frac{Q}{C}\right)^{1.85} \times \frac{L}{D^{4.87}} - USC$$

Q = volume flow rate, m³/s, (gpm)   C = Hazen-Williams friction coefficient
D = diameter of pipe, m (in) S_f = friction slope = h_f/L = $\Delta h/L$

The Hazen-Williams equation is valid when the flow velocity is less than 3 m/s and the pipe diameter is larger than 2 cm. The roughness coefficient C values for selected pipe materials are given in **Table 3.2**.

Example Problem 3.9

Water from a well is pumped through a 12 in diameter and 950 ft long transmission main. Calculate the head loss for a flow of 850 gpm assuming C =100 (rough) and C = 130 (smooth).

Given:
           D = 12 in h_f = ?   L = 950 ft   Q = 850 gpm   C= 100, 130

Solution:

$$h_f = 10.4 \times \left(\frac{Q}{C}\right)^{1.85} \times \frac{L}{D^{4.87}} = 10.4 \times \left(\frac{850}{100}\right)^{1.85} \times \frac{950\,ft}{12^{4.87}} = 2.87 = 2.9\,ft$$

TABLE 3.2

Hazen-Williams coefficient, C

Pipe Material	C
asbestos cement	140
cast iron	
• cement, lined	130–150
• new, unlined	130
• 5-year-old, unlined	120
• 20-year-old, unlined	100
concrete	130
copper	130–140
plastic	140–150
commercial steel	120
new riveted steel	110

$$h_f = 10.4 \times \left(\frac{Q}{C}\right)^{1.85} \times \frac{L}{D^{4.87}} = 10.4 \times \left(\frac{850}{130}\right)^{1.85} \times \frac{950 \, ft}{12^{4.87}} = 1.769 = 1.7 \, ft$$

Example Problem 3.10

Compute water flow rate in a 1 m diameter commercial steel pipe for an allowable head loss of 4 m per km length of pipeline using Darcy-Weisbach equation and Hazen-Williams flow equation.

Given:

   $D = 1.0$ m $S_f = 4$ m/km $= 4.0 \times 10^{-3}$

Solution:
   To solve this by applying the Darcy-Weisbach equation we have to resort to the trial and error technique. Absolute roughness for commercial steel is 4.5 μm. Since both v and f are unknown, assume f = 0.02 to start with.

$$Q = \sqrt{\frac{1.23 \, g \, S_f D^5}{f}} = \sqrt{\frac{1.23}{0.02} \times \frac{9.81m}{s^2} \times 0.004 \times (1.0 \, m)^5} = 1.55 \, m^3 / s$$

$$N_R = \frac{4Q}{\pi Dv} = \frac{4}{\pi} \times \frac{1.55 \, m^3}{s} \times \frac{1}{1.0 \, m} \times \frac{s}{1.05 \times 10^{-6} m^2} = 1.88 \times 10^6$$

$$\frac{D}{\varepsilon} = \frac{1.0 \, m}{4.5 \times 10^{-5} m} = 2220$$

$$f = 0.0055 + 0.0055 \times \sqrt[3]{\frac{20000}{2220} + \frac{1.0 \times 10^6}{1.88 \times 10^6}} = 0.0172$$

New trial

$$Q = \sqrt{\frac{1.23\, g\, S_f D^5}{f}} = \sqrt{\frac{1.23}{0.0172} \times \frac{9.81\, m}{s^2} \times 0.004 \times (1.0\, m)^5}$$

$$= 1.675 = 1.68\, m^3 / s$$

$$N_R = \frac{4}{\pi} \times \frac{1.675\, m^3}{s} \times \frac{1}{1.0\, m} \times \frac{s}{1.05 \times 10^{-6} m^2} = 2.03 \times 10^6$$

$$\frac{D}{\varepsilon} = \frac{1.0\, m}{4.6 \times 10^{-5} m} = 2220$$

$$f = 0.0055 + 0.0055 \times \sqrt[3]{\frac{20000}{2220} + \frac{1.0 \times 10^6}{2.03 \times 10^6}} = 0.0169$$

$$Q = \sqrt{\frac{1.23\, g\, S_f D^5}{f}} = \sqrt{\frac{1.23}{0.0169} \times \frac{9.81\, m}{s^2} \times 0.004 \times (1.0\, m)^5}$$

$$= 1.689 = 1.69\, m^3 / s$$

Using Hazen-Williams equation, solution is direct. For new pipes, C = 120

$$Q = 0.278\, C\, D^{2.63} S_f^{0.54}$$

$$= 0.278 \times 120 \times (1.0)^{2.63} \times (4.0 \times 10^{-3})^{0.54} = 1.69 = 1.7\, m^3 / s$$

3.7.3   MANNING'S FLOW EQUATION

The Manning's flow equation is also empirical in nature. It can be applied both to closed pipe and open channel flow systems, though it is more commonly used for open channel flow such as storm and sanitary sewers.

Manning's flow equation

$$v = \frac{1}{n} \times R_h^{2/3} \times \sqrt{S} \quad (SI) \quad v = \frac{1.49}{n} \times R_h^{2/3} \times \sqrt{S}\, (USC)$$

n = *Manning's coefficient of roughness and* R_h = *hydraulic radius*

Hydraulics radius is defined as water flow area divided by wetted perimeter. For Circular pipes flowing full or half full, hydraulic radius is equal to one-fourth of pipe diameter. Making this modification, Manning's equation for a circular pipe flowing full becomes:

Manning's flow equation for flow velocity

$$v_F = \frac{0.4}{n} \times D^{2/3} \times \sqrt{S} \quad (SI) \quad v_F = \frac{0.6}{n} \times D^{2/3} \times \sqrt{S} \, (USC)$$

Manning's flow equation for flow rate

$$Q_F = \frac{0.312}{n} \times D^{8/3} \times \sqrt{S} \quad (SI) Q_F = \frac{0.464}{n} \times D^{8/3} \times \sqrt{S} \quad (USC)$$

Example Problem 3.11

Calculate the minimum slope on which a rectangular channel with 1.2 m of width should be laid to maintain a velocity of 1.0 m/s while flowing at a depth of 0.90 m. The channel is made of formed, unfinished concrete.

Given:

$$b = 1.2 \text{ m} \qquad d = 0.90 \text{ m} \qquad v = 1.0 \text{ m/s} \qquad n = 0.014 \qquad S = ?$$

Solution:

$$R_h = \frac{A}{P_w} = \frac{1.2 \, m \times 0.90 \, m}{(1.2 \, m + 2 \times 0.90 \, m)} = 0.36 \, m$$

$$S = \frac{n^2 v^2}{R_h^{4/3}} = 0.014^2 \times \frac{1.0^2}{0.36^{4/3}} = 7.65 \times 10^{-4} = \underline{0.077\%}$$

Example Problem 3.12

A 12 in diameter sanitary sewer with n of 0.13 is laid on a grade of 0.10%. Determine full flow capacity of the sewer pipe.

Given:

$$n = 0.013 \qquad D = 12 \text{ in} = 1 \text{ ft} \qquad S = 0.1\% = 0.001$$

Solution:

$$Q_F = \frac{0.464}{n} \times D^{\frac{8}{3}} \times \sqrt{S} = \frac{0.464}{0.013} \times 1.0^{\frac{8}{3}} \times \sqrt{0.001} = 1.128 = 1.13 \, ft^3 / s$$

$$= \frac{1.128 \, ft^3}{s} \times \frac{7.48 \, gal}{ft^3} \times \frac{60 \, s}{min} = 506.5 = 510 \, gpm$$

Example Problem 3.13

A 300 mm diameter sanitary sewer with n of 0.012 is required to carry a flow of 22 L/s when flowing full. What is the minimum slope required?

Given:

    $n = 0.012$ $D = 300$ mm $= 0.30$ m $S = ?$ $Q_F = 22$ L/s $= 0.022$ m³/s

Solution:

$$S = \left(\frac{Qn}{0.312D^{8/3}}\right)^2 = \left(\frac{0.022 \times 0.012}{0.312 \times (0.3)^{8/3}}\right)^2 = 4.40 \times 10^{-4} = \underline{0.044\%}$$

Discussion Questions

1. List the important characteristics of hydrostatic pressure.
2. Compare the following:
    a. Pressure and pressure head
    b. Absolute pressure and gauge pressure
    c. Hydrostatic pressure and dynamic pressure
    d. Open flow and pipe flow
    e. Absolute roughness and relative roughness.
3. Briefly describe what is meant by continuity of flow.
4. Describe the principle on which Bernoulli's equation is based.
5. What does the term "head" mean? What does hydraulic head represent in a fluid flow system?
6. Define hydraulic grade line and compare it with energy grade line.
7. Which of the flow equations has a theoretical basis? Modify this equation to find friction slope.
8. Discuss the limitations of Bernoulli's equation.
9. Which flow equation is common for solving open channel flow problems? Modify this equation to find the full flow capacity of a circular sewer pipe of diameter D.
10. Modify the power formula for USC units: power in hp, Q in gpm, and h in ft.
11. Modify the Hazen-Williams formula to find friction slope.
12. What is the typical flow velocity in water distribution systems. Show that velocity head for this velocity is negligibly small.

Practice Problems

1. A water reservoir is filled with water to a depth of 18 m. Work out the pressure at the bottom. At what point above the bottom is pressure 100 kPa? (180 kPa, 7.8 m).

2. A chemical tank is filled with a liquid chemical of specific gravity (SG = 1.3). What is the liquid level when the pressure gauge at the bottom of the tank reads 12 psi? (21 ft).

3. What is the flow velocity in a 300 mm diameter water main carrying a flow of 45 L/s? (0.64 m/s).

4. What diameter pipe should be selected to carry 500 gpm without exceeding flow velocity of 2.0 ft/s? (10 in).

5. Water is flowing at a velocity of 2.0 ft/s in an 8 in pipe. The pipe size is reduced to 6 in by placing a reducer. What is the flow velocity in the narrow section? (3.6 ft/s).

6. What size of pipe is needed to carry a water flow of 50 L/s without exceeding a flow velocity of 2.0 m/s? (200 mm).

7. A 150 m long section of a 300 mm water main is made ready for a leakage test. During filling it is recommended not to exceed the flow velocity of 0.30 m/s. What is the rate of inflow? (21 L/s).

8. A 16 in water main is to be flushed maintaining a flow velocity of 5.0 ft/s. What should be the rate of flow to achieve this velocity? (3100 gpm).

9. Calculate the flow carrying capacity of a 400 mm diameter pipe at a flow velocity of 2.0 m/s (250 L/s).

10. Water is flowing at the rate of 12 L/s in a 50 mm diameter pipe. The water pressure in the pipe at a point 7.0 m above the datum is 280 kPa. Calculate the hydraulic head and total head (36 m, 37 m).

11. A pump is pumping water from reservoir A to reservoir B. Total head losses are estimated to be 11 ft when pumping water at the rate of 920 gpm. Determine the water power when difference in water levels in the two reservoirs is 33 ft (10 hp).

12. An 18 in sewer with n = 0.013 is placed on a grade of 0.15%. What is the flow capacity? (1800 gpm).

13. Pressure gauges attached to the inlet and outlet side of a centrifugal pump delivering water @ 12 L/s read 25 kPa vacuum and 150 kPa respectively. Determine the total head and power added by the pump (18 m, 2.1 kW).

14. Estimate the frictional losses in a 2.2 km long, 300 mm diameter water main with a friction coefficient C of 120 while carrying a flow of 50 L/s? (4.6 m).

15. Estimate the flow capacity of a 300 mm diameter main for an allowable friction slope of 0.1%. Assume friction factor f is 0.020 (38 L/s).

16. After cleaning a 8 in diameter water main, it was intended to check in the improvement in the coefficient by performing a pressure test. The pressure drop of 3.5 psi was observed over a length of 2000 ft when carrying a flow of 330 gpm. Compute the improved value of roughness coefficient C (97).

17. A water treatment plant gets its water from a lake. The intake is located below the water surface at an elevation of 230.4 m. The lake water is pumped to the plant influent at elevation 242.4 m. The total head losses are estimated to be 5.5 m when water is drawn at the rate of 340 L/s. Work

out water power and the pump power assuming pump is 72% efficient? (58 kW, 81 kW).

18. What grade is required for a 36 in sewer main, with n = 0.012 to carry a design flow of 11 MGD when flowing full? (0.02%)

19. Select sewer pipe size to convey 2500 gpm without exceeding flow velocity of 10 ft/s and allowable slope of 3.0% (12 in).

20. What friction factor f, will you use for a 300 mm water main with absolute roughness of 250 μm and Reynold's number, N_R of 1.5×10^5? (0.017).

4 Basic Chemistry

Chemicals like alum and chlorine are the backbone of water processing and so you need to know the basics of applied chemistry. Chlorine is commonly known for its disinfecting power. Chlorine is also used for odor control and the removal of iron and manganese. Chemical treatment is used to soften and fluoridate the water, control corrosion and in laboratory testing. Alum is used for coagulation for the removal of turbidity in water treatment and the removal of phosphorus in wastewater treatment.

4.1 STATES OF MATTER

Matter is anything which occupies space and can exist as liquid, solid, or gas. The term fluid is applied to both liquid and gases since both can flow or take the shape of the container in which they are poured. In the topic on hydraulics, above, the physical property of density was discussed. Like density, different types of matter have chemical properties which makes them unique, as indicated by the structure of the atom of a particular element.

4.1.1 STRUCTURE OF ATOM

An atom is the basic building block of all matter. Each element is unique in its properties and the atom is the smallest possible part possessing all these properties. It is the number and arrangement of subatomic particles, protons, electrons, and neutrons that provide uniqueness to each atom. An atom maintains a neutral charge and consists of a positively charged **nucleus** and negatively charged **atomic shell** that contains **electrons**. For an atom to maintain its neural charge, the number of protons and electrons must be equal. The nucleus is composed of positively charged **protons** and of **neutrons** that have no charge.

Atomic mass number or simply **mass number** represents the number of neutrons and protons in the nucleus. Electrons do not contribute to the mass of the atom but occupy very large spaces. **Atomic number** on the other hand, refers to the number of protons or electrons. The number of electrons determines how a given atom is going to interact with other atoms. All the 103 elements are distinguished by their atomic number. Atoms of the same element have the same number of protons or atomic number. However, the number of neutrons may vary to form isotopes. **Isotopes** are atoms of the same element with different numbers of neutrons or mass number. Hydrogen (H), for example, consists of one proton at its nucleus and one electron on its outside shell. An atom of helium (He) has two protons and two neutrons.

Atomic mass, also called atomic weight, is the mass of the nucleus or neutrons and protons combined. The atomic mass of an element is relative to the mass of carbon-12, which is the assigned atomic mass of 12 atomic mass units (amu). The atomic mass of carbon (chemical symbol C) is 12 and its mass number is 6.

DOI: 10.1201/9781003347941-5

4.1.2 PERIODIC TABLE

The periodic table is called the chemist's road map for a good reason. It lists all the elements indicating their chemical symbol, atomic mass, atomic number, and the arrangement of electrons in various orbits. The symbols and atomic mass of some common elements are shown in **Table 4.1**. The unique identity of an element is established by its atomic number and based on this number elements are arranged in the periodic table. Any introductory chemistry book will contain this table. The first element in the periodic table is hydrogen, followed by helium. Oxygen is down the list because its atomic number is eight. Some key points about the periodic table are as follows:

- The chemical properties of an element are more closely related to its atomic number rather than atomic mass.

TABLE 4.1
Atomic mass of common elements

Element	Symbol	Atomic Mass
aluminum	Al	27
arsenic	As	75
cadmium	Cd	112
calcium	Ca	40
carbon	C	12
chlorine	Cl	35
chromium	Cr	24
copper	Cu	64
fluorine	F	19
hydrogen	H	1
iron	Fe	56
lead	Pb	207
magnesium	Mg	24
manganese	Mn	55
mercury	Hg	201
nitrogen	N	14
oxygen	O	16
phosphorus	P	31
potassium	K	39
selenium	Se	79
silicon	Si	28
silver	Ag	108
sodium	Na	23
sulphur	S	32
zinc	Zn	65

- The properties of elements are a periodic function of their atomic number.
- The vertical column is called a group or a family, 1–18 under a new system.
- The elements in a group do not have consecutive atomic numbers.
- The elements in a group have the same number of electrons in the outer shell.
- Going down in a group of the periodic table, the size of the atom increases. In the halogen group, the atomic size increases from fluorine to iodine.
- The electron affinity of elements (tendency to combine) decreases on going down in a group. Fluorine is more reactive than chlorine.
- Rows in the periodic table are called periods or series (1–7).
- Each period represents an electron shell or orbit.
- Moving from left to right in a period, the metallic character of elements decreases. Left-side elements are metals and right-side non-metals.
- In the third period, sodium, potassium, and aluminum are metals and phosphorus, sulphur, and chlorine are non-metals. Moving from left to right in a period, the chemical reactivity of elements first decreases and then increases.

4.2 COMPOUNDS

The way an element behaves chemically depends primarily on the number of electrons in the outermost shell or orbital. As an example, chlorine is very aggressive to combine with other elements while helium is not very reactive. Certain groups of atoms act together as a unit in many different molecules and are called **radicals** (e.g., hydroxyl, carbonates, sulphates, and nitrates). Radicals themselves are not compounds but join with other elements to form compounds.

Atoms of one element combine with those of another in a definite ratio defined by their **valence**. Valence is the combining power of an element based on that of the hydrogen atom, which has been assigned a value of 1. Thus, an element with valence of 2+ can replace two hydrogen atoms in a compound. Compounds are formed by either the transfer or the sharing of the electrons among two or more atoms. For example, table salt, chemically called sodium chloride (NaCl), is formed by the transformation of one electron from the outermost shell of the sodium atom (valence +1) to the outermost shell of the chlorine atom (valence −1). On transformation of electrons, atoms lose their neutrality and are called **ions**. When table salt is dissolved in water, it exits in form of ions. As a result, sodium has a positive charge (cation) and the chloride has a negative charge (anion).

Molecular mass of a compound is the sum of the atomic masses of the combined elements and is known as one **mole**. A molecule of hydrogen gas (H_2) consists of two atoms of hydrogen – hence its molecular mass is 2 or 2 g/mol. Calculation of the molecular mass of another common compound, calcium carbonate, is shown below.

Molecular mass of calcium carbonate

$$CaCO_3 = Ca + CO_3 = \frac{40\ g}{mole} + \frac{12\ g + 3 \times 16\ g}{mole} = 100\ g/mole$$

Equivalent mass represents the combining mass of an ion, radical, or a compound. Equivalent weight is equal to molecular weight per unit valence of that radical or compound. Mass of a compound divided by its equivalent mass yields equivalents of the compound.

equivalent mass = molecular mass / valence

equivalent = mass / equivalent mass

Equivalent mass of calcium carbonate

$$CaCO_3 = \frac{100\ g}{mole} \times \frac{mole}{2\ equivalents} = 50\ g\,/\,eq = 50\ mg\,/\,meq$$

For example, if hardness of water is 200 mg/L as $CaCO_3$, it can be expressed as 2.0 mmol/L or 4.0 meq/L.

Example Problem 4.1

Determine the hardness of water that contains 15 mg/L of Ca^{++} and 12 mg/L of Mg^{++}.

Solution:

$$Hardness = \left(\frac{15\ mg}{L} \times \frac{meq}{20\ mg} + \frac{12\ mg}{L} \times \frac{meq}{12\ mg} \right) \times \frac{50\ mg\ CaCO_3}{meq}$$

$$= 87.5 = 88\ mg\,/\,L\ as\ CaCO_3$$

In other types of compounds, electrons are shared rather than transferred to form a molecule of a compound. Water, the most common compound, is a good example. In the case of water molecule, the two atoms of hydrogen tend to share their electrons with the oxygen atom. This is called **covalent bonding**. Where there is a limited number of elements, the number of compounds is too many to list. Some of the compounds are naturally occurring (e.g., table salt) while others are manmade: household bleach (sodium hypochlorite) is a good example.

ORGANIC AND INORGANIC COMPOUNDS

All organic compounds contain carbon in combination with other elements such as hydrogen, oxygen, nitrogen, phosphorus, and sulphur. Another important characteristic of organic compounds is that they are part of all living things: carbohydrates, proteins, fats, etc. Some of the organic compounds are so complex that scientists are still not able to fully understand them. Inorganic compounds tend to be simpler and usually do not contain carbon, exception being carbon dioxide (CO_2).

4.3   ACIDS AND BASES

Acids and bases are used in many of the water and wastewater treatment processes. In the technical sense, acids are substances that cause an increase of the hydrogen ion (H+) concentration in aqueous solutions. A substance that causes the hydroxyl ion (OH–) concentration to increase is called a base. An aqueous solution with equal amounts of hydrogen and hydroxyl ions is known to be neutral. Pure water, for example, is neutral.

Acids and bases may be characterized as weak or strong depending on the degree to which they cause an increase in hydrogen or hydroxyl ions. From a layman's point of view, substances acidic in nature are on the sour side and bases are usually bitter in taste. Acids turn the litmus paper red, and bases turn it blue.

4.3.1   NEUTRALIZATION

Sulphuric acid (oil of vitriol) and hydrochloric acid (muriatic acid) are strong acids, and acetic acid (household vinegar) is a good example of a weak acid. Sodium hydroxide, commonly called caustic soda, is a very strong base. Such substances are also called **alkaline**. The chemical reaction between acid and base is called neutralization. When this happens, salt and water are formed. Table salt (NaCl), for example, is a product of the chemical reaction between hydrochloric acid and sodium hydroxide. The symbolic representation of a chemical reaction is known as a **chemical equation**. The chemical equation representing the reaction between sodium hydroxide and hydrochloric acid is presented below.

$$NaOH + HCl \rightarrow NaCl + H_2O$$

Things on the left-hand side of a chemical equation are called **reactants** and those on the right are called **products**. In addition, a balanced chemical equation also indicates the number of molecules of each category. This information is used to decide the chemical dosage rate in chemical treatment processes. According to the above equation, one mole of acid reacts with mole of caustic soda to produce one mole of sodium chloride.

4.3.2   THE pH SCALE AND ALKALINITY

Alkalinity and pH are the two most commonly used terms to express water quality or to describe chemical treatment processes. Whereas alkalinity refers to the buffering capacity, pH describes the degree of strength of ionization. It may be helpful to think in terms of pressure and total volume of water available. There is no direct relationship between the pH and alkalinity. However, based on pH, you can know the nature of alkalinity present.

The pH Scale

The pH is a number ranging from 0 to 14 for aqueous solutions. It indicates the degree or strength of an acidic or basic solution. A pH value of 7 – in the middle of the range – is indicative of a neutral solution, or one that is neither acidic nor basic. Pure water is neutral because it contains an equal number of hydrogen and hydroxyl ions. Any little addition of an acid or a base can shift this balance and make the pH to increase or decrease. Rainwater, due to the presence of carbon dioxide, is acidic in nature (pH <7). Mathematically, pH is the negative logarithms of hydrogen ion concentration expressed as mol/L.

$$pH = -\log\left[H^+\right] = \log\left[\frac{1}{H^+}\right]$$

If the hydrogen ion concentration in a water sample is 10^{-6} mol/L, pH of the sample is:

$$pH = -\log\left[H^+\right] = -\log\left(10^{-6}\right) = 6$$

Similarly, if pH is known, the concentration of the hydrogen ion can be found by taking the antilog of the pH value. To remind you, the antilog is the exponential form that is the base raised to the power.

$$\left[H^+\right] = 10^{-pH} = 10^{-6} \; mol/L$$

Similar to pH, pOH is defined as the negative logarithm of hydroxyl ion concentration. When water gets ionized, it possesses both hydrogen ions and hydroxyl ions. In neutral water, the concentration of both hydrogen and hydroxyl ions equals 10^{-7} mol/L. Due to this fact, the following holds true.

$$pH + pOH = 14$$

Since pH is the logarithm of hydrogen ion concentration, each unit change in pH actually represents a tenfold change in the degree of the acidity or alkalinity of a solution. For example, a solution of water with a pH of 8 is 10 times more basic than neutral water. Thus, pH increase by 2 units indicates a solution has become 100 times more alkaline.

Some common liquids at home like vinegar, orange juice, and bleach respectively have approximate pH values of 3, 4, and 13.

Example Problem 4.2

What is the pH and pOH of an aqueous solution known to have a hydrogen ion concentration of $10^{-3.5}$M?

Solution:

$$pH = -\log\left[H^+\right] = -\log\left(10^{-3.5}\right) = 3.5$$

$$pOH = 14 - pH = 14 - 3.5 = 10.5$$

ALKALINITY

In simpler terms, the terms acidity and alkalinity indicate buffering capacity. For practical purposes, the alkalinity of water can be defined as the concentration of carbonates, bicarbonates, and hydroxyl ions. A solution with high alkalinity will resist changes in pH. The alkalinity of water is measured by titrating with a strong acid like H_2SO_4 to a pH near a pH value of 4.5. Acidity and alkalinity are measured in mg/L equivalent amounts of calcium carbonate. Water with low alkalinity will show rapid changes in pH due to the addition of an acidic solution. If natural alkalinity is low, chemicals like caustic, lime, or soda ash can be added to boost alkalinity. If the pH of water is below 8, alkalinity in water is due to bicarbonates that represents temporary hardness.

Example Problem 4.3

Determine the alkalinity of water that contains carbonates, bicarbonates, and hydroxyl concentrations of 20, 488, and 0.17 mg/L respectively.

Solution:

$$Alk = \frac{15 \, mg}{L} \times \frac{meq}{30 \, mg} + \frac{418 \, mg}{L} \times \frac{meq}{61 \, mg} + \frac{0.17 \, mg}{L} \times \frac{meq}{17 \, mg}$$

$$= \frac{7.36 \, meq}{L} \times \frac{50 \, mg \, CaCO_3}{meq} = 368.1 = \underline{370 \, mg \, / \, L \, \text{ as } CaCO_3}$$

4.4 SOLUTIONS

A solution is a uniform mixture of two or more substances existing in a single phase. One component is known as the solvent, which will dissolve the component(s). The other component, which is dissolved, is called the solute. Solutions in water are called aqueous solutions. When you dissolve salt (solute) in water (solvent), you get a brine solution. A **standard solution** is a solution whose exact concentration is known. **Normality**, N, relates to the mass of solute per unit volume of solution. A 1.0N solution contains 1 equivalent mass of a substance in one liter of solution. For example, 1.0N solution of NaOH requires 40 g of pure NaOH be weighed and made up to 1 L by adding distilled water. Similar to normality, **molarity** is defined as number of moles per liter of solution. A 0.1M solution of NaOH in concentration is 0.1 mol/L or 4.0 g/L.

Example Problem 4.4

What is the normality and mass concentration of a 0.25M solution of $CaCO_3$?

Solution:

$$Conc., \, C = \frac{0.25 \, mol}{L} \times \frac{2 \, eq}{mole} = 0.50 \, eq \, / \, L = 0.5 \, N$$

$$= \frac{0.5 \, eq}{L} \times \frac{50 \, g \, CaCO_3}{eq} = 25 \, g \, / \, L$$

4.5 EXPRESSING CONCENTRATIONS

The concentration of a solution refers to the amount of solute in a solution. A dilute or weak solution has a relatively small amount of solute. The properties of solutions like liquid bleach, and suspensions like municipal wastewater, depend to a large extent on their concentrations. Concentrations need to be expressed quantitatively, rather than qualitatively as weak or strong. Concentrations of dilute solutions are expressed in terms of mass per unit volume, or mass per unit mass as parts per million (ppm), or parts per billion (ppb). A concentration of strong solutions as in the case of liquid chemicals is commonly expressed as a percentage or parts per hundred (pph). For example, the chlorine content of commercial chlorine bleach (sodium hypochlorite) is typically 12%.

4.5.1 Mass per Unit Volume, $C_{M/V}$

One of the most common terms for concentration is milligrams per liter, mg/L. If it is found that the concentration of suspended solids (SS) in a wastewater sample is 250 mg/L, it will mean that each million liters (ML) of this wastewater contains 250 kg of dry SS. This equivalence of units needs to be clearly understood, as it will be greatly helpful when estimating dosage rates of chemicals fed to treat a given volume of water or wastewater at a desired concentration.

4.5.2 Mass per Unit Mass, $C_{M/M}$

Concentration of highly concentrated solutions like liquid chemicals and sludges are expressed as mass to mass, as parts per hundred (pph). The two types of concentrations are related by the density of solution as shown below.

Expressing concentration

$$\boxed{C_{m/v} = C_{m/m} \times \rho = C_{m/m} \times SG \times \rho_w}$$

For aqueous solutions below 10%, density of a solution is assumed to be the same as density of water. For such cases, SG can be safely assumed as 1.

$$C_{m/v} = 1\ ppm = \frac{1g}{10^6 g} \times \frac{1\ g}{mL} \times \frac{1000\ mL}{L} \times \frac{1000\ mg}{g} = 1\ mg/L$$

$$C_{m/v} = 1\ ppm = \frac{1}{10^6} \times \frac{8.34\ lb}{gal} = 8.34\ lb/mil\ gal = 8.34\ lb/MG$$

$$C_{m/v} = 1\% = pph = \frac{1}{100} \times \frac{8.34\ lb}{gal} = \frac{0.0834\ lb}{gal} = \frac{10\ g}{L}$$

Unit equivalents

$$\boxed{\frac{mg}{L} = \frac{g}{m^3} = \frac{kg}{ML} = 1\ ppm = \frac{8.34\ lb}{mil\ gal} = \frac{8.34\ lb}{MG}}$$

For highly concentrated liquids, like liquid alum and normal sulphuric acid, the density of the liquid must be considered. For example, commercial liquid alum is typically 48.5% with a specific gravity (SG) of 1.35. Expressed as mass concentration, it equals 655 g/L and not 485 g/L, because liquid alum is 1.35 times heavier than water.

$$C_{m/V} = C_{m/m} \times \rho = \frac{48.5\%}{100\%} \times \frac{1.35 \times 8.34 \ lb}{gal} = 5.46 = 5.5 \ lb / gal$$

$$C_{m/V} = C_{m/m} \times \rho = \frac{48.5\%}{100\%} \times \frac{1.35 \times 1 \ kg}{L} = 0.655 \ kg / L = 655 \ g / mL$$

Example Problem 4.5

100 kg of HTH containing 70 kg of chlorine are added to disinfect a reservoir containing 1.5 ML of water. What is the chlorine dosage applied?

Solution:

$$C = \frac{m}{V} = \frac{70 \ kg}{1.5 \ ML} \times \frac{mg / L}{kg / ML} = 46.6 = 47 \ mg / L$$

4.6 STOICHIOMETRY

Chemical reactions describe the transformation of one or more elements or compounds(reactants) into different products. A **chemical equation** is a shorthand way, through the use of chemical formulas, to write a chemical reaction. A balanced chemical reaction entails not only the reactants and products of a chemical reaction but also the numerical relationship. The number of atoms on each side must be equal. The numerical relationship is called **stoichiometry**. This technique is used to find how much of a one reactant is needed to react with the other reactant to complete the reaction. A common type of stoichiometric relationship is the mole ratio, which relates the amounts in moles of any two substances. In water softening, lime is added to precipitate carbonate hardness, as shown by the following chemical equation.

$$Ca(HCO_3)_2 + Ca(OH)_2 = 2CaCO_3 \downarrow + 2H_2O$$

The 2 in front of $CaCO_3$ is called the **coefficient**. The molar ratio of calcium carbonate hardness to lime is 1:1, although 2 moles of calcium carbonate are produced as precipitate for each mole of lime.

Example Problem 4.6

What dosage of lime as CaO is needed to react with 70 mg/L of calcium carbonate hardness (CaCH) as $CaCO_3$?

Solution:

As shown earlier, $CaCO_3$ is 100 g/mole

$$lime \ as \ CaO = Ca + O = \frac{40 \ g}{mole} + \frac{16 \ g}{mole} = 56 \ g \ / \ mole$$

$$\frac{CaO}{CaCH} = \frac{1 \ mole \ CaO}{1 \ mole \ CaCO_3} \times \frac{56 \ g}{mole \ of \ CaO} \times \frac{mole \ of \ CaCO_3}{100 \ g} = 0.56$$

$$CaO = CaCH \times 0.56 = \frac{70 \ mg}{L} \times 0.56 = 39.2 = 39 \ mg \ / \ L$$

4.7 CHEMICAL FEEDING

Most of the processes in water treatment require chemical feeding. The correct dosing of chemicals is very important in maintaining the efficiency of a process and to avoid overdosing. Chemical dosage and feed rate are based on a mass balance equation. The dilution formula is another expression for the mass balance of a chemical. The dilution formula is useful in performing the calculations for the volume needed to dilute a solution, making a solution of a desired strength and volume, and determining the volume of solution to be fed in order to apply a desired dosage of the chemical.

Dilution formula

$$\boxed{V_1 \times C_1 = V_2 \times C_2 \quad or \quad Q_1 \times C_1 = Q_2 \times C_2}$$

When diluting, subscript 1 refers to the liquid chemical or the original solution, and subscript 2 refers to the desired solution. When dosing a given volume of water, subscript 1 refers to the liquid chemical or its solution and subscript 2 indicates the water being dosed, as illustrated in **Figure 4.1**.

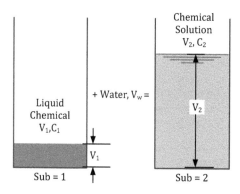

FIGURE 4.1   Making a solution of desired strength

Example Problem 4.7

The strength of the liquid polymer supplied to a water plant is 10%. How many liters of the chemical should be mixed to produce 200 L of 0.5% solution?

Solution:

$$V_l = V_s \times \frac{C_s}{C_l} = 200\ L \times \frac{0.5\%}{10\%} = 10.0 = \underline{10\ L}$$

10 L of polymer is mixed with 190 L of water to make up 200 L of solution.

Example Problem 4.8

What is the concentration in mg/L of 48.5% alum solution with a specific gravity of 1.35? What volume in mL of liquid alum is required to make 2.5 L of 0.10% alum solution?

Given:

$$C = 48.5\% \quad SG = 1.35 (\rho = 1.35\ g/mL) \quad C_2 = 0.1\% = 1.0\ g/L \quad V_2 = ?$$

Solution:

SG of liquid alum is 1.35, meaning thereby it is 1.35 denser than water

$$C = \frac{48.5\%}{100\%} \times \frac{1.35\ g}{mL} \times \frac{1000\ mL}{L} = 654.8 = 655\ g/L$$

$$V_1 = V_2 \times \frac{C_2}{C_l} = 2.5\ L \times \frac{1.0\ g}{L} \times \frac{L}{654.8\ g} \times \frac{1000\ mL}{L} = 3.816 = \underline{3.8\ mL}$$

FEED PUMP SETTING

Since water treatment involves the addition of chemicals, it is important to know how to set the chemical feed pump to apply a set dosage for treating a given flow rate (**Figure 4.2**). Based on the principle of conservation, the ratio of the feed pump rate to the water pump rate must be equal to the ratio of the dosage applied to the strength of the solution feed. In other words, the feed rate of strong solutions will be relatively lower to treat a given water supply at a fixed dosage. It is recommended to adjust the strength of the feed solution so that the feeder setting is about in the middle range (40%–60%). This allows the feeder setting to go up and down and ensures feed pump operation under optimal conditions.

Example Problem 4.9

If the average water flow through the plant is 10.2 ML/d, and the desired chlorine dosage is 7.7 mg/L, what should be the setting for chlorinator feed rate in kg/d?

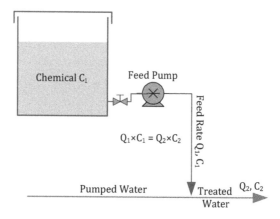

FIGURE 4.2   Liquid chemical feeding

Given:

$$C = 7.7 \text{ mg/L} \quad Q = 10.2 \text{ ML/d} \quad M = ?$$

Solution:
Chlorine feed rate

$$M = C \times Q = \frac{7.7 \text{ mg}}{L} \times \frac{kg/ML}{mg/L} \times \frac{4.2 \text{ ML}}{d} = 32.34 = 32 \text{ kg}/d$$

Example Problem 4.10

Determine the chemical feed pump rate in mL/min to treat flow rate of 4.3 MGD. Liquid alum strength is known to be 5.36 lb/gal and desired dosage is 10 mg/L.

Given:

$$C_1 = 5.36 \text{ lb/gal} \quad C_2 = 10 \text{ mg/L} \quad Q_2 = 4.2 \text{ MG/d} \quad Q_1 = ?$$

Solution:
Feed pump rate of liquid alum

$$Q_1 = \frac{C_2}{C_1} \times Q_2 = \frac{10 \text{ mg}}{L} \times \frac{8.34 \text{ lb}/MG}{mg/L} \times \frac{gal}{5.36 \text{ lb}} \times \frac{4.2 \text{ MG}}{d}$$

$$= \frac{65.35 \text{ gal}}{d} \times \frac{3.78 \text{ L}}{gal} \times \frac{1000 \text{ mL}}{L} \times \frac{d}{1440 \text{ min}} = 171.5 = 170 \text{ mL}/min$$

4.8 COMMON CHEMICALS

Table 4.2 lists many of the chemicals used in water industry. Some of the most commonly used chemicals in water and wastewater treatment are briefly discussed.

4.8.1 ACTIVATED CARBON

Activated carbon has been used for ages to treat taste and odor problems. It is available in powder or granular form. Activated carbon is a very porous material, which is used to adsorb particles from water. Adsorption is a process where particles are trapped on the surface of a material. This differs from absorption, which refers to the trapping of particles within the body of another material. Because of its high surface area, activated carbon is able to remove soluble and colloidal materials which sand filters cannot. It can be reactivated or cleared by intense heat and reused.

TABLE 4.2
Common chemicals in water industry

Chemical Name	Common Name	State	Formula
aluminum sulphate	alum	solid	$Al_2(SO4)_3$
ammonia		gas	NH_3
calcium carbonate	limestone	solid	$CaCO_3$
calcium hydroxide	slaked lime	solid	$Ca(OH)_2$
calcium hypochlorite	HTH	solid	$Ca(ClO)_2$
calcium oxide	lime	solid	CaO
carbon dioxide		gas	CO_2
carbon monoxide		gas	CO
chlorine		gas	Cl_2
copper sulphate	blue vitriol	solid	$CuSO_4$
iron oxide	rust	solid	Fe_3O_4
hydrogen sulphide		gas	H_2S
hydrochloric acid	muriatic acid	liquid	HCl
hypochlorous acid		liquid	$HClO$
nitric acid		liquid	HNO_3
nitrogen		gas	N_2
oxygen		gas	O_2
ozone		gas	O_3
sodium carbonate	soda ash	solid	Na_2CO_3
sodium chloride	table salt	solid	$NaCl$
sodium hydroxide	lye	solid	$NaOH$
sodium hypochlorite	liquid bleach	solid	$NaOCl$
sulphur dioxide		gas	SO_2
sulphuric acid	oil of vitriol	liquid	H_2SO_4

4.8.2 ALUM (ALUMINIUM SULPHATE)

Alum is common name for aluminum sulphate. This metallic salt is often used in the water and wastewater treatment as a chemical coagulant. Coagulants will react with the water and cause the destabilization of suspended particles. In water treatment, alum is used to coagulate turbidity particles, while in wastewater it is used to precipitate soluble phosphorus. This destabilization upsets the repulsive forces between the particles and allows them to stick together forming a **floc**. With gentle mixing, called flocculation of the water, the flocs will gather to gain strength and size and be easily removed through filtration or sedimentation. Although alum is the most common coagulant, many other chemicals, including the iron salts of ferrous sulphate and ferric chloride, are also used.

4.8.3 CHLORINE (Cl)

Chlorination is the most common method of disinfection currently used in North America. Chorine is available as a solution in water (sodium hypochlorite), a powder (calcium hypochlorite), or a gas. Commercial sodium hypochlorite solution is about 12% available chlorine, compared to household bleach, which typically is 5% in strength. One reason for its popularity with small plants is that it is relatively safe to handle. Chlorine gas, on the other hand, is extremely dangerous to handle. It is a severe respiratory irritant, which may result in death, and is flammable and explosive.

- Chlorine gas is a greenish/yellow color.
- It has a penetrating pungent odor
- It is toxic at concentrations as low as 0.1% by volume.
- It is about 2.5 times heavier than air and thus will concentrate in low-lying enclosed areas.
- Chlorine supplied to water and wastewater plants is highly compressed (liquefied) and comes in 68 kg (150 lb) and 1 ton cylinders
- These cylinders must be handled with special care.
- One volume of liquid chlorine equals 450 volumes of chlorine gas.

During chlorine handling, operators must ensure that they are aware of the mechanical procedures, safety procedures, and how to use emergency leak stoppage equipment. Leaks of chlorine gas can be detected by using commercial ammonia, which when passed below the leak produces a white smoke. To work on a chlorine gas leak, an operator must wear a self-contained breathing apparatus (SCBA). Chlorine, in both solution and gas forms, is an effective low-cost method of disinfection. When using chlorine, it is important to ask how much chlorine is required and how long the chlorine should be in contact with the water. The amount of chlorine used is called dosage, C. The time required for disinfection to occur is called contact time, T. Effectiveness of disinfection depends on the dosage and the time of contact, also known as CT factor.

4.8.4 LIME AND SODA ASH

Lime (CaO) and soda ash (Na_2CO_3) are commonly used to soften water. These compounds are also used to add alkalinity when natural alkalinity is low. Lime as CaO is called quick line and slaked lime is $Ca(OH)_2$ or hydrated lime. Soda ash (the commercial name) is available up to 98% purity.

Discussion Questions

1. Differentiate the following:
   a. Element and compound
   b. Atom and molecule
   c. Inorganic and organic
   d. Ionic bonding and covalent bonding
   e. Atomic mass and atomic number.
2. Which chemical is a common coagulant, the main purpose for which is its use in water and wastewater treatment?
3. Discuss the role of pH and alkalinity in chemical reactions.
4. Define a radical and give examples of three common radicals.
5. Define pH and discuss its relationship with pOH.
6. What does the pH of water say about alkalinity in water?
7. If two water samples of different pH values are mixed, the pH of the blended sample will not be the simple average of the two pH values. Discuss.
8. How are the elements in periodic table arranged and what information about each element is shown?
9. What role does density play in expressing concentrations? Show that concentration of 1.0% is equivalent to 10 g/L.
10. Prove that concentration in mg/L is the same as concentration in ppm. What is the relationship between ppb and µg/L?
11. Explain why vents in a chlorine room are provided in the floor?
12. Handling of chlorine requires special care and safety equipment. What other chemicals can be used to chlorinate water?
13. Write a balanced chemical reaction when lime (CaO) is slaked by adding water to yield $Ca(OH)_2$. Using principles of stoichiometry, find the ratio between CaO and $Ca(OH)_2$.
14. Explain neutralization. Write a balanced chemical equation when a solution of sulphuric acid is neutralized using a caustic (NaOH) solution.
15. What is the importance of lime and soda ash in water treatment?

Practice Problems

1. If the pH of a wastewater sample is 7.2, what is the hydroxyl ion concentration? ($10^{-6.8}$ mol/L)
2. After chlorination, the pH of the water is read to determine the efficiency of the chlorination. If the pH of the water is 6.8, determine the pOH and hydrogen ion concentration (7.2, $10^{-6.8}$ mol/L).

3. What is the pH of an acidic solution of sulphuric acid concentrate of 10 mg/L? (4).

4. A water well sample contains 25 mg/L of calcium and 12 mg/L of magnesium. What is the hardness of water as $CaCO_3$? (110 mg/L).

5. A water calcium hardness is 150 mg/L. What is the concentration of the Ca ion? (60 mg/L).

6. The alkalinity of water is 12 mg/L as carbonate ion and 120 mg/L as bicarbonate ion. What is the alkalinity expressed as $CaCO_3$? (120 mg/L).

7. A lab procedure calls for 3.5 moles of sodium bicarbonate ($NaHCO_3$). How many grams? (294 g).

8. What is the percentage of fluoride ion in NaF? (45%).

9. An alum dosage of 40 mg/L is applied in treating a water supply of 5.0 MGD. How many lb of alum are added per day? (1700 lb/d).

10. A chlorine dosage of 1.0 mg/L is applied to a well water. What should be the chlorine feed rate in kg/d if the well is pumped at the rate 25 L/s? (2.2 kg/d).

11. If 45 lb of pure sodium silicofluoride is pumped into 3.5 MG of water, what is the fluoride dosage? (0.93 mg/L).

12. During excess lime softening, carbon dioxide present in water reacts with hydrated lime $Ca(OH)_2$ to form $CaCO_3$ precipitate. Write a balanced equation of CO_2 reacting with hydrated lime $Ca(OH)_2$ to form $CaCO_3$. Calculate the dosage of lime as CaO to react with 15 mg/L of CO_2. If the commercial lime is 88% pure, what dosage of lime is required? (19 mg/L, 22 mg/L).

13. In the recarbonation process, CO_2 is added to convert calcium carbonate to soluble calcium bicarbonate. Write a balanced chemical equation and determine the dosage of CO_2 to convert 40 mg/L of $CaCO_3$ to bicarbonate. (18 mg/L).

14. In excess lime water softening process, lime slurry $Ca(OH)_2$ is added to precipitate $Ca(HCO_3)_2$ as $CaCO_3$. Calculate the dosage of as 85% CaO to react with calcium carbonate hardness of 75 mg/L expressed as $CaCO_3$. (49 mg/L).

15. To make a 2.5 L of a 0.10% solution, how many grams of the dry chemical should be weighed? (2.5 g).

16. Using analytical balance, 15.68 g of dry polymer are weighed out. To prepare a 0.10% solution of this polymer, what volume of solution will you make? (16 L).

17. What is the molecular mass of liquid alum, $Al_2(SO_4)_3.14.3H_2O$? (600 g/mol).

18. What volume of 13.7% concentrated ferric chloride solution (SG = 1.39) is required to prepare 1.0 L of 1.0% solution of ferric chloride? (53 mL).

19. A total chlorine dosage of 12 mg/L is required to treat a particular surface water. If the flow is 5.0 ML/d and the hypochlorite has 65% available chlorine, what feed rate in kg/d of hypochlorite will be required? (92 kg/d).

20. A sludge with 3.0% solids concentration has a volume of 450 000 gal. How many tons of dry solids are in the sludge? (56 tons).

21. It is required to dose a water supply at the rate of 2.0 mg/L chlorine by applying hypochlorite solution with 20% available chlorine. For a water supply rate of 90 L/s, calculate the required feeder setting in L/d. (78 L/d)

22. In water softening, if 25 lb of $Ca(OH)_2$ is added, how many lb of $Ca(HCO_3)_2$ would react with it? (55 lb).

23. A well chlorination pump is set to feed at the rate of 120 L/d. The strength of the feed solution (sodium hypochlorite) is 12.5%, with a SG of 1.21. If the well pumping rate is 16 ML/d, calculate the dosage of chlorine applied. (1.1 mg/L).

24. How many grams of $MgSO_4$ are required to make 5.0 L of 0.1 M solution? (60 g).

25. What volume of 14% concentrated ferric chloride solution (SG = 1.4) is required to prepare 5.0 L of 1% solution of ferric chloride? (260 mL).

5 Microbial Water Quality

Water quality is an important topic in both water and wastewater systems. Disease-causing microorganisms such as some bacteria and viruses are called **pathogens**. The absence of pathogens in drinking water determines its safety. Disinfection, an important process in water treatment, refers to the removal or inactivation of pathogens. In the natural purification and biological treatment of wastewater, microorganisms play an important role. Some basic concepts of microbiology as related to water and wastewater are discussed below.

5.1 BASICS OF MICROBIOLOGY

Microbiology is the branch of the science of biology that deals with microorganisms. These microscopic living things include both plants and animals. Many of us prefer to call them by other names, like critters and bugs, etc. To some of us they might be a source of nuisance, but the reality is they are an important part of many environmental processes responsible for breaking down waste. The biological treatment of wastewater and sludge digestion in a sewage treatment plant are excellent examples. Fermentation of sugars to alcohol, yogurt, and many more such things are all due to the wonders of microorganisms.

5.1.1 BACTERIA

Just as the atom is the basic element in chemistry, the cell is the base of all living things. Bacteria are single celled organisms, which cannot be seen with the naked eye. They are typically 1–2 μm in size and only can be seen with the aid of a microscope. Most living microorganisms are similar in cellular structure to the bacterial cell. They are found everywhere in the world and each particular location usually contains bacteria that are characteristic of that location, the so-called **normal bacterial flora**. The human intestines contain a normal flora of bacteria, which includes the coliform group.

The interior cellular mass contains the cytoplasm in addition to proteins and carbohydrates. The nuclear area contains deoxyribonucleic acid (DNA), which contains the blueprint of the key information needed in reproduction. Depending upon the environmental conditions for their growth and food habits bacteria are classified as aerobes or anaerobes.

Aerobes versus Anaerobes

One way of classifying bacteria depends on how they metabolize their food. Those require oxygen are called aerobic bacteria or aerobes. On the other hand, those living in the absence of free oxygen are called anaerobic bacteria or anaerobes. The end products of aerobic reaction are stable whereas the biochemical reactions of anaerobic

DOI: 10.1201/9781003347941-6

bacteria take longer and their end products are unstable and include acids, alcohols, methane, and carbon dioxide. There is another class of bacteria, which can survive in both types of environments. They are called **facultative**. This distinction is of greater significance in water pollution and wastewater treatment. For example, lagoons or ponds commonly employed in processing wastewater in rural areas are classified as aerobic, anaerobic, or facultative based on the presence or absence of dissolved oxygen.

Autotrophic or Heterotrophic

This distinction is based on the type of food the microorganism requires: organic or inorganic. Those that utilize simpler inorganic compounds are called autotrophic. Those that primarily use complex organic compounds as their source of food supply are called heterotrophic.

The nitrifying bacteria, for example, which use ammonia as food and convert it to nitrate, are autotrophs. Other examples of autotrophs are iron bacteria and sulphur bacteria. These two are nuisance bacteria since they cause many problems in drinking water supplies. Iron bacteria grow rapidly in iron-rich environments (red water) and cause taste and odor problems. Sulphur bacteria, which are also anaerobes, are active in sewers and cause corrosive conditions by converting hydrogen sulphide to sulphuric acid. In addition to the offensive odor, hydrogen sulphide is very toxic.

5.1.2   ALGAE

Algae fall into the category of plants and contain chlorophyll. They are autotrophic organisms and support themselves by converting inorganic materials into organic matter by using energy from the sun. This process of conversion in the presence of sunlight is known as **photosynthesis**. During the process of photosynthesis, like other plants, they take in carbon dioxide and give off oxygen. Algae are of great importance in wastewater treatment and a kind of nuisance in water supplies. In facultative lagoon systems, algae uptake inorganic chemicals and add oxygen to water to maintain aerobic conditions in the top layer. Algae play a role in the **eutrophication** or aging of lakes. When phosphorus and other nutrients are present, the excessive growth of algae, called algae bloom, is often unsightly.

In water supplies, algae are undesirable because they produce bad tastes and odors, and shorter filter runs through the premature clogging of filters. Certain types of algae, for example blue-green algae, are toxic to animals. Lake water rich in algae are commonly populated with cyanobacteria.

5.1.3   FUNGI

Fungi are aerobic, multicellular, non-photosynthetic microorganisms obtaining their food from dead organic matter. Along with bacteria, fungi are the principal organisms responsible for the decomposition of organics in the biosphere. Fungi have two main advantages over bacteria: they can grow in low moisture conditions and low-pH environments. Because of these characteristics, fungi play an important role in the biodegradation of organic matter. Fungi vary in size from microscopic organisms to mushrooms.

5.1.4 Protozoa

Protozoa are the simplest of animal species and consume bacteria and algae for food. Protozoa like ciliates and rotifers are an important part of the population of organisms used in the biological treatment of wastewaters. Rotifers are the simplest multicellular animals.

Some protozoa, including cryptosporidium (or "crypto") and Giardia lamblia, can cause waterborne diseases including dysentery. In water treatment, their removal before chlorination is important since they are highly resistant to disinfection by chlorination.

5.1.5 Viruses

Viruses are the smallest organisms – much smaller than bacteria that are capable of reproducing themselves. They are essentially parasites and require a host in order to live. For this reason, some people debate whether to call them living or non-living. Viruses are extremely small pathogens that can cause a variety of illnesses in humans, including chicken pox, rabies, yellow fever, polio, influenza, and of course the common cold. Viruses that can infect cells of the intestinal tract of humans are called **enteric viruses**.

5.2 MICROBIOLOGICAL CONTAMINANTS

Untreated or raw wastewater can contain any number of harmful microorganisms that can pose a serious health risk to the public. Bacteria are living organisms present in wastewater and may or may not be harmful. When bacteria are referred to as being pathogenic, it means they are "disease causing". Parasites are multi-cellular organisms and can form a tough dormant stage referred to as a cyst, which protects the parasite against disinfection methods. Viruses are non-living DNA chains. If a certain trigger is present, then the viruses comes alive and can cause diseases. Untreated wastewater can also contain worms, such as round, tape, hook, or whip worms, which can cause severe abdominal pain and illness. The **Table 5.1** indicates the types of organism and diseases associated with them.

TABLE 5.1
Diseases and causative organisms

Disease	Causative Organism	Type of Organism
typhoid	salmonella typhosa	pathogenic bacteria
bacillary dysentery	shigella dysenteries, escherichia coli (e. coli)	pathogenic bacteria
hepatitis a, b, c		virus
amoebic dysentery	entamoeba histolytic	parasite
cryptosporidiosis	cryptosporidium	parasite

Bacterial contamination of a drinking water source can lead to that drinking water source becoming unsafe for consumption without treatment. Bacterial contamination of surface water sources can also lead to unsafe drinking water conditions, as well as beach closures or conditions deemed unsafe for recreational uses.

NUTRIENT CONTAMINANTS

Untreated wastewater can contain high levels of **nutrients**, including nitrogen compounds and phosphorous. Nitrogen species that can present a public health or environmental risk include ammonia, which can be toxic to fish and other aquatic life. High levels of nitrate, another nitrogen species present in untreated wastewater, can cause a condition called **methemoglobinemia**, which can be very dangerous to infants. This condition is also called "blue baby syndrome".

Depending on the source of the wastewater, it may also be high in phosphorous. Although phosphorous in not considered a risk to public health, it can present environmental problems in surface waters, leading to excess algae and plant growth. Excess algae and plant growth lead to a lack of oxygen in receiving waters, which in turn can lead to severe degradation of water quality and possible reductions in fish and aquatic animal populations. This is sometimes called **eutrophication**. Excess phosphorous can also lead to blue-green algae outbreaks, some of which are toxic to humans.

5.3   MICROBIOLOGICAL TESTS

Microbiological composition is the most important aspect of drinking water quality because of the possible presence of disease-causing organisms. Most microorganisms are harmless and play an important role in the natural purification of water. Such organisms are called non-pathogens and are beneficial to plants, animals, and soil. Pathogen organisms include bacteria, protozoa, viruses, cysts, and parasitic worms.

5.3.1   INDICATOR ORGANISMS

One of the primary objectives of water treatment is to remove pathogen organisms. Water safe to drink is called hygienic water. As discussed earlier, the number of disease-causing organisms is usually very small compared to the number of non-pathogens. Testing for a specific pathogen such as salmonella (typhoid) or Giardia (dysentery) is very difficult, time-consuming, and almost an impracticable task. The population of these organisms is small enough to elude detection. That is where the role of indicator organisms comes in. The presence of such organisms will signal the presence of pathogens. A group of bacteria that fit into this role is called **coliforms**. Coliforms fit this role very well for the following reasons:

- Coliforms are generally not pathogenic.
- They are always present in contaminated water.
- They are easily detected.
- They are hardier than pathogens, and survive longer than pathogens.

Since coliforms exist in large numbers compared to pathogens and are sturdier, the absence of this group indicates water is hygienic and safe to drink. In simple terms, no coliforms equals zero pathogens. Based on their origin, coliforms are further classified into total coliforms and fecal coliforms. Total coliforms refer to all the members of the group regardless of origin. Fecal coliforms are those from the intestines of warm-blooded animals. E. coli are fecal coliform from humans.

A coliform test is particularly applicable to determine the safety of drinking water. Drinking water must be free of any of coliforms of any kind. A fecal coliform test is appropriate for monitoring the pollution of both surface and groundwaters.

The **coliform group**, whose name means rod-shaped, is made up of various types of bacteria. These are associated with human feces, those of warm-blooded animals, and with soil. Presence of fecal coliform bacteria in a water sample is a better proof of sanitary pollution than the presence of total coliform bacteria because the fecal bacteria are restricted to the intestinal tracts of mammals. **Escherichia coli (E. coli)** is the predominant member of the fecal coliform group and is present in large numbers but has a short life in the environment. *Detection of E. coli in drinking water, therefore, is an indication of recent pollution.*

5.3.2 MEMBRANE FILTRATION METHOD

The simplest and the most common method for microbiological testing for coliforms is the **membrane filter Test (MF).** An aliquot of sample is filtered through a sterile membrane of special design. This membrane has pores of size in the range of 5–10 nm, on which bacteria will be retained, if present. The filter is then rinsed with a sterile buffer solution, placed on a pad saturated with suitable nutrient medium or broth and incubated at an appropriate temperature. Bacteria if present will grow on the medium and form colonies, each colony representing one bacterium in the original sample.

When testing for total coliforms, M-endo broth is used with incubation at 35°C for 20 to 22 hours. Color and shine of the colony are a characteristic of this type of bacterium. Coliform colonies are pink to dark red, with a golden metallic sheen, often with a greenish tinge. Non-coliform bacteria which grow lack this sheen. For fecal coliform count, the medium is M-FC broth, incubated at temperature of 44.5°C for 22 hours. The coliform colonies here are blue and the other bacteria which grow on this medium are gray- to cream-colored. The coliforms are detected by counting the number of colonies on the filter. It can be assumed that each colony counted represents one coliform in the original sample. The results are presented as the number of coliforms per deciliter.

5.3.3 MULTIPLE-TUBE FERMENTATION METHOD

The coliform group of bacteria ferments lactose with formation of gas within a maximum period of 48 hours when incubated at a temperature of 35°C. This characteristic is used to indicate the presence of coliform in a water sample. This method is known as the **multiple-tube fermentation technique**. In this test, a broth containing lactose and other substances, which inhibit non-coliforms, is placed in a series of test tubes, five for each of the three dilutions. For the first dilution, 10 mL of the sample aliquot, 1 mL for the second set of five tubes, and 0.1 mL for the last set. After incubation at 35°C for 24 hours, the formation of gas in tubes is noticed. The presence of a gas a **positive presumptive test**. If no gas is formed, tubes are incubated for another 24 hours, thus making the total incubation period of 48 hours. If no gas is found after 48 hours of incubation, it indicates a negative test for the presence of coliforms.

The test tubes showing positive presumptive gas are further subjected to a **confirmatory test**, which will eliminate certain bacteria of non-sanitary importance. Based on the number of positive test tubes, the statistical methods are used to determine the bacterial density called the **most probable number** or **MPN**. Based on the number of positive test results from each series of test tubes, MPN can be read from tables. A given water supply is considered unsafe if any of the following conditions exist:

- Fecal coliforms are detected in any distribution sample by any analytical method.
- Total coliforms are detected in consecutive samples from the same site or in multiple samples taken as a single submission from a distribution system.

5.3.4 SAMPLE COLLECTION FOR MICROBIOLOGICAL TESTING

Microbiological samples are collected to determine the safety of drinking water supplies. They are manually collected grab samples at designated locations in the water distribution system. Here are some important tips when collecting a water sample for microbiological testing – usually referred as a **bacti test** (bacteriological) sample. This procedure is also suitable for collecting other samples. Bottles used for collecting bacteriological samples should not be rinsed as they contain **sodium thiosulphate** to neutralize any chlorine residual.

- The sampling tap should be free from leaks, have no attachments like an aerator, and able to provide a steady flow.
- If steady flow cannot be obtained, use that faucet to collect a sample.
- Run the faucet for 2 to 5 minutes at a steady flow before collecting a sample.
- Opening the faucet to full flow is not generally desirable.
- Remove the cap from the bottle and hold it with its threads down during sample collection to avoid any contamination
- Without touching the bottle, fill the sample leaving some head space.
- Do not rinse the bottle.
- Label the bottle and refrigerate at 4°C.
- The exception to the above procedure is sampling for lead and copper analysis, where first draw is used to fill bottle.

Preservation and Storage of Samples

Some samples require **preservation** to ensure the stability of the target compounds during transportation and storage or to eliminate substances that may interfere with the analysis. In some cases, preservation of the sample is optional, and, if selected, will allow for a longer storage period before analysis must be initiated. Other parameters require that the analysis be conducted immediately following sample collection.

Storage time is defined as the time interval between the end of the sample collection period and the initiation of analysis. All samples should be stored for as short a time interval as possible and under conditions that will minimize sample degradation.

It may be possible to analyze a number of parameters from one container, provided there is enough sample volume (i.e., sodium, ammonia, nitrate, nitrite, turbidity, alkalinity, pH, chloride, color, sulphate). For required sample volume, one should check with the laboratory. As a rule of thumb, if not indicated by the laboratory, a minimum of 1 L of sample should be collected.

5.4 BIOCHEMICAL OXYGEN DEMAND

Biochemical oxygen demand, or BOD, is defined as the oxygen required to biochemically oxidize biodegradable matter. This parameter is not a pollutant but indirectly indicates biodegradable matter in a sample. BOD indicates the strength of wastewater. In water bodies, it indirectly describes the amount of biodegradable organic matter. Oxidation of organic matter can cause dissolved oxygen (DO) stress due to the depletion of oxygen, and hence impair the quality of the water.

Theoretical oxygen demand, or ThOD, is the total oxygen required for the complete decomposition of pure materials. It can be estimated from stoichiometry. In the case of a hydrocarbon, which contains only carbon and hydrogen, or an alcohol, which also has oxygen, then the decomposition products are CO_2 and H_2O. Once the chemical reaction is known and balanced, the theoretical oxygen demand can be found, as shown in the following example problem. In the case of BOD, oxidation is brought biochemically, that is by the microorganisms to break down organic matter over a certain period.

Chemical oxygen demand (COD) is the oxygen equivalent of organic matter, irrespective of the fact whether or not it is biodegradable. In the majority of organic compounds, COD is very close to ThOD. Whether there is good correlation, COD is typically 0.95–0.98 of the ThOD.

Example Problem 5.1

What is the theoretical oxygen demand of 1.8 mmol/L solution of glucose, $C_6H_{12}O_6$?

Given:

Conc. = 1.8 mmol/L = 0.0018 mol/L   ThOD = ?

Solution:

$$chemical\ reaction: C_6H_{12}O_6 + 6O_2 = 6CO_2 + 6H_2O$$

6 moles of oxygen are required to completely oxidize I mol of glucose

$$ThOD = \frac{0.0018\ mol}{L} \times \frac{6\ mol\ O_2}{mol} \times \frac{32\ g}{mol\ O_2} \times \frac{1000\ mg}{g} = 345.6 = 350\ mg\ /\ L$$

BOD Test

A BOD test is carried out by observing the depletion of dissolved oxygen in a sample or diluted sample after incubation over a given period at 20°C. Dilution becomes necessary when expected BOD exceeds 4.0 mg/L. In most cases, a portion of sample (aliquot) is diluted with water saturated with oxygen and containing other chemicals necessary for growth of microorganisms. A portion of sample (or **dilution factor**) is chosen based on the expected BOD since water can have only a limited amount of dissolved oxygen at a given temperature. As a rule of thumb, the sample dilution factor can be chosen based on expected BOD of the sample.

$$dilution\ factor, DF = \frac{diluted\ sample}{sample\ portion} = \frac{expected\ BOD}{4}$$

To be on the safe side, actually more than one dilution is done. Only BOD dilutions with depletions in the range of 40% to 70% are considered valid. Results are considered invalid if the final DO falls below 2.0 mg/L.

Biochemical oxygen demand calculation

$$BOD = DF \times \left(D_i - D_f \right) = DF \times \left[\left(D_i - D_f \right) - f \left(B_i - B_f \right) \right]$$

D_i, D_f = initial and final DO readings of the diluted sample
B_i, B_f = initial and final DO readings of the seeded blank
f = volume of seed in dilution/volume of seed in blank.

Since BOD tests take five days, it cannot be used for process control purposes. In such cases, a COD test is preferred, which takes about 3 hours to get results and is not affected by the presence of toxins. BOD for most wastewaters is in the range of 60% to 75 % of COD.

Example Problem 5.2

A series of BOD tests were run on a wastewater sample using three different dilutions of 1.0%, 0.5%, and 0.25%. Initial DO in all three cases was 9.5 mg/L and final DO respectively were 2.3 mg/L, 5.9 mg/L, and 7.6 mg/L. What is the BOD of the sample?

Solution:
Observations and results are presented in the following table.

%Dil.	DF	DO_i mg/L	DO_f mg/L	ΔDO mg/L	%	BOD = ΔD × DF mg/L	Comment
1.0%	100	9.5	2.3	7.2	76	720	>70% depletion, Low DO_f
0.5%	200	9.5	5.7	3.8	40	760	DO_f > 2.0 mg/L, ΔDO=40%
0.25%	400	9.5	7.4	2.1	20	840	<40% depletion, High DO_f

For valid BOD tests, percent depletion should be in the range of 40%–70% and final, DO should be above 2.0 mg/L. Second dilution meets both the creteria, hence the result of 760 mg/L is considered to be the correct value.

Example Problem 5.3

A BOD test was carried out on a wastewater sample by making a 1% dilution. Initial and final DO readings in the diluted samples are 7.95 and 2.15 mg/L and the same readings for the seeded blanks are 8.15 mg/L and 7.95 mg/L. Calculate BOD of the sample?

Given:
$$DF = 100\%/1\% = 100 \quad f = 100\% - 1\% = 99\% \quad BOD = ?$$

Solution:
$$BOD = DF \times \left[(D_i - D_f) - f(B_i - B_f) \right]$$

$$= 100 \times \left[(7.95 - 2.15)\frac{mg}{l} - \frac{99\%}{100\%}(8.95 - 7.95)\frac{mg}{L} \right]$$

$$= 480.2 = 480 \; mg / L$$

5.5   NITROGEN (N)

Nitrogen is a vital element for the life processes of living organisms. There is great number of compounds, organic and inorganic, associated with nitrogen. Nitrogen is an important consideration for wastewater effluent since it acts as a nutrient, which stimulates plant growth. In terms of drinking water, nitrates are known to be toxic, particularly to infants, as it can lead to blue baby syndrome. In addition, the presence of nitrogen compounds is often an indication of sewage pollution.

5.6   SOLIDS

Solids occur in water either in solution or in suspension. These two types of solids are distinguished by passing the water sample through a glass fiber filter. Solids retained on top of the filter are known as **suspended** and those passing through the filter are termed as **dissolved** or filterable solids. In wastewater processing, the efficiency of various treatment operations and the quality of the final effluent is usually expressed in mg/L of total suspended solids, TSS.

In drinking water supplies, dissolved solids may cause taste and odor problems. Hardness and corrosion are usually due to an excess of total dissolved solids, TDS.

In wastewater treatment, suspended solids are further classified as settleable and non-settleable. Settleable solids are found by reading the volume of settleable solids in an Imhoff cone. Solids can also be categorized as **inorganic** and **organic**. Organic solids can be broken down by biological processes while inorganic solids cannot and hence are also called fixed solids. Organic solids are measured using the volatile portion of the solids. A **volatile solids test** is where organics are burned off at 550µC, leaving only the inorganic or the inert material. The strength of wastewater is usually expressed in concentration of SS and BOD. **Particulate** BOD refers to BOD contributed by SS.

5.7   HAZARDOUS CONTAMINANTS

Depending on the source, untreated wastewater can contain high levels of various chemicals, some of which may be hazardous. Hazardous chemicals can present serious public health and environmental risks if the collection system is not carefully installed, operated, and maintained. Some communities with high-risk industries will initiate sewer by-laws limiting the amount or concentrations and or discharges of certain chemicals that can be discharged to the public wastewater collection system.

As a rule, only industry effluents that are amenable to sanitary wastewater are allowed to enter into the sanitary sewer system, as is the practice in the developed world. Industries such as dairies and breweries, with high strength wastes (BOD exceeding 200 mg/L and or solids exceeding 300 mg/L), can be allowed provided a municipal plant has the capacity to treat extra wastewater. In such cases, a municipality collects a surcharge fee for high strength discharges.

High strength effluents, for example BOD exceeding 30 mg/L, if discharged to local water bodies will cause water pollution and make it unfit for other uses. In India, since wastewater treatment plants are limited in capacity, industries are not allowed to discharge effluents into the municipal system.

5.8   SAMPLING

One of the most common activities in any water and wastewater treatment systems is obtaining representative samples both in plant and in the field. In water distribution systems, Provincial regulation demands a minimum number and frequency of sampling based on the population served. Provincial laws and the facility's Certificate of Approval, or COA, set specific sampling requirements to ensure that the final product of the treatment facility meets the required health and environmental standards. In addition, sampling within a facility is an important tool for optimizing the operation of the treatment processes.

Experience has shown that the majority of the errors in water testing results are due to improper sampling. The use of highly sophisticated analytical methods is a waste if the sample being analyzed is not a representative one. The need for collecting a representative of the sample cannot be overemphasized. A sample that has been improperly collected, preserved, transported, or identified will result in invalid and useless test results. Whatever the purpose of the sampling, it is essential to sample in a consistent, accurate manner.

All necessary steps must be taken to eliminate any chance of contaminating the sample or the containers used in sampling. Sampling should be performed using a set routine to maintain consistency between sampling periods. Sample locations should also be consistent and must be chosen in order to obtain a water sample, which is representative of the actual flow conditions. It is important to remember that, depending on how the sample is going to be analyzed, different sampling procedures may be required. Finally, it is essential to label all samples indicating location, time and date, type of sample, type of analyses. The sampler should also take notes on weather conditions and other factors.

5.8.1 GRAB SAMPLES

A grab sample, as its name implies, is a single sample taken at one place and time. It is important to note that the test results from a grab sample only represent the condition of water at the particular time and location of the sample collection. Sampling at the same point at a given frequency will allow the identification of trends through time and can provide a measure of change over the long term. Grab samples are collected to

- allow a rapid assessment of source water conditions
- assess treatment performance
- determine parameters that change with time and temperature, like pH.

These samples may be appropriate where the water quality and quantity are not variable. Grab samples are most suitable when testing for chlorine residual, pH, coliform, and dissolved oxygen. They are usually collected manually.

5.8.2 COMPOSITE SAMPLES

Composite sampling is more appropriate when it is necessary to determine an overall or average rather than a snapshot of conditions of the flow stream. Composite samples are obtained by mixing a series of samples taken over a period of time. Composite sampling is not very common in water systems except for determining the average quality of sludges and waste wash water from filters. However, composite sampling is very common in wastewater sampling to determine BOD loading and overall plant effluent quality.

In cases where the flow is highly variable, it is recommended to mix the volume from individual grabs in proportion to flow at that time. Such a sample is called a **flow-proportioned sample**. A simple composite sample is made by mixing the same volume from each individual sample. The collection of composite samples using an automatic sampler is most common in wastewater collection and wastewater treatment plants.

Discussion Questions

1. Why are total coliforms groups used as indicator organisms?
2. Why is a positive test for fecal coliform in potable water considered more serious than a positive test for total coliforms?
3. Do protozoa fall into the plant kingdom or animal kingdom? What role do protozoans play in the biological treatment of wastewater?
4. What role do algae play in facultative stabilization ponds?
5. Name three waterborne diseases caused by pathogens.
6. Describe photosynthesis.
7. Which of the two kinds of protozoans are pathogens and must chemical assisted filtration be used to remove these in a public water supply?
8. Describe briefly two methods used for bacteriological testing of potable water supplies.
9. Describe the procedure for collecting a bacteriological sample from a public water supply?
10. What does the abbreviation BOD stand for and what does it indicate?
11. Differentiate the following terms:
    a. BOD test and COD test
    b. Aerobes and anaerobes
    c. Autotrophic and heterotrophic
    d. BOD_5 and BOD_u
    e. TSS and TDS.
12. When testing for BOD, explain why a wastewater sample is diluted and what does a valid BOD test indicate?
13. Describe two main types of samples and their suitability.
14. Define and describe lake eutrophication.
15. Since a BOD test takes five days to complete, it cannot be used for process control purposes. What are the alternatives?

Practice Problems

1. A dechlorinated sample of secondary effluent was seeded with 0.4% of stale sewage and seed control (blank) was incubated at 3.0% concentration. The DO depletions in the sample using 25% dilution were found to be 3.8

mg/L and that in the blank test was 4.9 mg/L. Calculate BOD of the sample. (13 mg/L).

2. A BOD test run on the primary effluent from the same plant using a 4% dilution yielded initial and final DO readings of 7.90 mg/L and 3.60 mg/L respectively. Determine BOD. (110 mg/L).

3. A 5 d BOD test was run on a raw wastewater sample by adding 6.0 mL of sample aliquot to 300 mL BOD bottle. Initial and final DO readings were observed to be 8.20 mg/L and 4.15 mg/L respectively. Calculate BOD of the raw wastewater. (200 mg/L).

4. A seeded BOD test is to be conducted on food processing waste with an expected BOD of 800 mg/L. The seed has an expected BOD of 200 g/L. What sample aliquots should be used for setting up the middle dilutions of the wastewater and seed tests? (0.5%, 2.0%).

5. A sample of final effluent is dechlorinated, and a BOD test is carried out resulting in the following data. Determine the BOD of the plant effluent. (8.3 mg/L).

Test	Dilution, %		DO, mg/L	
	Sample	Seed	Initial	Final
Sample	40	0.20	8.45	4.45
Blank	0.0	1.0	8.80	5.40

6. In a BOD test of a raw wastewater sample, a DO drop of 3.2 mg/L was observed when 1.5% dilution was chosen. What is the BOD of the sample? (210 mg/L).

7. Certain river water has a BOD_5 of 50 mg/L. The initial DO in the BOD bottle is 8.5 mg/L and the dilution is 1:10. What is the final DO in the BOD bottle? (3.5 mg/L).

8. Calculate the BOD_5 of a water sample. A BOD test was carried out using 3.5% dilution. Initial dissolved oxygen of the seeded dilution water was observed to be 9.1 mg/L and that of the sample aliquot and seeded dilution water, DO, was 8.9 mg/L. After a 5 d incubation period the final dissolved oxygen of seeded dilution water was 8.0 mg/L and the dissolved oxygen of the sample plus seeded dilution water was 2.4 mg/L. (160 mg/L).

9. Determine the theoretical BOD of a 150 mg/L C_4H_7OH solution. (370 mg/L).

10. In a BOD test, 4.0 mL of a raw wastewater sample was added to a 300 mL BOD bottle and filled by adding aerated dilution water. The initial DO of the diluted sample was observed to be 8.1 mg/L. After incubation for a period of 5 d at 20°C, the final DO was read to be 4.5 mg/L. What is the BOD of the sample tested? (270 mg/L).

11. For making a 2.0% dilution, 20 mL of sample aliquot was poured in a graduated cylinder. Dilution water was added to make it to 1000 mL. After mixing, the diluted sampled was transferred to BOD bottles. The initial DO

reading was 8.32 mg/L and the average final DO reading of the three BOD bottles was observed to be 4.20 mg/L. What is the BOD of the sample? (210 mg/L).

12. A suspended solids test carried out on 50 mL of a wastewater plant influent sample yielded the following data: weight of the filter disc = 0.2335 g Filter + dry solids = 0.2531 g Filter + ash = 0.2360 g. Calculate the concentration of suspended solids and fraction of volatile. (390 mg/L, 87%).

13. The weight of an evaporating dish is 38.820 g. A 50 mL volume of filtered sample is evaporated from the dish. The weight of the dish plus residue is found to be 38.843 g. What is the concentration of TDS in the water sample? (460 mg/L).

14. A sample of water from a stream is poured into a 300 mL BOD bottle. Initial DO is found to be 12.5 mg/L. After 5 d incubation at 20°C, the DO in the bottle dropped to 6.3 mg/L. What is the 5 d BOD of the stream water? (6.2 mg/L)

15. In a water pollution control plant daily, the average flow is 4.5 MGD and the peak hourly flow is 7.5 MGD. To prepare 2.5 L of flow proportioned composite sample, what portion of discrete sample collected at the peak hour should be mixed? (170 mL).

Section II

Water Treatment

6 Sources of Water Supply

To design a water supply system for a town, the engineer must know the various sources of water and their characteristics in the vicinity of the town. Unless a steady source of water is available the water supply system becomes unreliable. The origin of all sources of water is precipitation. It can be collected as it falls as rain before it reaches the ground; or as surface water when it flows over the ground surface; or is pooled in lakes or ponds; or as groundwater when it percolates into the ground and flows or collects as groundwater; or from the sea into which it finally flows. The quality of the water varies according to the source as well as the media through which it flows. Water source is either surface water in the form of lakes, reservoirs, and rivers or groundwater in the form of springs, infiltration galleries, and wells.

6.1 SURFACE WATER

Surface water, as the name indicates, is exposed to atmosphere, thus prone to pollution and contamination. However, for larger supplies, these are more reliable.

6.1.1 LAKES AND PONDS

These refer to naturally formed large depressions in the ground filled with water. Usually, lakes are designated in the mountainous region where water from springs and streams flow in whereas ponds are usually designated in the plains where water is collected from the surface runoff. Water from these sources would be more uniform in quality than water from flowing streams. Long storage and detention time permits sedimentation of suspended matter, bleaching of color and the removal of bacteria. Self-purification, which is an inherent property of water to purify itself, is usually less complete in smaller lakes than in larger ones.

The size of lakes and ponds varies greatly. For example, in Canada and USA, **Great Lakes** are large bodies of fresh water and serve very large populations as their water supply. Lake Superior, the largest of Great Lakes, has a detention time of more than one hundred years. Turbidity and color in this water is almost negligible whereas small lakes and ponds may be limited in capacity to serve smaller communities in hilly regions.

6.1.2 RIVERS AND STREAMS

Water from rivers and streams are generally dynamic and are of variable quality and less satisfactory than those from lakes and ponds. The quality of the water depends upon the character and area of the watershed, its geology and topography, the extent

DOI: 10.1201/9781003347941-8

and nature of development, seasonal variations, and weather conditions. Streams from relatively sparsely inhabited watersheds carry suspended impurities from eroded catchments, organic debris, and minerals. Substantial variations in the quality of the water may also occur between the maximum and minimum flows. In populated regions, pollution by sewage and industrial wastes will be direct. Natural and man-made pollution results in introducing color, turbidity, tastes and odors, hardness, and bacterial and other microorganisms in the water supplies. The quantity of water available depends on the type of river: perennial or seasonal. Rivers are the most important sources for public water supply schemes. In perennial rivers, water is available throughout the year whereas in non-perennial rivers it can be used only after constructing storage works such as reservoirs, dams, etc.

6.1.3   ARTIFICIAL RESERVOIRS

These are basins constructed in the valleys of streams. Reservoirs are formed by hydraulic structures thrown across river valleys and the quality of water is more or less the same as that of natural lakes and ponds. While the top layers of water are prone to develop algae, the bottom layers of water may be high in turbidity, carbon dioxide, iron, manganese, etc. The quantity of water depends on the capacity of the reservoir. However, silting of the reservoir may reduce capacity. In certain countries, reservoirs are necessary in order to store stream water during the monsoon season so that water supply works are not affected.

6.1.4   SEAWATER

Although plentiful, it is difficult economically to extract water of potable quality from seawater because it contains about 3% of salts, which involves costly treatment before supply. There are various desalination processes available for the conversion of brackish water into drinking water. Desalting or de-mineralizing processes involve the separation of salt or water from saline water. This is still a costly process and has to be adopted in places where seawater is the only source available and potable water has to be obtained from it, such as in ships on the high seas or in a place where an industry has to be set up and there is no other source of supply.

6.1.5   WASTEWATER RECLAMATION

Sewage or other wastewater of the community may be utilized for non-domestic purposes, such as water for cooling, flushing, lawns, parks, etc., firefighting and for certain industrial purposes, after giving the necessary treatment to suit the nature of use. The supply from this source to residences is prohibited because of the possible cross-connection with the potable water supply system. In places like California, highly treated wastewater is supplied to be used for flushing toilets and other similar purposes. This practice is becoming more common as the water supply in many parts of the world is dwindling.

6.1.6 STORED RAINWATER

At places where no water source is easily available, stored rainwater in cisterns or tanks can be used. Rainwater from roofs and paved courtyards is collected in water-tight tanks. The stored rainwater is of limited capacity and is more susceptible to contamination.

6.1.7 YIELD ASSESSMENT

A correct assessment of the capacity of the source is necessary to decide on its dependability for the water supply project. The capacity of flowing streams and natural lakes is decided by the area and nature of the catchment, the amount of rainfall, and allied factors. The safe yield of a subsurface source is decided by the hydrological and hydrogeological features relevant to each case.

6.2 INTAKE WORKS

Intake works comprise certain works wherein a structure is constructed on a surface water source to impound water, and arrangements are provided to withdraw water from this source and discharge it into an intake pipe. From this pipe, water is made to flow into the water works system. There are reservoir intakes, river intakes, and canal intakes. River intakes are further classified into weir intakes, intake wells, pipe intakes, and intake wells with approach channels.

6.2.1 RESERVOIR INTAKES

As shown in **Figure 6.1**, reservoir intake consists of a circular well of concrete. Its floor is well below the low level of the reservoir. A number of intake pipes provided with screens are installed at different depths, so as to facilitate drawing of clear

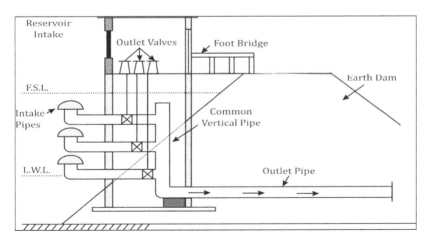

FIGURE 6.1 Reservoir intake

water into the well. It is better to draw water from upper levels, as water at lower levels contains more silt and other impurities.

There are valves provided for drawing off or shutting out water. The top of the dam is provided with a gangway or bridge so as to connect the dam and the valve tower reservoir intake. This is located inside the reservoir so as to get an adequate supply of water in all seasons. Water is drawn from the upstream side of the reservoir, where it is of comparatively better quality and is not affected by the pollutant discharged into the reservoir from the city.

6.2.2 TWIN TOWER RIVER INTAKE

A twin tower river intake (**Figure 6.2**) consists of collector well and a jack well, both well connected by an intake pipe. The **intake well** consists of a concrete masonry well located inside a river. The intake well is circular in shape or more preferably an oblong well. The intake well is located in the riverbed somewhat away from the riverbank. River water enters into the intake well through openings or ports fitted with bar screens. The area of openings is based on the inlet velocity through a screen which is kept in the range of 0.15–0.20 m/s (0.50–0.75 ft/s). Such openings, or ports (**penstocks**), are usually provided at two or more levels. The lower layer of ports permits direct entry during a low level of the river, as experienced during the summer.

The **intake pipe** connecting the intake well with the jack well is usually a non-pressure type and is laid on a gentle slope. Flow velocity in the intake pipe should not exceed 1.2 m/s (4.0 ft/s). The ends of the pipes are fixed with strainers. The water is drawn into a **jack well** and then pumped to purification works. This is a cheap and simple arrangement. Water entering the jack well is pumped to treatment works. A jack well should be founded on hard strata.

6.2.3 SINGLE WELL TYPE RIVER INTAKE

In alluvial rivers, water is usually ponded by constructing a weir across the river. This is done so that the river water level in the dry season is sufficient to provide

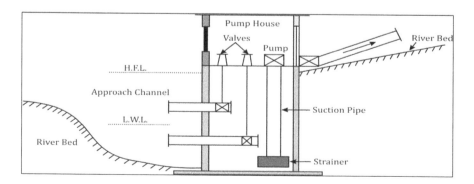

FIGURE 6.2 River intake

water. Intake water may be drawn from the weir through a channel into a sump well from which it can be pumped to the water supply. This arrangement will eliminate the construction of a separate inlet well and inlet pipe, as required in the twin tower case.

6.2.4 LAKE INTAKE

For obtaining water from lakes, generally submersible intakes are used. These intakes are constructed in the bed of the lake below the low water level. Essentially they consist of a pipe laid in the bed of the lake. The opening of the pipe is fitted with a bell mouth opening covered with a screen. The water enters the pipe through the bell mouth opening and flows under gravity to the sump well and is then pumped to the treatment plant.

6.3 WATER TRANSMISSION

The collected water is conveyed or transported from the source to the treatment works. After treatment, the water is conveyed to consumers via the water distribution system. Normally, the sources of supply are far away from towns and cities. Hence, structures called conduits have to be constructed. Conduits may flow under gravity or pressure. Pipes used for transmission are also called low pressure pipes, since they usually flow under low pressure conditions.

6.4 GROUNDWATER

Rainwater percolating into the ground and reaching permeable layers (**aquifers**) in the zone of saturation constitutes a groundwater source. Groundwater is normally beyond the reach of vegetation and is usually free from evaporation losses. As the water seeps down, it comes into contact with organic and inorganic substances during its passage through the ground and acquires chemical characteristics representative of the strata it passes through (hardness due to limestone, iron and manganese, fluoride, arsenic, etc.). Percolation into the sub-soil at times results in the filtering out of bacteria and other living organisms.

Generally, groundwater is clear and colorless but is harder than surface waters of the region in which it occurs. Groundwater is generally of uniform quality although changes may occur in the quality because of water logging, over-draft from areas adjoining saline water sources, and recycling of water applied for irrigation.

6.4.1 WATER WELLS

Water wells are structures made to tap groundwater from greater depths. In former times, wells used to be of large diameter and relatively shallow since modern methods of drilling did not exist. The larger size of the open well allows the groundwater to seep into the ground and acts as storage. Modern wells are small diameter holes dug into ground to tap groundwater. Such wells, called **tube wells**, consist of a conduit section to accommodate a pump. At the lower end of the pipe is a strainer that is the intake portion of the well.

6.4.2 Springs

Springs may be considered as outcrops of groundwater, which often appear as small water holes at the foot of hills or along riverbanks. Springs may be either perennial or intermittent. Springs are of two types: gravity and artesian. A **gravity spring** may result from the outcropping of an impervious stratum underneath a water-bearing formation. It may also be due to overflow of the water table by a continuous rise in it.

An **artesian spring** is formed when two impervious strata sandwich an aquifer or the water-bearing stratum being under pressure. Water flows through the weaker spots or fault or crack in the rock. The yield of the gravity spring varies with the position of the water table and of the rainfall, whereas the yield of the artesian spring is more uniform. Their usefulness as sources of water supply depend on the discharge and its variability during the year. It is a common practice to construct collecting tanks at the point of springs to ensure uniform supply and protection from contamination. Spring waters from shallow strata are more likely to be affected by surface contamination than deep-seated waters. Protected spring water contains minerals, which may supply the essential nutrients for organisms (mineral water).

6.4.3 Infiltration Galleries

An infiltration gallery is another way of tapping groundwater. It is essentially a shallow tunnel (3 m–5 m deep) dug along the banks of the river through the water-bearing strata. In the tunnel, pipes are laid under the ground to withdraw groundwater at moderate depths. The shape of the tunnel is generally circular, and they are covered by graded filter material to retain unwanted solids. The graded material is laid in layers around the perforated pipe collecting filtered water. The outer layer covering the pipe consists of relatively fine material typically in the range of 3 mm to 10 mm in size. The inner layer is of coarser material varying in sizes from 25 mm to 50 mm (1–2 inches). This make the gradation in the direction of flow from finer to coarser. The length of the gallery varies depending on water demand and permeability of the formation and ranges from 10 m to 100 m (30 ft–300 ft).

6.4.4 Collector Wells

Radial collector wells are also called **horizontal wells**. They consist of a concrete cylinder typically 5 m (15 ft) in diameter which is sunk into the aquifer by excavating the earth. When the desired level is reached, it is sealed at its bottom with a thick concrete plug. The concrete cylinder has precast ports at the bottom end. Through these ports, a number of radial collector pipes are jacked horizontally into the formation. Perforated pipes with proper screens are inserted in these horizontal pipes.

Collector wells extract relatively large supplies of groundwater from valley fills and alluvial aquifers of high permeability. Entrance velocity of water is low, thus chances of premature plugging and scale deposition are minimized.

6.5 WELL TYPES

Based on the type and depth of aquifers tapped, wells can be called shallow, deep well, artesian, or water table wells (**Figure 6.3**). **Shallow wells** get their water supply from the subsoil water table. **Deep wells** derive water from more than one aquifer. **Artesian wells** are formed when a porous aquifer is enclosed between two impervious strata. When the piezometric surface lies above the ground surface, it forms a **flowing well** and water flows without pumping due to piezometric pressure. Depending on the method of construction, wells are classified as dug well, driven, bored, and drilled. **Dug wells** (also called percolation wells) and driven wells are shallow wells which are usually confined to soft ground, sand, and gravel. Dug wells are highly susceptible to contamination. The diameter of these wells ranges from 1 m to 4 m and in depth to about 20 m (65 ft). Tube wells are deep wells specially driven to obtain more yield. Due to the low yield of dug wells or driven wells, they are useful for small localities, whereas tube wells can be used for supplying water to larger areas.

Municipal wells are usually drilled. Except in consolidated formations, the end portion of the well consists of a strainer or screen. Screen slots or openings are selected based on the gradation of the aquifer material. Screen open areas should be such that entrance velocity is less than 0.03 m/s. In fine and non-uniform aquifers, a **gravel pack** is poured around the screen to make intake portion hydraulically more efficient. Municipal wells are usually gravel packed.

Recently, **tube wells** are commonly employed to withdraw groundwater for municipal supplies. Tube well consists of the lowering of a pipe in a hole drilled deeper so as to tap artesian aquifers for high yield and reliable water supply. Tube wells are as deep as 300m to 400 m (1000–1300 ft). Unfortunately, due to excessive

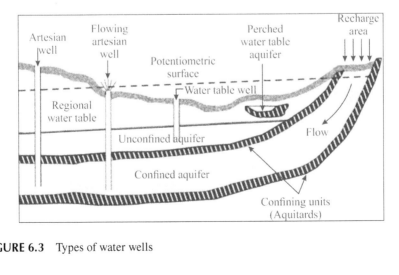

FIGURE 6.3 Types of water wells

draft, there is a significant lowering of the water table in many areas. This problem has made pumping uneconomical, and, in some cases, water quality has deteriorated to the extent that water is not suitable for drinking purposes. The main cause of this problem is that groundwater is extensively used for agricultural irrigation.

6.6 WELL HYDRAULICS

A **water well** is a hydraulic structure, which, when properly designed and constructed, permits the economic withdrawal of water from an aquifer. In the proper design and operation of wells, it important to understand some basic principles of well hydraulics. Darcy's Law combined with Laplace's equation allows us to study the groundwater movement under a given set of conditions. In following section, derivation, and application of equilibrium (steady state), well equations for artesian aquifers and water table aquifers are discussed. The application of a non-equilibrium well equation, pertaining to well testing and determination of aquifer parameters, is discussed briefly.

6.6.1 STEADY FLOW TO AN ARTESIAN WELL

Referring to **Figure 6.4**, pumping an artesian well under equilibrium conditions is achieved when there are no further changes in the drawdown (based on Darcy's Law). For steady state, this rate must be equal to the pumping rate Q and must equal

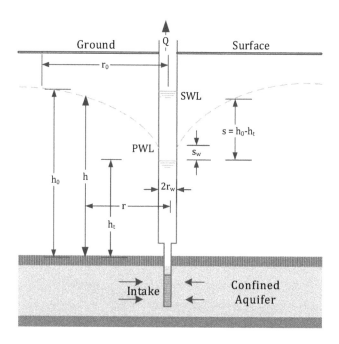

FIGURE 6.4 Pumping an artesian well

the water entering radially from the aquifer into the well. Understanding this, the following equation can be derived.

Equilibrium equation (artesian well)

$$Q = \frac{2\pi kb\left(h_2 - h_1\right)}{\ln\left(r_2 / r_1\right)}$$

Subscripts 1 and 2 are the two observation wells to observe drawdowns.

Artesian well yield

When point 1 refers to the outside of a well and point 2 lies at radius of influence, that is $r_1 = r_w$ and $r_2 = r_0$, the equilibrium well equation becomes:

Artesian well yield

$$Q = \frac{2\pi kb\left(h_0 - h_w\right)}{\ln\left(r_0 / r_w\right)} = \frac{2\pi T\left(h_0 - h_w\right)}{\ln\left(r_0 / r_w\right)}$$

$(h_0 - h_w) = s_w$ = drawdown in the pumped well

Specific capacity

Well yield is more commonly expressed as the specific capacity of the well. Specific capacity, SC, of a well is the yield per unit drawdown in the pumped well.

Specific capacity

$$SC = \frac{Q}{s_w} = \frac{2\pi kb}{\ln\left(r_0 / r_w\right)} = \frac{2\pi T}{\ln\left(r_0 / r_w\right)}$$

It can be said that the yield of the well is directly proportional to the drawdown. This is not strictly true since some drawdown is caused by well losses. In an ideal case, well losses are negligible. In real situations, well losses will be greater than zero. In other words, the drawdown inside the pumped well will be greater than the theoretical drawdown as given by the well equation. The difference in the actual and theoretical drawdown indicates the well losses. The well losses might increase over the years, thus causing a drop in well efficiency.

Well Efficiency

Well efficiency accounts for well losses as the water enters the well intake section. In a perfect well (ideal) loss would be zero. However, in the real world, there will always be some losses. The equilibrium equation accounts for drawdown due to formation only. Hence, drawdown in the pumped well will be more and hence well efficiency is less than 100%.

$$E_w = \frac{s_i}{s_a} = \frac{ideal}{actual} = \frac{Formation\ loss}{Formation\ loss + Well\ loss}$$

Example Problem 6.1

A well is installed in a 15 m thick sandstone artesian aquifer at a depth of 35 m from ground surface. The diameter of the intake section is 0.6 m. The piezometric surface is at a depth of 9.0 m and the radius of the cone of depression is 300 m. Given that the permeability of the sandstone is 0.075 mm/s, calculate the pumping rate that will lower the water level in the well to 12 m above the aquifer.

Given:

$$k = 0.075 \text{ mm/s} = 7.5 \times 10^{-5} \text{ m/s} \quad b = 15 \text{ m} \quad Q = ?$$

Parameter	Pumped Well	Radius of influence
Distance r, m	0.30	300
Head h, m	12.0	26.0
Static head h_o, m	26.0	26.0
Drawdown s, m	14.0	0.0

Solution:

$$Q = \frac{2\pi k b \left(h_0 - h_w\right)}{\ln\left(r_0 / r_w\right)} = 2\pi \times 15\,m \times \frac{7.5 \times 10^{-5}\,m}{s} \times \frac{(26-12)\,m}{\ln\left(300 / 0.30\right)}$$

$$= \frac{0.0143\,m^3}{s} \times \frac{1000\,L}{m^3} = 14.3 = \underline{14\,L/s}$$

Estimating T or K

Applying the equilibrium equation, the **transmissivity** of an aquifer can be determined by observing the data at two points when the well installed in that aquifer is pumped.

$$Q = \frac{2\pi T \left(h_2 - h_1\right)}{\ln\left(r_2 / r_1\right)} \quad or \quad T = kb = \frac{\ln\left(r_2 / r_1\right)Q}{2\pi\left(h_2 - h_1\right)}$$

All the parameters on the right-hand side of this equation can be determined from a pumping test. The well is pumped at a constant rate Q to reach the equilibrium conditions. Drawdowns in two observation wells located at distances r1, and r2 from the pumped well are observed.

Radius of Influence

When drawdowns (s_1, s_2) in two observation wells 1 and 2 ($r_2 > r_1$) are known, the radius of influence can be estimated by employing steady state equations.

Example Problem 6.2

A 2 ft diameter well is constructed in 48 ft thick confined aquifer. Before pumping started, pumping, the static water level was measured to be 60 ft above the top of aquifer. The well is pumped at the rate of 200 gpm and equilibrium conditions are achieved after 2 days of pumping. The observed data is given in the following table. Calculate the transmissivity, specific capacity, and radius of influence.

Given:

$$Q = 13 \text{ L/s} \quad b = 48 \text{ ft} \quad \Delta s = 12.1 - 7.9 = 4.2 \text{ ft} \quad s_{w\text{-actual}} = 32.5 \text{ ft}$$
$$K, T, SC, s_w, E_w = ?$$

Parameter	Pumped Well	Observation Well 1	2	3
r, ft	1.0 (r_w)	30	90	(r_0)
s, ft	31.5	12.1	7.9	0
h, ($h_0 - s$), ft	?	47.9	52.1	60

Solution:

Aquifer parameters

$$T = \frac{\ln(r_2/r_1)Q}{2\pi\Delta s} = \frac{\ln(90/30)}{2\pi} \times \frac{200 \, gal}{min} \times \frac{1}{4.2 \, ft}$$

$$= \frac{8.326 \, gal}{ft.min} \times \frac{ft^3}{7.48 \, gal} \times \frac{1440 \, min}{d} = 1602.9 = 1600 \, ft^2/d$$

$$K = \frac{T}{b} = \frac{1602.90 \, ft^2}{d} \times \frac{1}{48 \, ft} \times \frac{d}{1440 \, min} = 0.0231 = 0.023 \, ft/min$$

Well Efficiency

$$(s_w - s_1) = \ln\left(\frac{r_1}{r_w}\right)\frac{Q}{2\pi T} = \ln\left(\frac{90 \, ft}{1.0 \, ft}\right) \times \frac{200 \, gal}{min} \times \frac{ft.min}{2\pi \times 8.326 \, gal}$$

$$= 17.203 = 17.2 \, ft$$

$$s_w = s_1 + 17.2 \, ft = 12.1 \, ft + 17.2 \, ft = 29.30 = 29 \, ft$$

$$E_w = \frac{s_w}{s_{act}} = \frac{29.3 \, ft}{32.5 \, ft} \times 100\% = 90.15 = 90\%$$

6.6.2 UNCONFINED WELL EQUATION

The equilibrium equation for a water table well can be found by considering a cone of depression of radius r and thickness dr and height h such that gradient at point r is

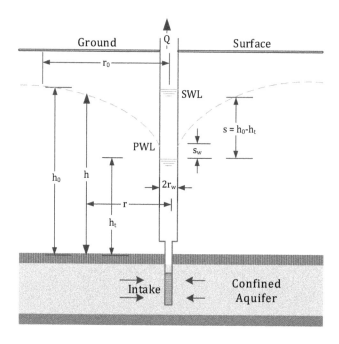

FIGURE 6.5   Pumping a water table well

dh/dr as shown in **Figure 6.5**. Under equilibrium conditions, the pumping rate must be equal to water entering the cylinder defined by the outside boundary of the cone of depression

Equilibrium equation (water table well)

$$Q = \frac{\pi k \left(h_2^2 - h_1^2 \right)}{\ln \left(r_2 / r_1 \right)}$$

Assumptions
Derivation of the well equilibrium equations are based on the following assumptions:

   i. Hydraulic conductivity is constant.
  ii. Aquifer thickness is constant.
 iii. Well losses are negligibly small.
  iv. The intake portion of the well penetrates the entire aquifer.
   v. The water table or potentiometric surface is horizontal.
  vi. Laminar flow exists throughout the aquifer.
 vii. The cone of depression has reached equilibrium so that there is no further enlargement of radius of influence as pumping continues.

These assumptions seem to limit the use of well equations because in the real world these assumptions are rarely found completely valid. In fact, these equations can yield reliable results provided good judgment is applied. If in a certain case there is a serious derivation from the above-mentioned assumptions, extra caution should be exercised in making decisions based on the results obtained using equilibrium equations.

Example Problem 6.3

The pumping test was conducted in an unconfined aquifer using a test well extending to the underlying impervious stratum at a depth of 20 m. From the observed data given below, calculate the coefficient of permeability.

Given:

$$Q = 35 \text{ L/s} \quad h_0 = 20 - 5 = 15 \text{ m}$$

	Observation Well	
Parameter	1	2
Distance r, m	0.30	300
Drawdown s, m	14.0	0.0
Head h = h_0 – s, m	11.96	14.2

Solution:
  Hydraulic conductivity

$$k = \frac{Q \ln(r_2 / r_1)}{\pi\left(h_2^2 - h_1^2\right)}$$

$$= \frac{35 \text{ L}}{\text{s}} \ln\left(\frac{120 \text{ m}}{20. \text{m}}\right) \times \frac{1}{\pi\left(14.2^2 - 11.96^2\right)\text{m}^2} \times \frac{\text{m}^3}{1000 \text{ L}} \times \frac{1000 \text{ mm}}{\text{m}}$$

$$= \frac{0.34 \text{ mm}}{\text{s}} \times \frac{3600 \text{ s}}{\text{h}} \times \frac{24 \text{ h}}{\text{d}} \times \frac{\text{m}}{1000 \text{ mm}} = 29.0 = \underline{29 \text{ m/d}}$$

6.6.3 Modified Non-Equilibrium Equation

Usually, the pumping conditions are unsteady. Under unsteady conditions, well drawdown goes on increasing with the increase in pumping duration, although at a lesser rate. When pumping conditions are unsteady, aquifer parameters can be found using Cooper and Jacob modified non-equilibrium equations. For example, an artesian aquifer with transmissivity T and coefficient of storage S is pumped at rate Q. When r is small and t is large, drawdown s at any radial distance r from the pumped well is given by the following equation.

Modified Non-Equilibrium Equation

$$s = \frac{Q}{4\pi T} \times \ln\left(\frac{2.25Tt}{r^2 S}\right) = \frac{0.183Q}{T} \times \log\left(\frac{2.25Tt}{r^2 S}\right)$$

Based on the modified equation, the following conclusions can be drawn:

i. Drawdown increases in direct proportion with increasing pumping rate.
ii. For the same pumping rate, drawdown will be high in an aquifer with poor transmissivity.
iii. For a given pumping rate, drawdown per log cycle of time, that is $(t_2/t_1) = 10$ remains constant. This allows us to predict drawdown at various continuous pumping times
iv. Drawdown decreases with logarithms of distance away from the pumped well. Put differently, the increase in drawdown shrinks as you move away from the pumped well.

6.7   FAILURE OF WELLS AND REMEDIATION

The clogging of wells by filling with sand or by corrosion or incrustation of the screen may reduce the yield very greatly. Wells may be readily cleaned of sand by means of a sand pump or bucket, but if the strainers are corroded, they must be pulled out, cleaned, and renewed or replaced. If the clogging is due to fine sand collecting outside the tube, it may be removed to some extent by forcing water into the wells under high pressure, or by use of a hose or by other suitable means. Sometimes, instead of the yield of a well becoming less through continued operation, it is increased owing probably to the gradual removal of the fine material immediately surrounding the well.

6.8   SANITARY PROTECTION

Water wells must be protected against pollution. Any activity or operation like land-fill that may impair groundwater must be prohibited within the **well head** area (50 m radius). Well casing should extend at least 0.4 m (1.5 ft) above ground surface and properly grouted. A **sanitary seal** should be provided to protect from the entry of contaminated water into the ground.

6.9   WELL ABANDONMENT

After its useful life, a well must be abandoned in a proper fashion. Well bore should first be filled with contaminated material and the top portion grouted with cement mixture. If not done properly, it can become a source of major pollution and a safety issue.

6.10   WATER QUANTITY

As the effect of rainfall is most direct on the surface sources of the water supply, the quantity of water available is abundant. However, since the rainfall may not be uniformly spread throughout the year, considerable variations in the flow of surface waters are likely. Thus, the flow in the streams or rivers may vary from a maximum during the rainy season, sufficient to result in floods, to a minimum during dry months, sufficient to cause long droughts. In the case of impounding reservoirs, in addition to the rainfall and runoff, the topography of the catchment area is important. It should be such as to drain off water from all remote points.

As regards the groundwater sources, the quantity of water available is usually less than that in the case of surface sources; the effect of rainfall now being most indirect. The quantity depends upon the available underground storage and the geological formations of the substrata, viz., permeable or non-permeable. In the case of shallow wells and springs, it is easier to get supplies by tapping the upper water-bearing strata, but such a storage may be temporary and fall off during the dry season, resulting in the failure of the source. The underground supplies drawn from greater depth, viz., deep wells, are more constant in their yield and hence more reliable.

6.11   WATER QUALITY

Impurities in water normally are of two types: suspended and dissolved. Surface waters are characterized by suspended impurities whereas groundwaters are generally free from suspended matter but are likely to contain a large amount of dissolved impurities, which they gather during the course of their travel in the underground strata comprising rocks and minerals. The suspended matter often contains the pathogenic or disease-producing bacteria and as such, surface waters are not considered to be safe for a water supply without the necessary treatment. Groundwaters are comparatively safer and fit for use with or without minor treatment only. A comparison of surface water and groundwater supplies are shown in **Table 6.1**.

TABLE 6.1
Surface water versus groundwater supplies

Feature	Surface Water	Groundwater
quality	low in minerals, soft water	high in minerals/hard water
changes in quality	low quality during lake turnovers and after storms	little variation; unaffected by storms
quantity	usually, a large quantity	low if well yield is low
taste and odor	taste and odor problems from algae	taste and odor problems from iron and H_2S
contamination	high	low
minimum treatment	chemically assisted filtration	disinfection only
costs to treat	medium to high	low to medium

River water varies in quality. This variation is caused by the great difference in maximum and minimum flows. Maximum flow is caused by high floods, resulting in an increase in turbidity and bacteria due to the surface wash brought into the river. Minimum flow is due to the flow of groundwater into the river, resulting in the decrease of turbidity but increase of dissolved impurities. River water is also usually found to be contaminated with sewage or industrial water from towns and cities. River water, therefore, must be thoroughly treated before supplying for public use.

An impounding reservoir stores water by the construction of a dam across a natural water course. The storage provided may be as much as 60 days. This long storage enables suspended matter to settle down and be removed. There is also considerable reduction in harmful bacteria and in colors present. Long storage is, however, unacceptable in one way and this is that it creates the growth of microscopic organisms in water, impairing its general quality. While top layers of water are prone to develop algae, bottom layers of water may be high in turbidity, carbon dioxide, iron, manganese, and even hydrogen sulphide. Aeration and chlorination are thus normally required before water is considered fit for supply.

The quality of groundwater is comparatively much better. This is due to the fact that water gets filtered during its passage through the porous underground strata. The geological formations with which water comes into contact also impart to it certain qualities like softness or hardness. In granite formations, groundwaters are soft, low in dissolved mineral and high in carbon dioxide, and are actively corrosive, while in limestone formations, they are very hard and tend to form deposits in pipes, but are relatively non-corrosive. The bacterial content of waters from springs, infiltration galleries, and deep wells is usually low due to the straining action involved. In general, groundwater is good in quality but may require some treatment to improve its chemical characteristics.

6.12   GROUNDWATER UNDER THE INFLUENCE (GUDI)

GUDI stands for groundwater under the direct influence of surface water. Groundwater is clean because the water gets naturally filtered as it moves through subsurface layers. However, sometimes this process of filtration is not complete as water takes shorter routes and is not filtered adequately, especially when surface water easily makes its way to shallow groundwater. When this happens, there is a concern that solids and microorganisms could get into groundwater. This results in a water supply that is under the direct influence of surface water. In many developed countries, the minimum treatment required for GUDI is the same as for surface water supply systems – that is chemical-assisted filtration. Here are some examples of GUDI systems.

- A drinking water system that gets its water from a well that is not drilled or a well that does not have watertight casing and that extends to a depth of at least 6 m (20 ft) below the ground surface.
- A water system where the source of water is from a infiltration gallery.
- A shallow well within the vicinity of surface water (<100 m).
- A drinking water system that has been contaminated by surface water.

6.13   CHOICE OF SOURCE OF WATER SUPPLY

Considerations in the selection of a particular source for the supplying of water are: quantity of water available, its quality, and cost of production. The quantity of water available from the source should be sufficient to cater for the needs of the town or city regarding domestic service, industrial demands, firefighting requirements, and other public uses. The quantity of water supplied should also include the design requirements, which means the calculated quantity would be somewhat higher than the bare needs. The quality of water should be wholesome, safe, and free from pollution of any kind. The health of the public should in no way be endangered due to epidemics associated with waterborne diseases.

The quantity and quality of water are prime considerations in the selection of any source of supply. Cost considerations regarding the development and operation of the water supply are also significant. The cost of supply would depend on whether the system of supply is such that the water flows by gravity from the source or it has to be pumped first before supplying. Cost would, naturally, be less in the first case. The cost shall also depend upon the distance between the source of supply and the distribution system. A longer distance means the greater cost of conduits and other appurtenances required. In short, the investment cost of water supply must be reasonable compared to the number of people served and must bear a fair relation to the value of property served so that by equitable taxes and reasonable charges for water the original cost of the system can be repaid at the end of the design period, which is usually 20 to 30 years.

Discussion Questions

1. Compare reservoir intake versus lake intake.
2. Describe two main types of river intakes.
3. Compare groundwater supplies versus surface water supplies.
4. List and describe various sources of surface water supplies.
5. Describe various methods of tapping groundwater.
6. What factors would you consider when choosing the source of water supply?
7. What are the reasons that for larger cities and towns, usually surface water supplies are used?
8. Why do GUDI systems require rigorous treatment to make water potable? Under what conditions and situations, should groundwater source be treated as GUDI?
9. Compare artesian wells and water table wells.
10. Define and explain the following terms related to wells?
    a. Transmissivity and hydraulic conductivity.
    b. Specific capacity, well efficiency.
    c. Shallow well, deep well.
    d. Specific yield, storage coefficient.
    e. Radius of influence, cone of depression.
    f. Dug well, infiltration gallery.
    g. Artesian well, flowing well.

11. "Most municipal wells are gravel packed." Discuss.
12. Write short notes on: sanitary seal, well abandonment, and sick wells.
13. Explain why the drawdown in the pumped well is more than the drawdown at the well face.
14. State the assumptions on which equilibrium equation are derived.
15. Search the internet to find if in your country there are any government regulations pertaining to groundwater wells (groundwater withdrawal, construction, design and abandonment).

Practice Problems

1. The specific capacity of a well is 43 m^3/m·d. What is the maximum yield of a well if the maximum available drawdown is 18 m? (9.0 L/s).
2. Before pumping started, the static water level in a well is observed to be at a depth of 60 ft. After 22 hours of continuous pumping @ 35 L/s, the water level stabilized at a depth of 50.2 m. Determine the specific capacity of the well. (1.1 L/s·m).
3. A 50 cm diameter well is drilled in an artesian aquifer with a transmissivity of 80 m^3/m·d. What is the theoretical specific capacity? Assume the radius of influence is 1200 m. (0.7 L/s·m).
4. A pumping test was conducted in a water table well penetrating to the clay layer at a depth of 60 ft. Two observation wells were located at a distance of 50 ft and 300 ft from the pumped well. Before pumping started, static water level was observed at a depth of 15 ft from the ground surface datum. After continuously pumping at the rate 550 gpm for 40 hours, drawdowns of 9.1 ft and 2.4 ft were observed in the two observation wells. Calculate the hydraulic conductivity of the unconfined aquifer. (110 ft/d).
5. A pumping test was conducted in an unconfined aquifer. The impervious stratum is found at the depth of 20 m and the static water level is 5 m below the ground surface. After 10 hours of continuous pumping, the drawdown in an observation well was observed to be 3.04 m. What is h and h_o in m? (12.0 m, 15 m).
6. The following measurements were made at a well that supplies water to a portion of a small community. SWL = 127.4 m, DWL = 152.1 m, Q = 15.0 L/s, r_0 = 290 m. What is the specific capacity of the well? (0.61 L/s·m).
7. After pumping continuously for 15 h @ 15 L/s, depth to water level in the pumped well is observed to be 78 m. Before the pumping started, depth to water level was 55 m. What is the specific capacity of the well? (0.65 L/s·m).
8. After pumping continuously for 15 h, water level in the pumped well drops by 25 m. If the SWL before the pumping was observed to be at a depth of 55 m, what is depth to DWL in the pumped well? (80 m).
9. An artesian aquifer has a transmissivity of 160 m^3/m·d. The piezometric readings in the two observation wells located at 10 m and 150 m are 12.45 m and 15.5 m respectively. What is the steady pumping rate? (13 L/s).

10. Determine the permeability of a 25 m thick artesian aquifer being pumped at a constant rate of 125 L/s. at equilibrium, the drawdown in an observation well 25 m away is 3.5 m and drawdown in the second observation well 200 m away is 0.25 m. (44 m/d).

11. A 30 cm well fully penetrates a 24 m thick confined aquifer. The well is pumped at the rate of 65 L/s. Two test wells are located at 15 m and 40 m respectively. After reaching equilibrium conditions, the drawdowns in wells 1 and 2 were observed to be 3.1 m and 1.8 m respectively. Estimate the transmissivity of the aquifer. Also find the theoretical drawdown in the pumped well and hence the theoretical specific capacity. If the actual draw-down in the well is observed to be 11.7 m, determine the efficiency of the well. (T = 670 m/d, s = 9.2 m, SC = 7.1 L/s.m, efficiency = 79%).

12. A 20 cm diameter well is pumping an unconfined aquifer with a coefficient of permeability of 40 m/d. After equilibrium conditions are achieved, the two observation wells located at distances of 10 m and 100 m indicate water height of 7.3 m and 8.1 m respectively. Determine the steady pumping rate. (7.8 L/s).

13. A 25 cm diameter water table well is pumped at a constant rate of 32 L/s. Before pumping started, the elevation of static water level was recorded to be 22.0 m above the impervious layer. After achieving equilibrium condi-tions, drawdowns in the pumped well and observation well at 50 m respec-tively were 2.5 m and 0.45 m. What is the hydraulic conductivity of the aquifer? (63 m/d).

14. An artesian well of intake diameter of 2.0 ft is installed in a confined aqui-fer. The depth of the aquifer is 90 ft from the ground surface. Based on the pumping test data, piezometric surface is 25 ft below the ground surface. Assuming aquifer transmissivity of 350 ft^2/d and radius of influence at a distance of 2000 ft, find the specific capacity of the well. (1.5 gpm/ft).

15. A 30 cm diameter artesian well is installed in a 30 m thick aquifer which has permeability of 22 m/d. Find the yield of the well under a drawdown of 4.0 m at the well face. Assume radius of influence at a distance of 500 m. (24 L/s).

7 Water Demand and Water Quality

The proper design and execution of a water supply scheme necessities an estimate of the total amount of water required for the community. The total quantity of water required by a community is called as the "**demand**". The demand is normally expressed as total annual demand in million liters (gallons) per annum or average daily demand in liters (gallons) per day, or average daily demand in liters (gallons) per capita per day. Municipal water use in the USA averages about 500–600 L/c.d. This includes residential, commercial, and industrial use.

7.1 DESIGN PERIOD

The design of a water supply scheme is an activity that should be aimed to meet the needs of the people for a specified period in the future. The useful life of the project is called the "**design period**". In a robust design, the life of the individual components and the distribution system should be more than the defined design period. Moreover, the design of the scheme will depend on the projected future population to be served and their probable demand. Commonly used design periods are shown in **Table 7.1**

This warrants a study of the expected population increase in the community and an understanding of the trends of development in the region, including industrial and commercial growth. The period should not be too short (which demands immediate replacement) or too long (the present population would make a huge investment for a long future). Water supply projects are planned, normally, with a design period of 30 years. The time lag between the design and completion of the project should also be accounted for in deciding the total design period, which should not exceed two to five years, depending on the size of the project.

7.2 FORECASTING POPULATION

It is essential to estimate the likely population of the town by the end of the design period. The cost of any water supply scheme depends on the total population to be served and the total demand per capita per day. Any water supply scheme is usually designed in two stages: for an anticipated population after 25–30 years and for a future population at the end of 50–60 years

Hence, it is essential to know how to project the population at the end of a few decades, based on the present trends in population growth. There are several methods of predicting population. Most of these methods try to project the future population

DOI: 10.1201/9781003347941-9

TABLE 7.1

Design periods of various components

Component	Design Period
storage reservoir/dams	50
infiltration works	30
pump house and civil works	30
electric motors and pumps	15
water treatment units	15
pipe connections and small appurtenances	30
raw and clear water conveyance mains	30
overhead and ground level service reservoirs	15
distribution system	30

based on the present trend in population growth. A brief description of commonly used methods for population growth follows.

7.2.1 ARITHMETICAL INCREASE

According to this method, population increase per decade, C, is assumed to be constant, thus indicating a linear relationship. Population at the end of n number of decades can be found as follows.

$$P_2 = P_1 + C(t_2 - t_1) = P_1 + nC$$

This method is adopted for old and well-settled cities where the trend in population increase is more or less constant.

7.2.2 GEOMETRICAL INCREASE METHOD

In this method, the percentage increase in population from decade to decade is assumed to be constant. Population of two or more previous decades is considered, the percentage increase calculated, and the average percentage adopted to calculate the population of future decades. Arithmetic mean or preferably geometric mean can be used to find the average growth rate. If the present population is P, and the average percentage growth is r, the population after n decades will be

$$P_n = P\left(1 + \frac{r}{100}\right)^n$$

Arithmetical increase is comparable to simple interest and geometrical increase to compound interest. The geometrical increase method of population growth is suitable for towns with rapid growth due to more employment opportunities and labor migration.

7.2.3 INCREMENTAL INCREASE METHOD

In this case scenario, the average increase per decade and the average incremental increase per decade are calculated and their sum is added to the present population. This process can be repeated for each successive decade till the population of the required decade is arrived at. This method will yield results, somewhere between the forecasts obtained using the arithmetical technique and the geometrical method. This method yields better results since it assumes that growth rate varies decade to decade.

Example Problem 7.1

In **Table 7.2**, the population of a town is shown for six decades. Based on this data, forecast the future population for decades 2021 and 2031 using the three techniques discussed above.

Solution:
In column 3 of **Table 7.2**, change in population, ΔP, for every successive decade is calculated. At the end of the column, average values based on data of six decades are shown.

Sample of Calculation:
For the first two decades:

$$\Delta P = 250 - 180 = 70 = \frac{70}{180} \times 100\% = 38.88 = 38.9\%$$

TABLE 7.2
Worksheet for population forecast

Year	Population	ΔP	ΔP	$\Delta(\Delta P)$
	×1000	×1000	%	×1000
1951	180			
1961	250	70.0	38.9	
1971	350	100.0	40.0	30.0
1981	460	110.0	31.4	10.0
1991	620	160.0	34.8	50.0
2001	820	200.0	32.3	40.0
2011	1100	280.0	34.1	80.0
	Average	153.3	35.3	42.0

Year	Population Forecast × 1000		
	Arithmetical	Geometrical	Incremental
2021	1253	1488	1295
2031	1407	2012	1491
2041	1560	2722	1686

Based on the data of first three decades:

$$\Delta(\Delta P) = 40 - 38.9 = 1.1$$

Expected population in the decade 2021:

$$Arithmetic, P_{2021} = P_{2011} + \overline{\Delta P} = 1100 + 153.3 = 1253.3 = 1253$$

$$Geometric, P_{2021} = P_{2011}\left(1 + \frac{r}{100}\right)^n = 1100(1 + 0.353)^1 = 1488.3 = 1488$$

$$Incremental, P_{2021} = P_{2011} + n\left(\overline{\Delta P} + \overline{\Delta(\Delta P)}\right)$$

$$= 1100 + 1 \times (153.3 + 42) = 1295$$

From **Table 7.2**, it is clear that the arithmetical increase method gives a very low value, the geometrical increase method a very high value, and the incremental increase method a reasonable value. Hence, the incremental increase method is generally adopted.

7.3   ESTIMATING WATER DEMAND

Before designing a water supply scheme, it is pertinent to evaluate water demand and water available.

It is not possible to assess the demand for water very accurately as it involves various factors that affect consumption. To arrive at a fair estimate, a few empirical formulae and rules of thumb are used. Demand can be classified as domestic, industrial and commercial, public use, in firefighting, for compensating various losses, and the specific requirement for various buildings.

7.3.1   DOMESTIC

This is the requirement of water for private buildings. It includes the water required for drinking, cooking, bathing, maintaining lawns and gardens, domestic sanitation, washing, and so on. The amount of water required by a person, called per capita consumption, varies over wide limits. In many developed countries, per capita water consumption is roughly twice that of developing countries, like India. As per a 1998 water use survey, per capita water consumption in the USA for residential water use is 280 L/c·d (75 gal/c·d), as shown in **Table 7.3**. In comparison, per capita water consumption in India is 135 L/c·d, as shown in **Table 7.4**.

The total domestic demand for a town will be equal to the product of the total population of the town and the per capita (domestic) consumption. This amounts to about 50% of total consumption.

TABLE 7.3
Residential water consumption in USA

Purpose	L/c·d
faucets	40
bathing	5
leaks	35
showers	50
toilets	75
washing of clothes	65
miscellaneous	10
Total	280

TABLE 7.4
Residential water consumption in India

Purpose	L/c·d
drinking	5
bathing	55
cooking	5
washing of utensils	10
washing of floors, etc.	10
washing of clothes	20
flushing of toilets	30
Total	135 (200) *

* For high income families water demand is 200 L/c·d.

7.3.2 INDUSTRIAL AND COMMERCIAL

Water required by offices, factories, industries, hotels, and so on comes under this category. This accounts for 10% of total consumption. For various industrial and commercial uses water demand in USA and India is presented in **Tables 7.5 and 7.6** respectively.

7.3.3 PUBLIC USE

This is the water required by public buildings such as hospitals, city hall, and jails, as also for washing streets, and maintaining municipal lawns and orchards. This accounts for 10% of the total demand.

7.3.4 FIREFIGHTING

Fire demands must be considered in a municipal water system design. Design requirements vary by municipality or country and must be according to these guidelines.

TABLE 7.5
Water consumption in USA

Type of structure	Unit	L/d
airport	Passenger	15
apartment building	Bedroom	480
gas station	Vehicle served	40
motel	Guest	230
hotel	Guest	270
	Employee	40
mobile home park	Unit	550
movie theatre	Seat	10
office building	Employee	50
restaurant	Customer	35
shopping centre	Parking Space	8
laundromat		
machine		1700
customer		180

TABLE 7.6
Water consumption in India

Type of structure	Unit	L/d
(a) factories with bathrooms	employee	45
(b) factories without bathrooms	employee	30
hostels	resident	135
offices	employee	45
hospitals	bed	400
restaurants	seat	70
schools: (a) day schools	student	45
(B) board	student	135
cinemas, theatres	seat	15
hostels	bed	135
airports and sea ports	passenger	180
Railway stations		
junction stations	passenger	70
terminal station	passenger	45
intermediate station	passenger	45

Although the annual water requirements for firefighting are relatively small, short-term demand rate might be large and is governed by the design of the distribution system, distribution storage, and pumping equipment. Firefighting requirements for residential areas vary from 30–1000 L/s (500–15 000 gpm), depending on the population density and land use.

Withdrawal of large quantities of water is not the preferred method of firefighting. For many buildings, an automatic sprinkler system is the most effective way to protect life and property. The minimum fire flow for buildings without sprinklers is 30 L/s (500 gpm) at a residual pressure of 140 kPa (20 psi).

This will be 5%–10% of total demand. Fire can be extinguished cheaply and quickly by water but entailing heavy demands for brief periods. It depends upon the population, the number of buildings, and the planning and design of buildings. Satisfactory results can be obtained by applying the following empirical formulae.

$$Q(Kuching) = 3182\sqrt{P} \quad Q(Freeman) = 1136.5 \times \frac{P}{5} + 10$$

Q = fire demand in L/min and P = population in thousands

However, these formulae are not perfect and only give an idea of the demand.

Duration

The required duration for fire flow is 2 h for up to 150 L/s (2500 gpm) and 3 h for fire flows exceeding 200 L/s (3000 gpm). The major components of a water system on which the reliability of fire flow depends, including pumps, supply mains, treatment plant, and power source, must have the capacity to deliver the maximum water use rate for several days plus fire flow for a specified duration at any time during that interval. The period may vary from 2 to 5 days, depending on the system components under consideration.

Pressure

The pressure in a distribution system must be high enough to permit pumps to withdraw an adequate supply from hydrants. A minimum residual pressure of 140 kPa (20 psi) during fire flow in used to analyze the adequacy of a water system.

Example Problem 7.2

At present, the population of a town is 28 000 and the average daily water demand is 4.1 ML/d. Design capacity of the water filtration plant is 6.0 ML/d and it is expected that the population of the town will grow to 44 000 in the next two decades. Applying the arithmetical rate of population growth method, find after how many years will the water demand of the town reach design capacity.

Solution:

$$Constant\ rate\ of\ growth, C = \frac{(44000 - 28000)}{2} = 8000\,/\,decade$$

$$Per\ capita\ demand, = \frac{4.1\,ML}{28000\,p} \times \frac{10^6\,L}{ML} = 146.4 = 150\,L\,/\,p.d$$

Population to reach design capacity

$$P_n = \frac{6.0\,ML}{d} \times \frac{p.d}{146.4\,L} \times \frac{10^6\,L}{ML} = 40981 = 41000\,p$$

$$n = \frac{P_n - P_0}{C} = \frac{(409813 - 28000)}{8000\,/\,decade} \times \frac{10\,y}{decade} = 16.2 = 16\,years$$

7.4   TOTAL DEMAND

The total quantity of water required by a community is estimated using the following two pieces of information:

1. The anticipated population of a particular town for a given design period, say 30 years, worked out by any of the methods of population forecasting.
2. The per capita demand, liters per person per day, which is the average amount of water required by a person every day, worked out by considering the water requirements for various purposes such as domestic use, commercial use, and so on, divided by total population and 365 days.

Based on these two data, the annual water requirement of the town can be worked out. For example, if the projected population of a town is 50 000 and the per capita demand 200 L/c·d, the total amount of water required per year would be:

$$Q = \frac{200\,L}{p.d} \times 50\,000\,p \times \frac{ML}{10^6\,L} = 1.0\,ML\,/\,d$$

7.5   FACTORS AFFECTING PER CAPITA DEMAND

The annual average demand for water is not the same for every country or city or town. In India, it varies from 100 to 300 liters per capita per day. Factors affecting the water consumption are further discussed in the following sections.

Size of the Town

The amount of water used in a community is directly related to the size, distribution, and composition of the local population. The bigger the town, the greater is the per capita demand of water. This is because in big cities, water requirements will be more, in order to cater for maintenance of health and sanitation and meet the needs of industry and commerce.

Climatic Conditions

Water consumption in a hotter country will be more than in colder regions, as more water will be required for bathing, cleaning, air coolers, air conditioners, and the maintenance of gardens, lawns, etc.

Lifestyle

The quantity of water consumed varies from rich people to poor people. Highly affluent sections of society usually use more water than their poorer counterparts. The amount of water consumed usually depends upon the status of people. As mentioned earlier, In India, per capita water consumption for high income families is 30% more than that of average families.

Industrial and Commercial Activities

The industrial and commercial activities in a particular area definitely increase demand for water. Many industries require very large quantities of water. Demand due to industries and commerce depends on the number of industries and business activities and not on the population or size of the town.

System of Supply

In most developed countries, water is supplied continuously; however, this is not the case in underdeveloped and developing countries, where water is supplied intermittently. In many cases, intermittent supply of water can result in reduced demand because there is saving in water consumption due to reduction of wastage and loss. But in some places, intermittent supply may produce increased demand for the following reasons:

   i. Water stored by consumers during non-supply periods may be thrown away during the next supply period, resulting in losses and wastage of water.
   ii. If taps are kept open during the non-supply period, and the consumer is not present when water is supplied, then a lot of water is wasted.

Quality of Water Supplied

Good-quality water is welcomed by domestic consumers and factories and the demand for this can only increase. If the quality of water supplied is good, the consumer will never think of alternatives. This in turn results in an increase in demand for the public water supply.

Pressure in the Distribution System

Higher pressures in the distribution system results in an increase of water consumption and thereby demand. If pressures are high, those living in the upper floors (or in flats) can get water. Naturally, this results in increased demand. Further, higher distribution pressures produce higher losses in water mains and household plumbing fixtures, which also increase demand. This is especially true in developing countries where an increase of pressure from 200 kPa to that of 300 kPa can produce an increase of loss of 30%.

Development of Sewerage Facilities

The conservancy system demands less water than the water carriage system. This is so because in the latter system water is used liberally to flush toilets, urinals, and so on.

Policy of Distribution

If the supply is unmetered and consumers are required to pay a fixed amount every month, they are more liable to waste water. But if the supply is metered, people are more cautions, as they have to pay for the extra consumption.

Cost of Water

The cost of any material influences consumption and thereby demand. If the cost of water is made high, people will use it carefully and thus reduce demand.

Environmental Protection

Social attitudes towards environmental protection strongly affect water allocation and use. Water use forecasts must consider the amount of water that is to be dedicated to environmental protection.

Conservation

With depleting water resources and increases in water demand, more emphasis is put on the adoption of conservation practices and the reuse of wastewater and storm water.

Management Practices

Water management practices, including inter-basin transfers, saline water conversion, water reclamation and reuse, and many other practices influence water use trends. The impact of technological change on water use can be significant.

Tourism

Places such as Florida, Niagara Falls, and Agra (India) have annual tourist populations that significant exceed their resident populations. The impact of such occurrences must be recognized when forecasting future water demands.

7.6 VARIATION IN DEMAND

Average water demand represents the average daily demand over a period of one year. There are peak periods as well as lean periods. There are wide variations in seasonal, daily, and hourly water demands. Generally, during holidays and festivals, demand is high; hot and dry days have more demand than wet and cold days; within a day, the demands in the mornings and evenings are higher. Accordingly, variations in demand are broadly categorized as seasonal, daily, and hourly.

7.6.1 SEASONAL VARIATION

The variation in demand from season to season is indicated in this category. Normally, demand peaks during the summer. Fire breakouts are generally greater in summer, increasing demand. In winter, people generally consume less water.

7.6.2 DAILY VARIATION

Water demand can depend on the variation in daily activity during the week. People draw out more water on Sundays and on festival days, thus increasing demand on these days.

7.6.3 HOURLY VARIATION

This is very important as there is a wide range. During active household working hours, i.e., from 6:00 to 10:00 in the morning and 4:00 to 8:00 in the evening, the bulk of daily consumption is made. During other hours, the requirement is less or even negligible. The maximum day demand represents the amount of water required during the day of maximum consumption in a year. Peak hour demand represents the amount of water required during the maximum consumption hour in a given day. Hence, in order to design a water supply project, one must consider demand during the peak hours of the day and the peak periods in a year.

The design should ensure an adequate quantity of water to meet the peak demand. To meet all the fluctuations, the supply pipes, service reservoirs, and distribution pipes must be properly sized. If water is supplied by pumping directly, then the pumps and distribution system must be designed to meet the peak demand. The effect of monthly variation influences the design of storage reservoirs, and the hourly variation influences the design of pumps and service reservoirs. As the population decreases, the fluctuation rate increases. Therefore, a careful study to understand these fluctuations must be made for each city from the past data. If such water demand data is not available, the equation given below can be used to estimate the maximum monthly, weekly, daily, and hourly demand.

$$Percent\ Demand, P = 180\,t^{-0.10}$$

where P = percent of annual average demand for time, t
t = time in days. The time t in days varies from 2/24 to 365
For example, for hourly demand (t = 1/24 d) the multiplication factor is:

$$= 180 \times \left(\frac{1}{24}\right)^{-0.10} = 247\% = 2.5 \times$$

Some empirical relationships to work out peak demand are as follows.
maximum daily demand = 1.8 × annual average daily demand
maximum hourly demand of maximum day, i.e., peak demand
= 1.5 × average hourly demand
= 1.5 × maximum daily demand
= 1.5 × (1.8 × annual average daily demand)
= 2.7 × annual average daily demand
= 2.7 × average hourly demand
Hourly demand does not have to have the units of /h; it can be any units, ML/d or L/s, etc. If the average demand for a town is 10 ML/d, then maximum daily demand is 15 ML/d and peak demand is 27 ML/d. **Figure 7.1** shows a typical hourly variation of water consumption. The maximum demand is observed to be at about 8:30 in the evening and the minimum is at midnight.

The many factors affecting municipal water use preclude generalizations that could apply to all areas. General trends and representative figures are useful, but it should be understood that local usage may vary considerably from the reported

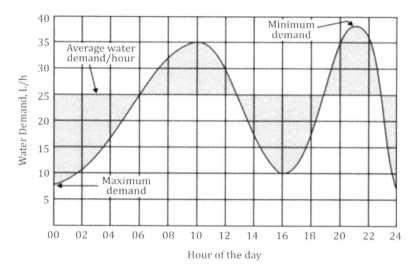

FIGURE 7.1   Hourly variation (diurnal variation)

averages. For design purposes, past records of the type and pattern of community use, the physical and climatic characteristics of the area, expected trends in development, population forecast, and other pertinent factors must be thoroughly studied.

7.7   WATER QUALITY STANDARDS

Ideally, water delivered to the consumer should be clear, colorless, odorless, and, more importantly, safe to drink, that is **potable**. It should not contain concentrations of chemicals that may be physiologically harmful or aesthetically objectionable. In most jurisdictions, the quality of drinking water is regulated by legislation. These standards pertain to physical, chemical, and bacteriological characteristics of the treated water. The WHO provides guidelines for the minimum level of quality for the public supply of water. However, the standards for quality control are evolved locally considering the limitations imposed by local factors. Normally, the check against transmitting diseases is comparatively more important than physical or chemical qualities, which are prescribed within a particular range.

The quality standards of water provide a description of the various inorganic and organic chemicals that may be found in drinking water. These standards should be used as a reference by those responsible for drinking water quality. The purpose of these standards is to ensure a safe and palatable drinking water and the operators of treatment systems have a legal obligation to meet these requirements. If the results of any analysis indicate that the level of any adverse parameter exceeds its quality standards, it indicates **adverse water quality** and must be reported.

In many jurisdictions, monitoring and reporting of the quality of water is mandated by legislation. For example, in the USA and Canada, regular monitoring and data submission are required by law. In addition, municipal water systems are required by legislation to put summaries of the data on the web. As an example, in Ontario, Canada drinking water quality is legislated under the Safe Drinking Water Act (SDWA). Under SDWA, two regulations specifically relate to water quality. In Ontario, Regulation 169, drinking water quality standards (ODWQS) are described. These standards pertain to physical, chemical, and bacteriological characteristics of treated water. Regulation 170 details the information related to sampling requirements, reporting of adverse water quality, and operational checks. As an example, the turbidity of filtered water should be tested with a minimum frequency or preferably monitored continuously. To comply, turbidity of filtered water must not exceed 1 nephelometric turbidity unit (NTU). This requirement is to make sure that, during disinfection, turbidity does not interfere with inactivation of pathogens. As regards aesthetic objectivity, however, turbidity up to 5 NTU is acceptable.

7.8 WATER QUALITY PARAMETERS

Analysis of water quality supplements any existing information on it. The parameters for such analysis depend on the specific use the water is to be put to. The treatment plant must be designed and operated on the basis of the information on impurities present in the water. Therefore, analysis of water is essential for designing the water treatment system. Similarly, after the treatment of water, its analysis is undertaken again to ascertain that the water has been purified and now meets the prescribed quality standards. Since quality is a variable with respect to time and space, it is imperative to conduct periodic/routine tests to ascertain regular maintenance of the quality of water. In the following sections, tests routinely done to determine the quality of water are described.

7.9 PHYSICAL PARAMETERS

Some of the main physical parameters used to monitor the quality of drinking water supplies include turbidity, color, temperature, and taste and odor.

7.9.1 TURBIDITY

The term turbidity is applied to water containing suspended matter that tends to scatter or absorb light rays. This gives the water a cloudy or murky appearance. It can be caused by a wide variety of suspended materials ranging in size from colloidal to coarse particles. Colloids are very small (<1.0 μm) particles suspended in water. It is very difficult to remove colloidal particles by plain sedimentation or filtration without the use of chemicals. Turbidity can result from erosion or disturbance of river bottoms, and material originating from domestic and industrial wastewaters. Turbidity itself may not be dangerous to human health but may provide an indication of pathogens or other harmful material.

From a public perspective, turbidity has two main effects. The first is aesthetic: a customer of drinking water expects to receive water with no noticeable cloudiness. Second, turbidity allows harmful microorganisms to become enveloped within solids and therefore protect themselves from disinfection. Turbidity removal normally requires both physical and chemical processes. For water treatment, an instrument called a nephelometer or simply a turbidimeter is used to measure the turbidity at various stages of treatment. These devices measure turbidity based on the amount of light scattered. Measurements made by such instruments are expressed in terms of NTU. Turbidity exceeding 5 NTU is noticeable to the average person. Water filtration plants can routinely produce water with turbidity less than 1 NTU. Water is tested for turbidity to determine the removal and adjust the operation of various processes like chemical dosing and backwashing of filters.

7.9.2 Color

The color of water is usually due to the presence of colloidal organic matter or to mineral and dissolved organic and inorganic impurities. Color is thus hard to remove by simple sedimentation or filtration. Like turbidity, color is an aesthetic parameter. Turbidity in water imparts some color to water, hence this measurement is called **apparent color**. On the other hand, **true color** is the color of water without the presence of the turbidity. To read the true color, the water sample should be filtered first to remove turbidity.

The color of the water is compared with standard color solution or color discs. The color produced by one milligram of platinum in a liter of distilled water has been fixed as the unit of color (CU). The permissible color for domestic water is 20 mg/L on the platinum-cobalt scale. Mostly, even the slightest color in water is objectionable as it directly affects the palatability of water. For esthetic purposes, color in water should be below 5 color units, also called Hazen units.

7.9.3 Temperature

From the treatment point of view, water temperature is not significant since it is not practicable to change the temperature of water during treatment. However, the temperature of water does affect the rate of chemical reactions and particle settling velocity. For the same chlorine residual, disinfection is more powerful at high temperatures. The temperature of water is measured by means of ordinary thermometers. The temperature of surface water is generally the same as the atmospheric temperature, while that of groundwater is generally more or less than atmospheric temperature. For public supply, the most desirable temperature is between 4°C and 10°C (50°F). Temperatures exceeding 25°C are undesirable and above 35°C (95°F) water is unfit for public supply, because of its lesser acceptability and the danger of it becoming more prone to bacterial contamination.

7.9.4 Taste and Odor

Like color and turbidity, taste and odor are important for aesthetic reasons. Although they do not cause any direct effect on health, most people object strongly to water that offends their sense of taste and smell. Hydrogen sulphide, characteristic of rot-ten-egg smell, is a common cause of problems in water supplies. Groundwater with an excessive amount of sulphur sometimes has this problem. The odor of water also changes with temperature. Water having a bad smell or odor is objectionable and should not be supplied to the public.

Odor is measured by **threshold odor number** (TON). The threshold number is the ratio by which the sample has to be diluted for the odor has to become virtually unnoticeable. Algae blooms cause serious odor problems. The major implication of tastes and odors in drinking water is customer dissatisfaction. Changes in drinking water taste and odor may also indicate:

- Possible contamination of the raw water supply.
- Excessive biological activity.
- Loss of chlorine residual in the distribution system.

7.9.5 SOLIDS

Total solids include suspended, colloidal, and dissolved solids. The quantity of sus-pended solids is determined by filtering the sample of water through a fine filter with a pore size of 1μm, drying and weighing. The quantity of dissolved and colloidal solids is determined by evaporating the filtered water (obtained from the suspended solid test) and weighing the residue. The total solids in a water sample can be directly determined by evaporating the water and weighing the residue. In drinking water supplies, dissolved solids may cause taste and odor problems. Hardness and corro-sion are usually due to an excess of total dissolved solids (TDS). Total solids can be estimated from the electric conductivity of the water. The maximum limit of total solids in water is 500 mg/L. **Table 7.7** gives a summary of various suspended and dissolved impurities in water.

Electrical Conductivity

Electrical conductivity or the specific conductance of water is related to the dissolved salts in water. It is a measure of the ability of the water to conduct electric current. It is determined by means of a di-ionic electrode and is expressed in micro-siemens per cm (μ S/cm) at 25°C or the specific conductivity of water multiplied by a coefficient (ranging from 0.5 to 0.9; generally, 0.65) to directly obtain the TDS content in mg/L.

7.9.6 WATER DENSITY

Density is a physical characteristic, a measure of the mass per unit of volume. In the case of a liquid, the density is given by the mass of the liquid contained in a constant volume. Density can be expressed as specific gravity, which is defined as density of a liquid with respect to density of water. Solids with specific gravity (SG) less than unity would float in water. This is why ice stays on top. Changes in water tempera-ture at the surface cause lake turn over. This would cause changes in raw water qual-ity during fall and winter in Northern climates.

TABLE 7.7

Suspended and dissolved impurities in water

Type	Constituent	Effect
suspended matter	bacteria	may cause disease
	algae	odor, color, turbidity
	silt	turbidity
dissolved solids	salts	
	calcium, magnesium	alkalinity, hardness, corrosion
	sodium	
	carbonates	alkalinity
	sulphates	foaming in boilers
	fluoride	dental fluorosis
	chloride	taste, cardiovascular
	metals and compounds	
	iron	taste, red color, corrosion
	manganese	black or brown stains
	lead	slow poisoning
	arsenic	toxicity, poisoning
	barium	toxicity (heart)
	cadmium	toxic
	cyanide	fatal
	boron	effects nervous system
	selenium	toxic to animals, fish
	silver	discoloration of skin
	nitrates	blue baby syndrome
	gases	
	oxygen	corrosion
	carbon dioxide	acidity, corrosiveness
	hydrogen sulphide	odor, acidity, corrosiveness

7.9.7 VISCOSITY

Viscosity is the physical property of a liquid which prevents it from flowing. For example, corn syrup is more viscous than water. In general, viscosity and density decrease with an increase in temperature. Water is unique in the sense that density of water is greatest (1.0 kg/L) at 4°C. Water density and viscosity are important parameters for treatment processes (i.e., coagulation, sedimentation, and filtration).

In general, the efficiency of sedimentation and other water treatment processes is reduced when water is cold (denser and more viscous). Changes in water density have a dramatic effect on the settling of particles. For example, when warm influent water enters a settling tank that has colder water, the warmer water rises to the surface of the tank and flows directly to the effluent channels without providing enough time for the particles in the water to settle in the tank. And when cold water flows into a tank containing warm water, the influent water sinks to the bottom of the tank because is heavier.

7.10   CHEMICAL PARAMETERS

The word chemical is very upsetting and alarming to some people when referring to water quality. This may be for a good reason. Many organic and inorganic chemicals affect water quality. In drinking water supplies, these effects may be related to public health or to aesthetics and economics. Several chemical parameters are also of concern in wastewater. In surface waters, chemical quality can affect the aquatic environment. In this section, the most common parameters of water quality are discussed.

7.10.1   HYDROGEN ION CONCENTRATION (pH)

As explained in the unit on basic chemistry, pH indicates whether a water is acidic or basic in nature. The parameter pH plays an important role in the chemistry of a given sample of water. For example, the strength of the free chlorine residual is very well dependent on the pH value. The desirable range of pH in drinking water supplies is 6.5 to 8.5. Water with low pH value has a sour taste and indicates the water is aggressive and will cause corrosion. On the other hand, water with a high pH has a bitter taste and causes the scaling of water pipes and plumbing fixtures.

7.10.2   ALKALINITY

Much like pH, alkalinity is an important parameter to describe the chemical nature of a given water. In simpler terms, alkalinity indicates the buffering capacity of water and is due to the presence of carbonates, bicarbonates, and hydroxyl ions. Since it can be due to various ions, alkalinity is expressed as mg/L of equivalent $CaCO_3$. In some ways, pH and alkalinity are related terms. Knowing the pH, it is possible to know the nature of ions causing alkalinity but not the concentration. Water with high alkalinity will resist changes in pH. The alkalinity of water is measured by titrating with a strong acid like H_2SO_4 to a pH near the pH value of 4.5. Water with low alkalinity will show rapid changes in pH due to the addition of an acidic solution. If natural alkalinity is low, chemicals like caustic lime or soda ash can be added to boost alkalinity. It is recommended that residual alkalinity of treated water is >30 mg/L.

7.10.3   HARDNESS

The hardness or softness of water is dictated by the concentration of mineral salts dissolved in the water. Hard water is not harmful from a health point of view but may restrict its use by industrial users like breweries and laundries. Hard water contains an excess of these salts and requires a considerable amount of soap to produce a foam or lather. Hard water will also leave scale in hot water heaters and boilers, as you might have noticed in a tea kettle.

Hardness is primarily caused by the presence of dissolved calcium and magnesium salts. Hardness is normally found in waters running through or over limestone. Groundwater, in general, is found to be hard. Again, as in case of alkalinity, hardness is expressed in milligrams per liter as calcium carbonate (mg/L as $CaCO_3$). Hardness due to the presence of carbonates associated with bivalent ions is called temporary or carbonate hardness (CH). Since alkalinity is also caused by carbonates, the relationship between the two parameters is as follows.

$$TH < Alk, NCH = 0$$

$$TH > Alk, CH = Alk$$

TH = total hardness   NCH = non carbonate hardness   alkalinity = Alk

Waters containing less than 50 mg/L of hardness are considered soft and are aggressive in nature, and water exceeding 300 mg/L is classified as very hard water. Hardness in the range of 80–120 mg/L is most acceptable. Hardness can be removed using softening chemicals or using an ion exchange process, like the one mostly used in homes. Hardness in the lab is generally determined by the titrimetric method using ethylenediaminetetraacetic acid (EDTA).

7.10.4 IRON AND MANGANESE

Like hardness, Iron (Fe), and manganese (Mn) are generally present in groundwater as a result of mineral deposits and chemical reactions that occur in aquifers. Anaerobic bacteria (bacteria which do not require oxygen) reduce solid iron and manganese oxides to soluble forms. This can also happen at the bottom of lakes when low dissolved oxygen conditions occur. Some of the problems due to excess of iron and manganese present in water supplies are as follows.

- Iron and manganese in drinking water can cause aesthetic problems and makes water less palatable.
- Growth of iron bacteria forms slime in wells and on pipe walls and contributes to taste and odors and corrosion.
- Staining of laundry and plumbing fixtures.
- Red, brown, or black water.

If iron in water exceeds 0.3 mg/L, the water is not suitable for bleaching, dyeing, and laundry purposes. The presence of iron and manganese in water gives it a brownish red color, causes growth of microorganisms, and corrodes water pipes. The presence of manganese produces black stains on clothes and fixture. Iron and manganese also give it undesirable taste and odor. The quantity of iron and manganese is determined by colorimetric methods.

7.10.5 FLUORIDES

Fluorides in natural water are either due to leaching of rocks (geological origin) or discharge of industrial wastes containing fluorides. Up to 1 mg/L fluorides in water are considered beneficial, especially for dental health. However, a concentration

TABLE 7.8
Maximum permissible limit (MPL) of chemicals

Chemical	MPL, mg/L	Chemical	MPL, mg/L
magnesium	125.0	cadmium	0.01
zinc	15.0	chromium	0.05
silver	0.05	sulphate	250
arsenic	0.05	phenolic (as phenol)	0.001
lead	0.05 to 0.1	cyanide	0. 2
fluoride	1.5	iron	0.3
copper	1.0 to 3.0	manganese	0.05
barium	1.0	selenium	0.05

exceeding 1.5 mg/L causes "fluorosis" (a disease characterized by mottling or staining of teeth and disfigurement of bones). **Table 7.8** gives the maximum permissible level of chemicals in water.

7.10.6 NITROGEN

Though initially present in trace amounts in natural water, the concentration of nitrogen increases due to discharge of industrial wastes. The presence of nitrogen indicates the presence of organic matter. Nitrogen may be present in water as nitrites, nitrates, free ammonia, and albuminoid nitrogen. Groundwater supplies, due to the absence of oxygen, may contain free ammonia due to the reduction of nitrates in water. The albuminoid is normally derived from animal and plant life normal to an aquatic environment. Its presence gives an indication of organic pollution.

The presence of nitrites in water due to partly oxidized organic matters is very dangerous. Therefore, under no circumstances should nitrites be allowed in water. Nitrites are rapidly and easily converted to nitrates by full oxidation of the organic matter. Concentration of nitrates in water should not exceed 45 mg/L, since it can cause **blue baby syndrome** in children (infant methemoglobinemia). The presence of free ammonia indicates recent pollution or decomposition, the presence of nitrites indicates partial decomposition, and the presence of nitrates indicates fully oxidized organic matter. The presence of free ammonia in water should not exceed 0.15 mg/L. Any sudden change in nitrate level indicates the recent pollution of a water source. Nitrate is measured by reduction to ammonia.

7.10.7 DISSOLVED GASES

The various gases which get dissolved in water may be nitrogen, oxygen, carbon dioxide, ethane, and hydrogen sulphide. The nitrogen content is not important.

Dissolved Oxygen

Dissolved oxygen (DO) of water is dependent on atmospheric pressure and water temperature. In a way, DO content affects the palatability of water. Too much DO may lead to corrosion in water distribution pipes. Low DO content in raw water indicates the presence of biodegradable organic matter and is indicated by **biochemical oxygen demand** (BOD). Though BOD is of greater significance in determining the characteristics of wastewater, raw water should not have BOD exceeding 5.0 mg/L and BOD of treated drinking water must be nil.

Carbon Dioxide

The carbon dioxide content of water may make water aggressive at low pH. Carbon dioxide and other undesirable gases like hydrogen sulphide can be removed by stripping.

Hydrogen Sulphide

The presence of hydrogen sulphide, even in low concentrations, can lead to strong odors and make water esthetically unacceptable. It is found mostly in groundwater and can be removed by stripping or aeration, as mentioned earlier.

Discussion Questions

1. What factors determine per capita demand? Compare the water demand of developed versus developing countries, like India.
2. Describe with the help of a hydrograph, diurnal variation in water demand.
3. Explain briefly various factors that affect population growth.
4. Describe the incremental method for population forecasting. Also state its superiority over arithmetical and geometrical methods.
5. What physical tests are conducted for the examination of drinking water supplies?
6. Compare the following terms:
   a. True color, apparent color.
   b. Potable, palatable.
   c. Hardness, alkalinity.
7. List water treatment processes in which alkalinity and pH play important roles.
8. What causes taste and odor problems in drinking water supplies? Describe how taste and odor in water are measured.
9. List dissolved gases generally present in drinking water supplies. Also discuss the problems caused due to their presence.
10. In the treatment of some waters, fluoride is added, while in other cases water is defluorinated. Comment.
11. What types of problems are caused by an excess of iron and manganese?
12. When source water is high in color, it may lead to the formation after disinfection of by-products. Discuss.
13. A water which looks clear may not be always safe to drink, and water that is cloudy in appearance does not necessarily mean it is not safe to drink. Explain.
14. What processes are the components and conventions of a water treatment process scheme.
15. Neither very soft nor very hard water is desirable. Discuss.

Practice Problems

1. The following data shows the variation in population of a town from 1941 to 1991.

Year	1941	1951	1961	1971	1981	1991
P, ×1000	72	85	110.5	144	184	221

Estimate the population in the year 2021 using:
a. arithmetical method (310 400).
b. geometrical method (435 000).
c. incremental increase method (346 400).

2. The following data shows the variation in population of a town from 1960 to 2000.

Year	1960	19760	1980	1990	2000	2010
P, ×1000	25	28	34	42	47	?

Estimate the population in the year 2010 using:
a. arithmetical method (57 000)
b. geometrical method (60 000)
c. incremental increase method (67 500).

3. Based on the data of three decades shown in the table below, estimate the population of the town in the year 2011.

Year	1981	1991	2001
P, ×1000	258	495	735

4. Based on demand of 120 gal/c·d, estimate the maximum daily demand and the peak hourly demand for a town with a projected population of 80 000. (17, 26 MGD).

5. Average annual demand for a city is estimated to be 450 L/c.d. For a projected population of 110 000, estimate the peak hourly demand. (130 ML/d).

6. For a certain region, an empirical formula shown below is used to find the multiplication factor in order to determine maximum demand.

$$Multiplication\ factor, F = 2 \times t^{-0.10}$$

Where t is the time period in days, based on this formula, determine the maximum daily and peak hourly demand when average demand is 25 ML/d. (50, 69 ML/d).

8 Coagulation and Flocculation

The main purpose of water treatment is to provide a **potable** and **palatable** water supply. Few raw water sources can supply water meeting these objectives. Even groundwater, which is usually of a good quality, requires at least **disinfection** to provide residual protection against possible contamination in the water distribution system. Surface water should never be used for drinking without proper treatment. In most advanced countries like the USA and Canada, there are regulations that require a minimum level of treatment depending on the source of water supply. Water treatment plants with surface water as the source of supply, or groundwater under the direct influence (GUDI), must have chemical assisted filtration followed by disinfection as part of their processes.

8.1 SOURCE OF SUPPLY AND TREATMENT

Surface water usually requires a higher degree of treatment than groundwater. Surface waters are open to a variety of contaminants from domestic and industrial wastes to the accidental spillage of harmful chemicals. River water is subject to extreme fluctuations and therefore is more difficult and costlier to treat. Nevertheless, surface waters are the mostly widely used water supplies because of their high yield and other advantages.

The quality of water in a lake or a reservoir varies a great deal from season to season. To protect the water supply, the **watershed** feeding the river or lake must be properly managed. The same thing applies to the well fields and the recharge zone of groundwater aquifers. For good reasons, the concept of watershed management in drinking water supplies is gaining popularity. With increasing pollution and stringent requirements for water quality, a **multi-barrier** approach is the way to meet the need. In **Figure 8.1**, the conventional scheme of surface water treatment is shown. It consists of coagulation, flocculation, sedimentation, followed by filtration and disinfection.

Groundwater quality problems are mostly concerned with high levels of hardness and the presence of iron and manganese and dissolved gases. In general, groundwater is the preferred source of water supply, from the quality and ease of treatment point of view. However, suitable aquifers with sufficient yields are not always available. The simplest treatment system consists of disinfection, as in case of groundwater with a medium level of hardness. Excess iron and manganese are removed from

DOI: 10.1201/9781003347941-10

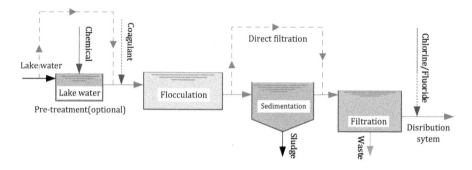

FIGURE 8.1   Process schematic of a filtration plant

TABLE 8.1
Main treatment processes

Main Processes	Purpose
aeration	removes odors and dissolved gases, adds aeration to improve taste
coagulation	converts non settleable particles to settleable particles
sedimentation	removes settleable particles
softening	removes hardness-causing chemicals
filtration	removes finely divided particles, suspended flocs and most microorganisms
adsorption	removes organics and color
stabilization	prevents scaling and corrosion
fluoridation	adds fluoride to harden tooth enamel
disinfection	kills disease-causing organisms

groundwater by adding chlorine or potassium permanganate followed by filtration. Groundwater containing excessive minerals needs additional treatment including **softening.** Aeration of groundwater is performed to strip out dissolved gases and adds oxygen.

In drinking water supplies, physical and chemicals treatments are commonly employed. The most commonly used treatment processes are listed in **Table 8.1.** In the majority of source waters, especially surface water, there is non-settleable matter consisting of a combination of microorganisms, including pathogens and organic and inorganic solids. The term applied to all suspended matter in water is turbidity. Removal of turbidity is thus a primary goal of water treatment.

8.2   PRELIMINARY TREATMENT

In river water supplies, the first step is termed the **preliminary treatment**. Screening, pre-sedimentation, and micro straining are preliminary treatment processes intended to remove grit, algae, and other small debris that can damage or clog the plant equipment. Chemical pre-treatment conditions the water for the further removal of algae and other aquatic plants that cause taste and odor problems.

8.3 CONVENTIONAL TREATMENT

The treatment scheme most commonly used to remove turbidity is known as the **conventional treatment**. Conventional treatment consists of coagulation, flocculation, sedimentation, and filtration. If turbidity is not removed, not only is the appearance not pleasing, but turbidity may also hide pathogens that come into contact with disinfectant and can result in unsafe water.

8.4 COAGULATION

Coagulation is the chemical reaction that occurs when a coagulant (chemical) is rapidly mixed into water **Table 8.2**. The chemical reaction destabilizes the suspended particles and forms tiny particles called **microfloc**. Because many of the microfloc particles are now positively charged, they begin to attract more negatively charged turbidity-causing and color-causing particles. This reaction happens in microseconds. During **flocculation**, suspended particles are stimulated to come together and grow and to make large floc. This makes it easier to remove the suspended materials from the water by the following sedimentation and filtration processes.

Coagulation is a chemical process and takes place in less than 30 seconds. In fact, much of coagulation is complete in the first couple of seconds. Thus, it becomes necessary to quickly disperse the coagulant uniformly into water. That is accomplished by high-speed mixing, also called **flash mixing**. Coagulated water starts forming invisible microfloc. Microfloc still carries positive charge from the coagulant. Flocculation must follow coagulation to allow the growth of microfloc to macrofloc. Flocculation is basically gentle or slow mixing for about 20 to 60 minutes. Whereas coagulation is basically a chemical process, flocculation is mainly a physical process. By agglomeration of microfloc, floc grows and becomes dense enough to be removed by settling and or filtration.

TABLE 8.2
Common coagulants and doses

Coagulant	Name	Formula	Dose, mg/L
aluminum sulphate	alum	$Al_2(SO_4)_3.14. H_2O$	15–100
copper sulphate		$CuSO_4$	5–20
ferric sulphate	ferrosol	$Fe_2(SO_4)_3$	10–50
ferric chloride	ferrichlor	$FeCl_3.6H_2O$	10–50
ferrous sulphate	copperas	$FeSO_4.7H_2O$	5–25
sodium aluminates		$NaAlO_2$	5–50
polymers		various	0.1–1

8.5   COAGULATING CHEMICALS

Coagulants refer to chemicals causing coagulation. When added to water, they form insoluble and gelatinous precipitates. These gelatinous precipitates absorb and entangle non-settleable solids during their formation and descent through the water.

8.5.1   PRIMARY COAGULANTS

Primary coagulants are responsible for destabilizing the particles and making them clump together to form floc. The coagulants more commonly used in the water industry consist of positively charged ions. Trivalent cations, as in the case of aluminum and iron salts, are more effective since they carry triple the charge of monovalent cations. The coagulants having trivalent ions like aluminum and iron are 50–60 times more effective compared to bivalent ions like calcium. Table 8.2 lists the coagulants and doses used in the treatment of water.

Alum is the most common and universal coagulant. Alum is found effective in the pH range of 6.5 to 8.5. Alum is preferred over other coagulants since it also reduces taste and odor problems. It is cheap and makes tough floc more easily. One of the disadvantages is that the sludge formed is difficult to dewater.

Hydrated ferrous sulphate, more commonly known as copperas, has too high solubility to act as a satisfactory coagulant at the usual pH range. Hence it is first chlorinated to form trivalent ions and is known as **chlorinated copperas.**

$$6FeSO_4.7H_2O + 3Cl_2 = 2Fe(SO_4)_3 + 2FeCl_3 + 7H_2O$$

Chlorinated copperas is effective in removing color also. Ferric sulphate and ferric chloride can also be used independently. **Ferrous sulphate or copperas** can react with natural alkalinity, but it is a slow reaction. To expedite reaction, lime is added.

$$6FeSO_4.7H_2O + 3Cl_2 = 2Fe(OH)_2 + CaSO_4 + 7H_2O$$

The ferrous oxide thus formed is efficient floc. However, it is oxidized to a ferric state by the dissolved oxygen, as shown below.

$$4Fe(OH)_2 + O_2 + 2H_2O = 4Fe(OH)_3$$

Floc thus formed is gelatinous in nature and heavier than alum floc. Since copperas is effective above a pH of 8.5, it is unsuitable for the treatment of soft colored water. Iron coagulants makes stronger floc and are effective over a wide range. However, iron coagulants leave the water more corrosive, and some iron residual can promote growth of iron bacteria and cause staining.

Sodium aluminate, though an effective coagulant, is less commonly used due to its high cost compared to alum. Its reaction with natural alkalinity is as follows:

$$Na_3Al_2O_4 + 3Ca(HCO_3)_2 = CaAl_2O_4 \downarrow + Na_2CO_3 + 6CO_2 + H_2O$$

$$Na_3Al_2O_4 + CaSO_4 = CaAl_2O_4 \downarrow + Na_2SO_4$$

As shown above, sodium aluminate reacts with both temporary (carbonate) hardness and permanent (non-carbonate) hardness. It is effective in the pH range of 6 to 8.5.

8.5.2 COAGULANT AIDS

Coagulation and flocculation aids are chemicals or substances that when added in small quantities increase the effectiveness of the process. Aids help in overcoming temperature drops that slow coagulation. In addition, coagulant aids reduce chemical costs and the amount of sludge produced. Coagulant aids and doses commonly used in water industry are listed in **Table 8.3**.

Activated Silica

The key chemical in activated silica is sodium silicate. Activated silica will improve coagulation by strengthening the floc and widen the pH range for effective coagulation. Improved color removal and better floc formation in colder temperatures can also result. The main disadvantage is the precise control required in preparation and feed rate. The dosage of **activated silica** is usually less than 10% of the coagulant used. The normal dosage range of cationic and anionic polymers is 0.1 to 1.0 mg/L. Overdosing is not only wasteful, it can upset the process and allow the discharge of chemicals in the finished water. Overdosing can also upset settling and filter operation.

Weighting Agents

As the name indicates, the primary role of weighting agents is to add weight to the floc to improve settleability. **Bentonite clay** is a very common weighting agent. Weighting agents are used to treat water high in color, low in turbidity, and low in mineral content. Typically, dosages of 10–50 mg/L are used. However, this will add to the solids in the sludge.

Polymers/Polyelectrolyte

Polymers are long-chained synthetic organic compounds, commonly referred to as polyelectrolyte. Molecular mass of polymers may vary from 10^2 to 10^7, as mentioned

TABLE 8.3
Recommended doses of coagulant aids

Coagulant Aid	Typical Dose, mg/L
activated silica	7%–10% of the coagulant
cationic polyelectrolyte	0.1–1
anionic polyelectrolyte	0.1–1
non-ionic polyelectrolyte	1–10
bentonite (clay)	10–50

earlier, based on electrical charge; polymers are classified as anionic-negative charge, cationic-positive charge, and anionic-neutral, or no charge.

Cationic polymers can be used both as primary coagulants or coagulant aids. The primary advantage of polymers is in reducing the quantity of alum sludge produced. The sludge produced is also easily dewatered. Alum sludge is difficult to dewater. Another advantage is that using cationic polymers as a coagulant aid, the dosage of the primary coagulant can be significantly reduced.

It is recommended to prepare a fresh solution of polyelectrolytes. The polyelectrolyte is first added to the mixing tank to achieve the right concentration. It is then added to the treatment process a short distance downstream of the coagulant rapid-mixing unit.

8.6   CHEMISTRY OF COAGULATION

Coagulants like alum react with the alkalinity of the water and form the insoluble floc. Sufficient chemical must be added to water to exceed the solubility limit of the metal hydroxide. The optimum pH range is between 5 and 8. There should be sufficient natural alkalinity present to react with the coagulating chemical and to serve as a buffer. If the source water is low in alkalinity, it may be increased by the addition of lime or soda ash.

8.6.1   CHEMICAL REACTIONS

Coagulation reactions are quite complex. The chemical reactions presented below are hypothetical but are useful in estimating the quantities of reactants and products. The chemical reaction of alum with alkalinity is shown below:

Alum reacting with natural alkalinity

$$Al(SO_4)_3 \cdot 14.3H_2O + 3Ca(HCO_3)_2 = 2Al(OH)_3 \downarrow + 3CaSO_4 + 6CO_2 + 14.3H_2O$$

Stoichiometric requirements

$$\frac{Alkalinity}{Alum} = \frac{3\,mol \times 100\,g\,/\,mol}{1\,mol \times 600\,g\,/\,mol} = 0.50$$

Alum reacting with added alkalinity(lime)

$$Al(SO_4)_3 \cdot 14.3H_2O + 3Ca(OH)_2 = 2Al(OH)_3 \downarrow + 3CaSO_4 + 14.3H_2O$$

As per stoichiometry, the molar ratio of alum to alkalinity is 3 to 1. As shown above, one unit of alum reacts with a half unit of the alkalinity, expressed as $CaCO_3$. When alum reacts with alkalinity, carbon dioxide is produced, and sulphate ions are added to finished water. Thus, the addition of alum results in reducing the pH, the alkalinity, or both. In addition, some permanent hardness such as calcium sulphate is added and the production of carbon dioxide makes the water corrosive. Despite these drawbacks, alum is still the most used coagulant. Alum is available both in solid and dry forms. Filter alum is the name used for solid alum. Liquid alum comes in three strengths. The strongest liquid alum is about 48.5%, with a specific gravity of 1.32.

8.6.2 ALUM FLOC (SLUDGE)

In the hypothetical coagulation equation, aluminum floc is written as $Al(OH)_3$. The quantity of sludge produced as $Al(OH)_3$ can be estimated as follows:

$$\boxed{\frac{Alum\ Floc}{Alum} = \frac{2\ mol \times 78\ g\ /\ mol}{1\ mol \times 600\ g\ /\ mol} = 0.26}$$

Experiments have shown that the actual production of alum floc is twice as much. In addition to aluminum hydroxide, turbidity will be removed. The total solids produced in alum coagulation can be estimated by using the following empirical relationship.

$$Total\ SS = 0.44 \times Alum\ dose + 0.74 \times Turbidity\ removed\ in\ NTU$$

For a known concentration of dry solids produced as sludge, the volume of sludge can be estimated. Remember, sludge with a solids content of 1% will contain 10 g of dry solids in every liter of sludge. Consistency of alum sludge ranges from 1%–3%.

Example Problem 8.1

A surface water is coagulated by adding 50 mg/L of alum and an equivalent dosage of lime. How many kg of coagulant are used to treat one million liters of water processed? How many kg of lime are required at a purity of 88% CaO?

Given:

     Alum = 50 mg/L   Lime = Equivalent dosage

Solution:

$$Alum = \frac{50\ mg}{L} \times \frac{kg\ /\ ML}{mg\ /\ L} = 50\ kg\ /\ ML = \frac{50\ mg}{L} \times \frac{meq}{100\ mg} = 0.5\ meq\ /\ L$$

$$Lime = \frac{0.50\ meq}{L} \times \frac{56\ mg\ CaO}{meq} \times \frac{1}{0.88} = 31.8 = 32\ mg\ /\ L = 32\ kg\ /\ ML$$

Example Problem 8.2

The optimum alum dose as determined by jar testing is 10 mg/L. The chemical feed pump has a maximum capacity of 4.0 gal/h at a setting of 100% capacity. The liquid alum delivered to the plant is 48.5%, with a SG of 1.35. Determine the setting on the liquid alum feeder for treating a flow of 4.5 MGD? How long will 100 gal of liquid alum last?

Given:

Parameter	Liquid Alum = 1	Treated Water = 2
concentration	48.5%	10 mg/L
SG	1.35	1.0
pump rate	?	4.5 MGD (mil gal/d)

Solution:

Strength of commercial liquid alum

$$C = C_{m/m} \times \rho = \frac{48.5\%}{100\%} \times \frac{1.35 \times 8.34\,lb}{gal} = 5.46 = 5.5\,lb/gal$$

Alum feeding

$$Q_1 = Q_2 \times \frac{C_2}{C_1} = \frac{4.5\,mil\,gal}{d} \times \frac{10\,mg/L \times 8.34\,lb/mil\,gal}{mg/L} \times \frac{gal}{5.46\,lb}$$

$$= \frac{68.73\,gal}{d} \times \frac{d}{24\,h} = 2.86 = \underline{2.9\,gal/h}$$

$$S_2 = S_1 \times \frac{Q_1}{Q_2} = 100\% \times \frac{2.9\,gal/h}{4.0\,gal/h} = 71.5 = \underline{72\%}$$

$$t = \frac{V}{Q} = \frac{1000\,gal.h}{2.86\,gal} \times \frac{h}{24\,h} = 14.56 = \underline{15\,d}$$

It is recommended to adjust the strength of the feed solution so that the feeder setting is about in the middle range (40%–60%). This allows the setting to go up and down and ensures feed pump operation under optimal conditions.

Example Problem 8.3

Raw Safwan River water has an alkalinity of 35 mg/L as $CaCO_3$. The jar test indicated an optimum alum dosage of 30 mg/L to reduce the turbidity to less than 1 NTU. To assure complete precipitation, a minimum of 30 mg/L of residual alkalinity is recommended. Work out the dosage of alkalinity to be added to maintain the required level of residual alkalinity. Also, find the dose of 85% hydrated lime in mg/L that will be needed to complete the reaction. What should be the setting on the lime feeder in kg/d when the flow is 10 ML/d? Also work out the production of carbon dioxide.

Given:

Raw Alk = 35 mg/L   Residual Alk = 30 mg/L   Alum dose = 30 mg/L
Q = 10 ML/d

Solution:

$$Alum\,Dose = \frac{30\,mg}{L} \times \frac{mmole}{600\,mg} = 0.05\,mmole/L$$

$$Alk\ consumed = 0.50 \times Alum = 0.5 \times 30\ mg/L = 15.0 = 15\ mg/L$$

$$Alk\ added = (15 + 30 - 35)\frac{mg}{L} \times \frac{mmole}{100\,mg} = 0.10\,mmole/L$$

$$lime\ added = \frac{0.1\,mmol}{L} \times \frac{74\,mg}{mmol} = 7.40 = \underline{7.4\ mg/L\ of\ Ca(OH)_2}$$

$$Lime\ feed = \frac{7.4\,kg}{ML} \times \frac{10\,ML}{d} \times \frac{100\%}{85\%} = 87.05 = \underline{87\ kg/d}$$

$$CO_2 = \frac{6.0\,mmol\ CO_2}{mmol\ alum} \times \frac{0.05\,mmol\ alum}{L} \times \frac{44\,mg}{mmol\ CO_2} = 13.2 = \underline{13\ mg/L}$$

8.7 FLOCCULATION PHENOMENON

Flocculation is a process of slow, gentle mixing of water to encourage the tiny particles to collide and clump together to become settleable floc. Slow mixing is a physical process, which helps to transform the microfloc formed during coagulation into macrofloc, which is easy to settle. As the floc particles grow, they become more fragile, so that the speed of the mixers needs to be controlled. A typical floc size is 1–2 mm and varies depending on the characteristics of raw water. If algae are present, they too can be trapped or caught up in the floc particles. When algae are present in large numbers, the flow will have a stringy appearance.

8.7.1 MIXERS

Mixing is the most important part of the coagulation/flocculation process. Because the reaction takes place in less than a minute, flash mixers are used in coagulation (**Figure 8.2**). The most commonly used are mechanical mixers and in-line mixers or blenders. A minimum detention time of 30 seconds should be provided.

A different kind of mixing is needed for flocculation. Here we are attempting to move particles around so that they collide and clump together to make settleable floc.

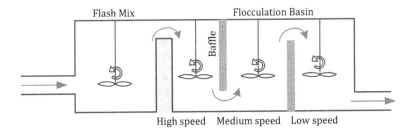

FIGURE 8.2 Coagulation and flocculation mixers

Since 95% of the floc is water, it is fragile and has to be treated gently. High-speed flocculation can be damaging. Since floc becomes more fragile as it grows, **mixing intensity** decreases in the direction of flow. Each successive compartment in the flocculation unit receives less mixing energy than the previous compartment. In the final stage, peripheral speed of the pedals should be in the range of 25–35 cm/s (50–70 ft/min). The minimum detention time recommended for flocculation ranges from 5–10 min for direct filtration systems and up to 30 min for conventional filtration. Flow through velocity shall be kept in the range of 2.5–7.5 mm/s (0.5–1.5 ft/min).

8.7.2 FLOCCULATION TANKS

Flocculation tanks for horizontal flocculators are usually rectangular; tanks for vertical flocculators are generally square. The depth of the flocculation tanks is comparable to sedimentation tanks. Best flocculation results are obtained when the mixing intensity decreases in the direction of flow. As water flows through the tank, floc size grows, and reduced mixing intensity reduces the chances of floc break up. The desired intensity is achieved by varying the speed of the flocculators.

A parameter called a **G-factor** or simply G is often used to express the intensity of mixing in terms of power input per unit of flow. Typical values for G in a flocculation tank decreases from a maximum of 75 to 30/s. In the case of rapid or flash mixing, G values can be as high as 1000/s. Factor G is actual velocity gradient and has the units of 1/s. Mathematically, velocity gradient G (/s) can be expressed as follows:

velocity gradient

$$G = \sqrt{\frac{C_D A v_r{}^3}{2 V \nu}}$$

A = area of cross section of paddles, V = volume of the tank
ν = kinetic viscosity = μ/ρ, C_D = coefficient of drag, v_r = relative velocity

A mechanically operated flocculation tank seems to perform better than a baffle-type flocculator (hydraulic mixing), due to better efficiency of mixing and negligible head loss. For reliability of the operation, either provide multiple tanks or a spare electric motor.

Example Problem 8.4

Size a flocculation basin to treat 20 ML/d. The detention time is 30 min. Assume the basin width is 15 m, consisting of 3 equal width units separated by perforated walls.

Given:

Q = 20 ML/d   W = 20 m t_d = 30 min   d = 4.0 m (assumed)

Solution:

$$V = t_d \times Q = \frac{20\,ML}{d} \times \frac{1000\,m^3}{ML} \times 30\,min \times \frac{d}{1440\,min} = 416.6\,m^2$$

$$A_X = \frac{V}{d} = \frac{416.6\,m^2}{4.0\,m} = 104.15 = 104\,m^2$$

$$L = \frac{A_X}{W} = \frac{104.15\,m^2}{15\,m} = 6.94 = 7.0\,m$$

Three units of 7 m × 5 m × 4 m will do the job

Example Problem 8.5

A water plant is to process water at the rate of 55 ML/d. It has a flocculation basin measuring 22 m × 12m × 4 m. It is fitted with four horizontal shafts each supporting four paddles that are 13 m × 0.20 m, centered 1.5 m from the shaft, and rotated at 1.5 rpm. Assume viscosity of water 1.2 × 10⁻⁶ m²/s. Making appropriate assumptions, calculate the velocity gradient, and flocculation time

Given:

Q = 55 ML/d, Tank = 22 m × 12m × 4.0 m, Paddle = 13 m × 0.20 m, r = 1.5 m, N = 1.5 rpm Assumed C_D = 1.9 and mean flow velocity as 30% of paddle peripheral velocity.

Solution:

Flow/paddle velocity

$$v_P = 2\pi rn = \frac{2\pi \times 1.5\,m}{revolution} \times \frac{1.5\,revolutions}{min} \times \frac{min}{60\,s} = 0.235 = 0.24\,m/s$$

$$\Delta v = 0.7v_P = 0.7 \times 0.2356\,m/s = 0.164 = 0.16\,m/s$$

$$G = \sqrt{\frac{C_D A (\Delta v)^3}{2Vv}} = \sqrt{\frac{1.9}{2} \times \frac{4 \times 4 \times 13\,m \times 0.2\,m}{22\,m \times 12\,m \times 4.0\,m} \times \left(\frac{0.16\,m}{s}\right)^3 \times \frac{s}{1.2 \times 10^{-6}m^2}}$$

$$= 11.3 = 11/s$$

Flocculation time

$$t_d = \frac{V}{Q} = \frac{22\,m \times 12\,m \times 4.0\,m.d}{55\,ML} \times \frac{ML}{1000\,m^3} \times \frac{1440\,min}{d} = 27.6 = \underline{28\,min}$$

Example Problem 8.6

A water plant is to process water at the rate of 12 MGD. It is planned to have two rectangular flocculation tanks each with an operating depth of 10 ft. Select the size of each tank to provide minimum detention time of 20 min. Assume length is twice that of width.

Given:

$$Q = 12/2 = 6 \text{ mil gal/d} \quad d = 10 \text{ ft} \quad L = 2 \times W \quad t_d = 20 \text{ min}$$

Solution:

$$t_d = \frac{V}{Q} = \frac{L \times W \times d}{Q} = \frac{2W \times W \times d}{Q} = \frac{2W^2 \times d}{Q} \quad \text{or}$$

$$W = \sqrt{\frac{t_d \times Q}{2d}} = \sqrt{20 \text{ min} \times \frac{d}{1440 \text{ min}} \times \frac{6 \times 10^6 \text{ gal}}{d} \times \frac{ft^3}{7.5 \text{ gal}} \times \frac{1}{2 \times 10 \text{ ft}}}$$

$$= 23.5 = 24 \text{ ft} \quad hence$$

$$L = 2 \times 24 = 48 \, ft$$

8.7.3 FACTORS AFFECTING FLOCCULATION

The purpose of all flocculators is to provide gentle mixing that will produce a quick settling floc. The main factors affecting flocculation are as follows.

Degree of Mixing

If mixing is too gentle or too fast, it will prevent the formation of large floc. Very slow mixing fails to bring the suspended particles into contact with each other, while mixing too fast mixing tears the floc particles. Mixing energy is reduced in the direction of flow to achieve better results.

Duration of Mixing

A minimum time of mixing is necessary for flocculation to be completed. In actual plant operations, depending on the temperature of the raw water, a period of 20 to 40 minutes is usually sufficient. To provide the required detention time, short-circuiting should be prevented. For this reason, at the entrance to the flocculator, the flow is directed downwards by placing a baffle.

Number of Particles

Relatively clear water is harder to flocculate than turbid water containing a lot of suspended matter. A large number of particles allows more collision thus resulting in large size floc.

Degree of Coagulation

Coagulation destabilizes the particles causing turbidity, and flocculation clumps these particles together forming a settleable floc. Improper chemical dosage, change in the quality of raw water, and temperature can affect the coagulation and flocculation processes. Since there are so many factors affecting the coagulation/flocculation process, whenever there is a change in the type of coagulant or the raw water characteristics, a laboratory procedure called the **jar test** is performed.

8.8   JAR TESTING

Jar testing is basically a laboratory simulation of coagulation and flocculation and settling to determine the optimum dosage of a coagulant and coagulation aids. The apparatus (**Figure 8.3**) consists of a set of six paddle mixers driven by a variable-speed motor. A glass beaker acting as a batch reactor is placed under each paddle mixer and a different dosage is added to each beaker. The set of dosages to be tried will depend upon the characteristics of the raw water. During jar testing, chemicals should be added in the same sequence as in plant operations. The working solutions for jar testing should preferably be made from the commercial chemical used by the plant. When adding a chemical, all jars should be dosed at the same time and paddle speed should be kept at more than 200 rpm to simulate flash mixing.

The main purpose of jar testing is to determine the **optimum dosage** or the minimum dosage that will do the job. If the coagulant dose is low, floc formation will not be as effective as it could be. On the other hand, overdosing not only costs more, but it may also cause problems later in the treatment processes and leave residual coagulants in the finished water.

When the characteristics of the raw water change rapidly because of storm conditions or other causes, jar tests may have to be conducted frequently. A shortcut to jar testing is to collect a water sample in a glass beaker directly from the outlet of the rapid mix chamber. By gently stirring the water sample for a few minutes, you can

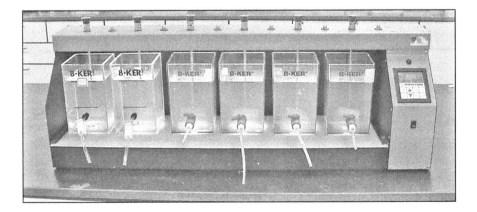

FIGURE 8.3   Jar testing apparatus

sometimes get an idea of how well the water is coagulating. The jar test will give a good indication for the chemical dosage required, although the actual dosage used in the plant operation may be a bit higher or lower. Compared to jar testing, the results are usually better when the same dosage is applied in the actual plant operation.

There have been some improvements in jar testing procedure. Since coagulant dose does affect the pH, in addition to coagulant dosage, pH also needs to be optimized. Another modification is to observe turbidity of the settled water by collecting a filtered sample. A filter media of crushed glass with effective size of 0.45 mm is packed in a small column. This filter column has provision for withdrawing a sample from the jar.

Example Problem 8.7

For jar testing, you need to prepare 0.10% solution of alum using 45% liquid alum of SG of 1.3. What volume of liquid alum would you require preparing 1 L of solution?

Solution:

Strength of commercial liquid alum

$$C = C_{m/m} \times \rho = \frac{45\%}{100\%} \times \frac{1.3\ kg}{L} \times \frac{1000\ g}{kg} = 585.0 = 585\ g/L$$

$$V_1 = V_2 \times \frac{C_2}{C_1} = 1.0\ L \times \frac{1.0\ g}{L} \times \frac{L}{585\ g} \times \frac{1000\ mL}{L} = 1.70 = \underline{1.7\ mL}$$

8.9   OPERATIONAL CONTROL TESTS

A brief discussion of tests used to optimize the coagulation and flocculation processes follows.

8.9.1   ACIDITY TESTS

It is well established that coagulation flocculation is very much affected by the chemical environment indicated by pH and the alkalinity of the water. If changes in pH are noticed due to natural changes in raw water or changes in other treatment chemicals, the doses of coagulants need to be adjusted accordingly.

8.9.2   TURBIDITY TESTS

Turbidity of the raw water and settled water should be observed and compared with the jar test results to see the actual removal after sedimentation. The turbidity removal should be comparable. Any drastic difference from the jar testing should be investigated and corrected. For a well-operating plant, turbidity of the settled water should not be less than 10 NTU. Turbidity levels exceeding this level may prematurely seal the filter and shorten the filter runs.

8.9.3 FILTERABILITY TESTS

Filterability tests are done using a pilot filter to determine how efficiently the coagulated water can be filtered. The pilot filter contains the same media as the actual filter and is equipped with an online turbidimeter. The amount of water passing through the filter before turbidity breakthrough can be used to judge the filter run under the same coagulant dosage. The water used in this test is the actual coagulated water from the plant.

8.9.4 ZETA POTENTIAL

Zeta potential is a measure of the negative charge or excess of electrons on the surface of particulate matter. The degree of this charge determines efficacy of the coagulation process. To induce the particle to settle properly, the ZP should be reduced to close to zero. Zeta potential below −10 indicates very poor settling due to strong repelling forces. A zeta meter is used to make zeta potential or ZP measurements.

8.9.5 STREAMING CURRENT MONITORS

Stream current in a flow stream of coagulated water is monitored using a stream current detector (STD). This measurement is very similar to the zeta potential discussed above. The optimal STD reading varies with the pH of the raw water. The advantage of the stream current monitor is that it provides a continuous reading and thus can be used effectively to control the process. If the pH of the water remains constant, it can be used for automatic control.

8.9.6 PARTICLE COUNTERS

The monitoring of turbidity at various points in water treatment yields good information about the removal of particulate matter. However, it fails to recognize the size and number of particles removed. This is becoming increasingly important when we want to ensure the removal of Crypto and Giardia cysts, two commonly found pathogens in drinking water supplies. Since particle counters can provide size distribution, that is size and concentration, they can be used for optimizing the coagulation filtration process.

Particle counters are based on passing a beam of laser light through the water and the reflected or diffused light is sensed by photodiode. Computer software interprets this information. These instruments are quite expensive and need to be handled by specially trained personnel. However, it has a very promising future in water treatment optimization.

Discussion Questions

1. Draw a schematic diagram of a water treatment plant that uses river water as the source. Discuss the purpose of each of the unit operations or unit processes in the process scheme.

2. Draw a schematic diagram of a water treatment plant that uses groundwater as the source. Discuss the purpose of each unit operation or unit process shown in the process schematic.

3. To precipitate turbidity, alum is added that reacts with any alkalinity present in water. Show that during coagulation, alkalinity as $CaCO_3$ consumed is equal to half that of alum dosage.

4. What do you understand by coagulation and flocculation? Explain why they are necessary in potable water treatment.

5. Why is chemical-assisted filtration considered necessary for the treatment of surface water supplies and GUDI sources?

6. Describe the various type of coagulants and coagulant aids used in the water industry.

7. Discuss the role of coagulant aids and flocculation aids.

8. Compare rapid mixing with slow mixing.

9. Describe the various steps to perform jar testing. What are the disadvantages of applying too high a dosage?

10. Explain the factors that affect flocculation.

11. What process control test can be used to optimize the coagulation/flocculation process?

12. Why is alum the most commonly used coagulant in water treatment?

13. Under what conditions would you need frequent jar testing?

14. Define the term "velocity gradient" and the factors affecting it. What range of values are recommended for rapid mixing and slow mixing?

15. What improvements have been made in jar testing?

Practice Problems

1. The daily flow to a water plant is 3.2 MGD. Jar test results indicate that the best polymer dosage is 0.50 mg/L. How much polymer will be used in a week? (93 lb).

2. During a jar testing, one of the trial dosages is 25 mg/L. What volume of 1% alum solution should be added to 1 L of water in the jar to achieve this dosage? (2.5 mL).

3. A water plant has two flocculation basins, each measuring 12 m × 9 m × 2.5 m. What is the theoretical detention time for a flow of 10 MGD? (21 min).

4. What volume of 46% liquid alum (SG = 1.3) is needed to feed 1 kg of alum? (17 L)

5. In a small water filtration plant, 15% alum solution is fed at a rate of 50 gal/24h when treating a flow of 120 gpm. Find the alum dosage applied. (43 mg/L).

6. Raw water of a certain river has an alkalinity of 40 mg/L as $CaCO_3$. Based on jar testing, optimum alum dosage of 40 mg/L is required to reduce the turbidity to less than 1 NTU. To assure complete precipitation, a minimum of 30 mg/L of residual alkalinity is recommended.

a. What alkalinity must be added to maintain the desired level of residual alkalinity? (10 mg/L as $CaCO_3$).

b. Determine the dose of 80% hydrated lime in mg/L that will be needed to add alkalinity? (9.3 mg/L).

c. What should be the setting on the lime feeder in kg/d while treating a flow of 12 ML/d? (110 kg/d).

7. In a water treatment plant, due to the increase in water demand, the water flow rate is increased to 11 ML/d. For achieving alum dosage of 30 mg/L, work out the feed pump rate of 48% liquid alum with SG of 1.3? (370 mL/min).

8. A dosage of 60 mg/L of alum is added in coagulating river water.

a. How much natural alkalinity is consumed if 20 mg/L of alkalinity as $CaCO_3$ is added? (10 mg/L).

b. What is the increase in SO_4 ion in the treated water? (29 mg/L).

c. What is the reduction in alkalinity as bicarbonate? (12 mg/L).

d. How is the pH of the water affected? (drop).

9. An alum dosage of 30 mg/L and equivalent dosage of soda ash is added in coagulation of a surface water. What changes take place in the ionic character of water? (Addition of 14 mg/L of SO_4 ion, 7 mg/L of Na ions and decrease in pH due to CO_2).

10. Coagulation of soft water requires 40 mg/L of alum plus lime to supplement natural alkalinity for proper coagulation and floc formation. If it is desired to react not more than 10 mg/L of natural alkalinity, what dosage of lime as 85% pure CaO is required? (6.6 mg/L).

11. A river water is coagulated dosing ferrous sulphate at the rate of 30 mg/L and equivalent dosage of lime. Chemical reaction is as follows:

$$2FeSO_4.7H_2O + 2Ca(OH)_2 + 0.5O_2 = 2Fe(OH)_3 + 2CaSO_4 + 13H_2O$$

a. How many kg of hydrated lime are needed per ML of water? (8.0 kg/ML).

b. How many kg of 88% CaO? (6.9 kg/ML).

c. How many kg of sludge as $Fe(OH)_3$? (12 kg/ML).

12. Write the chemical reaction between ferric chloride and lime slurry that results in the precipitation of $Fe(OH)_3$. At a conventional plant, surface water at the rate of 24 ML/d is treated applying coagulant dosage of 50 mg/L.

a. What is the required dosage rate of lime as 85% CaO? (730 kg/d).

b. Find dry solids as $Fe(OH)_3$ produced? (790 kg/d).

c. Assuming wet sludge is 2.0% solids, what is the daily production of inorganic sludge? (40 m^3/d).

13. Determine the settings in percent stroke on a chemical feed pump to apply a dosage of 2.5 mg/L. The water is pumped at the rate of 500 gpm and the strength of the chemical solution fed is 5.0%. The chemical feed pump has a maximum capacity (100% setting) of 5.0 gal/h. (30%).

14. Determine the amount of natural alkalinity consumed for the treatment of lake water with alum dosage of 30 mg/L. For a daily flow rate of 1.5 MGD, estimate the amount of sludge produced as $Al(OH)_3$ assuming sludge is 2.5% solids. (15 mg/L, 470 gal/d).

15. Optimum alum dose as determined by jar testing is 15 mg/L. The label on the liquid alum drum indicates a strength of 45%, with a density of 1.3 kg/L.
    a. Determine the setting on the liquid alum feeder in L/24h for treating a flow of 15 ML/d? (390 L/24h).
    b. How long would a container holding 1.5 m³ of liquid alum last? (3.9 d).

16. The natural alkalinity in muddy river water is found to be 35 mg/L as $CaCO_3$. The alum dose required, as determined by jar tests, is 50 mg/L. Calculate the lime dose required in mg/L as 80% CaO if 30 mg/L of alkalinity as $CaCO_3$ is required for complete precipitation. (14 mg/L)

17. Jar tests indicate that 40 mg/L is the optimum alum dose for treating Rainy River water. If 30 mg/L of residual alkalinity expressed as $CaCO_3$ is required to promote complete precipitation, what is the total alkalinity required for chemical coagulation? (50 mg/L)

9 Sedimentation

In conventional water treatment schemes, sedimentation follows coagulation and flocculation to allow the gravitational settling of flocs and particulate matter. Settled water has low turbidity – usually less than 10 NTU – which is removed by filtration. Sedimentation is necessary for the treatment of turbid waters. However, when the turbidity level in the source water is less than 10 NTU, the sedimentation process may be omitted. This process scheme is called **direct filtration**.

In gravitational settling, heavier matter in water is allowed to settle out under quiescent conditions. For effective settling, water flow velocity needs to be kept low – – typically in the range of 0.1–0.5 mm/s (1–1.5 ft/h). Sedimentation will happen when the **settling velocity** is more than the **surface overflow rate** and the sedimentation tank provides sufficient **hydraulic detention time.** The parameters important for design and operation of sedimentation tanks include detention time, overflow rate, and weir loading. For properly designed and operated sedimentation basins, solids removal efficiencies greater than 95% can be achieved. In a well operated plant, the turbidity of the settled water should be less than 10 NTU.

9.1 THEORY OF SEDIMENTATION

Sedimentation is the separation of solids from water using gravity settling under quiescent conditions. In water treatment, the sedimentation processes used are of two types.

Type I: to settle out discrete non-flocculant particles in dilute suspension. This may arise due to the plain settling of surface waters prior to treatment by sand filtration

Type II: to settle out flocculant particles in dilute suspension. This is done to settle the floc created during coagulation and flocculation.

9.1.1 PLAIN SEDIMENTATION

Plain settling involves the settling of discrete particles like sand and grit. When plain settling is done before the water enters the plant, it is called **presedimentation**. This may be is practiced where the surface water bodies are subjected to heavy loads of solids, such as grit and sand. A lake is natural presedimentation tank. In the case of turbid waters, presedimentation is very important since water loaded with grit can damage the mechanical equipment and pipes. **Plain sedimentation** tanks can be rectangular or circular, though rectangular tanks are more common. Plain sedimentation is becoming obsolete these days since it is not able to remove colloidal particles.

DOI: 10.1201/9781003347941-11

9.1.2 DISCRETE SETTLING

Figure 9.1 shows type I settling. Particles settle out individually and it is assumed that there is no flocculation between the particles. This kind of settling is usually the first physical process for surface waters and for grit removal in wastewater treatment.

The settling velocity of the discrete particles is determined by Stokes' Law. The velocity of flow, viscosity of water, size and shape, and more importantly density or specific gravity of the solid, influence the settling velocity of discrete particles in water. When buoyancy and drag forces counterbalance gravity forces, solid particles reach terminal velocity, as given by Stokes' Law.

Settling velocity of a discrete particle

$$v_s = \frac{gD^2}{18v}(G_s - 1) \quad Modified \ v_s = 418D^2(G_s - 1)\left[\frac{3T + 70}{100}\right]$$

Where settling velocity in mm/s and diameter in mm and temperature T in °C, Hazen observed that when the diameter of the particle of size >0.1 mm, settling velocity is proportional to the first power of the diameter and not the second power.

Assuming type I settling is applicable, with flow entering and leaving the tank uniform, there are three zones in the settling tank, and particles reaching the bottom stay there until scraped off. For a rectangular tank, as shown in Figure 9.2, the following relationship will apply.

Horizontal flow velocity and overflow rate

$$v_H = \frac{Q}{A_X} = \frac{Q}{L \times d_w} = \frac{d_w}{v_O} \quad and \quad v_o = \frac{Q}{A_S} = \frac{Q}{L \times W} = \frac{d_w}{t_d}$$

Detention time

$$t_d = \frac{V}{Q} = \frac{L \times W \times d_w}{Q} = \frac{d_w}{v_O} = \frac{L}{v_H}$$

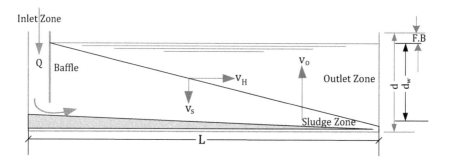

FIGURE 9.1   Discrete settling

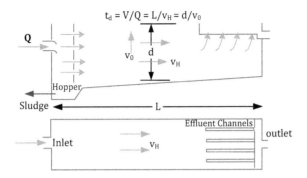

FIGURE 9.2 Plan and sectional view of a rectangular basin

Example Problem 9.1

A spherical particle with a diameter of 0.07 mm and a specific gravity of 2.8 settles in water at a temperature of 15°C. Use Stokes' Law equation to calculate the settling velocity.

Given:

$$D = 0.07 \text{ mm} \quad G_s = 2.8 \quad T = 15°C \quad v = 1.14 \times 10^{-6} \text{ m}^2/\text{s} = 1.14 \text{ mm}^2/\text{s}$$

Solution:

Settling velocity

$$v_s = 418D^2 (G_s - 1)\left[\frac{3T + 70}{100}\right] = 418 \times (0.07)^2 \times (2.8 - 1)\left(\frac{3 \times 10 + 70}{100}\right)$$

$$= 4.23 = 4.2 \text{ mm} / s \text{ or}$$

$$v_s = \frac{gD^2}{18 v}(G_s - 1) = \frac{9.81 m}{s^2} \times \frac{(0.07 \text{ mm})^2}{18} \times \frac{s}{1.14 \text{ mm}^2} \times 1.8 \times \frac{1000 \text{ mm}}{m}$$

$$= 4.21 = 4.2 \text{ mm} / s$$

For 100% removal of this size particle, the overflow rate must be equal to or less than 4.2 mm/s.

Example Problem 9.2

Size a square type I settling tank to treat 18 ML/d of raw water, with a surface rate of 12 m³/m²·d and detention time of 6 h. What minimum size particle can be 100% removed? Assume specific gravity of 2.5 and water temperature of 15°C.

Given:

$$D = ? \quad G_s = 2.5 \quad Q = 18 \text{ ML/d} \quad t_d = 6.0 \text{ h} \quad T = 15°C \quad v = 1.14 \text{ mm}^2/\text{s}$$

Solution:

Surface area

$$A_s = \frac{Q}{v_o} = \frac{18\,ML}{d} \times \frac{d}{12\,m} \times \frac{1000\,m^3}{ML} = 1500\,m^2$$

Assuming L = 2W

$$W = \sqrt{\frac{1500\,m^2}{2}} = 27.3 = 28\,m \;\; say, \;\; hence \; L = 2W = 56\,m$$

Side water depth

$$d_w = v_0 \times t_d = \frac{12\,m}{d} \times 6h \times \frac{d}{24h} = 3.0\,m \;\; d = 3.0 + 0.5(f.b) = 3.5\,m$$

Minimum size particle that will be 100% removed ($v_o = v_s$)

$$D = \sqrt{\frac{18v \times v_o}{g(G_s - 1)}} = \sqrt{\frac{18s^2}{9.81m} \times \frac{1.14\,mm^2}{s} \times \frac{12m}{d} \times \frac{d}{24 \times 3600\,s} \times \frac{1}{(2.5 - 1)}}$$

$$= 0.0139 = 0.014\,mm$$

Particle smaller than 0.014 mm will be partially removed

Example Problem 9.3

A sedimentation basin is designed to carry all particles larger than 0.04 mm of SG = 2.65. The effective depth of the basin is 3.0 m and the length 40 m. What horizontal velocity should be maintained to achieve that removal. Assume kinetic viscosity at the prevailing temperature is 1.0×10^{-6} m²/s.

Given:

$$D = 0.04 \; mm \quad G_s = 2.65 \quad v = 1.0 \times 10^{-6} \; m^2/s = 1.0 \; mm^2/s$$

$$d_w = 3.0 \; m \quad L = 40 \; m \quad v_H = ?$$

Solution:

$$v_s = \frac{gD^2}{18v}(G_s - 1) = \frac{9.81m}{s^2} \times \frac{(0.04\,mm)^2}{18} \times \frac{s}{1.0\,mm^2} \times 1.65 \times \frac{1000\,mm}{m}$$

$$= 1.438 = 1.4\,mm/s$$

$$v_H = v_s \times \frac{L}{d_w} = \frac{1.438\,mm}{s} \times \frac{40\,m}{3.0\,m} = 19.18 = 19\,mm/s$$

9.2 SEDIMENTATION AIDED WITH COAGULATION

As mentioned earlier, plain sedimentation is unable to remove fine suspended particles unless detention times in terms of days and months are provided. Moreover, surface charge on colloidal particles keeps them apart and gives them long-term stability. As discussed in the previous unit, coagulation is used to destabilize these particles. It is followed by flocculation for floc to grow and strengthen. Sedimentation follows coagulation and flocculation to allow floc to settle and produces effluents with turbidity less than 10 NTU. This helps to make filtration efficient and achieve long fitter runs. Sedimentation of coagulated water is type II settling as floc grow during sedimentation process. The detention time for sedimentation tanks following coagulation is typically 2–4 h – significantly less than in the case of plain sedimentation. Overflow rates are typically 24–32 $m^3/m^2 \cdot d$ (600–800 $gal/ft^2 \cdot d$), which is about twice what can be afforded in plain sedimentation tanks. This is another reason that sedimentation following coagulation is the norm of the day.

9.3 SEDIMENTATION BASINS AND TANKS

For efficient operation, sedimentation basins are designed so that water can enter the tank, pass through, and leave without creating much turbulence, while preventing any short-circuiting. The operation of a settling tank can be better understood by visualizing it in four distinct zones. In the **inlet zone**, water entering the tank is distributed across the section of the tank by the baffle and slows to a uniform velocity. The **settling zone** is the main part of the tank. Here the water flows slowly through the tank and the suspended solids settle. The settled matter, or sludge, accumulates at the bottom of the tank, referred to as the **sludge zone.** In the **outlet zone**, the clarified water flows over the weirs into effluent channels and leaves the tank.

9.3.1 RECTANGULAR BASINS

Rectangular-shaped basins are more popular in water treatment plants. In **Figure 9.2**, plan and sectional views of a rectangular sedimentation tank are shown. The main reasons for their popularity include predictable performance, comparable cost, low maintenance and minimal short circuiting, which lends itself to sudden increases in loading.

A modified design of a rectangular basin is the double deck version. **Double deck** basins provide twice the effective sedimentation area of a single basin of equivalent land area. Due to high operation and maintenance costs, they have failed to gain popularity.

9.3.2 CIRCULAR AND SQUARE BASINS

In circular and square basins, the flow radiates from the center towards the periphery. They are generally more likely to short circuit. The bottom of a circular clarifier is graded towards the center (typically 1:12) to make efficient the collection and withdrawal of sludge. In a circular clarifier with a diameter D, the volume of water and sludge can be expressed in terms of D. Total volume is the sum of cylinder of height H equal to side water depth and cone made by the sloped part of the clarifier.

9.3.3 TUBE SETTLERS

Conventional sedimentation basins are modified by placing tubes at an upward angle in the direction of flow through the tank. Water is directed upwards and each tube acts as an individual settler. **Tube settlers** are basically shallow depth settlers. Together, they provide a high ratio of settling surface area per unit volume of water. High-rate settlers are particularly useful for water treatment applications where the site area is in constraint, in packed-type units, and to increase the capacity of existing units. Since the depth of settling is reduced significantly, the time for a particle to settle is also reduced. Due to increased overflow rates, coagulant aids should be used to strengthen the floc.

9.3.4 SOLIDS CONTACT UNITS

Solid contact units combine the coagulation, flocculation, and sedimentation and sludge removal into a single compartmented tank as shown in **Figure 9.3**. For this reason, they are also called flocculator-clarifiers. Such units are common for industrial plants and where softening and turbidity removal are performed simultaneously. The flow is generally in an upward direction through a properly controlled sludge blanket; hence the name "solids contact unit" or "upflow clarifier".

Referring to **Figure 9.3**, raw water and coagulant enter the draft tube just below the mixers. The mixer draws the raw water and the chemical up the draft tube through the rapid mix zone. This phase represents the coagulation process. Water then flows from the top of the draft tube down into a slow mix zone under the skirt where flocculation takes place. The contact with the sludge solids helps to build

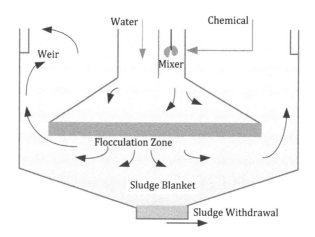

FIGURE 9.3   A solids contact unit

better floc. Floc settles to form a layer of sludge at the bottom of the clarifier. From the bottom of the flocculation zone the water flows up to the launders. The upward flow velocity or rise rate decreases in this zone to allow the floc to settle. From the launders, the clarified water flows to the next process, usually filtration.

During the operation of such units, some of the settled sludge is recycled with the incoming raw water and the remainder periodically drawn off. There must be a proper balance between the incoming raw water turbidity and the amount of sludge to be drained. The maintenance of the sludge blanket is critical to the operation of the upflow clarifier. Reactor/clarifiers are efficient and economical, especially for the softening processes. However, they are very sensitive to changes in influent flow and turbidity.

9.3.5 Pulsator Clarifier

The pulsator clarifier is a vertical type of sludge tank. A pulse of upward flow is generated every 30 s to give rapid flow for 5–10 s. This causes the sludge blanket to rise and fall alternately.

9.3.6 Ballasted Flocculation

Ballast is defined as a heavy substance that imparts weight. Injecting water with heavy sand enhances flocculation and sedimentation. Usually, micro sand, in size range 60–200 μm, is used as ballast. This is also known as the **actiflow** process. Typically, sand is applied after coagulation along with a flocculation aid like polymer. Both are added to initiate floc formation for a couple of minutes. In the maturation tank, polymer allows the floc to gain size and strength. Typically, hydraulic detention time is about 6 minutes. Clean sand is separated from the sludge slurry and recycled to the injection tank.

Ballasted flocculation is appropriate in treating surface water with fluctuating quality, especially turbidity. This may happen due to algae blooms and naturally occurring organic matter from land drainage.

9.4 DESIGN PARAMETERS

When designing a sedimentation process, engineers must use basin guidelines. It is important to understand these guidelines in order to communicate effectively with design engineers and to identify the cause of any operational problems.

9.4.1 Detention Time

Hydraulic detention time (HDT), or retention time, refers to the time required by a parcel of water to pass through a sedimentation basin at a given flow rate. It can be thought of as the time to fill up the basin at a given flow rate. Detention time is calculated by dividing the volume of the basin by the rate of flow and is commonly expressed in hours. Commonly used values of HDT are 2–4 hours. The actual flow-through time may be less than the calculated value due to short-circuiting.

9.4.2 SURFACE OVERFLOW RATE

Overflow rate (v_o), or surface loading, represents the upward velocity with which water is going to rise in a sedimentation basin (**Figure 9.4**). The particles with settling velocities less than the overflow rate will not settle out. Overflow rate is determined by dividing basin flow rate by the basin surface area.

Typically, surface loading ranges from 20 to 32 $m^3/m^2 \cdot d$ (500–800 $gal/ft^2 \cdot d$). In colder temperatures, settling velocities reduce due to an increase in density and the viscosity of water. As a result, plant capacity is lowered during the winter months. To compensate for a reduced settling rate, weighted agents may be applied, especially in light density floc.

9.4.3 OVERFLOW RATE AND REMOVAL EFFICIENCY

Since overflow rate represents velocity with which water rises, particles with settling velocity less than or equal to the overflow rate will surely settle out in an ideal plug flow system. Put differently, particles with settling velocity, $v_s \geq v_o$ will be 100% removed. However, particles with lower settling velocities will also settle if introduced at a lesser height. This will only be true in the case of a plug flow reactor and not an upflow clarifier. Following this logic, if particles are uniformly distributed, 50% of the particles with settling velocity of 0.5 v_o should also settle out. Likewise, 25% of particles with settling velocity of 0.25 v_o will also be removed.

9.4.4 EFFECTIVE WATER DEPTH

Ideally, sedimentation basins should be shallow, with a large open area. However, practical considerations, such as depth of sludge blanket, current and wind effects, and desired flow velocity must be considered in the selection of an appropriate basin

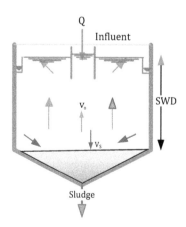

FIGURE 9.4   Detention time and overflow rate

depth. The side water depth (SWD), overflow rate (v_0), and hydraulic detention time (HDT) are interrelated parameters.

$$HDT \text{ or } t_d = \frac{V}{Q} = \frac{side\,water\,depth}{overflow\,rate} = \frac{d_w}{v_O} = \frac{L}{v_H}$$

Since v_0 represents the upward traveling rate, distance divided by velocity gives you the time it takes for a given amount of water to rise vertically for discharge through effluent channels.

9.4.5 MEAN FLOW VELOCITY

Mean flow velocity is the horizontal flow velocity along the length of the basin. In an ideal basin, it should be uniform. However, due to the non-uniform accumulation of sludge, wind action, and density currents, it is not completely uniform throughout the length of the basin. High flow velocities will have the tendency to scour the settled sludge. It is recommended to limit the mean flow velocities to 2.5 mm/s at the design flow rate.

Example Problem 9.4

Design a rectangular sedimentation tank to treat a flow of 4.0 ML/d so as to provide a detention time of 3.5 h and without exceeding a flow velocity of 1.5 mm/s.

Given:

$$Q = 4.0 \text{ ML/d} = 4000 \text{ m}^3/\text{d} \quad v_H = 1.5 \text{ mm/s} \quad t_d = 3.5 \text{ h}$$

Solution:

$$L = t_d \times v_H = 3.5\,h \times \frac{1.5\,mm}{s} \times \frac{3600\,s}{h} \times \frac{m}{1000\,mm} = 18.9 = \underline{20\,m\,say}$$

$$A_X = \frac{Q}{v_H} = \frac{4000\,m^3}{d} \times \frac{s}{1.5\,mm} \times \frac{1000\,mm}{m} \times \frac{d}{24 \times 3600\,s} = 30.86\,m^2$$

Assume a working depth of 3.0 m

$$B = \frac{A_X}{H} = \frac{30.86\,m^2}{3.0\,m} = 10.28 = \underline{10\,m\,say}$$

1. To provide room for sludge accumulation and free board, select the depth of the tank as 3.0 +1.0 + 0.5 m = 4.5 m. Final design is a rectangular basin of 20 m × 10 m × 4.5 m.

2. Short-circuiting with proper baffles and inlet design must be prevented in order to achieve high displacement efficiency and uniform flow velocity.
3. Displacement efficiency refers to the ratio of actual detention time to theoretical detention time. In an ideal plug flow, displacement efficiency is 100%

Example Problem 9.5

Size a circular sedimentation tank to provide a detention time of 2.5 h for a design flow of 1.0 MGD.

Given:

$$Q = 1.0 \text{ mil gal/d} \quad t_d = 2.5 \text{ h} \quad SWD = 10 \text{ ft (assumed)}$$

Solution:

$$V = t_d \times Q = 2.5h \times \frac{10^6 \, gal}{d} \times \frac{d}{24h} \times \frac{ft^3}{7.48 \, gal} = 1.393 \times 10^4 \, ft^3$$

$$D = \sqrt{\frac{4V}{\pi H}} = \sqrt{\left(\frac{4}{\pi} \times \frac{1.393 \times 10^4 \, ft^3}{10 ft}\right)} = 42.1 = 42 ft$$

A 42 ft circular tank with a total depth of 10 + 1.5 + 1.5 (free board) = 13 ft will do the job

9.4.6 Weir Loading Rate

Weir loading rate is the flow rate per unit length of the effluent weir. If the weir loading rate becomes too high, floc will be carried out of the basin. Weir loading is more important in shallow basins than in deeper basins. The maximum weir loading rate is 250 m³/m·d.

$$WLR = \frac{Flow \, Rate}{Weir \, Length} = \frac{Q}{L_W}$$

9.5 FACTORS AFFECTING OPERATION OF SEDIMENTATION

Some key factors are listed below:

- Large and dense floc will settle faster, so a lower tank overflow rate is preferred.
- Inlet arrangements should distribute incoming flow over the full cross section of the tank. The outlet design should collect settled water near the top and uniformly across the width of the tank.
- Current caused by wind action and density differences can cause short-circuiting.

- Particles settle faster in warm water than cold water.
- Detention time of 2–4 hours residence time is sufficient.
- Sudden changes in raw water quality will affect the sedimentation process.
- Changes in source water alkalinity and pH caused by storms seriously affect the performance of sedimentation because of poor coagulation or flocculation.
- Sudden increases in settled water turbidity will cause premature clogging of filters and may increase the turbidity of the filter effluent.

9.6   VOLUME OF SLUDGE

Sludge collected from the bottom of a water treatment clarifier is mainly inorganic. Based on the dosage of the coagulant and the turbidity reduced, the mass of solids removed can be found. Knowing the consistency of wet sludge, the volume of sludge produced can be estimated. Though inorganic solids are heavier, sludge solid concentration is not more than 2%. The solid concentration of alum sludge is typically in the range of 1%–3%. Alum sludge is difficult to dewater.

Example Problem 9.6

In a water treatment sedimentation tank, the sludge produced has solids concentration of 0.80%, and the specific gravity of dry solids in the sludge is 1.5. Determine the specific gravity of the wet sludge.

Given:

$$SS \text{ (sludge)} = 0.80\% \quad \text{Water content, } w = 99.2\% \quad G_{ss} = 1.5$$

Solution:
   Specific gravity of wet sludge

$$\frac{100\%}{G_{sl}} = \frac{SS\%}{G_{ss}} + \frac{w\%}{G_w} = \frac{080\%}{1.5} + \frac{99.2\%}{1} \quad or$$

$$G_{sl} = \left(\frac{0.008}{1.5} + .992 \right)^{-1} = 1.002$$

This further validates the point earlier that for low solid concentrations, the specific gravity of wet sludge can be safely assumed to be unity. Thus $0.8\% = 8.0 \text{ kg/m}^3$.

Example Problem 9.7

In a flocculation clarifier, alum dose is 30 mg/L to treat river water at the rate of 4.0 ML/d. Estimate the volume of wet sludge produced as aluminum hydroxide, assuming the solids concentration in the sludge is 0.80%. Also, find the quantity of sludge produced if 15 NTU of turbidity is removed.

Given:

Q = 4.0 ML/d, alum dose = 30 mg/L, SS = 0.80% = 8.0 kg/m³
wet sludge, Q_{sl} = ?

Solution:

$$Q_{sl} = \frac{M_{SS}(removed)}{SS_{sl}} = \frac{Q \times 0.26\, Alum}{SS_{sl}} = \frac{4.0\,ML}{d} \times \frac{0.26 \times 30\, kg}{ML} \times \frac{m^3}{8.0\, kg}$$

$$= 3.90 = \underline{3.9\, m^3 / d}$$

In the second case, 15 NTU of turbidity is removed, which would increase SS in the sludge:

$$SS_{rem} = 0.44\, Alum + 0.74 \times NTU\ removed$$

$$= 0.44 \times \frac{30\, mg}{L} + 0.74 \times 15 = 24.3\, mg / L$$

$$Q_{sl} = \frac{Q \times SS_{rem}}{SS_{sl}} = \frac{4.0\,ML}{d} \times \frac{24.3\, kg}{ML} \times \frac{m^3}{8.0\, kg} = 12.1 = \underline{12\, m^3 / d}$$

That is roughly three times the wet sludge produced without considering solids removed due to turbidity.

Example Problem 9.8

A 35 ML/d capacity water plant removes turbidity causing solids by alum coagulation. The alum dose is 20 mg/L and raw water contains 30 mg/L of suspended solids. What is the production of dry solids due to alum floc and SS removal as sludge (1.8%) in the sedimentation tank? Assume 65% removal of SS and that sufficient alkalinity is present to react with alum.

Given:

Q = 35 ML/d, Alum dose = 20 mg/L, SS_{raw} = 30 mg/L, Removal = 65%

Sludge: SS_{sl} = 1.8% = 18 kg/m³, Wet sludge, Q_{sl} = ?

Solution:

$$SS_{rem}\ as\ Al(OH)_3 = \frac{2\, mol \times 78g / mol}{1\, mol \times 600\ g / mol} \times \frac{20\, mg}{L} = 5.2\, mg / L$$

$$SS_{rem} = 0.65 \times 30\, mg / L + 5.2\, mg / L = 24.7\, mg / L$$

$$M_{SS} = \frac{24.7\, kg}{ML} \times \frac{35\,ML}{d} = 864.5\, kg / d$$

$$Q_{Sl} = \frac{M_{SS}(removed)}{SS_{sl}} = \frac{864.5\ kg}{d} \times \frac{m^3}{18\ kg} = 48.02 = 48\ m^3\ /\ d$$

9.7 SLUDGE DISPOSAL

In many plants, the most common practice is to discharge the sludge into the sanitary sewer to be treated at the municipal wastewater plant. Recently, water plant residue has found a number of other applications: soil conditioner is the major one. Recent practices of disposal include cement and brick manufacturing, turf farming, composting with yard waste, road sub grade, forest land application, citrus grove application, landfill, and land reclamation.

Discussion Questions

1. Based on the alum reacting with alkalinity, show that alum floc produced as aluminum hydroxide is roughly about one-fourth of the alum dosage.
2. List the conditions suitable for direct filtration.
3. Describe, with a neat sketch, the functional zones of a sedimentation basin.
4. Derive the relationship between detention time and surface overflow rate.
5. Prove that, theoretically, the surface loading and the not the depth is a measure of the removal of particles in a sedimentation tank.
6. Explain the working of a circular sedimentation tank with the help of a neat sketch.
7. Define the terms: detention time, flow-through period, overflow rate, and weir loading.
8. List the advantages and limitations of solid contact units.
9. Describe the working of a solids contact unit, with the help of neat sketch.
10. Explain how the use of tube settlers can help to increase the plant capacity without expanding its footprint.
11. How does the flow scheme for direct filtration differ from the conventional flow scheme? What limitations of raw water quality are recommended for the adoption of direct filtration?
12. Write a note on the disposal of sludge from a water treatment plant.
13. What is the nature of sludge from a water treatment plant? What operating parameters affect the quality of sludge produced?

Practice Problems

1. The solids removal in a sedimentation basin is estimated to be 35 mg/L. How many m^3 of sludge with solids concentration of 2.0% is formed for every 1.0 ML of flow treated? (1.8 m^3).
2. If settling velocity of alum floc is 25 mm/min, how long would it take for the alum floc to settle through a depth of 3.0 m when the overflow rate is 20 $m^3/m^2 \cdot d$? (4.5 h).

2. What should be the water depth of a circular clarifier to provide a detention time of 3.0 h and an overflow rate of 20 m³/m²·d? (2.5 m).

3. How deep must a circular clarifier be to provide a detention time of 3.0 h? Design overflow rate is 500 gal/ft²·d. (8.3 ft).

4. What should be the diameter of a circular clarifier to provide an overflow rate of 20 m³/m²·d when treating a flow of 15 ML/d? (31 m).

5. The settling velocity of alum floc is known to be 1.0 in/min. How long will it take for the floc to settle through a depth of 10 ft when hydraulic loading is maintained at the rate of 500 gal/ft²·d? (4.5 h).

6. What detention time is achieved in a circular clarifier with side water depth of 8.5 ft? The clarifier achieves an overflow rate of 550 gal/ft²·d at the design flow. (2.8 h).

7. The settling velocity of alum floc is known to be 0.40 mm/s. What is the minimum detention time needed to allow 100% removal of this size floc in a sedimentation tank with a depth of 3.0 m? (2.1 h).

8. Calculate the settling velocity of a discrete particle with a diameter of 45 μm, assuming laminar flow conditions. Assume specific gravity of the solid to be 2.62 and water temperature of 20°C ($v = 1.0 \times 10^{-6}$ m²/s). (1.8 mm/s).

8. A rectangular sedimentation basin 16 m long, 10 m wide, and 3.2 m deep treats a flow of 4800 m³/d. Determine detention time, the overflow rate, and mean flow velocity through the basin. (2.6 h, 30 m³/m²·d, 1.7 mm/s).

9. In a water filtration plant, a sedimentation basin 50 ft long, 32 ft wide, and 10 ft deep treats a flow of 1.3 MGD. Determine the detention time, the overflow rate, and mean flow velocity through the basin. (2.2 h, 810 gal/ft²·d, 0.38 ft/min).

10. Size a rectangular sedimentation basin to treat a flow of 5.5 ML/d so as to provide a detention time of 3.5 h and without exceeding a flow velocity of 1.5 mm/s. Assume the side water depth of 3.3 m. (19 m × 13 m × 4 m)

11. A sedimentation tank is 3.0 m deep and 50 m long. What flow velocity would you recommend if you want to effectively remove 25 μm particles and the water temperature is 20°C? Assume the specific gravity of the solid particles is 2.7. (9.6 mm/s).

12. In a water filtration plant, water is processed at the daily rate of 5.5 ML/d and turbidity is removed applying alum at a dosage of 35 mg/L. Estimate the volume of wet sludge produced as aluminum hydroxide, assuming the solids concentration in the sludge consistency is 0.90% solids. (5.6 m³/d).

13. In a water plant, coagulated water at the rate of 4 ML/d is fed to two circular settling basins operating in parallel. Each basin is 2.5 m deep and has a single peripheral weir attached to the outer wall. What diameter basin is required based on surface loading of 18 m³/m²·d? Compute detention time and weir loading. (12 m, 3.3 h, 53 m³/m·d).

14. Size a circular clarifier for treating a flow of 2.5 MGD. The maximum overflow rate is 500 gal/ft²·d and a detention time of 3.0 h must be achieved. (D = 80 ft, 8.5 ft deep).

15. What diameter circular clarifier and side water depth is needed for treating a flow of 15 ML/d? Maximum overflow rate not to exceed $16 \ m^3/m^2 \cdot d$ and a detention time of 4.0 h must be achieved. (35 m, 2.7 m).

16. A spherical particle with a diameter of 0.02 mm and a specific gravity of 2.8 settles in water having a temperature of 20°C. Use Stokes' Law equation for calculating the settling velocity. (0.39 mm/s).

17. Estimate the sludge solids produced in coagulating a surface water with turbidity of 14 NTU and alum dosage of 45 mg/L when treating a flow of 5.0 MGD. What is the volume of sludge if the settled sludge is concentrated to 1500 mg/L of solids in a clarifier thickener? (30 mg/L, 100 000 gal/d).

18. Two rectangular clarifiers, each 27 m × 5 m × 3.8 m, settle coagulated water at an average daily flow rate of 6.0 ML/d. Calculate the detention time, overflow rate, and mean flow velocity. (4.1 h, $22 \ m^3/m^2 \cdot d$, 1.8 mm/s).

10 Filtration

Filtration is an important process in surface water treatment. In fact, many such facilities are called filtration plants, even though filtration is just one of the processes. In most developed countries, chemical-assisted filtration is regulated for surface water sources and groundwater under direct influence (GUDI). When raw water turbidity is <10 NTU, the process of sedimentation can be eliminated. When the coagulants are added directly to the filter inlet pipe and flocculation and sedimentation are eliminated, it is called **in-line filtration**.

10.1 FILTRATION MECHANISMS

The process of filtration removes particulate matter and floc by passing the water through a porous bed of sand, coal, or other granular material. Filtration is not just straining but a combination of physical chemical process and in some cases biological processes. Although straining plays a significant role in the overall removal process, it is important to realize that most of the particles removed during filtration are considerably smaller than the pore openings in the filter media. The entire removal process is a complex combination of straining, sedimentation, adsorption, and biological action.

Straining is a physical process causing the entrapment of particles by the pore openings in the filter media. Sedimentation occurs when particles settle out on the surface of the media as water flows through the openings or pores of the media. Some media like carbon have absorptive properties. Biological action is important when filtration rates are relatively slow as in the case of slow sand filtration. Biological growth on the media retains the impurities from the water, especially organics.

10.2 TYPES OF FILTERS

Filtration rate, defined as flow rate per unit surface area of filter, was extremely low in early filters, hence the name "slow sand filters" because of the relatively low rate of filtration. Modern muti media filters perform at much higher filtration rates and are cleaned by backwashing. **Backwashing** refers to reversing the flow of water through the bed at a rate higher than the filtration rate.

10.2.1 SLOW SAND FILTERS (SSFs)

A schematic of a slow sand filter is shown in **Figure 10.1**. Slow sand filters consist of uniform sand media with no provision for backwashing. Removal mechanisms in slow sand filters are straining, adsorption, and biological action. The main advantage

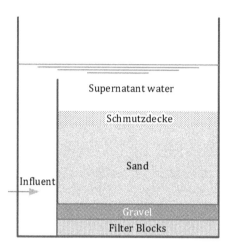

FIGURE 10.1   Slow sand filter

of this method is low maintenance and simplicity of operation. Due to low hydraulic loading, most of the removal takes place in the top portion of the bed. When the bed becomes clogged with turbidity, the top 2–3 cm (1 in) is scraped off and replaced with new media. Due to high labor and area requirements combined with its unsuitability for treating turbid waters (>10 NTU), slow sand filtration is not common in larger plants.

10.2.2   Rapid Gravity Filters (RGFs)

In the middle of the twentieth century, rapid gravity filters completely took over from slow sand filters, except in rural areas. In the latter case, slow sand filtration is commonly used without any prior coagulation. The popularity of rapid gravity filtration is because of the enhanced filtration rate of 5–20 m/h (15–65 ft/h) compared to 0.1–0.2 m/h for slow sand filters (50–100×). Rapid gravity filtration is commonly used to filter coagulated water and thus produces high-quality water filtration. Coarse sand is used as the filter media. Thus, filtration rates are 30 to 40 times greater than that of slow sand filtration. This is achieved by maintaining 2–3 m (7–10 ft) of head above the media. In addition to straining, the removal mechanisms include sedimentation, adhesion, and adsorption. Because of the increased rate of filtration, filter runs on RGF range from 20 h to 60 h. Cleaning of the bed is achieved by agitating it either mechanically or with compressed air and washing water upwards through the bed to the surface and out through the troughs.

Rapid gravity filters may be of three possible media types:

1. Single media, usually sand or anthracite.
2. Dual media, usually sand and anthracite.
3. Multimedia, usually garnet, sand, and anthracite.

TABLE 10.1

Comparison of various types of filters

Characteristic	Slow Sand	Rapid Gravity Filters (Media)		
		Single	Dual	Multi
filtration rate L/s·m² (gpm/ft²)	0.03 (0.05)	1.5 (2.0)	2–3 (3–5)	2–5 (3–8)
filter media	sand	sand	A, S	A, S, G*
effective size (mm)	0.15–0.3	0.6	1.0, 0.5	1,0.5, 0.2
uniformity coefficient	<3	<2	<2	<2
depth, m (ft)		0.7 (2.3)	0.6, 0.15	0.5, 0.2, 0.1
media distribution	unstratified	fine to coarse	coarse to fine	coarse to fine
filter run	20–60 d	12 -36 h	12–36 h	24–48 h
initial loss of head, m (ft)	0.03 (0.1)	0.3 (1)	0.3 (1)	0. 0.3 (1)
terminal head loss, m (ft)	1.1 (3.6)	2.5 (8)	2.5 (8)	2.5 (8)
backwash water used (%)	n/a	2%–4%	5%–6%	5%–6%

A = anthracite, S = sand G = garnet

10.2.3 HIGH-RATE FILTERS

Dual media and multimedia filters are high-rate filters. In high-rate filters, filtration rates 100 to 150× that of slow sand filters can be achieved. This is afforded by providing gradation from coarse to fine. Gradation employs the entire depth of media in removal rather than just the top portion, as is the case with rapid sand filters. **In-depth filtration** enables the achievement of higher filter runs. A **Filter run** is defined as the period between successive back washings. A comparison is shown in **Table 10.1**.

10.2.4 PRESSURE FILTERS

As the name indicates, filters operate under positive pressure. Filter media and other components are contained in a steel tank. Filtration rates are in the range of 1.5 –3 L/s·m². Pressure filters are more common in industrial applications and smaller municipalities. Other less common type of filters includes **diatomaceous earth** filters and biofilters, which use granular activated carbon. In diatomaceous earth filters, the filter media is a thin layer of diatomaceous earth, a natural powder-like material made from the shells of microscopic organisms. These types of filters are commonly used for industrial or swimming pool applications.

10.3 COMPONENTS OF A GRAVITY FILTER

Figure 10.2 shows the main components of a gravity filtration system. The main components of gravity filter shown in this figure are: Filter Box, Filter Media,

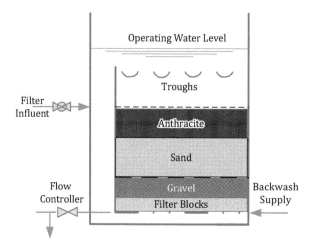

FIGURE 10.2 Components of a gravity filter

Underdrain System, Gravel Bed, Surface Wash System, Wash Water Troughs and Control Equipment.

10.3.1 FILTER BOX

A **filter box** is usually made of concrete and contains all the system components except for the control equipment. In the majority of cases, the filter box is a rectangular section. A typical filter box is 3.0 m (10 ft) deep, but the media is only 0.75 m or 2.5 ft deep. Located above the surface of the filter bed are **wash-water troughs**, which carry away the dirty water during backwashing. During filtration, water enters above the filter media through an inlet flume. After passing though the granular media and the supporting gravel bed, it is discharged through an underdrain pipe to the **clear well** storage.

10.3.2 FILTER MEDIA

The type and depth of filter media greatly affects the removal efficiency of a filter. As indicated above, filter media can be single, dual, or multimedia. However, only multimedia can produce **in-depth filtration**. An ideal filter media should have a uniform decrease in pore size with increasing filter depth. Media depth should be adequate to allow relatively long **filter runs.**

After backwashing, the gradation in single media filters is from fine to coarse, thus reducing the filter run due to surface plugging. Dual media beds of coarser anthracite overlying the sand filter provide coarser to fine gradation. Mixed media beds can almost be considered the ideal filter since the gradation is from coarse to fine with finer material at the bottom of the media. The various types of filter media are usually classified by the size of grains and specific gravity, as shown in **Table 10.2**.

TABLE 10.2
Filter media characteristics

Material	Grain Size		Specific Gravity
	mm	in/1000	
conventional sand	0.5–0.6	20–25	2.6
coarse sand	0.7–3.0	28–120	2.6
anthracite coal	1.0–3.0	40–120	1.5–1.8
garnet	0.2–0.4	8–16	3.1–4.3
gravel	1.0–50	40–200	2.6

After backwashing, the gradation in single media filters is from fine to coarse, thus reducing the filter run due to surface plugging. Dual media beds of coarser anthracite overlying the sand filter provide a coarser to fine gradation. Mixed media beds can almost be considered the ideal filter since the gradation is from coarse to fine with finer material at the bottom of the media. The various types of filter media are usually classified by the size of grains and specific gravity, as shown in **Table 10.2**. For a given material, size is indicated by effective size and uniformity of material by uniformity coefficient

Effective size is defined as tenth percentile (P_{10}), which means only 10% of the media grains by mass are smaller, or 90% are greater, than the effective size. In a way, effective size represents the finest size in a graded media. Since larger particle size will produce larger pore openings, coarser media will allow a longer time to reach the terminal head loss but shortens the time for turbidity to break through.

Uniformity coefficient (C_u) is defined as the ratio of grain sizes comprising 60th percentile (D_{60}) and 10th percentile (D_{10}). A lower uniformity coefficient indicates more uniform materials. In other words, grain sizes are closer to each other. Generally, the more uniform the material, the slower the head loss builds up. In non-uniform (C_u >2–4) materials, larger pores are filled by smaller particles, thus reducing the hydraulic effectiveness.

10.3.3 UNDERDRAIN SYSTEM

Underdrain is the bottom portion of the filter supporting filter media. Its main function is to evenly distribute the backwash water without disturbing the gravel bed and the media, and collecting the filtered water and the feed in of the backwash water. Several different kinds of manufactured filter bottoms are available, including porous plates, filter blocks, and nozzles.

Layers of gravel have been used as the top portion of the underdrain system. The number and depth of gravel layers, both to prevent loss of media during filtration and to allow uniform distribution of backwash water, depend on the size of the opening in the underdrain. Large openings require a deep gravel layer, hence a deeper filter box, but have reduced head loss. Smaller openings, such as slots in nozzles, require a fine gravel size, hence higher head loss. The gravel bed is typically 200–300 mm (8–12 in) in thickness, with a gradation in the range of 3–20 mm.

10.3.4 SURFACE WASH SYSTEM

The purpose of surface wash equipment is to assist in removing particles trapped in the filter media. The surface washer breaks up the surface of the filter and causes particles to collide and helps release the impurities. The surface wash equipment is positioned directly above the top surface of the filter media.

10.3.5 WASH-WATER TROUGHS

The **wash-water troughs** are located above the surface wash equipment. During backwash, the level of water in the filter rises and enters the trough. During filtration, the water trough remains submerged and as such plays no role.

10.3.6 CONTROL EQUIPMENT

A **flow controller** is required to maintain a constant rate of flow. The differential pressure across a venturi or an orifice meter installed in the effluent piping activates the controller. Depending on the impurities caught in the media, head loss across the filter increases during the filter run. In **declining rate filter operation**, filtration rate declines as filtration progresses. However, in a **constant rate filtration** operation, filtration rate is kept constant by the control valve attached to a venturi meter. When the filter is close to end the run, the valve is fully opened since flow is controlled by the head loss across the filter.

10.4 FILTRATION OPERATION

Filtration has three steps: filtering, backwashing, and filtering to waste. A brief discussion follows.

10.4.1 FILTERING

During filtration, water passes downward through the filter due to gravity. The maximum operating head is the difference between the water levels above the filter and in the clear well, commonly 3–4 m. During filtration, the depth of water above the filter is 0.9–1.2 m (3–4 ft). To provide additional head, the underdrain pipe is submerged in the clear well. At the start of the filter run, the media is clean and head loss is very low (about 30 cm), thus resulting in high operating head conditions (**Figure 10.3**).

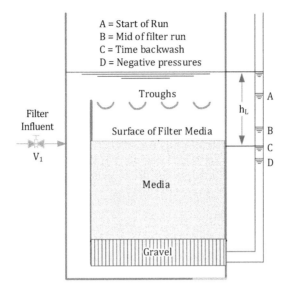

FIGURE 10.3   Head loss in a filter

10.4.2   DECLINING RATE CONTROL

Flow rate is allowed to vary with head loss. Each filter operates at the same water surface level. The filtration rate is highest at the beginning and drops as the filter gets clogged. This system is relatively simple but requires an effluent weir to provide adequate media submergence.

10.4.3   SPLIT FLOW CONTROL

As the name indicates, a control weir divides the flow to each filter. With this system, equal flow is automatically distributed to each filter. The filter effluent position is controlled by the water level in the filter. Each filter operates within a narrow water level range.

10.4.4   BACKWASHING

During filtration, the voids in the filter media become clogged with the filtered-out material. The media grains also become coated with the floc and become very sticky. To clean the filter bed, the media grains must be agitated and allowed to rub against each other. Therefore, the backwash rate must be high enough to expand the filter media by about 20%. Surface wash is recommended to provide extra agitation, especially in high-rate filters due to in-depth filtration. A filter is backwashed when:

1. head loss approaches a set value, usually about 2–2.5 m (6–8 ft) or
2. floc breakthrough of the filter causes an increase in turbidity or
3. filter run reaches a set value, usually 36–48 hours, whatever comes earliest.

The decision to backwash the filter should be made based on all these factors. As an example, when the source water is very clean, as in Lake Superior, filter runs based on head loss or turbidity breakthrough can be very long – of the order of four days. However, long filter runs can cause a gradual build-up of organic materials and bacterial populations within the filter. For this reason, filter runs rarely are allowed to exceed four days.

Filter Ripening

After the backwash operation, the filter needs to be rested to allow the media and other particulate matter to settle down. If filter ripening is not done, filtered water will be of poor quality and turbidity levels may exceed the regulatory limits.

10.4.5 BACKWASH OPERATION

Referring to **Figure 10.4**, a typical backwash begins by closing the influent (V1) and effluent (V4) valves and opening the drain valve (V2), allowing the water to drop to a level about 15 cm above the media. The surface washers are turned on and allowed to operate for 1 to 2 minutes. If air agitation is used, compressed air is introduced through the underdrain for a scouring. By opening the wash water supply valve (V5), clean water enters the underdrain and passes up through the filter, which is the reverse direction. Before opening V5, valve V3 should be opened to allow wash water to go to waste.

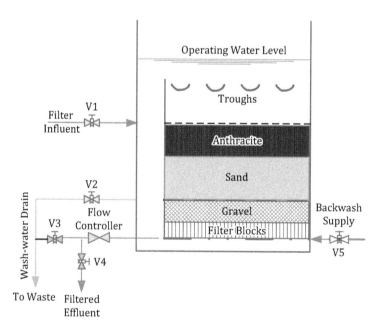

FIGURE 10.4 Backwashing of a gravity filter

During reverse flow, filter media is expanded hydraulically, and dirty wash water is collected by troughs and conveyed to disposal or for treatment. The expanded bed is washed for 5 to 15 minutes depending on how dirty the filter is. The surface washers are usually turned off for about 1 minute before the backwash flow is stopped. To avoid any damage to the underdrain or loss of media, the *backwash supply valve should be opened slowly.*

10.4.6 FILTERING TO WASTE

At the completion of the backwash, the first few minutes of the filtered water should be wasted. This is accomplished by opening the filter to the waste valve (V3) when the influent valve (V1) is also open. All the other valves remain shut. As the turbidity of the effluent drops to an acceptable level, opening the effluent valve (V4) and closing the filter to waste valve (V3) starts the next filter run. The three main types of backwashing aids are hydraulic surface agitators, mechanical rakes, and compressed air.

Circular filters are often equipped with mechanical rakes suspended above the media surface and rotated horizontally about a vertical axis during backwashing operations. Compressed air is sometimes introduced into the underdrain system before or at the same time the wash water is turned on. Air scouring provides better cleaning.

10.4.7 BACKWASHING KEY POINTS

Some general recommendations for backwashing of a filter are as follows.

- If surface washers are available, use them at the start of the backwash operation.
- It is recommended to ramp the backwash rate in stages so as to backwash each layer of the media.
- Too much backwashing can remove some bed ripening. This would result in waste and lengthen the ripening process.
- If air scour is available, it would greatly help to break up floc adhering to the media.
- Based on the historical data, the majority of gravity filters use 4–6 m^3/m^2 of backwash rate.
- If possible, allow the filter to rest after backwash. This would help to ripen it better and produces a better quality water.
- Though the backwash operation is mostly automatic, it is strongly recommended to observe the media during backwash.

10.5 DESIGN AND PERFORMANCE PARAMETERS

10.5.1 FILTRATION RATE

Filtration rate is defined as flow per unit surface of the filter and is a measure of the hydraulic loading rate. Thus, the filtration rate basically indicates the velocity of the flow (v_F) through the filter. It can be directly observed by noting the water drop rate after closing off the influent valve. Keep in mind, though, that it is not the velocity of water as it moves through the filter media. Actual velocity would be significantly more as the open area is less since some space is occupied by the solids. Higher filtration rates can be afforded in multimedia filters. The filtration rate is one measure of filter production. Along with filter run time, it provides valuable information for the operation of the filter. Problems can develop when design filtration rates are exceeded. Typical values fall in the range of 2–6 L/s.m²(3–9 gpm/ft²).

Example Problem 10.1

Six slow sand filter beds are used to treat a maximum flow of 15 ML/d with a filtration rate of 5.0 m³/m²·d. What should be the size of the rectangular filter box of length twice that of width? Also, assume that one unit out of six is kept as standby.

Given:

$$Q = 15.0 \text{ ML/d} \quad v_F = 5.0 \text{ m/d} \quad L = 2B \quad v_s = ?$$

Solution:
  filter surface required

$$A_F = \frac{Q}{v_F} = \frac{15 \, ML}{d} \times \frac{d}{5.0 \, m} \times \frac{1000 \, m^3}{ML} = 3000.0 = 3000 \, m^2$$

Since one unit is to be kept standby, 5 units should provide the required surface.

$$B = \sqrt{\frac{A_F}{2 \times 5}} = \sqrt{\frac{3000 \, m^2}{10}} = 17.3 = \underline{17.5 \, m \, say}$$

Six filter units each measuring 35 m × 17.5 m will meet the requirements.

Example Problem 10.2

Design a dual media sand filter unit to produce 1.2 MGD of water supply with all its principles components. Design filtration rate is 4.0 gpm/ft². Make appropriate assumptions.

Given:

$$Q = 20 \text{ mil gal/d} \quad \text{backwash water} = 6\% \text{ (assumed)}$$
$$\text{downtime} = 0.5 \text{ h (assumed)} \quad L = 1.5 \times B$$

Solution:
  flow to the filters

$$Q = \frac{1.2 \times 10^6 \, gal}{d} \times 1.06 \times \frac{d}{23.5h} = \underline{5.41 \times 10^4 \, gal \, / \, h}$$

filter surface area required

$$A_F = \frac{Q}{V_F} = \frac{5.412 \times 10^4 \; gal}{h} \times \frac{min.ft^2}{4.0 \; gal} \times \frac{h}{60 \; min} = 225.5 = 230 \; ft^2$$

Assuming two filter units for flexibility

$$B = \sqrt{\frac{A}{2 \times 1.5}} = \sqrt{\frac{225.5 \, ft^2}{3}} = 8.66 = 10 \, ft \; say$$

Thus 2-filter units each measuring 10 ft × 15 ft will suffice.

10.5.2 UNIT FILTER RUN VOLUME

Unit filter run volume (UFRV) is a term commonly used by operating personnel to compare and evaluate filter runs. This parameter is a good measure of filter performance. UFRV is defined as the volume of water filtered per unit surface of the filter in a filter run. For the majority of filters, UFRV falls in the range of 200–400 m³/m² (5000–10 000 gal/ft²) of the filter area.

$$\boxed{UFRV = \frac{V_{WF}}{A_F} = \frac{Q \times t_{FR}}{A_F} = V_F \times t_{FR}}$$

Example Problem 10.3

A filter with a media surface of 90 m² produces a total of 76 ML during a 74 hour filter run. What is the average filtration rate and unit filter run volume (UFRV)?

Given:

$$A_F = 90 \; m^2 \quad t_F = 74 \; h \quad V = 76 \; ML \quad v_F = ? \quad UFRV = ?$$

Solution:
  average filtration velocity

$$v_F = \frac{Q}{A_F} = \frac{76 \, ML}{74 \, h} \times \frac{1}{90 \, m^2} \times \frac{1000 \, m^3}{ML}$$

$$= \frac{11.4 \, m}{h} \times \frac{1000 \, L}{m^3} \times \frac{h}{3600 \, s} = 3.20 = 3.2 \, L / s.m^2$$

unit filter run volume

$$UFRV = \frac{V_{WF}}{A_F} = \frac{76 \, ML}{90 \, m^2} \times \frac{1000 \, m^3}{ML} = 844 = 840 \, m^3 / m^2$$

Example Problem 10.4

A multimedia filter with a media surface of 40 ft × 25 ft is operated maintaining a filtration rate of 5 gpm/ft². What should be the minimum filtration run to get a UFRV of 2000 ft³/ft²?

Given:

$$OA_F = 25 \text{ ft} \times 40 \text{ ft} = 1000 \text{ ft}^2 \quad t_{FR} = ? \quad UFRV = 2000 \text{ ft}^3/\text{ft}^2$$

Solution:

$$t_{FR} = \frac{UFRV}{V_F} = \frac{2000 \, ft^3}{ft^2} \times \frac{min.ft^2}{5.0 \, gal} \times \frac{7.48 \, gal}{ft^3} \times \frac{h}{60 \, min} = 49.86 = \underline{50 \, h}$$

$$V_{WF} = UFRV \times A_F = \frac{2000 \, ft^3}{ft^2} \times 1000 \, ft^2 \times \frac{7.48 \, gal}{ft^3}$$

$$= 1.496 \times 10^7 \, gal = 15 \, mil \, gal$$

10.5.3 FLOW RATE AND VOLUME OF WATER FILTERED

In the calculation of filtration rate, the flow rate must be known. Flow rate can be directly read from the flow meter. The flow rate can also be determined if the total volume of water filtered per filter run is known. Total flow volume produced divided by the length of the filter run (time) would yield the average flow rate during filter operation. If the flow rate is known, the expected volume of water filtered (V_{WF}) can be estimated by multiplying it by the length of the filter run. It is important to note that this equation will yield the average values over the filter run period. Filtration rate can also be estimated by measuring the water drop in a filter when the influent valve to the filter is closed. Water drop rate is essentially the filtration rate. The filtration rate calculated by measuring the drop rate will correspond to the time when the drop rate is observed. For a given filter, it may vary depending upon the operation time after the backwash cycle. It makes sense to expect higher values in the beginning and a drop off towards the end of the filter run.

10.5.4 BACKWASH RATE

Backwash rates can be calculated in a similar fashion as filtration rates. Backwash rate represents the rise rate. Backwash rates are 5 to 10 times that of filtration rates. The volume of water used to backwash is typically in the range of 4%–6%.

Example Problem 10.5

For the filter unit described in Example Problem 4.1 ($A_F = 90 \text{ m}^2$), determine the backwash-pumping rate in L/s if the desired backwash rate is 15 L/s.m².

Given:

   A = 90 m²   v = 15 L/s.m²   t = 10 min

Solution:

$$Q_{BW} = v_{BW} \times A_F = \frac{15\ L}{s.m^2} \times 90\ m^2 \times \frac{m^3}{1000\ L} = 1.35 = \underline{1.4\ m^3\ /\ s}$$

$$V_{BW} = Q_{BW} \times t_{BW} = \frac{1.35\ m^3}{s} \times 10\ min \times \frac{60\ s}{min} = 810.0 = \underline{810\ m^3}$$

$$= \frac{810\ m^3}{76\ ML} \times \frac{ML}{1000\ m^3} \times 100\% = 1.06 = \underline{1.1\%}$$

10.6   OPERATING PROBLEMS

Sudden changes in water quality indicators, such as turbidity, pH, alkalinity, threshold odor number, temperature, chlorine residual, or color are an indication of problems in the filtration process or processes preceding filtration. During a normal filtration run, the operator should watch for sudden changes in head loss and turbidity breakthrough.

FILTER BREAKTHROUGH

Probably the most common filter operation problem is filter breakthrough. Filter breakthrough can be defined as a steady increase in the turbidity of the filtered water. Normally, the turbidity of filtered water will stay relatively low and constant.

MUD BALLS

Mud balls look like small irregularly shaped balls of mud resting near the coal-sand or sand-gravel interfaces. Mud balls are formed due to inadequate backwashing, and they grow with time if the problem is not corrected. These mud balls, if allowed to remain, will cause clogged areas in the filter. Generally, proper surface washing will prevent mud ball formation.

AIR BINDING

Air binding is a result of a negative head condition due to excessive head loss. This is caused by the release of dissolved air in saturated cold water due to a decrease in pressure below atmospheric. During backwash, there can be violent agitation as the air is released from the filter media. This can damage the filter and result in the loss of filter media.

Media Breakthrough

When media breakthrough occurs, the media starts appearing in the filter effluent or even in the distribution system. When media breakthrough occurs, filtration has to be stopped quickly.

Gravel Mounding

When this happens, gravel is blown up and is mixed with the filter media. Mounding is often caused by hydraulic surges due to improper backwashing.

Media Boils

Media boils appear during backwash when there is uneven distribution. In filters with nozzle-type underdrains, boils are often the result of nozzle failure.

10.7   OPTIMUM FILTER OPERATION

Getting the best from your filter units depends on understanding the factors responsible for causing poor efficiency and keeping the system in good working order. Here are some ways to achieve optimum performance.

- Checking media cleanliness frequently by core sampling or visual inspection.
- Effective backwashing is very important in operating the filter efficiently and getting long filter runs. If the nature of the floc is sticky, it is advisable to use surface wash or air scouring to remove the sticky material from media grains during backwashing.
- Keeping filter media clean and maintaining media depth are essential
- Aggressive backwash with air–water combination helps to wash the media clean, with low flow rates and less frequent backwashing.
- Maintaining good records of water quality, filter run length, backwash frequency, and any changes made helps in taking appropriate action in time.

Discussion Questions

1. What is meant by in-depth filtration? How it can be accomplished?
2. Compare a slow sand filter with a rapid sand filter.
3. With the help of a neat sketch, describe the working of a dual media filter.
4. Compare declining rate filtration with constant rate filtration.
5. In a gravity filter, the depth of water above the filter surface is typically 1.5 m. How can the head loss gauge record head loss of 2.5 m?
6. A filter run is terminated as a result of the occurrence of a number of conditions. Cite these conditions.
7. What are the advantages of using a multimedia filter over a single media sand filter?

8. What is the function of a rate of flow controller in the filter operation?
9. Describe the common operating problems encountered in filter operation.
10. What would be the effect of inefficient coagulation-flocculation in filter removal efficiency and operation?
11. Describe the steps to backwash a gravity filter.
12. Describe various mechanisms that take place during filtration.
13. What is meant by direct filtration? In what situations would you recommend it?
14. Explain why head loss is not allowed to exceed 2.5 m.
15. What are the possible cause of short filter runs?

Practice Problems

1. What UFRV is achieved in a filter operation with an average filter run of 28 hours when operating @ 3.0 L/s.m²? (300 m³/m²).
2. A filter is designed to achieve an average UFRV of 400 m³/m² when operating at the rate of 2.8 L/s.m². What is the average filter run? (40 h).
3. A high-rate filter has a surface of 10 ft × 15 ft and it treats a flow of 1.2 MGD. Compute the filtration rate and express it as gpm/ft² and ft/min. (5.6 gpm/ft², 0.74 ft/min).
4. Design five slow sand filter beds to treat a maximum flow of 12 ML/d with a filtration rate not to exceed 6.0 m³/m²·d. The filter box is rectangular with length twice that of its width. Also assume that 1 unit out of 5 will be kept as standby. (32 m × 16 m).
5. A filter with a media surface of 11 m × 7.5 m produces a total of 68 000 m³ of filtered water during a 3 d filter run. What is the average filtration rate and unit filter run volume? (11 m³/m²·h, 820 m³/m²).
6. A filter unit has a surface of 80 m². What should be the backwash pumping rate if the desired backwash rate is 15 L/s·m²? Find the change in depth of water in the backwash tank after the completion of the backwash cycle. The diameter of the tank is 25 m and the filter is backwashed for a duration of 12 minutes. (1.2 m³/s, 1.8 m).
7. A filter with a media surface of 30 ft × 25 ft is operated maintaining a filtration rate of 5.0 gpm/ft² during a 2 d filter run. What is the unit filter run volume? (14 000 gal/ft²).
8. A filter unit described in Practice Problem 7 is backwashed at 5× the rate of filtration. What must the backwash pumping rate be? (19 000 gpm).
9. A new water treatment plant is to be considered for treating 80 ML/d.
   a. Estimate the number of filters required, if each filter area is not to exceed 60 m² and one filter is to be kept as standby. Assume design filtration rate is 12 m/h. (6 filters).
   b. What are the dimensions of each filter if the length to width ratio is 4:1? (15 m × 3.9 m).
10. A filter run is terminated by closing the influent valve. It was observed that it took 3 min and 30 seconds for the water level on top of the media to drop

by 0.50 m. What is the filtration velocity and flow through the filter if the media surface measures 5.0 m × 5.0 m? (8.6 m/h, 5.1 ML/d).

11. A filter measuring 14 m × 7 m produces a total of 72 ML during a 72 hour filter run. What is the average filtration rate in L/s m²? (2.8 L/s.m²).

12. Calculate the backwash pumping rate in L/s for a filter measuring 10 m x 10 m if the desired backwash rate is 11 L/s.m². What volume of backwash water is required if the filter is to be backwashed at this rate for 10 min? (1100 L/s, 660 m³).

13. Compute the length required of a square filter box that will treat a flow of 2.1 MGD without exceeding a filtration rate of 4.0 gpm/ft². (19 ft).

14. If the filter designed in Practice Problem 13 is backwashed every couple of days at a rate of 20 gpm/ft² for a duration of 10 min, what percentage of total flow is used to clean the filter? (1.7%).

15. The average filtration rate during a particular filter run was determined to be 7.8 m/h. If the filter run time was 42.5 hours, calculate UFRV. (330 m³/m²).

16. After closing the influent valve, it is observed that it took 6 min and 30 seconds for the water level to drop by 0.50 m. What is the filtration rate in mm/s and flow through the filter in ML/d if the filter measures 8.0 m × 8.0 m? (1.3 mm/s, 7.1 ML/d)

17. Calculate the backwash pumping rate for a filter measuring 10 m × 10 m if the desired backwash rate is 11 L/s m². What volume of backwash water is required if the filter is to be backwashed at this rate for 10 min? (1.1 m³/s, 660 m³).

18. A backwash pumping rate of 13 000 gpm is desired for a total backwash time of 9 minutes. Calculate the depth of water required in the 65 ft diameter backwash water tank to supply this amount of water. (4.7 ft).

19. During a filter run, 7.4 ML of water was filtered. At the end of the run, the filter was backwashed at a flow rate of 0.90 m³/s for 7 minutes. Calculate the percentage of the product water used for backwashing. (5.1%).

20. After closing the influent valve, it is observed that it took 4 min and 30 seconds for the water level on top of the filter surface to drop by 1 ft. What is the filtration rate and flow through the filter if the filter surface measures 25 ft × 25 ft? (1.7 gpm/ft², 1000 gpm).

11 Disinfection

Disinfection is one of the cleansing processes to make water hygienic and suitable for consumption. In cases such as good quality groundwater, disinfection is the only process of water treatment. This is usually the last process before water is pumped into the distribution system. Disinfection kills or inactivates **pathogens**. These organisms are very small – usually microscopic, ranging from 1–5 microns (μm). Pathogenic organisms can be carried and transmitted by water and cause diseases like cholera, dysentery, giardiasis, and typhoid. Dealing with pathogenic organisms is a major concern of water treatment.

11.1 DEFINITION

Disinfection is defined as the selective destruction/inactivation or removal of pathogenic organisms. **Sterilization**, on the other hand, is the complete destruction of all living organisms. Sterilization is not necessary in water treatment and is also very expensive. Though disinfection is necessary to make water safe, it can also create objectionable tastes and excessive levels of **disinfection by-products** (DBPs). Untreated water from domestic water sources may contain the following organisms:

- viruses, which could cause diseases like infectious hepatitis or poliomyelitis;
- bacteria, which could cause diseases like cholera, typhoid fever, dysentery, or Legionnaires' disease;
- intestinal parasites, like giant roundworm and others, which could cause amoebic dysentery, giardiasis, or cryptosporidiosis.

These three types of organisms differ in size and in their resistance to disinfection. Of the above-mentioned pathogens, Giardia and Cryptosporidium are the hardest to destroy.

11.1.1 PRIMARY DISINFECTION

Primary disinfection refers to the removal or inactivation of pathogens by various treatment methods. Thus, pre-chlorination, coagulation, settling, filtration, and post-chlorination all contribute to primary disinfection.

11.1.2 SECONDARY DISINFECTION

Secondary disinfection refers to leaving a residual disinfectant to provide a defense against any possible contamination on its way to the consumer. Thus, for secondary disinfection, only those disinfectants can be used which create a persistent residual, for example chlorine, chlorine dioxide, and chloramines.

DOI: 10.1201/9781003347941-13

   i. Secondary disinfection is used to protect water from microbiological recontamination, to reduce bacterial regrowth, control **bio-film** formation, and serve as an indicator of distribution system integrity.

   ii. The loss of residual disinfectant may indicate that the system integrity has been compromised.

   iii. Chlorine, chlorine dioxide, and monochloramine provide a persistent residual disinfectant and can be used for secondary disinfection. UV and ozone do not provide a residual and cannot be used for secondary disinfection.

   iv. In larger water distribution systems, the maintenance of the minimum required residual may not be possible without rechlorination facilities at one or more points within the distribution system. Such chlorination units are called **chlorine booster** stations.

   v. The rapid decay of a disinfectant residual may occur as a result of a number of other causes, such as heavy incrustation, sediment accumulation, and biofilm activity.

   vi. Free chlorine residual is more powerful, but decays faster compared with combined chlorine residual that lasts longer.

11.2 DISINFECTION METHODS

Pathogens can be removed or inactivated by applying a combination of physical and chemical treatment processes.

11.2.1 REMOVAL PROCESSES

Conventional filtration consists of chemical coagulation, rapid mixing, flocculation, and sedimentation followed by rapid rate sand filtration. **Direct filtration** eliminates the step of sedimentation from conventional filtration. Direct filtration is suitable for plants with raw water sources of high quality. Turbidity level in raw water is usually less than 10 NTU.

    Membrane filtration involves the passing of water through membranes consisting of very fine pores. Based on the pore sizes, this is known as micro or ultrafiltration. Membrane filtration can remove bacteria, Giardia, and some viruses. They are more suitable for polishing water that has already been treated by other methods. Algae and other solids can quickly clog cartridge filtration. Little information is available concerning the effectiveness of cartridge filters for virus removal. The effectiveness of removal processes is indicated by the turbidity of the filter effluent. For disinfection credits, filter effluent turbidity must be less than 0.5 NTU 95% of the time.

11.2.2 INACTIVATION PROCESSES

Inactivation of pathogens involves those methods that create a harsh environment for the organisms. Strong oxidizing chemicals are commonly used. Such chemicals destroy or impair pathogens by diffusing the cell wall and impairing or destroying the organism. Some of the methods that could be used to disinfect water are as follows.

UV Light

Ultraviolet light can be used to destroy pathogens, but this process is expensive. However, UV light is gaining popularity since it does not produce any harmful by-products and is more effective in the control of cysts. In the past, this type of disinfection was limited to small or local systems, but nowadays many municipal plants are being retrofitted with this process. Because UV radiation leaves no residual disinfectant, chlorination is still required. Multiple barrier systems incorporating UV and chlorination are effective in disinfecting both virus and Cryptosporidium parvum.

Heat

Heat can disinfect water. Boiling water for about five minutes will destroy all microorganisms. This method is expensive because of the energy it requires and thus is usually used only in disaster and emergency situations ("boil water advisory").

Chemicals

Chemicals that could be used to disinfect water have problems associated with them. For example, iodine can be used but is expensive and has possible physiological effects. There are difficulties in the handling of bromine. Alkaline chemicals such as sodium hydroxide and lime leave a bitter taste in the water. Also, excess lime has to be removed before supplying water for public use.

Potassium Permanganate

Potassium permanganate is more commonly used for controlling taste and odor problems. It is also used as a disinfectant, especially when chemicals such as trihalomethanes (THMs) are formed due to the presence of humic acid and fulvic acid in raw water. Feeding permanganate as the initial oxidant allows chlorine to be applied later in the treatment process when the precursors have been reduced. When added, this chemical imparts a pink color to water. It is commonly used as a disinfectant in rural areas. In places like India, where sources of water supply are open wells, well water is dosed with this chemical to disinfect the water, especially in the rainy season. Potassium permanganate is found to be very effective in the prevention of Cholera.

Ozonation

Ozone is a bluish, toxic gas with a pungent odor. Ozone is a powerful oxidizing agent and is used as a disinfectant in some countries. In North America, the process is gaining popularity in conventional water treatment. However, ozonation is too expensive for small systems. The other disadvantage of ozonation is that it fails to provide a measurable residual. Despite its disadvantages, there appears to be renewed interest in ozonation as it does not form THMs and has little pH effect. However, it needs to be produced at site THMs are considered carcinogenic.

Chlorination

Chlorination offers several advantages for water treatment. It is reliable and relatively low in cost. Also, the slight chlorine residual that stays in the water after purification serves as a tracer that can be used to indicate the presence of the disinfecting agent at any point in the system. The chlorine residual also protects against contamination in the distribution system (secondary disinfection). In addition to disinfection, oxidant chemical can serve other purposes during the disinfection process such as:

- control of biological growth in tanks and pipes and water mains;
- control of tastes and odors;
- color removal and aid to flocculation;
- precipitation of iron and manganese.

11.3 CHLORINE COMPOUNDS

Chlorination is by far the most effective way of disinfecting drinking water as both a primary and a secondary disinfectant. Chlorination involves the use of chlorine or chlorine compounds to disinfect water.

i. Chlorine itself could be added to the water to be treated. This method of chlorination is called **gas chlorination**. Chlorine cylinders containing liquefied gas come in various sizes.

ii. Chlorine compound solutions, like sodium hypochlorite, could be used. This method of chlorination is called **hypochlorination**. Since only part of the chemical is chlorine, it is relatively safer to use than chlorine gas. This practice is, however, limited to small water systems requiring lesser amounts.

iii. Another method of chlorination that may be used involves the use of **chlorine dioxide**. This form of chlorine is becoming more common due to its disinfectant efficiency and lower production of DBPs.

11.3.1 GAS CHLORINATION

One method of chlorination, especially in larger installations, involves the use of chlorine gas as the disinfectant. This method of chlorination is referred to as gas chlorination. Due to safety considerations and the initial cost of a gas feeder, gas chlorination is not used in small, seasonal drinking water systems. Chlorine gas is about 2.5 times heavier than air. It has a pungent odor and greenish yellow color. The gas is highly toxic. Its odor can be detected at concentrations as low as 0.3 ppm.

Chlorine liquid is compressed gas at high pressure. The liquid is about 99.5% pure chlorine, amber in color, and 50% denser than water. Most municipal plants use the liquid chlorine. One volume of liquid chlorine yields about 450 volumes of gas. Chlorine is non-explosive but reacts violently with greases, hydrocarbons, ammonia, and other flammable materials. Chlorine will not burn, but it supports combustion.

11.3.2 Chlorine Safety

Chlorine is very toxic and hazardous. A few breaths of 0.1% chlorine in air can cause death even at low concentration; it can also cause ill health effects. Handle chlorine carefully and use all the necessary precautions. Due to the dangers associated with the handling of chlorine, it is also called "the green goddess of water". Some key points related to chlorine safety are as follows.

- Do not enter a chlorine-containing atmosphere without wearing protective gear.
- All apparatus, cylinder lines, and valves should be checked regularly for leaks.
- Since chlorine is 2.5 times heavier than air, store chlorine at the lowest point possible. For the same reason, do not stoop down if you notice a chlorine smell.

11.4 HYPOCHLORINATION

A common way in which chlorine is used for water disinfection in a small water system is called hypochlorination. Because chlorine is so dangerous and requires special handling, many smaller water plants use a liquid chlorine compound called hypochlorite instead of chlorine gas. Another advantage of hypochlorination is that it does not need elaborate equipment or a separate room. It is important to note that hypochlorite compounds, which contain algaecide, should not be used as a disinfectant in potable water systems. Adding chlorine using hypochlorite would raise the pH whereas gaseous chlorine suppresses the pH.

11.4.1 Calcium Hypochlorite

Calcium hypochlorite ($Ca(OCl)_2$) is a dry, white, or yellow-white granular material produced from the reaction of lime and chlorine. Calcium hypochlorite is available commercially in granular powder or tablet forms. A commercially available compound, known as **high test hypochlorite** (HTH) typically contains 65% available chlorine. That is to say, for every 1 kg of chlorine needed, you will require 1.5 kg of 65% HTH. These products are dissolved in water to form a liquid solution before they are used to disinfect water. While calcium hypochlorite is not as dangerous as chlorine gas, it should be handled according to the recommended procedures.

11.4.2 Sodium Hypochlorite

Sodium hypochlorite (NaOCl), commonly called bleach, is a water-based solution of sodium hydroxide and chlorine. This form of chlorine is purchased in containers of various sizes in varying concentrations of anywhere from 9 to 15% by mass. **Household bleach** is sodium hypochlorite in a 3%–5% concentration. Large systems usually purchase liquid bleach in carboys, drums, and railroad tank cars. If needed in very small quantities, one gallon (4 L) jugs can be purchased. Sodium hypochlorite is alkaline, with a pH ranging from 9 to 11, depending upon the strength. Due to poor stability, liquid bleach can lose 2%–4% of its available chlorine every month. It is therefore recommended not to store it for more than three months.

11.4.3 CHLORINE DIOXIDE DISINFECTION

Chlorine dioxide is a greenish yellow gas at room temperature and is odorous like chlorine. Its use as a disinfecting chemical instead of chlorine is of interest to water treatment operators for several reasons.

- Some of the undesirable chemical compounds (Trihalomethanes) formed when chlorine gas or hypochlorite are employed are not formed when chlorine dioxide is used.
- Chlorine dioxide is about two to three times as effective compared with chlorine in killing bacteria and it is many times more effective in killing viruses and Cryptosporidium. When Giardia or Cryptosporidium is the problem, treatment by chlorine dioxide followed by chlorine or chloramines is very effective.
- However, because of its higher cost, chlorine dioxide is not extensively used as a disinfectant. Instead, it is used to treat difficult taste and odor problems in water and for oxidizing iron and manganese in difficult-to-treat water.
- The difficulty with chlorine dioxide is that is must be generated on site by reacting sodium chlorite solution with chlorine solution. Another problem is the handling of sodium chlorite, which is very combustible around organic materials. It is explosive with concentrations exceeding 10% in the air.

$$2NaClO_2 + Cl_2 = 2NaCl + 2ClO_2 \uparrow$$

11.5 CHEMISTRY OF CHLORINATION

When chlorination is performed properly, it is a safe, effective, and practical way to destroy pathogens. It also provides a stable residual to prevent regrowth in the distribution system. When chlorine is added to distilled water, it forms Hypochlorous acid, which, depending on the pH, can ionize to the hypochlorite ion. When pH is below 7, the bulk of the HOCl remains un-ionized.

$$Cl_2 + H_2O \rightarrow HCl + HOCl$$

$$HOCl \overrightarrow{pH > 8} \ H^+ + OCl^-$$

$$Ca(OCl)_2 + H_2O \rightarrow Ca^+ + H_2O$$

Chlorine existing in water as Hypochlorous acid and hypochlorite ion are two forms of **free chlorine residual**. Hypochlorous acid as a disinfectant is 100 times more effective compared to hypochlorite ion. In other words, the pH plays an important role in determining the effectiveness of disinfection by chlorination. The effectiveness of disinfection increases as the pH drops.

BREAKPOINT CHLORINATION CURVE

When chlorine is added, the amount of free chlorine residual formed is less than the amount of chlorine added since some of the chlorine is used up in reacting with impurities in water. When chlorine is added to natural water containing ammonia and other impurities, the residual that develop yield a curve similar to that shown in **Figure 11.1**. Breakpoint chlorination curve represents chlorine residual, corresponding to various dosages, remaining after a specified contact time. The reactions that take place can be divided into four groups.

Stage I
No chlorine residual since all of the chlorine added is used up in reactions with reducing agents such as iron and manganese, nitrites, sulphides, and dissolved organics. No disinfection happens.

Stage II
As more chlorine is added, it reacts with ammonia and organic water forming mono-chloramine and chloro-organic compounds. These compounds have some disinfectant properties and are called *combined chlorine residual.*

$$NH_3 + HOCl = NH_2Cl + H_2O$$

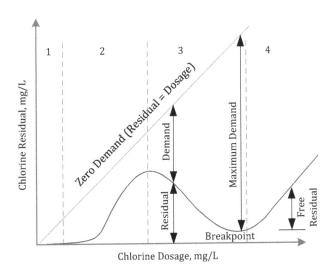

FIGURE 11.1 Breakpoint chlorination curve

Stage III

Adding more chlorine to water decreases the residual. The decrease results because the additional chlorine oxidizes some of the chloro-organic compounds and ammonia.

$$2NH_3 + 3Cl_2 = N_2 + 6HCl$$

The additional chlorine also changes some of the monochloramine to dichloramine and trichloramine. Chloramines residual declines as more chlorine is added until it reaches a minimum value, referred to as the **breakpoint.** The amount of chlorine added (dosage) minus the chlorine residual is the *chlorine demand* of water. Chlorine demand is at a maximum at the breakpoint (demand is 100% satisfied).

Stage IV (4 Onwards)

Beyond breakpoint, any chlorine addition will produce free chlorine residual. It is for this reason: the breakpoint chlorination curve beyond breakpoint is a straight line with a one-to-one slope. This is since all the demand of the water has been satisfied and there is nothing left for chlorine to react with. Breakpoint chlorine dosage is unique for each sample of water tested. Depending on the number of constituents in water like ammonia and other reducing agents, some stages may be absent.

Example Problem 11.1

A water pump delivers on average 200 gpm against typical operating heads.

    a. If the desired chlorine dosage is 2.0 mg/L, what should be the setting on the rotameter for the chlorinator in lb/d?

    b. If the pump is operated 18 hours per day, how many lb of chlorine will be used up over a period of one week?

Given:

    Q = 200 gpm   C = 2.0 mg/L   M =?

Solution:

$$M = Q \times C = \frac{200\,gal}{min} \times \frac{2.0 \times 8.34\,lb}{10^6\,gal} \times \frac{1440\,min}{d} = 4.80 = 4.8\,lb / d$$

$$Weekly\,Use = \frac{4.8\,lb}{24\,h} \times \frac{18\,h}{d} \times \frac{7\,d}{wk} = 10.5 = 25.2 = 25\,lb / wk$$

Example Problem 11.2

A chlorinator is set to feed filtered water at a dosage rate of 12 kg/d. This dose results in a chlorine residual of 0.55 mg/L when the average 24 hour flow is 5.0

ML/d. Determine the chlorine demand of water. When the chlorine setting is increased to 13 kg/d, determine the expected increase in free chlorine residual.

Given:

Dosage rate M = 12 kg/d   Residual = 0.55 mg/L   Q = 5.0 ML/d   Demand = ?

Solution:

$$Dosage, C = \frac{M}{Q} = \frac{12\ kg}{d} \times \frac{d}{5.0\ ML} \times \frac{mg/L}{kg/ML} = 2.40 = 2.4\ mg/L$$

$$Demand = dosage - residual = (2.4 - 0.55) = 1.85\ mg/L$$

$$C_{increase} = \frac{(13-12)kg}{d} \times \frac{d}{5.0\ ML} \times \frac{mg/L}{kg/ML} = 0.20\ mg/L$$

Since demand is not changed, the increase in residual will also be 0.2 mg/L.

Example Problem 11.3

Different chlorine dosages were applied to a water sample and the chlorine residual were tested after 12 min of contact time. For the data shown in the table below, plot the breakpoint chlorination curve and find the maximum chlorine demand, breakpoint dosage and residual, and dosage rate for treating a flow of 12 ML/d and for achieving a residual of 0.65 mg/L.

Dosage, mg/L	0.20	0.4	0.60	0.80	1.0	1.2	1.4	1.6	1.8	2.0
Residual, mg/L	0.05	0.15	0.30	0.45	0.30	0.20	0.35	0.62	0.78	1.0

Given:

Q = 12 ML/d   residual = 0.65 mg/L   dosage =?   feed rate,   M =?

Solution:
The chlorine demand for each dosage is calculated by subtracting the chlorine residual from chlorine demand, as shown in the table below.

Dosage, mg/L	0.20	0.40	0.60	0.80	1.00	1.20	1.40	1.60	1.80	2.00
Residual, mg/L	0.05	0.15	0.30	0.45	0.30	0.20	0.35	0.62	0.78	1.00
Demand, mg/L	0.15	0.25	0.30	0.35	0.70	1.00	1.05	0.98	1.02	1.00

A plot of chlorine residual against chlorine dosage is shown in **Figure 11.2**. From this plot, the following values are read.

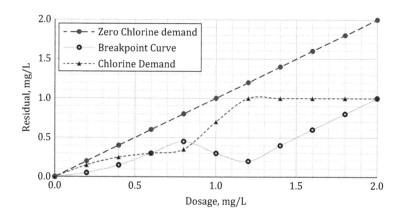

FIGURE 11.2   Breakpoint chlorination (Ex. Prob. 11.3)

$$Breakpoint\ dosage = 1.2\,mg\,/\,L \quad Max.\ Demand = 1.0\,mg\,/\,L$$

$$Dosage = Demand + residual = (1.0 + 0.65)\,mg\,/\,L = 1.65\,mg\,/\,L$$

$$Dosage\ Rate = \frac{12\,ML}{d} \times \frac{1.65\,kg}{ML} = 19.8 = \underline{20\,kg\,/\,d}$$

11.6   CHLORINE PRACTICES

There are a number of ways chlorine is fed into water. Some of these practices are discussed here.

11.6.1   CHLORAMINATION

Chloramination produces combined chlorine residual. In this method, chlorine is combined with natural or added ammonia. This type of residual is less effective but lasts longer. This practice is more suitable when free residual is going to cause taste and odor problems as in the case of presence of phenolic compounds. This practice is more commonly used for secondary disinfection.

The main reason for chloramination is the prevention of the formation of THMs and chlorophenols as DBPs. The only limitation is that chloramines residual is significantly weaker. The typical dosage of ammonia is one part to three parts of chlorine. Higher dosages of chlorine form dichloramine and will cause taste and odor problems and may cause the problem of nitrification at high temperatures. Due to the slower bactericidal action of combined residual, 25 times combined residual is required compared to free chlorine. Hence, combined residual of 2.5 mg/L or greater should be maintained for secondary disinfection.

11.6.2 BREAKPOINT CHLORINATION

Free chlorine residual is produced beyond breakpoint. Any dosage over and above the breakpoint dosage would result in an equal amount of free chlorine residual. This practice is more common, and the dosage depends on the chlorine demand and the level of free chlorine residual to be attained. Free chlorine residual is 25 times more effective compared to combined residual and needs less contact time. To obtain an equivalent bactericidal action with an equal amount of combined and free chlorine, a contact time approximately 100 times longer is required.

11.6.3 SUPERCHLORINATION

In a way, superchlorination can be thought to be breakpoint chlorination with high levels of free residual. This practice is used in emergencies, like outbreaks of epidemic, breakdowns, water main repairs, or heavily polluted water. Heavy doses of chlorine (10–15 mg/L) are used to effectively destroy the resistant organisms and cysts. Excess chlorine will then have to be removed by dechlorination.

11.6.4 DECHLORINATION

After superchlorination, water contains a high chlorine residual. These residuals must be brought down to acceptable levels before water exits the plant. When superchlorination, is used to disinfect tanks and pipes, water must be dechlorinated before discharge. Reducing compounds like sulphur dioxide, sodium thiosulphate, and sodium bisulphate are commonly used for neutralizing chlorine. Prolonged storage and adsorption on charcoal or activated carbon is also effective.

11.7 POINTS OF CHLORINATION

Chlorine is used at various stages of a water supply system, starting from raw water pumping to the distribution system. Terms like pre-, post-, and re-chlorination are often used to indicate the points of application.

11.7.1 PRE-CHLORINATION

Pre-chlorination is the application of chlorine to raw water before any other treatment process. It is also called source water chlorination and it is mainly done to achieve the following:

- Start the process of disinfection early.
- Control biological growth on filters, pipes, and basins.
- Promote improved coagulation.
- Prevent the formation of mud balls and slime in filters.
- Oxidize iron and manganese for removal in following processes.
- Reduce taste, odor, and color problems, minimize the post-chlorination dosage.

- In heavy polluted waters, pre-chlorination should be used with caution as chlorine reacts with organic compounds to form THMs, as discussed in section 11.8.

11.7.2 POST-CHLORINATION

Post-chlorination is the application of chlorine to treated water to achieve residual for water exiting the plant and is called secondary disinfection. This provides protection against any possible contamination in the distribution system or customer plumbing system. Water regulations provide specific guidelines for the type and level of chlorine residual before water leaves the plant. A free chlorine residual of 0.20 mg/L is recommended.

11.7.3 RE-CHLORINATION

In large and complex water distribution systems, it may not be possible to maintain the minimum residual of 0.2 mg/L, especially in remote parts of the system. In such cases, chlorine needs to be applied in stages to boost the chlorine residual. This is known as re-chlorination.

11.8 FORMATION OF TRIHALOMETHANES

Organic compounds like humic and fulvic acid originate from decaying vegetation and are present in almost all waters. They can cause color, taste, and odor problems in water. In addition, they react with chlorine to form THMs, which are considered carcinogenic. For this concern, the maximum contaminant level (MCL) for THMs is 80 µg/L (80 ppb). Any increase in dosage or residual increases DBP formation. Warmer temperatures increase the formation of DBPs. Lower pH favors the formation of haloacetic acids (HAAs), while higher pH favors THMs.

The removal of THMs is very difficult, thus prevention is the best solution. Most THMs are formed during pre-chlorination. One way of reducing the formation of THMs is by the elimination of pre-chlorination or by moving the injection point to just ahead of the filters. The other methods for reducing THM levels include the use of alternative disinfectants, such as ozone or chlorine dioxide, and the application of activated carbon to adsorb Trihalomethanes and humic substances. However, the risk of incomplete disinfection, that can accompany disinfectant losses, must be avoided.

11.9 FACTORS AFFECTING CHLORINE DOSAGE

It has already been noted that waters with high turbidity cannot be adequately disinfected. The particles of turbidity inhibit disinfection by providing hiding places for bacteria where chlorine cannot reach.

CHLORINE CONCENTRATION

The higher the concentration of free available disinfection chlorine, the more effective and faster is the disinfection.

CONTACT TIME BETWEEN THE ORGANISM AND CHLORINE

The longer the contact time, the more effective is the disinfection. If the chlorine concentration is decreased, then the **contact time** (t) must be increased. The disinfecting power, often referred to as the kill, is directly related to these two factors: disinfectant residual and time of contact, known as the **CT factor**.

$$Inactivation = C \times T = CT \; Factor$$

C = chlorine residual at the end of contact time = strength in mg/L
T = t_{10} = 10th percentile contact time in minutes

Contact time, t_{10}, in the above equation, is the length of time during which no more than 10% of water passes through a given disinfection process. It is calculated using an estimate of baffling in the tank or through a tracer test. For the plug flow system, as in the case of the pipe flowing full, contact time can be assumed to equal theoretical detention time. This product factor is more commonly known as the CT concept and is used to assess inactivation for chemical agents employed for disinfection. The effectiveness of disinfection is indicated by the cumulative value of the CT.

LOG REMOVAL

Log removal rather than percentage removal is more commonly used to indicate inactivation of pathogens. If the concentration is reduced by a factor of 10 – that is 90% removal – it amounts to 1 log removal, since the log of 10 is 1. If subscripts i and e refer to influent and effluent of a process, or a combination of processes, log removal and percent removal can be found as follows.

$$Log \; Removal, LR = \log(C_0 / C_e) \quad PR = \left(1 - 10^{-LR}\right) \times 100\%$$

For example, 2 log removal means reduction by a factor of 2, or 99% removal. Regulating authorities specify minimum values of CT factor to achieve a given level of disinfection (1–4 log removals). For surface treatment water systems, minimum CT values must be provided for peak flow conditions to achieve 4 log removal for viruses, 3 log removal for Giardia, and 2 log removal for Cryptosporidium. If disinfection follows conventional filtration, 2 log removal of viruses is achieved. Thus, only 2 log removal need be achieved by the disinfection method. In ground water systems, virus removal is only 2 log since filtration through the ground layers significantly contributes to pathogen removal.

TYPE OF CHLORINE RESIDUAL

Free chlorine is a much more effective disinfectant than combined chlorine. Combined chlorine residual requires a greater concentration acting over a longer period to do the same job of killing the pathogens. When contact time is short, only a free residual will provide effective disinfection.

TEMPERATURE OF THE WATER

Usually, the higher the temperature, the more effective is the disinfection. If the water system operator cannot control the temperature, then they must increase the contact time or dosage at a lower temperature.

THE pH OF THE WATER

In general, disinfection is more effective at a low pH since HOCl is 100 times more effective than the OCl ion. So, with combined residuals, monochloramine is least effective. Disinfection is most effective in the pH range of 4–7. As an example, the contact time required for 99.9% (3 log) Giardia removal, applying 0.6 mg/L of chlorine residual, is 100 minutes at a pH of 6 and twice as many minutes at a pH of 8. The pH also effects corrosivity.

SUBSTANCES IN THE WATER

Turbidity in water can provide shelter to organisms. Therefore, for chlorination to be effective, turbidity must be reduced to the greatest extent possible by the preceding water treatment methods.

Example Problem 11.4

A well pumping system maintains a free chlorine residual of 0.8 mg/L. Contact time is provided by a 120 m³ capacity tank. Baffling in the tank is such that effective contact time is 40% (baffling factor) of the theoretical time. Maximum flow through the tank is 800 L/min. Determine if the system meets the requirement for 2 log removal for virus. Water temperature is 10°C and pH of water is 7.5.

Given:

$$C = 0.80 \text{ mg/L} \quad V = 12 \text{ m}^3 \quad Q = 800 \text{ L/min} \quad T = 10°C \quad pH = 7.5$$
$$BF = 40\% = 0.40$$

Solution:

$$t_{eff} = BF \times \frac{V}{Q} = 0.40 \times 12 \ m^3 \times \frac{min}{800 \ L} \times \frac{1000 \ L}{m^3} = 6.0 \ min$$

$$CT = \frac{0.8\,mg}{L} \times 6.0\,min = 4.8\,mg.min\,/\,L$$

Since it is a groundwater system, 2 log virus removal is required. CT value for 2 log virus removal with free chlorine residual is 3 mg.min/L. Actual value is more than required, hence disinfection for virus removal is satisfactory.

Example Problem 11.5

A small community is served by a ground water supply consisting of a 15-in diameter main with a length of 4760 ft till it reaches the first consumer. The peak hourly pumping rate is 2100 gpm, and the temperature of the water is 5°C. The groundwater being under the influence of surface water, regulations demand 3 log virus inactivation. What free chlorine residual is required at the outlet of the pipeline?

Given:

$$C =?\quad L = 4760\ ft \quad Q = 2100\ gpm \quad D = 15\ in = 1.25\ ft$$

Solution:

$$t = \frac{A.L}{Q} = \frac{\pi(1.25\,ft)^2}{4} \times \frac{4760\,ft \times 7.48\,gal}{ft^3} \times \frac{min}{2100\,gal} = 20.8 = 21\,min$$

From the table in appendix, the CT factor for 3 log virus removal at 5°C is 6.0 mg.min/L, so to meet the desired requirement for free chlorine residual is

$$C = \frac{CT}{t} = \frac{6.0\,mg.min}{L} \times \frac{1}{20.8\,min} = 0.2880 = 0.29\,mg\,/\,L$$

Example Problem 11.6

A filtration system is required to provide 0.5 log Giardia and 2 log virus removal by post chlorination. The t_{10} time evaluated for the clear well is 43 min and the free residual of 0.45 mg/L is maintained. How much removal is achieved for Giardia and virus?

Given:

At 10°C, CT = 19 mg.min/L /0.5LR for Giardia   3.0 mg.min/L/ 2 LR for virus

$$C = 0.50\ mg/L \quad t_{10} = 43\ min$$

Solution:

$$For\ Giardia = \frac{CT}{CT/LR} = \frac{0.45\,mg \times 43\,min}{L} \times \frac{LLR}{19\,mg\,/\,min} = 0.5\,LR$$

$$For\ Virus = \frac{CT}{CT/LR} = \frac{0.45\,mg \times 43\,min}{L} \times \frac{LR}{3.0\,mg.min\,/\,L} = 13\,LR$$

11.10   GAS CHLORINATION EQUIPMENT

Because of the toxic nature of chlorine, gas chlorination systems require a different type of chlorine feed system, as well as specialized equipment for health and safety.

CYLINDERS

Chlorine cylinders hold 68 kg (150 lb) of chlorine. The handling of chlorine cylinders needs special care. They should not be rolled and should be kept in an upright position secured with a chain. The **ton container** is a reusable welded tank that holds 910 kg (2000 lb) of chlorine. Ton containers are designed to rest horizontally and are equipped with two valves – an upper valve for chlorine gas and a lower valve for liquid chlorine. Chlorine cylinders can deliver up to 18 kg/d (42 lb/d) of chlorine gas and ton containers can deliver up to 180 kg/d. Excess withdrawal from cylinders can cause freezing and plugging of lines. Ton containers equipped with evaporators can deliver up to 4400 kg/d.

WEIGHING SCALE

A weighing scale is used to determine the amount of chlorine used or remaining in a container or a cylinder. By recording weights at regular intervals, the chlorine dosage rate can be calculated.

VALVES AND PIPING

Except when a direct chlorinator is used, an auxiliary valve is connected to the container valve. The auxiliary valve can be used as a shut off to the supply during emergencies. The valve assembly is connected to the chlorine supply piping or manifold by flexible tubing. The tubing is usually 10 mm copper rated at 3500 kPa.

CHLORINATOR

Typical chlorine feed equipment is shown in **Figure 11.3**. A chlorinator can be a simple direct mount unit or it can be a freestanding cabinet. A variable orifice inserted in the feed line controls the feed rate of chlorine. The reduced pressure downstream of the orifice control allows a uniform gas flow, accurately metered by a rotameter or feed rate indicator. This also acts as a safety feature in case a leak develops in the vacuum line.

INJECTOR

Injector or ejector is basically a venturi. Water flowing through the venturi creates a vacuum that draws in gas to mix with water to create a strong chlorine solution. This solution can be safely piped to various points in the treatment plant.

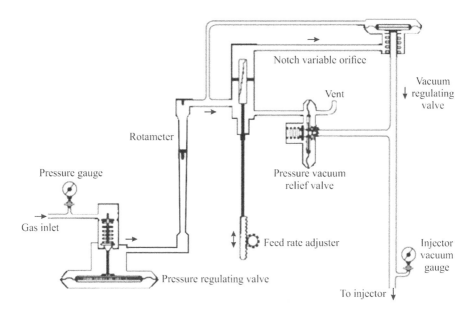

FIGURE 11.3   Chlorine feed system

DIFFUSER

The purpose of the diffuser is to disperse the chlorine solution into the main flow of water. To disperse the solution uniformly and quickly, the diffuser pipe is usually perforated.

Example Problem 11.7

On a Monday morning, a chlorine gas cylinder weighs 84 lb. At the same time on Tuesday, after chlorination, the cylinder weighs 58 lb. Calculate the chlorine dosage applied in mg/L if the amount of flow treated for the same period is 890 000 gallons.

Given:

$$m = (84 - 58) = 26 \text{ lb} \quad V = 890\ 000 \text{ gal} \quad C = ?$$

Solution:

$$C = \frac{m}{V} = \frac{26\ lb}{8.9 \times 10^5\ gal} \times \frac{10^6\ gal}{8.34\ lb} \times \frac{mg}{L} = 3.50 = \underline{3.5\ mg/L}$$

11.11   CHLORINE FEED CONTROL

Chlorine feeders can be operated manually or automatically based on flow, chlorine residual, or both.

11.11.1   Manual Control

The chlorine is fed at a constant rate irrespective of flow and chlorine demand of the water. Adjustments must be made each time the flow rate changes. This type of control is satisfactory when flow and demand are relatively constant.

11.11.2   Automatic Proportional Control

This control is recommended when the chlorine demand of the water remains constant. Proportional control adjusts the feed rate to provide a constant pre-established chlorine dosage. This is accomplished by transmitting the flow signal to the feeder which responds to the transmitted signal by increasing or decreasing the feed rate.

11.11.3   Automatic Residual Control

Automatic residual control provides a set residual rather than dosage, as in the case of proportional control. This control is desirable when there are fluctuations in the chlorine demand of the water. In this control, there is a chlorine residual sensor in addition to a flow sensor. The feeder receives signals from both the flow meter and the chlorine analyzer.

11.11.4   Hypochlorination Facilities

Hypochlorination is well suited to smaller water supply systems with chlorine uses of less than 1.5 kg/d. Both calcium and sodium hypochlorite systems use hypochlorite solution feeders called hypochlorinators to meter the liquid stock volume into water. Calcium hypochlorite also requires mixing and storage tanks for making up the stock solution from powder.

11.11.5   Hypochlorinators

Types of hypochlorinators available include positive displacement feeders, aspirator feeders, suction feeders, and tablet hypochlorinators. Hypochlorinators provide a reliable method for applying chlorine to disinfect water. The hypochlorinators either pump or inject a chlorine solution into the water. When injecting the chlorine solution, a hydraulic device such as a venturi is used to create a negative pressure. In this arrangement, the hypochlorite solution is pumped through an injector, which draws in additional water for dilution of the hypochlorite solution. Most hypochlorinators are controlled electrically from a master control panel. The stopping and starting of the hypochlorinator is synchronized with the pumping unit. A flow switch or other sensing device can be used for this purpose. Having a flow switch avoids operating problems associated with chemicals being fed when the raw water pumps fail to start.

Example Problem 11.8

Water pumped from a well is disinfected by a hypochlorinator. A chlorine dosage of 1.3 mg/L is applied to maintain the desired level of chlorine residual. During a one-week period, the flow totalizer indicated 8 830 m³ of water was pumped. A 2.5% sodium hypochlorite solution is stored in a 1.0 m diameter tank. Find the expected drop in the level of the hypochlorite tank.

Given:

$$C_1 = 2.5\% = 25 \text{ g/L} \quad C_2 = 1.3 \text{ mg/L} \quad V_1 =? \quad V_2 = 8830 \text{ m}^3 \quad D = 1.0 \text{ m}$$

Solution:

$$V_1 = \frac{C_2}{C_1} \times V_2 = \frac{1.3\,mg}{L} \times \frac{L}{25\,g} \times 8830\,m^3 \times \frac{g}{1000\,mg} \times \frac{1000\,L}{m^3} = 459 = 460\,L$$

$$\Delta d = \frac{\Delta V}{A} = \frac{0.459\,m^3}{\pi/4(1.0\,m)^2} \times \frac{100\,cm}{m} = 58.4 = 58\,cm$$

Example Problem 11.9

How many lb of hypochlorite (65% available chlorine) are required to disinfect 4500 ft of a 24 in water main at a chlorine dosage of 50 mg/L?

Given:

$$D = 24 \text{ in} = 2.0 \text{ ft} \quad L = 4500 \text{ ft} \quad C = 50 \text{ mg/L}$$

Solution:

$$m = V \times C = \frac{\pi}{4}(2.0\,ft)^2 \times 4500\,ft \times \frac{7.48\,gal}{ft^3} \times \frac{50\,mg}{L} \times \frac{8.34\ lb/MG}{mg/L}$$

$$= 44.096\,lb\,Cl_2 \times \frac{lb\,hypo}{0.65\,lb\,Cl_2} = 67.84 = 68\ lb\ of\ hypochlorite$$

Discussion Questions

1. Differentiate between the following:
   a. disinfection and sterilization;
   b. primary disinfection and secondary disinfection;
   c. free chlorine residual and combined chlorine residual;
   d. gas chlorination and hypochlorination;
   e. pre-chlorination and post chlorination.

2. Name and explain various types of disinfectants.
3. Explain the breakpoint chlorination curve, with the help of a neat sketch.
4. Describe chlorination practices along with application points.
5. What types of safety measures are taken where chlorination is practiced?
6. What are the main components of chlorination equipment? Discuss briefly.
7. Define CT factor and explain how it can be used to determine the disinfection efficiency?
8. What is the most important water treatment process to prevent the spread of water borne disease? Is there and any potential of harmful effect?
9. What are the characteristics of a good disinfectant?
10. Briefly describe the factors that affect chlorine dosage.
11. What is ozonation? What are its advantages and disadvantages?
12. What are the characteristics of a good disinfectant?
13. Define the meaning of C × t product. In addition to C and t, what other factors influence the efficiency of disinfection. What kind of pathogens are readily inactivated by free chlorine and what are the most difficult to inactivate?
14. Why is heavy pre-chlorination of surface water undeniable?
15. In what situations may dechlorination of water be required?

Practice Problems

1. A chlorine feeder operates on a proportional control and is set to provide a chlorine dosage of 2.0 mg/L. At a given hour, the rotameter indicates a reading of 24 kg/d. Calculate the hourly flow rate in L/s. (140 L/s).
2. In certain well water systems, the chlorine feed is controlled manually. What should be the setting on the feeder when the water pumping rate is 880 gpm and the desired chlorine dosage is 2.5 mg/L? (26 lb/24 h).
3. Chlorine use at a water treatment plant averages 41 kg/d. Assuming that the average daily flow is 52 ML/d, what dosage of chlorine is applied? (0.80 mg/L).
4. Water from a well is being disinfected using a hypochlorinator containing a 1.0% chlorine solution. If the feed pump rate is 7.0 gal/24h and the well pumping rate is 13 gpm, what is the dosage of chlorine? (3.7 mg/L).
5. How many kg of chlorine would you have to apply to a reservoir containing 20 ML of water to obtain 5.0 mg/L residual when the chlorine demand is 2.0 mg/L? (140 kg).
6. Chlorine at the rate of 4.0 mg/L is added continuously to a water flow that averages 5.2 MGD. How much chlorine will be used in 30 days? (5200 lb/month).
7. Your chlorinator breaks down and you decide to feed a sodium hypochlorite solution containing 15% chlorine as a temporary measure. Chlorine dosage of 6.0 mg/L is required for 7.2 ML pumped in 18 h. What should be the feed rate of hypochlorite solution? (16 L/h).

8. How many grams of 65% HTH are needed to disinfect 60 m of a repaired 6 in line section by applying a chlorine dosage of 25 mg/L? (41 g).

9. In a provincial park facility, a 2% solution is fed at the rate of 10 L/24h to treat a water flow of 50 kL/d. If the chlorine residual is 0.6 mg/L, what is the chlorine demand of lake water? (3.4 mg/L).

10. A chlorinator is set to feed filtered water at a dosage rate of 15 kg/d. This dose results in a chlorine residual of 0.65 mg/L when the average 24 hour flow is 270 m³/h. Determine the chlorine demand of water. When the chlorine feed rate setting is increased to 18 kg/d, what is the expected increase in free chlorine residual? (1.7 mg/L, 0.46 mg/L).

11. A well water pump delivers on average 240 gpm against typical operating heads. If the desired chlorine dosage is 3.2 mg/L, what should be the setting on the rotameter for the chlorinator? If the pump is operated 18 hours per day, how many 100 lb chlorine cylinders should be ordered per month? (9.2 lb/24h, 3/month).

12. Water pumped from a well is disinfected by a hypochlorinator. A chlorine dosage of 1.3 mg/L is applied to maintain the desired level of chlorine residual. During a one-week period, the flow totalizer indicated 8 830 m³ of water was pumped. A 2.5% sodium hypochlorite solution is stored in a 1 m diameter tank. Determine the expected drop in the level of the hypochlorite tank. (58 cm)

13. How many kg of hypochlorite (65% available chlorine) are required to disinfect 800 m of a 400 mm water main at a chlorine dosage of 100 mg/L? (15 kg).

14. How many liters of chlorine bleach with 12% available chlorine should be used to add to a 2.5 m diameter well to achieve a dosage of 50 mg/L? Water stands to a height of 3.0 m in the well. (6.1 L).

15. A small community is served by a ground water supply consisting of a 450 mm diameter main of length 1400 m till it reaches the first consumer. The peak hourly pumping rate is 11 m³/min and the temperature of water is 10°C. The groundwater being under the influence of surface water, regulation demands 3 log virus inactivation (CT required 4.0 mg.min/L). What free chlorine residual is required at the outlet of the pipeline? (0.20 mg/L).

16. The filtered water at peak hourly flow from a plant with direct filtration has a pH of 7.0 and temperature of 15°C. The effective detention time in the clear well is 24 min followed by 1.3 km of transmission line at flow velocity of 1.2 m/s before entering the distribution system. Assuming 2.0 log removal of Giardia before chlorination, what chlorine residual is required at the outlet of the clear well and the pipeline? Assume no loss of residual in the pipeline. The required value of CT for 1 log giardia removal at the prevailing conditions is known to be 25 mg.min/L. (0.59 mg/L).

17. A new main is disinfected with water containing 50 mg/L of chlorine by feeding 1% chlorine solution to the water entering the pipe. How many kg of dry hypochlorite powder, containing 70% chlorine, must be added to prepare 200 L of 1.0% solution? At what rate should 1.0% solution be

applied to water entering the pipe to achieve a dosage of 50 mg/L? (2.9 kg, 5.0 L/m³).

18. It is desired to provide a chlorine residual of 0.5 mg/L to a flow of 1.0 MGD. What should be the chlorinator setting in lb/24 h if the chlorine demand of treated water is 2.5 mg/L? (25 lb/24 h)

19. For proper disinfection, a water supply is to be fed at a dosage rate of 2.5 mg/L of chlorine. How much chlorine will be left in a 68 kg chlorine tank after a week if the average daily water flow is 1.5 ML/d? (42 kg).

20. A well water supply is disinfected by feeding 2.5% sodium hypochlorite solution. Over a period of 12 hours, the total volume of water pumped is 16 580 m³. During the same period, the hypochlorinator level drops by 57 cm in the 0.8 m diameter solution tank. The hypochlorinator is operated continuously over the 12 hour period and a residual of 0.2 mg/L is maintained. Calculate the chlorine demand of well water and the feed pump rate in mL/min. (0.23 mg/L, 400 mL/min).

21. A hypochlorinator is used to disinfect the water pumped from a well. A chlorine dosage rate of 4.5 kg/d is required for adequate disinfection of the well water. What should be the strength of the chlorine solution if the feed pump is set to feed at 180 mL/min. A volume of 20 L of 12% liquid hypochlorite is added to a 300 L capacity graduated container. How many liters of water must be added to the container to produce 250 L of solution of this strength? (1.74 %, 120 L).

22. You are going to test water for chlorine demand. You have got 15% liquid chlorine. What volume of liquid chlorine should be pipetted to make 1.5 L of 0.1% solution? How many mL of this solution to 500 mL of water is required to achieve a dosage of 1.0 mg/L? (10 mL, 0.5 mL).

23. A hypochlorite solution with 3.0% available chlorine is used to chlorinate water. A chlorine residual of 0.5 mg/L is required. The chlorine demand of the water is 1.3 mg/L. If the water pumping rate is 300 gpm, what should be the solution feeder setting in gal/24h. (26 gal/24h).

24. Chlorine dosage rate in the treatment of 5.5 MGD of water is 19 lb/d. The chlorine residual after 12 min contact time is 0.22 mg/L. Find the chlorine dosage and chlorine demand of the water. (0.41 mg/L, 0.19 mg/L).

25. A flow of 3.5 MGD is disinfected using powdered hypochlorite containing 65% available chlorine. If the powder was fed at the rate of 75 lb/d, calculate the amount of chlorine dosage applied. (1.7 mg/L).

12 Water Softening

In layman's terms, hard water does not make good lather with soap and causes the scaling of pipes. The **hardness** of water is due to its calcium and magnesium content. Hardness due to other bivalent and trivalent cations is usually insignificant. Total hardness is the total of calcium hardness and magnesium hardness and is expressed in terms of equivalent calcium carbonate. Water hardness varies considerably in different geographical areas. Since groundwater is in contact with geological formations containing calcium (limestone) and magnesium (dolomite) salts, groundwater is normally harder than surface water.

Though hardness is not harmful to health, its excess makes it unsuitable for uses such as heating, cooking, washing, bathing, and laundering. Excessive hardness, for example, may deposit scale in water heaters and water piping. Hardness can waste large proportions of the soap used in laundering. Owing to these disadvantages, many municipalities soften their water supplies. **Table 12.1** shows a comparative classification for softness and hardness in water.

Hardness in the range of 80–120 mg/L is quite acceptable. Although relatively soft water is preferable, excessively soft water is corrosive.

12.1 TYPES OF HARDNESS

Hardness can be categorized based on cations and anions content, and can be calcium and magnesium hardness or carbonate and non-carbonate hardness. Hardness caused by calcium is calcium hardness regardless of the salts associated with it, which may include carbonates, sulphates, and chlorides. Likewise, hardness caused by magnesium is called magnesium hardness. Magnesium hardness is relatively hard to treat. Since hardness is found in various forms, it is expressed as equivalent of $CaCO_3$.

12.1.1 CARBONATE HARDNESS

Carbonate hardness (CH) is primarily due to bicarbonates. On boiling the water, bicarbonates break down into carbonates and settle out of the water. Because it can be removed by heating, carbonate hardness is also referred to as **temporary hardness**.

12.1.2 NON-CARBONATE HARDNESS

Non-carbonate hardness (NCH) is a measure of calcium and magnesium salts other than carbonates and bicarbonates, including sulphates, nitrates, and chlorides. Since non-carbonate hardness cannot be removed by prolonged boiling, it is also called **permanent hardness**. The sum of carbonate hardness and non-carbonate

DOI: 10.1201/9781003347941-14

TABLE 12.1

Classification of hardness

Classification	mg/L as CaCO$_3$
extremely soft	0–50
soft	0–75
moderately hard	75–100
hard	100–125
very hard	125–175
excessively hard	175–250
too hard	>250

TABLE 12.2

Milliequivalent table (Ex. Prob. 12.1)

Component	mg/L	mg/meq	Hardness meq/L	mg/L as CaCO$_3$
CO$_2$	25	22	1.14	57
Alk	300	61	4.92	246
Ca	135	20	6.75	338
TH	360	50	7.20	360

hardness is the **total hardness** (TH). In most water, hardness due to other cations, including trivalent, is negligible.

Example Problem 12.1

The chemical analysis of a water sample is as follows. Express each concentration as meq/L and determine the various types of hardness See (**Table 12.2**).

Given:

Ca = 135 mg/L Total hardness,   TH = 360 mg/L as CaCO$_3$
Alk = 300 mg/L as HCO$_3$ CO$_2$ = 25 mg/L

Solution:

$$MgH = TH - CaH = 360 - 338 = 22\,mg\,/\,L$$

$$NCH = TH - CH = 360 - 114 = 110\,mg\,/\,L$$

12.2   SOFTENING METHODS

The two common methods of softening water are: lime-soda ash and ion-exchange.

Ion-exchange softening is more suitable for treating water high in non-carbonate hardness and when the total hardness is below 350 mg/L. It is possible to achieve zero hardness with ion-exchange softening. The end products of lime treatment – calcium carbonate and magnesium hydroxide – are soluble in water to some degree. Hence, it is not practical to reduce the hardness of the finished water to less than 35 mg/L by lime-soda ash treatment.

12.2.1   LIME-SODA ASH SOFTENING

In the lime-soda ash process, lime and soda ash are added to the water to form insoluble precipitates, which are removed by sedimentation and filtration processes.

Lime Reactions

Lime added to water increases pH and reacts with carbonate alkalinity to precipitate as calcium carbonate.

$$Ca(HCO_3)_2 + Ca(OH)_2 \rightarrow 2CaCO_3 \downarrow + 2H_2O$$

$$CO_2 + Ca(OH)_2 \rightarrow CaCO_3 \downarrow + H_2O$$

If sufficient lime is added to reach a pH >10, magnesium hydroxide is precipitated.

$$Mg(HCO_3)_2 + Ca(OH)_2 \rightarrow CaCO_3 \downarrow + MgCO_3 + 2H_2O$$

$$MgCO_3 + Ca(OH)_2 \rightarrow CaCO_3 \downarrow + Mg(OH)_2 \downarrow$$

Soda-Ash Reactions

Soda ash is used to remove the non-carbonate hardness.

$$MgSO_4 + Ca(OH)_2 \rightarrow Mg(OH)_2 \downarrow + CaSO_4$$

$$CaSO_4 + Na_2CO_3 \rightarrow CaCO_3 \downarrow + Na_2SO_4$$

In removing magnesium non-carbonate hardness, both lime and soda ash are needed. The lime-soda ash process increases the sodium content of water and the pH of the water is high due to residual (insoluble) carbonates and hydroxides.

12.2.2   CHEMICAL DOSAGES

The amount of chemical required depend on the degree of hardness removal with various types of wastes. Regarding the chemical reactions shown above, the chemical requirements expressed in equivalents for treating various types of hardness are shown in **Table 12.3**.

TABLE 12.3

Lime and soda ash requirement

Type of Hardness	Component	Lime (eq/eq)	Soda Ash (eq/eq)
carbonate hardness, CH	$Ca(HCO_3)_2$	1	0
	$Mg(HCO_3)_2$	2	0
non-carbonate hardness, NCH	$CaSO_4$	0	1
	$MgSO_4$	1	1
carbon dioxide	CO_2	1	0

Referring to **Table 12.3**, the lime requirement for the removal of carbonate hardness associated with magnesium is twice as much. Also, the non-carbonate hardness associated with magnesium requires one each of lime and soda ash. Based on this, lime and soda ash requirements when expressed in terms of carbonate and non-carbonate hardness are as follows:

$$Lime = CO_2 + Alk + Mg + Excess$$

$$Soda\ Ash = NCH = Ca + Mg - Alk$$

Excess lime is required to raise the pH for precipitation of magnesium carbonate hardness. It is assumed here that all the alkalinity is due to bicarbonates, which is mostly true in sources of fresh water. Lime dosage is commonly expressed in terms of quicklime, CaO. The quicklime has a purity ranging from 70% to 96%, with a typical value of 88%. The purity of commercially available soda ash is as high as 98% to 99%.

Example Problem 12.2

How much soda ash is required (lb/d) to remove 40 mg/L of non-carbonate hardness as $CaCO_3$ from a flow of 6.0 MGD?

Given:

$$Q = 6.0\ mil\ gal/d \quad NCH = 40\ mg/L$$

Solution:

$$NCH = \frac{40\,mg}{L} \times \frac{meq}{50\,mg} = 0.800 = 0.80\ meq/L$$

$$Soda = \frac{0.80\,meq}{L} \times \frac{53\,mg}{meq} \times \frac{8.34\,lb/MG}{mg/L} \times \frac{6.0\,MG}{d} = 2121 = \underline{2100\ lb/d}$$

12.3 TYPES OF LIME-SODA ASH PROCESSES

As indicated by the chemistry of softening, different processes can be used depending on the type of hardness, the degree of hardness, and the quality of the finished water. For example, carbonate hardness or alkalinity can be removed simply by the addition of lime. However, both lime and soda ash are required for the removal of total hardness.

12.3.1 SELECTIVE CALCIUM REMOVAL

This process is used when magnesium hardness is less than 40 mg/L as $CaCO_3$. Magnesium hardness exceeding this limit will form scale. The usual process scheme is lime clarification with single-stage recarbonation followed by filtration. Because magnesium is not removed, no excess lime is needed. Soda ash may or may not be required depending on the amount of non-carbonated hardness.

Example Problem 12.3

Water with constituents shown below is to be softened by selective calcium hardness removal.

a. Calculate the lime and soda ash requirements.
b. Calculate the hardness and alkalinity in the softened water before and after recarbonation.

Given:

Cations: $Ca = 4.0$ $Mg = 1.0$ $Na = 2$ meq/L
Anions: $HCO_3 = 3.5$ $SO_4 = 4.5$ $CO_2 = 1.0$ meq/L

Solution:
Hypothetical Combinations

	Ca	Mg	Na	Σ
HCO_3	2.5	–	–	2.5
SO_4	1.5	1.0	2.0	4.5
Σ	4.0	1.0	2.0	7.0

$$CH = \frac{2.5\,meq}{L} \times \frac{50\,mg}{meq} = 125.0 = \underline{120\,mg/L}$$

$$NCH = \frac{(1.5+1.0)\,meq}{L} \times \frac{50\,mg}{meq} = 125.0 = \underline{120\,mg/L}$$

$$TH = CH + NCH = 125 + 125 = 250.0 = \underline{250\,mg/L}$$

$$CaH = \frac{(2.5+1.5)meq}{L} \times \frac{50\,mg}{meq} = 200.0 = \underline{200\,mg\,/\,L}$$

$$MgH = \frac{1.0\,meq}{L} \times \frac{50\,mg}{meq} = 50.0 = \underline{50\,mg\,/\,L}$$

Selective Calcium Removal

$$Lime = CO_2 + CaCH = (1.0+2.5) \quad meq\,/\,L$$

$$= \frac{3.5\,meq}{L} \times \frac{28\,mg}{meq} \times \frac{1}{0.85} = 115 = \underline{120\,mg\,/\,L}$$

$$Soda = CaNCH = \frac{1.5\,meq}{L} \times \frac{53\,mg}{meq} = 79.5 = \underline{80\,mg\,/\,L}$$

12.3.2 EXCESS LIME TREATMENT

To reduce magnesium hardness, excess lime is required to raise the pH above 10.6. The amount of excess lime used is typically 35 mg/L CaO (1.25 meq/L). When this treatment process is used, soda ash is used to remove non-carbonate hardness and recarbonation is used to stabilize the water. A schematic of the two-stage excess lie treatment is found in **Figure 12.1**.

Excess lime treatment can be performed in a single-stage or double-stage process. In the first stage, carbon dioxide is added to lower the pH to about 10.3 and convert the excess lime in to settleable $CaCO_3$ for removal by flocculation and sedimentation. In the second stage, further recarbonation reduces the pH to the range of 8.4–9.5 to

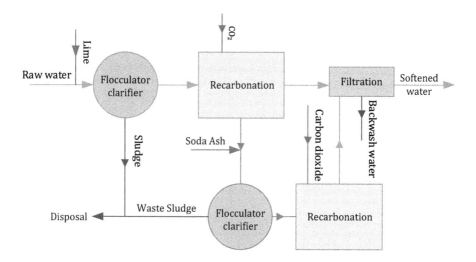

FIGURE 12.1   Excess lime softening treatment

convert the residual carbonate ion to bicarbonate ion to prevent scale formation during filtration and in the distribution of pipelines.

Example Problem 12.4

Calculate the lime (88% pure CaO) and soda ash (98% Na_2CO_3) dosage using excess lime treatment @ 0.7 meq/L of excess lime for treating water described in Example Problem 12.1.

Given:

$$CO_2 = 1.14 \text{ meq/L} \quad Alk = 4.92 \text{ meq/L} \quad Mg = 0.45 \text{ meq/L}$$

$$\text{non-carbonate hardness (NCH)} = 2.28 \text{ meq/L} \quad CH = Alk$$

Solution:

$$Lime = CO_2 + Alk + Mg + Excess = 1.14 + 4.92 + 0.45 + 0.70$$

$$= \frac{7.21 \, meq}{L} \times \frac{28 \, mg \, CaO}{meq} \times \frac{comm.}{0.88 \, pure} = 229.4 = \underline{230 \, mg/L}$$

$$Soda = \frac{2.28 \, meq}{L} \times \frac{53 \, mg}{meq} \times \frac{comm.}{0.98 \, pure} = 123.2 = \underline{120 \, mg/L}$$

12.3.3 SPLIT TREATMENT

In split treatment, only a portion of water is treated with excess lime, thus substantially reducing the lime and carbon dioxide requirement (**Figure 12.2**). Excess lime added to precipitate magnesium in the first stage reacts with bypassed water to remove calcium hardness. The quantity of flow split around the first stage depends on the hardness in the water and the desired quality of the finished water. While saving money, the reduction of magnesium to <40 mg/L is still possible.

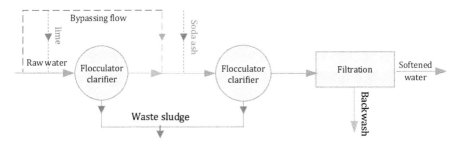

FIGURE 12.2   Split treatment softening plant

12.4   ION-EXCHANGE SOFTENING

Using zeolites, hardness is removed by replacing the bivalent ions of calcium and magnesium with sodium. Natural zeolites (boiling rock) are minerals that contain negative charge and have affinity for metal ions. Currently, synthetic zeolites (molecular sieves) are replacing natural zeolites. Since synthetic zeolites are porous, they have a greater exchange capacity. Synthetic resins are long chain polymers, microporous in nature, carrying exchangeable ions. With use, the ion-exchange capacity of the resins is exhausted. Passing brine solution through the unit regenerates the resins. During this process, Na^+ ions present in the zeolite get exchanged with Ca^{++} and Mg^{++} ions causing hardness.

$$Na_2Ze + CaCl_2 = CaZe + 2\ NaCl$$

$$Na_2Ze + MgSO_4 = MgZe + Na_2SO_4$$

The ion-exchange process is especially suitable for small points and when the major portion of the hardness is in non-carbonate form. With the ion-exchange method, water with hardness as low as 10 mg/L can be produced. An ion-exchange unit is very compact and does not require special skills to operate. During the ion-exchange operation, no precipitate or sludge is produced. However, during ion-exchange, sodium ions are added to the treated water. Water with a high sodium content is not recommended for heart patients and cannot be used in high-pressure boilers due to caustic embrittlement. This method is also not suitable for treating turbid waters as turbidity can clog pores of the zeolite. This method is not recommended when iron and manganese salts are present.

12.4.1   REMOVAL CAPACITY

Removal capacity indicates the total amount of hardness that can be removed by a unit volume of resin. Knowing the removal capacity, the exchange capacity of a softener can be calculated. It can be expressed as g or kg/m^3. Typically, the removal capacity of resins ranges from 15 to 100 kg/m^3.

12.4.2   WATER TREATMENT CAPACITY

Water treatment capacity is the volume of water that can be softened before the resin must be regenerated. Knowing the exchange capacity and the hardness of water, you can work out the water treatment capacity. Water treatment capacity indicates the volume of water that can be softened with one cycle of operation. Dividing the capacity by the average flow rate, hours or days of operation can be calculated. Pumping a concentrated brine solution of 5% to 20% onto the resin regenerates exhausted resin. The NaCl dosage required for regenerating resin ranges from 80 to 60 kg/m^3 of resin. The feed rate of the solution is typically 40 L/m^2.min (1 $gal/ft^2 \cdot min$). After regeneration, the medium should be flushed with softened water to remove excess brine.

Example Problem 12.5

An ion-exchange water softener is filled with 17.3 m³ of resin with a hardness removal capacity of 15 kg/m³. How many liters of water with a hardness of 180 mg/L can be treated before the resin is exhausted. How many kL of 1.5% brine (SG = 1.1) will be required to regenerate the resin assuming 3 kg of salt is required for every kg of hardness.

Given:

$$V = 17.3 \text{ m}^3 \quad \text{removal capacity} = 15 \text{ kg/m}^3 \quad \text{hardness} = 180 \text{ mg/L}$$

$$\text{NaCl required} = 3 \text{ kg/kg brine} = 1.5\% \text{ (SG} = 1.1) = 15 \times 1.1 \text{ g/L} = 165 \text{ g/L}$$

Solution:

$$Capacity = \frac{15 \ kg}{m^3} \times 17.3 \ m^3 = 259.6 = \underline{260 \ kg}$$

$$Volume \ of \ water = 260 \ kg \times \frac{ML}{180 \ kg} = 1.44 = \underline{1.4 \ ML}$$

$$Brine \ required = 259.6 \ kg \times \frac{3 \ kg \ salt}{kg \ hardness} \times \frac{m^3}{165 \ kg} = 4.72 = \underline{4.7 \ m^3}$$

Example Problem 12.6

A domestic ion-exchange water softener is able to remove 90% of the hardness of 5.7 m³ of water before exhaustion. It requires 120 L of 1.5% brine solution to regenerate zeolite. What is the hardness of the treated water?

Given:

$$\text{water} = 1 \quad V_1 = 5.7 \text{ m}^3 \quad \text{hardness removed } C_1 = ?$$

$$\text{brine} = 2 \quad C_2 = 1.5\% = 15 \text{ g/L} \quad V_2 = 120 \text{ L}$$

Solution:

$$C_1 = C_2 \times \frac{V_2}{V_1} = \frac{15 \ g}{L} \times \frac{eq \ of \ NaCl}{58.5 \ g} \times \frac{120 \ L}{5.7 \ m^3} = 5.39 = 5.4 \ eq/m^3$$

$$Hardness = \frac{5.39 \ eq}{m^3} \times \frac{0.10}{0.90} \times \frac{50 \ g}{eq \ of \ CaCO_3} = 29.98 = 30 \ g/m^3$$

Example Problem 12.7

An ion-exchange water softener is filled with 13.4 m³ of zeolite with a hardness removal capacity of 15 kg/m³. How much equivalent brine is required for regeneration? Assume strength of brine solution is 15% (SG = 1.1).

Given:

zeolite = 13.4 m³   removal capacity = 15 kg/m³

brine Solution C = 15% (SG = 1.1) = 150 × 1.1 g/L = 165 g/L   V = ?

Solution:

$$Capacity = \frac{15\,kg}{m^3} \times 13.4\,m^3 \times \frac{eq\ of\ CaCO_3}{50\,g} \times \frac{1000\,g}{kg} = 4020\ eq$$

$$Eq.\,brine, m = 4020\,eq \times \frac{58.5\,g}{eq\ of\ NaCl} \times \frac{kg}{1000\,g} = 235.17\,kg = 235\,kg$$

$$Volume, V = \frac{m}{C} = 235.17\,kg \times \frac{m^3}{165\,kg} = 1.425 = 1.4\,m^3$$

Discussion Questions

1. What causes hardness in water? Why is it necessary to remove hardness in water?
2. What is the relationship between hardness and alkalinity?
3. What is water softening? Name and explain various types of hardness in water.
4. Explain various water softening methods.
5. Explain lime ash softening methods and associated chemical reactions.
6. What are the advantages and disadvantages of ion-exchange water softening?
7. Under what situations is selective calcium removal recommended? Explain.
8. What is excess lime softening? Why it is necessary to add excess of lime and what happens to that in the finished water?
9. In what ways is split treatment more economical?
10. In lime-soda ash softening, solids contact units are commonly used. Explain.
11. If hard water causes problems, then why not use soft water?
12. Why is hardness in drinking water supplies not regulated? What is the preferred range of hardness in drinking water supplies?
13. If municipal water is hard, it may be unfit for some industrial uses. Explain. Give three examples of such industries.

14. Explain why the ion-exchange method is not suitable for municipal water supplies.
15. Write balanced chemical equations for lime and soda ash to remove hardness.

Practice Problems

1. A water sample contains 60 mg/L of calcium, 60 mg/L of magnesium, and 25 mg/L of sodium. What is the total hardness in mg/L as $CaCO_3$? (400 mg/L).
2. In a water sample, both Ca and Mg are tested to be 20 mg/L each. What is the hardness of the water expressed as $CaCO_3$? (130 mg/L).
3. A water sample bicarbonate component is 26 mg/L. What is the alkalinity expressed as $CaCO_3$? (21 mg/L).
4. A groundwater sample is analyzed and the hypothetical combinations are shown in the following table. Determine TH, CH, NCH, Ca-H, and Mg-H. (235, 110, 125, 185, 50 mg/L as $CaCO_3$).

	Ca	Mg	Na	Σ
HCO_3	2.2	–	–	2.2
SO_4	1.5	1.0	2.5	5.0
Σ	3.7	1.0	2.5	7.2

5. The results of the chemical analysis of a water sample are shown below. Make a hypothetical combination table and determine CH, NCH, and Mg-H. (246, 104, 38 mg/L as $CaCO_3$).
   Ca = 125 mg/L total hardness, TH = 350 mg/L as $CaCO_3$
   Alk = 300 mg/L as HCO_3 CO_2 = 25 mg/L
6. The results of analysis of a water sample are as follows:
   cations: Ca = 100 mg/L   Mg = 41 mg/L   Na = 25 mg/L
   anions: Alk = 180 mg/L SO_4 = 96 mg/L   Cl = 136 mg/L
   CO_2 = 8.8 mg/L
   Make hypothetical combinations. ($Ca(HCO_3)_2$ = 3.6, $CaCl_2$ = 0.40, $MgCl_2$ = 2.4, $MgSO_4$ = 1.0, $CaSO_4$ = 1.0, NaCl = 1.1 all in meq/L).
7. For the data of Practice Problem 6, calculate the quantity of lime as CaO with 88% purity to treat 4.0 MGD using excess lime treatment. (9200 lb/d).
8. For Practice Problem 6, calculate the quantity of 95% pure soda ash to treat 4.0 MGD of water. (8900 lb/d).
9. Calcium hardness is removed by coagulating with lime (85% CaO) @ 130 mg/L. What is the concentration of dry solids produced as $CaCO_3$ precipitate? (400 mg/L).
10. A groundwater supply has the following analysis:
    cations: Ca = 110 mg/L   Mg = 25 mg/L   Na = 12 mg/L

anions: HCO_3 = 240 mg/L SO_4 = 110 mg/L   Cl = 35 mg/L

Find the dosage of lime as CaO for selective calcium removal. (130 mg/L).

11. The analysis of a water sample (all components measured as meq/L) is as follows:

Ca	Mg	Na	K	HCO₃	SO₄	Cl
3.7	1.0	1.0	0.5	4.0	1.2	1.0

Calculate the chemical doses of lime as CaO and soda ash as Na_2CO_3 for excess lime softening. Assume 1.25 meq/L of excess lime. (175 mg/L as CaO, 37 meq/L as Na_2CO_3).

12. An ion-exchange water softener has a diameter of 3.0 m. It is filled with a resin to a depth of 1.8 m. If the removal capacity of the resin is 21 kg/m³, what is the total exchange capacity of the softener? (270 kg).

13. The meq/L concentrations of various components in a water sample are reported as follows:

Ca	Mg	Na	K	HCO₃	SO₄	Cl
3.5	0.8	0.6	0.1	4.0	0.6	0.4

Calculate the lime dosage in mg/L of $Ca(OH)_2$ required for the selective removal of calcium hardness. (130 mg/L).

14. For the data of Practice Problem 13, calculate the hardness and alkalinity of the softened water. (70 mg/L, 55 mg/L).

15. $Ca(OH)_2$ (lime slurry) reacts with calcium bicarbonate ($Ca(HCO_3)_2$) in solution to precipitate $CaCO_3$. Calculate the amount of 78% CaO required reacting with 185 mg/L of calcium hardness as $CaCO_3$. (133 mg/L).

16. A water sample was analyzed for soluble constituents. Partial results of the chemical analysis are shown below. Determine total hardness (TH), carbonate hardness (CH), and non-carbonate hardness (NCH). (300, 150, 150 mg/L as $CaCO_3$).

Constituent	Ca	Mg	HCO₃	Cl
meq/L	5	1	3	2

17. A domestic water softener is able to remove 95% of the hardness of 5.5 m³ of water before exhaustion. It requires 100 L of 1.5% brine solution to regenerate zeolite. What is the hardness of the treated water? (12 mg/L)

18. The hardness of 1000 L of a sample of water was removed by passing it through a zeolite softener. The zeolite softener then required 30 L of 1.5% sodium chloride solution to regenerate. Find the hardness of the water sample. (380 mg/L).

19. An ion-exchange water softener is filled with 15 m^3 of zeolite with a hardness removal capacity of 12 kg/m^3. How much equivalent brine is required for regeneration? The strength of the brine solution is 12% (SG = 1.1). (1.6 kL).

20. Given the following water analysis data (meq/L), determine the missing values in mg/L. (24, 23, 180 mg/L).

Ca	Mg	Na	K	HCO$_3$	SO$_4$	Cl	Alk	NCH
2.0	?	?	1.0	?	2.0	1.0	3.0	1.0

13 Miscellaneous Methods I

In addition to water treatment methods already discussed there are others which are used for specific purposes like fluoridation and iron and manganese removal. Such treatment is not part of conventional water treatment but is more specific to plants where the problem exists.

13.1 FLUORIDATION

It has been documented that fluoride in limited concentrations reduces tooth decay. Fluoride at a level of 1 mg/L can reduce the incidence of tooth decay among children by 65%. Controlled fluoridation at less than 1.5 mg/L is a safe, effective, and economical process. Mottling of the teeth occurs when the fluoride level exceeds 1.5 mg/L. Fluoride occurs naturally in water and concentrations more than 0.1% have been found in water from volcanic regions. Water with fluoride concentrations of 1.4 to 2.4 mg/L should be **defluoridated** to reduce the concentration to an optimum level of 1 mg/L. The maximum contaminant level (MCL) is 4.0 mg/L. Some facts about fluoride are as follows.

- Research has proven that fluorides dramatically reduce dental cavities in children.
- There are no harmful side effects from the use of fluorides within allowable limits.
- In excessive amounts, fluoride can cause fluorosis, a discoloration of the teeth (mottling).
- Optimum fluoride concentration in water is about 1 mg/L and concentrations greater than 2.4 mg/L are considered harmful.
- Deposits of fluoride get dissolved in the earth's water naturally.
- Some communities get drinking water with fluorides already in it. If raw water has an excess of fluoride, it needs to be defluoridated.

13.1.1 FLUORIDE CHEMICALS

The most used fluoride chemicals in the water industry are sodium fluoride (NaF), sodium silicofluoride (Na_2SiF_6), and hydrofluosilicic acid (H_2SiF_6). All these chemicals are refined from minerals found in nature and yield fluoride ions, which dissolve in water. A summary of fluoride chemicals is given in **Table 13.1**. When selecting a fluoridation chemical, safety, ease of handling, and cost must be given serious consideration. Hydrofluosilicic acid in liquid form is easy to feed and is popular with small and large plants. However, acid produces toxic fumes that must be vented. Large water works use gravimetric dry feeders to apply sodium silicofluoride that is commercially available in various gradations.

DOI: 10.1201/9781003347941-15

TABLE 13.1
Common fluoride compounds

Item	Sodium Fluoride	Sodium Silicofluoride	Fluosilicic Acid
chemical formula	NaF	Na_2SiF_6	H_2SiF_6
commercial form	powder or crystal	fine crystal powder	liquid
molecular mass, g/mol	42	188	144
commercial purity, %	90–98	98–99	22–30
fluoride ion, %	45	61	79
kg /ML for dosing @ 1.0 ppm F at indicated purity	2.26 (98%)	1.67 (98.5%)	4.21 (30%)
pH of saturated solution	7.6	3.5	1.2 (1.0%)
sodium ion contributed at 1.0 ppm F, ppm	1.17	0.40	0.00
solubility at 25°C, g/L of water	41	7.6	liquid
specific gravity	1–1.4	0.9–1.2	1.3

Example Problem 13.1

Determine the fluoride content of 30% pure commercial fluosilicic acid, H_2SiF_6.

Given:

$$H = 1 \text{ g/mol} \quad Si = 28 \text{ g/mol} \quad F = 19 \text{ g/mol} \quad Purity = 30\%$$

Solution:

$$H_2SiF_6 = 2\times1 + 28 + 6\times19 = 144\,g\,/\,mol$$

$$F = \frac{114\,g\,F}{144\,g\,H_2SiF_6} \times \frac{30\%}{100\%} = 0.2375 = 0.24 = \underline{24\%}$$

13.1.2 FLUORIDATION SYSTEMS

Depending on the natural fluoride level in the source water, there may be four different fluoridation systems.

1. When the natural fluoride level is zero and all the fluoride has to be added to the water supply.
2. Raw water source may have adequate or excessive ions. When water contains excessive fluoride, it must be defluoridated.
3. When the natural fluoride level is less than the optimum, a situation more commonly found, fluoride ions are added to bring the total to the desired level.

4. When there are two sources of water supply with different levels of fluoride, the degree of blending must be considered in selecting the fluoride dosage.
5. Both overdosing and underdosing are undesirable.

Chemical Feeding

The equipment used for fluoridation is like that used for feeding other water treatment chemicals. Fluoride ions can be added to water either by solution feeders or dry feeders. Solution feeders are most economical for small water systems. Whatever the type of feeding system may be, it is very important to maintain accurate feeding and prevent overfeeding and siphonage. A given feeding system should provide a means of measuring the back-fluoride level in the finished water. It is preferred to add fluoride after filtration to avoid any losses that may occur because of reactions with other chemicals, notably alum and lime.

Example Problem 13.2

A liquid feeder applies a 4.0% saturated sodium fluoride solution to treat a flow of 6.0 ML/d with a fluoride dosage of 1.0 mg/L. What must be the feeder pump setting (L/d) to feed at the desired dosage?

Given:

Parameter	Solution Fed	Water Treated
C	4.0% NaF	1.0 mg/L
Q	?	6.0 ML/d

Solution:

Solution conc. As F- ion

$$C_1 = \frac{40\,g\,NaF}{L} \times \frac{0.45\,F}{NaF} = 18\,g/L\,as\,F$$

Feed pump rate

$$Q_1 = \frac{C_2}{C_1} \times Q_2 = \frac{1\,kg}{ML} \times \frac{L}{18\,g} \times \frac{6.0\,ML}{d} \times \frac{1000\,g}{kg} = 333 = \underline{330\,L/d}$$

Example Problem 13.3

The hydrofluosilicic acid is pumped from a shipping drum placed on a platform scale. The recorded weight loss of the drum is 160 lb in processing 4.2 mil gal of water. Calculate the fluoride dosage applied.

Given:

$$m = 160\,lb\,of\,H_2SiF_6 \quad V = 4.2\,mil\,gal \quad C = ?$$

Solution:

$$C = \frac{m}{V} = \frac{160\,lb}{4.2\,MG} \times \frac{0.24\,lb\,F}{lb\,H_2SiF_6} \times \frac{mg/L}{8.34\,lb/MG} = 1.09 = \underline{1.1\,mg/L}$$

Example Problem 13.4

A flow of 8.5 ML/d is to be treated with a 20% solution of hydrofluosilicic acid, with a fluoride content of 79%. The relative density (SG) of the acid is known to be 1.2. Calculate the feed pump rate in L/h to apply 1.5 mg/L dose of fluoride.

Given:

Parameter	Solution Fed = 1	Water Treated = 2
C	20%, 79% F	15 mg/L
SG	1.2	10
Q	?	8.5 ML/d

Solution:

Acid F-ion concentration

$$C = \frac{20\%}{100\%} \times \frac{0.79\,F}{acid} \times \frac{1.2\,kg}{L} \times \frac{1000\,g}{kg} = 189.6 = 190\,g/L$$

Acid feed pump rate

$$Q_1 = \frac{C_2}{C_1} \times Q_2 = \frac{1.5\,kg}{ML} \times \frac{L}{189.6\,g} \times \frac{1000\,g}{kg} \times \frac{8.5\,ML}{d} \times \frac{d}{24\,h} = \underline{2.8\,L/h}$$

13.2 DEFLUORIDATION

Though a fluoride dose at the optimum level prevents dental caries in children, excessive fluoride in drinking water supplies can cause dental fluorosis (mottling of teeth) and hardening of arteries and bones in older people. The maximum contaminant level of fluoride is 4.0 mg/L and water with fluoride concentrations in the range of 1.4–2.4 mg/L should be defluoridated to bring it up to optimum level. The following are the commonly used chemical treatments to reduce fluoride in drinking water.

13.2.1 CALCIUM PHOSPHATE

Bone has great affinity for fluoride and can be used in the filter media to remove fluoride. The bones are calcinated at high temperatures followed by mineral treatment. It is then pulverized to 40–60 mesh and is used in the filter bed. When exhausted, the filter is regenerated with alkali and acid.

13.2.2 Tri-Calcium Phosphate

Bone charcoal is basically a mixture of tri-calcium phosphate and carbon. This material is very successfully used to reduce fluoride content. Calcium triphosphate can also be synthesized by adding phosphoric acid to lime. This chemical has been used in contact filters for the removal of fluoride. The regeneration of the exhausted calcium phosphate is done by adding 1.0% caustic solution followed by dilute hydrochloric acid or carbon dioxide to neutralize any excess of caustic.

Fluorex is a trade name. This is a special mixture of tri-calcium phosphate and hydroxyapatite and can be used as filter media. It can be regenerated by using caustic solution, followed by water rinse, and the excess of caustic soda can be neutralized with carbon dioxide (carbonic acid).

13.2.3 Ion Exchange

The ion exchange method can also be used for the removal of fluoride – for example with a cation exchanger of the sulphonated coal type and an amin resin. Alum-treated cation exchange resin from Avaram bark can be used as an effective material for fluoride removal.

13.2.4 Lime

During the softening process, fluoride can be removed along with magnesium. However, due to residual caustic alkalinity, it must be followed by recarbonation. This process is applicable when treating hard water with fluoride concentrations less than 4.0 mg/L.

13.2.5 Aluminum Compounds

Alum has a high absorption capacity for fluoride, which can be further enhanced with coagulant aids such as activated silica and clays. Another method utilizing aluminum salts involves the use of contact beds of insoluble materials impregnated with aluminum compounds. Dehydrated aluminum oxide can be used in contact beds.

13.2.6 Activated Carbon

Activated carbon has very high absorptive properties that can be used to remove fluoride. This removal takes place at a low pH. At pH 8.0 or above, no removal takes place. Exhausted carbon can be regenerated with a weak acid and alkaline solution.

13.3 IRON AND MANGANESE CONTROL

Iron and manganese are frequently found in groundwater and in some surface waters. They do not cause health problems but are objectional because they may cause aesthetic problems. Severe aesthetic problems may cause consumers to avoid a safe water supply in favor of water of questionable quality. Both elements are more commonly found in well water supplies and occasionally in the anaerobic bottom waters of deep lakes. Due to a lack of molecular oxygen, these elements exist in their reduced ferrous or manganous states. In supply reservoirs, dissolved oxygen is also low, particularly during the winter when reservoir surfaces are covered with ice and snow and aeration cannot occur.

PROBLEMS DUE TO HIGH FE, MN

In the soluble state, these minerals are colorless in the concentrations generally found in water supplies. However, when in contact with air during a pumping or aeration process the iron and manganese are oxidized and converted to insoluble forms. In concentrations as low as 0.3 mg/L, these compounds will be noticeable and will stain clothes and fixtures. In addition, a high content of iron and manganese may add a bitter taste to water and stimulate growth of nuisance microorganisms.

Perhaps the most troublesome consequence of iron and manganese in water is that they promote the growth of a group of microorganisms known as **iron bacteria**. Esthetic indications of the presence of iron bacteria include a white or reddish-brown slime on well pipes, sudden iron staining, and sometimes taste and odor problems. When the iron content of water is high, the tannic acid in tea or coffee may combine with the iron to darken the beverage like ink.

Iron and manganese in water can be detected easily by observing the color of the inside walls of the filters and the filter media. If the raw water is pre-chlorinated, there will be black stains on the walls below the water level and a black coating over the top portion of the sand filter bed. The black color usually indicates a high level of manganese while a brownish-black stain develops when water is high in both iron (Fe) and manganese (Mn). The generally acceptable limits for Fe and Mn are 0.3 mg/L and 0.05 mg/L respectively. Even if they were available in beneficial amounts, their presence is objectionable. Iron bacteria can pit and corrode pipe systems and lower the pH of water

13.4   CONTROL METHODS

Several methods are available to control iron and manganese in water. Preventive measures are successful when the concentration is low. However, for large concentrations, the iron and manganese must be removed by precipitation followed by filtration.

13.4.1   PHOSPHATE TREATMENT

Sequestering with phosphate may be a simple solution when water contains up to 0.3 mg/L of manganese and less than 0.1 mg/L of iron. Chlorine must usually be fed along with the polyphosphate to prevent the growth of iron bacteria. The chlorine dose for phosphate treatment should be sufficient to produce a free-chlorine residual of 0.25 mg/L after 5 min contact time. A minimum of 0.2 mg/L free-chlorine residual should be maintained throughout the system.

13.4.2 FEED SYSTEM

In a well water supply system, the polyphosphate and chlorine are fed through a polyethylene hose discharging below the suction bowls of the pump. Polyphosphate solutions stronger than 0.5 g/L are very viscous. Stale solutions (>48 hours) are not good because polyphosphates are hydrolyzed to form orthophosphates, which are less effective.

Polyphosphate treatment to control iron and manganese is usually most effective when the polyphosphate is added before the chlorine. They can also be fed together, but chlorine should never be fed before polyphosphate because chlorine will oxidize iron and manganese.

13.5 REMOVAL METHODS

The main techniques used for the removal of iron and manganese include oxidation followed by filtration. In other cases, both these elements can be removed by ion exchange with zeolites.

13.5.1 OXIDATION BY AERATION

Aerating the water to form insoluble ferric hydroxide can oxidize iron. Aeration followed by sedimentation and filtration or filtration alone is used to remove the precipitates. The oxidation by plain aeration is accelerated by an increase in pH. Lime is generally added to raise the pH. If the water contains any organic substances, the oxidation reaction is slower.

$$4Fe(HCO_3)_2 + 10H_2O + O_2 \rightarrow 4Fe(OH)_3 \downarrow + 8H_2CO_3$$

Plain aeration is not effective in oxidizing manganese. Hence this method is not suitable for treating water with high manganese concentrations. An advantage of this method is that no chemicals are used, except lime or soda ash, which may sometimes be used to raise pH.

13.5.2 OXIDATION WITH CHLORINE

Chlorine is a strong oxidant. Maintaining free chlorine residual throughout the treatment process can easily oxidize iron and manganese. The higher the chlorine residual, the faster the oxidation occurs.

$$2Fe(HCO_3)_2 + Cl_2 + Ca(HCO_3)_2 \rightarrow 2Fe(OH)_3 \downarrow + CaCl_2 + 6CO_2$$

This reaction works better in the neutral range of pH. Removing iron requires 0.64 mg/L of chlorine. The reaction with manganese works similarly. To completely oxidize manganese, 1.3 mg/L of free chlorine is required per mg/l of manganese. In some small plants the water is dosed to maintain a chlorine residual of 5–10 mg/L, filtered, and dechlorinated by using sodium bisulphite or other reducing agents. When the water is high in hardness, the calcium carbonate tends to form a coating on the valves in the solution feeder. To avoid this, use softened water to dilute the commercially available hypochlorite solution.

13.5.3 OXIDATION WITH PERMANGANATE

Much like chlorine, potassium permanganate ($KMnO_4$) is a strong oxidizing agent. For some water, $KMnO_4$ oxidation may be advantageous since pH adjustments are not required. The specific dosage required depends on the concentration of metal ions, pH, mixing conditions, and other factors. Bench tests should be performed to determine the optimum dosage. A dose which is too small will not oxidize all the manganese in the water and a dose which is too large will leave residual $KMnO_4$ and may produce a pink color.

$$Fe(HCO_3)_2 + KMnO_4 \rightarrow Fe(OH)_3 \downarrow + MnO_2$$

$$Mn(HCO_3)_2 + KMnO_4 \rightarrow MnO_2$$

Stoichiometric requirements

$$\frac{KMnO_4}{Fe} = \frac{1\,mol \times 158\,g\,/\,mol}{3\,mol \times 55.6\,g\,/\,mol} = 0.95$$

Potassium permanganate is a dark purple crystal or powder available commercially at 97.99% purity. Theoretically, the dosages of $KMnO_4$ required for oxidizing each 1 mg/L of iron and manganese respectively are 0.95 and 1.92 mg/L. In practice, the amount needed is often less than this. Effective filtration following chemical oxidation is essential. This method of filtration is effective when Fe and Mn are under 1.0 mg/L.

13.5.4 ION EXCHANGE WITH ZEOLITES

The continuous generation greensand filter process is another commonly used filtration technique for the removal of iron and manganese. Manganese zeolites or manganese greensand is a granular material. Greensand is coated with manganese dioxide that removes soluble iron and manganese. The greensand also acts like a filter media. After the zeolite becomes saturated with Fe and Mn oxides, it is regenerated using $KMnO_4$ to remove the insoluble oxides. In the continuous flow system, $KMnO_4$ solution is fed into the water. Water containing $KMnO_4$ passes through a pressure filter that contains a dual media anthracite and manganese zeolite bed. The upper filter layer removes the metal oxides formed by the chemical reaction with KMnO4.

The underlying zeolite layer captures any metal ions not oxidized. Any excess amount of $KMnO_4$ applied is used to regenerate the greensand. If too much $KMnO_4$ is applied, the effluent may be colored pink. A good operating zeolite system can remove 95% of both iron and manganese. However, if the iron content exceeds 20 mg/L, the efficiency of the process drops quickly. A residual of $KMnO_4$ must be present in the effluent water from the greensand for the zeolite to be effective.

Example Problem 13.5

Calculate the theoretical $KMnO_4$ dosage in mg/L to treat well water with 2.6 mg/L of iron before aeration and 0.3 mg/L after aeration. The manganese concentration is 0.8 mg/L before and after aeration.

Given:

$$Fe = 0.30 \text{ mg/L} \quad Mn = 0.80 \text{ mg/L} \quad KMnO_4 = ?$$

Solution:

$$C = 0.95Fe + 1.92Mn = (0.95 \times 0.3 + 1.92 \times 0.8) = 1.82 = \underline{1.8\, mg\,/\,L}$$

Actual dosage required may be less than 1.8 mg/L. Better to check by bench testing.

Example Problem 13.6

A 2.5% solution of potassium permanganate is fed into a manganese zeolite pressure filter. The desired dosage is 2.0 ppm. What should the feed pump rate be in gal/h if the water is being pumped at rate of 230 gpm?

Given:

$$KMnO_4 \text{ Solution} = 1 \quad \text{well water} = 2$$
$$Q_1 = ? \text{ gal/h} \qquad Q_2 = 230 \text{ gpm}$$
$$C_1 = 2.5\% = 2.5 \text{ pph} \quad C_2 = 2.0 \text{ ppm}$$

Solution

$$Q_1 = \frac{C_2}{C_1} \times Q_2 = \frac{2.0}{10^6} \times \frac{100}{2.5} \times \frac{230\,gal}{min} \times \frac{60\,min}{h} = 1.10 = 1.1 \ gal\,/\,h$$

13.6   ARSENIC REMOVAL

Arsenic is a naturally occurring contaminant in some water supplies, particularly groundwater where water is in contact with arsenic-bearing material. It may also be released to the aquatic environment by mining and smelting operations. Arsenic is a contaminant of concern near waste remediation sites. The maximum permissible limit in water supplies is 10 µg/L. Water supplies exceeding this limit can use a variety of methods to reduce the level of arsenic in their supplies. For most people, the most significant exposure to arsenic is through food. Although found in some surface waters, it is mainly found in groundwater supplies. In soluble form, arsenic primarily occurs as arsenite (AsO_3) and arsenates (AsO_4).

The process of precipitation has been the most frequently used method for the removal of arsenic. Proper coagulants can transform soluble arsenic into insoluble arsenic, which then can be removed by sedimentation and filtration. However, inorganic sludge thus produced would contain arsenic, so sludge must be treated properly before disposal.

Another effective way to reduce arsenic is by adsorption on activated carbon. After pH adjustment of the water to 5.5, a granular media of activated alumina can provide the efficient removal of inorganic arsenic, which is the more predominant form in water. Without pH adjustment, arsenic removal will be relatively low.

Other treatment options for arsenic removal include ion exchange and membrane filtration. Soluble forms of arsenic can be removed by strong anion exchange resins. Membrane filtration is discussed in the next chapter.

13.7 NITRATE REMOVAL

The maximum limit for nitrate in drinking water is 45 mg/L. Excessive amounts of nitrates are damaging to babies. In some groundwaters, nitrates can be excessive and need to be removed. One common method employed for nitrate removal is anion exchange, much like cation exchange in the case of iron and manganese. In anion exchange, selective nitrate resins are used. These resins are called selective because they are less selective for multivalent anions like sulphates. Sodium chloride or brine is used in the regeneration of both ordinary and selective resins.

$$\overset{\text{Nitrate removal}}{RCl + NO_3^- \quad \Rightarrow \quad RNO_3 + Cl^-}$$

Resins beds are susceptible to plugging caused by hardness, particulates, and bacterial growth that results in poor removal. To prevent this, water fed to a nitrate exchange unit must be softened to remove excessive hardness. If this is not done, backwashing will not be complete since metal oxides and other contaminants have greater resistance to fluidization.

Discussion Questions

1. What is fluoridation and explain when it is beneficial?
2. What are the chemicals used to add fluoride in potable water?
3. What are the problems associated with excess fluoride in drinking water?
4. Define defluoridation. What are the methods of defluoridation to bring fluoride to a safe level?

5. Briefly explain various methods of the removal of excess iron and manganese.
6. What problems are caused by excess of iron and manganese in potable water?
7. What problems are associated with the growth of iron bacteria?
8. The most common process scheme for the removal of iron and manganese from groundwater is aeration, chemical oxidation, and filtration.
   a. What is the purpose of aeration?
   b. Name the oxidation chemicals used.
   c. Why is the majority of iron oxide and manganese dioxide removed in the filter rather than during sedimentation?
9. What is the health concern associated with excessive nitrate in municipal water supplies? What is the common method used for removal of excessive nitrates?
10. Arsenic in water supplies is not desirable. What at are the common methods for its removal (a) in larger plants and (b) in smaller plants?

Practice Problems

1. What should be the feed pump setting when the well pump capacity is 2.0 m^3/min and the well pump is operated 12 h/d? The feed solution is 10% as fluoride and the desired dosage of fluoride is 0.80 mg/L. (12 L/d).
2. How many liters of diluted fluosilicic acid with 10% available fluoride must be added to 1 ML of water to achieve a fluoride dosage of 1 mg/L? (10 L).
3. A liquid feeder applies 25 gal of 4% saturated NaF solution to treat 0.5 MG of water. What is the dosage of fluoride? (0.9 mg/L).
4. Hydrofluosilicic acid containing 30% F-ion (SG = 1.26) is fed at the rate of 100 mL/min in a water plant processing water @ 48 ML/d. What is the fluoride dosage? (1.1 mg/L).
5. What volume of 25% hydrofluosilicic acid having a density of 1.2 kg/L is required to treat 1.5 ML of water at a fluoride dosage of 1 mg/L? (6.3 L).
6. Determine the quantity of $KMnO_4$ required per week to remove 0.50 mg/L of iron while treating a flow of 10 ML/d assuming removal efficiency of 95%. (35 kg/wk).
7. What is the theoretical dose of $KMnO_4$ to treat well water containing 1.5 mg/L of iron and 0.5 mg/L of manganese? (2.4 mg/L).
8. A well is being pumped at the rate of 340 gpm. How many gallons of 5% $KMnO_4$ solution should be fed per hour to provide a dosage of 2.5 mg/L? (1.0 gal/h).
9. How many mL of 0.1% polyphosphate standard solution should be added to dose 1.5 L of a water sample at the rate of 2.0 mg/L? (3.0 mL).
10. Hydrofluosilicic acid (30%) is pumped from a shipping drum placed on a platform scale. The recorded weight loss of the drum is 162 lb in processing 4.0 MG of water. What is the fluoride dosage applied? (1.2 mg/L).

11. A flow of 8.5 MGD is to be treated with a 20% solution of hydrofluosilicic acid with a fluoride content of 79%. The relative density of the acid is known to be 1.2. Calculate the feed rate in gal/h to apply 1.5 mg/L of fluoride. (3.4 gal/h)

12. What is the fluoride concentration of 25% H_2SiF_6, SG = 1.2? (240 g/L).

13. Calculate kg/d of commercial sodium silicofluoride to be fed if the natural level of fluoride in the water is 0.2 mg/L and the flow to be treated is 56 L/s. The level of fluoride desired is 1.0 mg/L. (6.5 kg/d).

14. A Na_2SiF_6 solution is prepared by dissolving 5.0 kg of 98% pure commercial salt to make 100 L of solution. Calculate the feed rate of this solution in L/ML of water flow to increase the fluoride content of water by 0.9 mg/L. (30 L/ML).

15. A flow of 24 ML/d is to be treated with a 20% solution of hydrofluosilicic acid with 79% F-content and SG of 1.2. What should be the acid feed rate of the acid in mL/min to increase the fluoride content of water from 0.2 to 1.0 mg/L? (70 mL/min).

16. A flow of 0.50 MGD is treated with 10 lb/d of commercial salt of NaF. The commercial purity is 98% and fluoride content is 45%. Determine the fluoride dosage. (1.1 mg/L).

17. What dosage of $KMnO_4$ is needed to treat city well water with 3.5 mg/L iron before aeration and 0.2 mg/L after aeration? The manganese content of 1.2 mg/L remains the same before and after aeration. (2.5 mg/L).

18. What dosage of $KMnO_4$ is required to oxidize 5.5 mg/L of iron and 1.2 mg/L of manganese? (7.5 mg/L)

19. 0.675 grams of polyphosphate are weighed out. What volume of solution should be made by adding water such that the solution concentration is 0.1% or 1 g/L. How many mL of this solution should be added to 1.5 L of water sample to dose water at the rate of 4.0 mg/L. Also calculate how many liters of this solution will be required to dose 1 m³ of water at the rate 2.5 mg/L. (675 mL, 6 mL, 2.5 L).

20. Polyphosphate is fed to sequester iron and manganese. What is the required dosage rate of polyphosphate to treat a flow of 0.6 MGD with a dosage of 3.5 mg/L.? If the polyphosphate solution strength is 2.0%, what will be the daily use of the solution in gallons? (18 lb/d, 110 gpd).

14 Miscellaneous Methods II

A major purpose of water treatment is to produce palatable and aesthetically appealing water. The importance of supplying odor-free water cannot be overemphasized. It should be clear, good tasting, and free of any objectionable odors.

14.1 TASTE AND ODOR CONTROL

Consumers can judge the quality of water by what they can see, smell, and taste. People will refuse to drink water if it is turbid or if it has an unpleasant taste or odor. On the other hand, they will happily drink toxin-loaded water if it looks clean and is odor free.

Those who have some experience working at a water plant very well know that most complaints relate to taste and odor problems in the water supply. To add to this, any such problems are more likely to become headlines in the local newspapers. All this may be ludicrous, but a good plant operator would be foolish to ignore it. Any neglect on the part of plant personnel may erode public confidence and force them to seek alternative sources of drinking water. These alternative sources including bottled water and water from private wells that may not be safe. Being a water professional you need to pay serious attention to consumer complaints, especially complaints related to taste and odor. At no cost should you lose public confidence regarding the safety and quality of water.

Adsorption has been used for years to remove organic-causing taste, odor, and color problems. This process has added importance since it also removes toxic and carcinogenic organics from drinking water.

14.1.1 ORGANICS IN RAW WATER

Organic substances like humic and fulvic acid are produced naturally because of decaying vegetation. For this reason, surface water usually contains greater concentrations and color than groundwater. In addition to humic substances, a great variety of organics, mainly man-made, are introduced when domestic, agricultural, and industrial wastes are discharged into received water bodies.

The presence of organics in water can interfere with other treatment processes such as coagulation and flocculation. During chlorination, THMs are formed when chlorine reacts with organics. Because of this, the practice of pre-chlorination is becoming unpopular and alternative disinfectants, including ozone and chlorine dioxide, are gaining in popularity.

DOI: 10.1201/9781003347941-16

The man-made organics include solvents, hydrocarbons, cleaning compounds, pesticides, and herbicides, many of which are toxic to humans. The best solution is the prevention of such substances entering the water systems. Emphasis should be given to watershed management and the implementation of sewer use control programs. In USA and Canada, source water protection programs have been initiated to protect source waters.

Algae are the most common cause of taste and odor problems. Their metabolic activities produce odorous compounds. In a eutrophic lake or reservoir, regular copper sulphate applications to the impounded water are effective in controlling algae blooms.

14.1.2 CHEMICAL DOSING

The chemical dose of copper sulphate pentahydrate ($CuSO_4.5HO_2$) partly depends on the characteristics including alkalinity, turbidity, and the temperature. At high levels of alkalinity, citric acid may have to be added to prevent precipitation of copper. The typical dosage of copper is 2.0 mg/L. When treating large reservoirs, sometimes only a depth of perhaps 6 m (20 ft) or a depth down to the thermocline may be used in the calculation of the volume of water to be treated. The desired copper sulphate dosage may also be specified as kg/ha or lb/acre. This format is generally used for water with relatively high alkalinity. Under such conditions, algae control is limited to the upper volume of water due to interference of other ions. When using such procedures, the phenomenon of die off causing odor problems must be considered. In many cases, it may be better to remove the microorganisms by plant processes rather than the chemical dosing of lakes and reservoirs.

14.2 TASTE AND ODOR REMOVAL

Taste- and odor-causing organic compounds can be removed by chemical oxidation or by absorption using activated carbon or synthetic resins.

14.2.1 OXIDATION

Some taste- and odor-causing compounds can be removed by using chlorine and potassium permanganate to oxidize the compounds by aeration (oxidation and air-stripping), by coagulation/sedimentation, and by filtration. As discussed earlier, heavy chlorination should be avoided because of THM formation.

14.2.2 AERATION

Aeration brings water into contact with air and allows volatile compounds to escape into the atmosphere. This method of the removal of volatile compounds is also called **degasification**. Aeration is also responsible for oxidation. However, oxidation by aeration is more effective for treating inorganic compounds rather than organic compounds. Therefore, for taste and odor problems, aeration can be considered a physical rather than a chemical process.

Aeration can be achieved by passing diffused air into water or cascading water through the air. A process called air stripping does both. In chemical oxidation, aeration may help prepare the water for treatment thus reducing chemical dosages required for oxidation.

14.2.3 CHEMICAL OXIDATION

Chlorine, potassium permanganate, and chlorine dioxide are some of the oxidizing chemicals commonly employed for taste and odor control. Ozone is another strong oxidant. When chlorination is used for taste and odor control, dosages are adjusted. Higher dosages than those required for disinfection are usually needed. It is recommended to add chlorine in the early stages of treatment to obtain the best results. When super chlorination is used for odor control, water must be dechlorinated to adjust the residuals. When water contains phenols or similar compounds, chlorine may be the wrong choice for taste and odor control since it will impart phenolic taste to water.

Potassium permanganate has been used very successfully for taste and odor control. This is a strong chemical oxidant and produces a purple color when mixed with water. After oxidation, the color changes from purple to yellow or brown. Any permanganate that has not reacted or reduced will result in purple finished water. Permanganate should be added to water as early as possible to maximize the contact time. Any excess of permanganate dosage can be treated with activated carbon.

Dosage of permanganate for odor control is typically in the range of 1–3 mg/L. Since permanganate is an expensive chemical, a cost comparison with alternative methods must be undertaken before making the final choice. It is very important to note that permanganate must never be stored in the vicinity of activated carbon. The two make a combustive mixture. Like most other chemicals, permanganate can be metered using chemical feed pumps. Or dry feeders can be used to apply the solid crystals to water. Since permanganate is very corrosive, provision for dust control must be made.

Chlorine dioxide and ozone are the other two disinfectants which can be used for taste and odor control. Chlorine dioxide is gaining popularity especially where phenols are the major cause of odor problems and there is danger of the formation of THMs. Due to the instability of chlorine dioxide, it must be generated on site and applied immediately. Ozone decreases the formation of THMs and is a strong oxidizing agent acting as a disinfectant as well as deodorant.

Example Problem 14.1

The volume of a reservoir is estimated to be 18 ML. The desired dose of copper is 0.5 mg/L. How many kg of copper sulphate will be needed?

Given:

$$C = 0.50 \text{ mg/L as Cu} \qquad V = 18 \text{ ML}$$

Solution:

Chemical CuSO₄

$$Molar\ mass = (63.5 + 32 + 4 \times 16 + 5 \times 18) = 249.5 = \underline{250\,g\,/\,mol}$$

$$Copper\ content = \frac{63.5\,g\,/\,mol}{249.5\,g\,/\,mol} \times 100\% = 25.4 = \underline{25\%}$$

$$Quantity\ required = 18\,ML \times \frac{0.50\,kg}{ML} \times \frac{100\%\ salt}{25.4\%} = 225 = \underline{230\,kg}$$

14.2.4 ADSORPTION

In adsorption, the organic compounds adhere to the surface of the adsorbing media, such as activated carbon. For adsorption to be effective, the adsorbent must provide an extremely large surface for the trapping of the organics. Activated carbon is an excellent adsorbent due to its porous structure and affinity for organics. These pores are created during the manufacturing process by steaming the carbon at temperatures as high as 800 C or 1500°F. This is known as activation. Each carbon particle consists of many small- and large-size pores that adsorb the odorous compounds. The exhausted carbon must then be replaced or reactivated by essentially the same process as activation.

14.2.5 FORMS OF ACTIVATED CARBON

Although activated carbon can be made from a variety of materials, such as wood, nutshells, peat, and petroleum residue, the activated carbon used in water treatment is usually made from bituminous or lignite coal. The two common forms of activated carbon are powder and granular. A comparison of the two is shown in **Table 14.1**.

Powder Activated Carbon

Powder activated carbon (PAC) is preferred for use in small plants and for controlling periodic taste and odor problems. Powdered carbon can be fed dry or as a slurry. Treatment plants that consistently use PAC use the slurry method.

TABLE 14.1
Comparing PAC with GAC

Property	PAC	GAC
particle diameter, mm	< 0.1	1.2–1.6
specific gravity	0.32–0.72	0.42–0.48
surface area, m²/g	500–600	650–1150

Point of Application

PAC is usually added ahead of coagulation to ensure no residual carbon enters the distribution system. The carbon added as powder ends up as sludge in the sedimentation basin and backwash from filters. It is not practicable to reuse the carbon.

Granular Activated Carbon

Granular activated carbon (GAC) is typically used when carbon is needed continuously to remove organics. Rather than feeding as in the case of PAC, water to be treated is passed through a bed of GAC. The bed can be part of the media of the conventional filter or be in separate units called **contactors**.

Conventional Filter

Partial or complete replacement of the filter media with GAC is economical and suitable for most situations. Granular carbon, being the same density as anthracite, is an equally effective filtering material. The minimum recommended depth of GAC media is 600 mm.

Contactors

Contactors are essentially pressure filters containing deep columns of GAC. Contactors, though expensive, have the following advantages.

- They provide longer contact time and longer filter runs and bed life.
- GAC is easier to replace and its operation has greater flexibility.
- Because contactors are primarily for adsorption and not filtration, they are placed after regular filtration.

14.3   MEMBRANE FILTRATION

Membrane filtration has recently become more common. First, the technology has improved and second, because of its smaller footprint, the capacity of a plant can be increased when the space is limited for expansion. **Membrane filtration** is a process in which hydrostatic pressure is applied to force the water through semipermeable membranes to filter out the contaminants. To avoid premature plugging and fouling of membranes, feed water must be treated adequately by conventional methods. There are four basic type of membrane filtration processes, including microfiltration (MF), ultrafiltration (UF), nanofiltration (NF), and reverse osmosis (RO).

14.3.1   MICROFILTRATION AND ULTRAFILTRATION

The terms "micro" and "ultra" indicate the size of particles that can be removed. MF membranes, generally considered to have an average size of 0.1 μm, can filter out suspended solids and some bacteria. UF membranes, which have pores of a smaller size, typically 0.01 μm, can remove colloidal particles, most bacteria, viruses, and large molecules. Membranes in these two cases are usually made of a hollow plastic fiber material. Several such tubes are wrapped together in a fiberglass tube, several of which are assembled in the form of modules. UF and MF membrane filters are operated at feed pressure of 700 kPa (100 psi) or less. Membranes systems are normally completely automated to control feed rate and frequency of backwashing.

14.3.2 NANOFILTRATION AND REVERSE OSMOSIS

Nanofiltration (NF) and reverse osmosis RO systems can remove small molecules and dissolved solids like those causing color and hardness. Pore sizes in NF membranes are on average I nm and the RO process has even smaller pore sizes than this. In these processes, semi-permeable membranes are used. Because of the smaller pore sizes, the operating pressure for reverse osmosis is in the range of 1–5 MPa (140–700 psi), depending on the recovery of product water. RO membranes are more susceptible to plugging, scaling, and fouling by dissolved compounds, chemical precipitates, and bacterial growths.

14.4 DESALINATION

Water with high level of dissolved minerals or salts is unfit for domestic, industrial, or agricultural use. Whereas brackish water salt concentration is typically 1.0 g/L or 0.1%, seawater has 3 to 4 times that concentration. To make these waters fit for various uses, salt need to be removed by a process called **desalination.** Reducing the salt content to less than 500 mg/L or less makes it potable.

14.4.1 MEMBRANE TECHNOLOGY

Reverse osmosis and **electrodialysis** are two membrane treatment processes used for desalination. However, electrodialysis is limited to treating brackish water. In electrodialysis, a voltage rather than pressure is applied across the salty water, causing ions to migrate towards electrodes of opposite charge. Plastic membranes that are selectively permeable to either cations or anions are used to separate fresh water from salty water.

In a reverse osmosis process, a semipermeable membrane separates salty water of two different concentrations. The normal osmotic flow of water through a semipermeable membrane allows the low-salt concentration water to pass though the membrane to dilute saline water of a higher concentration. The pressure created due to the difference in concentrations is called **osmotic pressure**. Reverse osmosis is the forced passage of water through a semipermeable membrane against natural osmotic pressure, hence the name reverse osmosis. The osmotic pressure of seawater is about 1 MPa (140 psi). However, brackish water having a lower concentration of salts has significantly lower osmotic pressure. A basic reverse osmosis system consists of pre-treatment units, high-pressure pumps, post treatment tanks and appurtenances for cleaning and flushing, and a disposal system for reject brine.

14.4.2 DISTILLATION OF SEAWATER

Multistage flash distillation is the most common process used for desalination. In distillation, fresh water is separated from seawater by heating, evaporation, and condensation, much like nature separates fresh water from the ocean to fall as rain. Water starts boiling at a lower temperature if the air pressure is lowered. This fact is used in flash distillation to save energy. This process is carried out in a series of closed vessels, set progressively at lower pressures. When preheated water enters a vessel that is at low pressure, some of that water rapidly boils into vapor. The water vapors are condensed into fresh water in heat-exchange tubes. The remaining saltwater flows to the next stage set at an even lower pressure. Some facilities can have as many as 40 stages.

14.5 WATER STABILIZATION

One of the several objectives of water treatment is the production of stabilized water that is neither corrosive nor scale forming. Unstable water can cause problems related to health, aesthetics, and economics. Some of the problems due to unstable water are color, taste and odor, and higher costs of operation. The build-up of corrosion products (**tuberculation**) or uncontrolled scale deposits can seriously reduce pipeline capacity and increase resistance to flow. Scaling can also cause the operation of hot water heaters and boilers to become more expensive due to reduced volume and heating capacity. Corrosive water can cause the leaching of toxic metals like lead and copper into a water supply.

14.5.1 CLASSIFYING WATER STABILITY

The stability of treated water can be classified into the following three categories: scaling, neutral, and corrosive. A brief discussion follows.

Scaling typically indicates hard water that is supersaturated with calcium carbonate. Other scale-forming compounds include magnesium carbonate, calcium sulphate, and magnesium chloride. Scaling tendencies are easily noticed in hot water heaters.

Neutral water is in equilibrium or stabilized. It is neither scale forming nor corrosive. However, any change in the chemical composition or temperature can make it aggressive or scale forming. By adjusting the water to slightly scaling, a protective eggshell crust forms on the interior pipe surface.

Corrosive water tends to dissolve pipe material. Deposits of tuberculation in a cast iron system are typical by-products. It is important to remember that the same water can be scaling at one point of use and be aggressive at another point due to changes in physical or chemical characteristics.

14.5.2 CHEMISTRY OF CORROSION

Localized corrosion occurs due to galvanic corrosion. It is called cell corrosion as corrosion generates an electric current, which flows through the metal being corroded. Due to minor impurities and variations in metal, one spot on the pipe starts acting like an **anode** in relation to another spot that acts like a **cathode**. At the anode, atoms of iron break away and go into the solution. As each atom breaks away, it ionizes by losing two electrons, which travel to the cathode. The formation of Fe $(OH)_2$ leaves an excess of H^+ near the anode, and the formation of H_2 leaves an excess of OH^- near the cathode. This imbalance in the H^+ and OH^- is the reason for the localized corrosion that causes pitting and tuberculation. Fe^{++} is further oxidized to Fe^{+++} in the presence of oxygen to form Fe $(OH)_3$ or rust.

$$2Fe^{++} + 5H_2 + 0.5O_2 \rightarrow 2Fe(OH)_3 \downarrow + 4H^+$$

The rust precipitates, forming deposits called **tubercles**. As the reaction indicates, the rate of corrosion is controlled by the concentration of dissolved oxygen. Scale-forming water can protect the pipe from corrosion by depositing a thin scale that bars the flow of the current. However, uncontrolled scale deposits can significantly reduce the flow-carrying capacity due to a reduction in the effective flow area and an increase in flow resistance.

14.5.3 STABILITY INDEX

The corrosive or scaling tendency of given water can be determined knowing the stability index like the Langelier Index or the Aggressive Index. In many cases, pH and alkalinity can be adjusted to make the water non-aggressive. Low alkalinity combined with low pH makes the water very corrosive.

LANGELIER INDEX

Any corrosivity index is a way of determining the tendency of water to form scale or to corrode. The Langelier Saturation Index is the most common and is applicable in the pH range of 6.5 to 9.5. The Langelier Index (LI) is defined as follows:

$$LI = pH - pH_S; \quad pH_S = A + B - \log(CaH \times Alk)$$

Constant A depends on temperature and constant B depends on total dissolved solids. The values of constants A and B can be read from **Table 14.2** and **Table 14.3**.

Alk is alkalinity and Ca-H is the calcium hardness, both expressed in mg/L as $CaCO_3$. This calculation is accurate enough up to a pH_S value of 9.3. A positive (>1)

TABLE 14.2

Values of constant A

TDS, mg/L	0	50	100	200	400	800	1000
A		9.63	9.72	9.75	9.80	9.86	9.94	10.04

TABLE 14.3
Values of constant B

T°C	0	5	10	15	20	25	30
B	2.34	2.27	2.20	2.12	2.05	1.98	1.91

LI indicates that the water is supersaturated with $CaCO_3$ and will tend to form scale. If the actual pH is less than the pH_S, (LI < 1), the water is corrosive. If the pH and pH_S are equal (LI = 0), the water is stable.

Aggressive Index, AI

The aggressive index is equal to pH plus the logarithm of the product of alkalinity and calcium hardness as $CaCO_3$. The aggressive index of 12 corresponds to LI of zero (**Table 14.4**).

$$AI = pH + \log(CaH \times Alk)$$

Example Problem 14.2

Calculate the Langelier Index of water at 10°C having a total dissolved solids concentration of 210 mg/L, alkalinity, and calcium content of 110 and 40 mg/L as $CaCO_3$ respectively.

Given:

CaH = 40 mg/L   Alk = 110 mg/L   TDS = 210 mg/L   pH = 8.4 at 10°C

Solution

Constant A, from Table 14.2 T = 10°C, A = 2.20
Constant B, from Table 14.3 TDS = 210 mg/L, B = 9.80

$$pH_S = A + B - \log(CaH \times Alk) = 2.2 + 9.8 - \log(40 \times 110) = 8.35 = 8.4$$

$$LI = pH - pH_S = 8.4 - 8.35 = 0.05 = \underline{0.1}$$

TABLE 14.4
Comparison of various indices

Corrosivity	LI	AI
high	< -2.0	< 10.0
moderate	-2.0 <LI <0.0	10 <LI <12
non-aggressive	>0.0	>12
stable	0.0	12

Example Problem 14.3

Calculate the aggressive index of water at 10°C having a TDS of 100 mg/L and alkalinity of 80 mg/L as $CaCO_3$. Actual pH at the prevailing temperature is 8.4.

Given:

$$TDS = 100 \text{ mg/L} \quad T = 10°C \quad Alk = 80 \text{ mg/L} \quad CaH = 40 \text{ mg/L as } CaCO_3$$

Solution:

$$AI = pH + \log(CaH \times Alk) = 8.4 + \log(40 \times 80) = 11.9 = \underline{12}$$

14.5.4 CORROSION CONTROL

Basically, there are two ways to inhibit, or slow down, corrosion. These are adjusting the pH or adding a corrosion inhibitor. Sometimes both are used.

Adjusting pH

If the water is corrosive and lead leaching is a problem, consider raising the pH to make the water more likely to form scale. **Table 14.5** lists several chemicals that can be used to raise pH. Using this chart in combination with the Baylis Curve or the Langelier Saturation Index, an operator can see how to change the water quality.

While each chemical listed raises the pH, each provides different sources of hydroxide (OH) or carbonate (CO_3) alkalinity. For example, only lime provides calcium, and it is the least expensive chemical per kg. It is also the most troublesome to feed. Liquid caustic soda can be fed easily and safely. However, long-term shortages of caustic soda will make it expensive. Before adjusting the pH, other factors may need to be considered, such as THM formation at higher pH levels and the fact that disinfection by chlorination is most efficient between pH 7.2 and pH 7.5.

Sodium phosphates have been used for film formation, sequestration, and dispersion for more than 100 years. They can also be used to stabilize and buffer a pH-adjusted water. These phosphates generally work over a wide range of pH and under a variety of conditions. Products such as sodium hexametaphosphate (SHMP) and sodium tripolyphosphate have been two traditional mainstays. Other powders and

TABLE 14.5
Proper use of pH adjustment chemicals

Chemical	Dosage mg/L	Alkalinity Increase/Unit	Equipment
−50% NaOH	1–29	1.25	chemical feed pump
Na_2CO_3	1–40	0.94	feeding pump system
$Ca(OH)_2$	1–20	1.35	lime slaker
$NaHCO_3$	5–30	0.59	feed pump, tank

powdered blends are also available. Recently, many liquid-blended phosphates have shown the best performance and ease of application.

For water below pH 8, zinc phosphates can be added for film formation. Zinc phosphates can be made of zinc orthophosphate or zinc polyphosphate; they come in liquid and powder forms. Maximum contaminant levels for heavy metals, such as for zinc in treatment sludge, have limited the use of zinc phosphates in some applications. Silicates, more commonly known as **water glass**, are known for the glass-like coatings they form in distribution systems. They can be blended with phosphates for corrosion reduction.

Corrosion Inhibitors

Chemical inhibitors may be used alone or with a pH adjustment program. Besides forming a film on the inside of distribution piping, some inhibitors also sequester minerals and buffer the water to help stabilize pH changes.

Discussion Questions

1. How does cathodic protection prevent internal corrosion of a steel water tank?
2. Explain how unprotected iron pipes are corroded? What is the most common method to protect pitting of ductile iron pipes?
3. Compare tuberculation with scaling?
4. What you understand by water stabilization?
5. Chlorine can be used for taste and odor control. However, if water contains phenols, chlorination may worsen matters. Explain.
6. Describe briefly the common methods used for taste and odor control.
7. Describe briefly various types of indices to measure the stability of water.
8. Why is the passage of water through a semi-permeable membrane for the removal of dissolved salts called reverse osmosis. What other process is used for desalination and how does it work?
9. How does the use of acid prevent scale formation on reverse osmosis membranes?
10. How does cathodic protection prevent internal corrosion of pipes and tanks?
11. What are corrosion inhibitors. List chemicals commonly used as corrosion inhibitors.

Practice Problems

1. A sample of PAC slurry was collected from a slurry-mixing tank. The one liter sample was allowed to settle. After six hours of settling, the volume of settled carbon was recorded to be 55 mL. An aliquot of 10 mL of settled carbon was dried and found to weigh 2.2 g. Calculate the concentration of carbon in the slurry (12 g/L).
2. Determine the volume of carbon slurry (Practice Problem 1) to dose water with 5.0 mg/L of carbon. (410 L/ML).

3. The volume of water in a reservoir is estimated to be 5.6 MG. The desired dose of copper is 0.5 mg/L. How many lb of copper sulphate will be needed? (92 lb).

4. Calculate the Langelier Index of water at 10°C having a total dissolved solids concentration of 400 mg/L, alkalinity, and calcium content of 150 and 70 mg/L as $CaCO_3$ respectively and pH of 8.3. (0.26)

5. Calculate the aggressive index of water that has calcium hardness of 35 mg/L and alkalinity of 100 mg/L as $CaCO_3$. Actual pH at the prevailing temperature is 8.4. (12).

6. For Practice Problem 5, calculate LI given that TDS are 200 mg/L and temperature is 15°C. (0.024).

7. For the data of Practice Problem 4, calculate the aggressive index. (12).

8. Calculate the Langelier Index and aggressive index of water having a total dissolved solids concentration of 400 mg/L, alkalinity of 130 mg/L as $CaCO_3$, and calcium content of 20 mg/L as calcium. At a temperature of 10°C, pH is 8.4. (0.21, 12).

9. An inland lake is spread over an area of 88 ha. It is planned to dose the top 1.0 m depth of water with copper dosage of 2.0 mg/L. Compute kg of $CuSO_4$ required. (6900 kg).

10. You have 2000 lb of $CuSO_4$. How many gallons of water can be treated by applying a copper dose of 1.5 mg/L? (41 MG).

Section III

Water Distribution

15 Water Distribution

Water distribution is the second most important step in potable water supply systems. In many cases, poorly operated and maintained water distribution is responsible for the degradation of water quality. It may be bacterial contamination due to lack of chlorine residual, or taste and odor problems caused by the poor circulation of water in the distribution system. Under severe water demand, water pressures can fall below atmospheric pressure causing **backflow** that can lead to contamination and/or collapsing of the water mains.

Operating personnel of a water distribution system must ensure that their clients have access to clean, safe water, while maintaining sufficient pressure and volume to meet all of their users' demands. It should be capable of meeting water demand during emergencies, including fighting fires. The design and operation of the system must provide safeguards against backflow contamination. To protect the entry of contaminated water, the water pressure in a distribution system should not drop below 250–300 kPa (35–40 psi). However, in developing countries, water pressures are usually less than 200 kPa (30 psi). Maximum pressures are usually kept below 550 kPa (80 psi) and not allowed to exceed 700 kPa (100 psi) to reduce the chances of leaks and water breaks.

A well-designed system must put in place the correct devices to protect the system against backflow. All distribution operating personnel should always be aware of the consequences of **cross connections** between a water distribution system and an unapproved source of water. Cross connections have led to serious public health problems in many cities. In summary, properly operated and maintained water distribution system must be able to sustain the following objectives.

- Meet normal water demand and during emergencies (e.g., firefighting).
- Always provide a positive pressure of more than 250 kPa (35 psi), except in emergencies and when firefighting (with a minimum pressure of 150 kPa (20 psi)).
- Maintain water quality and protect it from contamination.
- Ensure that interruptions are as infrequent as is reasonably possible.
- Minimize the costs of maintenance and operation.

15.1 SYSTEM COMPONENTS

Water distribution consists of pipelines of various sizes for carrying water to the streets, valves for controlling the flow and pressure, water hydrants for firefighting, service connections, pumps for lifting and pressurizing water, and reservoirs for storing the water supply. Typical water distribution systems consist of the water mains, pumping stations, storage reservoirs, valves, hydrants, meters, and other

DOI: 10.1201/9781003347941-18

ancillary devices. In smaller towns and cities, treated water from the water plant is pumped directly into the distribution mains and distributes the water into submains, branches, and laterals.

Personnel responsible for the operation have to be aware of the strengths and weaknesses of their system so as always to be on guard to maintain the safety of the water supply. The operation of a distribution system includes such things as the operation of pump stations, the location and repair of water main breaks, the flushing and cleaning of distribution lines, chlorination of water distribution lines, reservoirs, or storage containers and the thawing of distribution lines during winter.

As discussed above, distribution systems consist of piping, pump stations, and storage facilities such as reservoirs, elevated tanks, and standpipes. In addition, there are numerous appurtenances such as valves, fire hydrants, water meters, and service connections. It is essential that distribution systems are operated to minimize supply interruptions, protect the water from contamination, and maintain a positive pressure in the system at all times. Lack of positive pressure can cause contaminated groundwater to enter pipes through cracks. Moreover, negative pressure may draw contaminated water from an unapproved source of water. This phenomenon is known as backflow due to **back siphoning**.

Water distribution systems must also be prepared to operate under unusual conditions. These can include excessive demands caused by firefighting or broken or frozen water mains. Pipe sizes are based on the peak hourly demand, or the fire flow demands, whichever is larger. In addition, pump failures can create serious problems. Plans need to be prepared and taught to all staff before an emergency occurs.

15.2 METHODS OF WATER DISTRIBUTION

Depending on the location of the source and the water plant and topography of the area, treated water may be supplied by gravity, pumping, or by a combination of the two.

15.2.1 GRAVITATIONAL SYSTEM

In the gravity system, water from a high elevation source is distributed to consumers located at low elevations. For the gravity system to operate successfully, there must be sufficient elevation difference to maintain adequate pressure at the delivery points. Since no pumping is required, this method is the most economical. However, this system is only possible where sufficient gravity head is available, such as a water source on a hill with consumers located in the foothills.

15.2.2 PUMPING SYSTEM

By the use of high lift pumps, water from a clear well at a water plant can be directly pumped into the distribution system. To meet the variation in demand, either variable speed pumps are used or a combination of pumps may be used. Since there is no storage, this kind of system will not be able to deal with emergencies such as fire and power failures. This method is, therefore, generally not used.

15.2.3 COMBINED GRAVITY AND PUMPING SYSTEM

This type of system can be used for any situation and is thus very common. Water distribution system consists of pumping as well as storage. During low demand periods, water in excess of demand is used to fill up storage tanks. During peak demand, water from the elevated storage tanks feeds distribution in addition to water pumped directly from the plant.

15.3 EQUALIZING DEMAND

Water distribution systems are designed to meet peak flows and firefighting needs while maintaining uniform pumping and a treatment rate equal to the maximum daily demand. This is accomplished by providing equalizing or balancing storage to meet peak hourly demand. Generally, the volume of water needed to equalize the peak demand is about 20% of the average daily demand. The location of the reservoir is selected based on the following considerations.

- Where the areas of high demand are most likely to occur.
- The comparative cost of pumping and pipelines.
- Equalizing pressure in the water system, which will be determined by the topographic features of the area served.
- Hydraulic grade line developed by the reservoir and associated pumping facilities.
- If the topography permits, storage is located at a central location and at high elevation.

15.3.1 EQUALIZING STORAGE CAPACITY

An extremely important element in water distribution systems is water storage. System storage has far-reaching effects on a water system's ability to provide adequate and reliable water supplies for domestic and commercial use and especially for firefighting. The main function of storage is to equalize demand and storage during emergencies and for fighting fires. However, the primary function of storage is to balance the distribution to meet the fluctuating demand with a constant rate of water supply from the treatment plant.

Balancing storage can be worked out knowing the maximum cumulative deficit (supply minus demand) using either the mass curve or by using an analytical method. The latter is gaining more popularity as it becomes easier to set up a table using a spreadsheet computer program like MS Excel. Using a tabular method, the hourly rate supply and demand are entered, and cumulative values are calculated. A column indicating the difference in the cumulative supply and cumulative demand is then created. A negative value would indicate excess demand and a positive value excess of supply. The total equalizing storage capacity is then worked out by adding maximum excess demand and maximum excess supply. This procedure is further illustrated by the example problems shown below.

Example Problem 15.1

A medium-sized city has variable demand, as shown in **Table 15.1**. Determine the storage capacity required to equalize the demand against a constant rate of pumping @ 24 ML/d.

TABLE 15.1
Excel worksheet (constant supply)

1	3	4	5	6	7	8	9	10	11
hour	Demand, Q_o				Supply, Q_i			Q_i-Q_o	
	Rate	Volume	Σ	Rate	Volume	Σ	Volume	Excess	
	ML/d	m^3	m^3	ML/d	m^3	m^3	m^3	m^3	
1	12.0	499	120	26	1067	1000	1067	1067	Supply
2	11.4	476	596	26	1067	2067	1471	1471	Supply
3	9.8	408	1004	26	1067	3133	2129	2129	Supply
4	7.6	317	1321	26	1067	4200	2879	2879	Supply
5	7.1	295	1616	26	1067	5267	3651	3651	Supply
6	6.5	272	1888	26	1067	6333	4445	4445	Supply
7	10.9	453	2341	26	1067	7400	5059	5059	Supply
8	19.0	793	3135	26	1067	8467	5332	5332	Supply
9	27.2	1133	4268	26	1067	9533	5265	5265	Supply
10	32.6	1360	5628	26	1067	10600	4972	4972	Supply
11	34.8	1451	7079	26	1067	11667	4588	4588	Supply
12	38.1	1587	8665	26	1067	12733	4068	4068	Supply
13	35.9	1496	10161	26	1067	13800	3639	3639	Supply
14	34.8	1451	11612	26	1067	14867	3255	3255	Supply
15	34.3	1428	13040	26	1067	15933	2893	2893	Supply
16	34.8	1451	14491	26	1067	17000	2509	2509	Supply
17	34.8	1451	15941	26	1067	18067	2125	2125	Supply
18	36.4	1519	17460	26	1067	19133	1673	1673	Supply
19	40.3	1677	19137	26	1067	20200	1063	1063	Supply
20	50.0	2085	21223	26	1067	21267	44	44	Supply
21	45.7	1904	23127	26	1067	22333	-793	793	Demand
22	27.2	1133	24260	26	1067	23400	-860	860	Demand
23	17.4	725	24985	26	1067	24467	-519	519	Demand
24	15.2	635	25620	26	1067	25533	-87	87	Demand
Σ	624	25999		614					

Max. excess supply = 5332 m^3.

Max. excess demand = 860 m^3.

Required storage capacity = 6192 m^3.

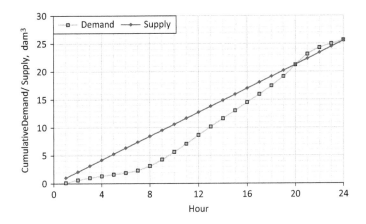

FIGURE 15.1   Mass curve (Ex. Prob. 15.1)

Solution:

CONSTANT SUPPLY

As seen in **Table 15.1**, the rate of demand in column 3 is converted to volume in column 4 and the cumulative value is shown in column 5. The same is done for the supply and the difference in cumulative supply and cumulative demand is shown in column 9. Storage capacity is calculated by adding the maximum of excess supply and the maximum of excess demand.

15.3.2   OTHER PURPOSES OF STORAGE

Distribution storage tanks and reservoirs are used to provide a community with emergency storage capabilities. It also allows the treatment facility to produce excess water during slow periods for use during periods of high demand. In this way the water treatment facility can be designed for the average demand rather than peak demand. Storage facilities improve system flows and pressures. Another advantage of storage is that it provides a location for **blending** water from various sources and locations for **extended chlorination**. Pumps operate at lower and uniform pumping rates resulting in cost savings.

15.3.3   TYPES OF STORAGE

Common types of storage facilities include ground storage, underground reservoirs, elevated tanks, and standpipes. Underground storage reservoirs built as part of a water treatment plant are called **clear wells**. The main purpose of an elevated tank is to maintain system pressure. Level switches control flow in and out of the tank. **Hydro pneumatic tanks** are used for small-scale and individual household water systems. Storage reservoirs need to be protected against corrosion, contamination, and vandalism. To maintain the quality of water, about two-thirds of the volume of the storage should be renewed every day.

Procedures for operating reservoirs depend on the type of facility and the magnitude of system demands. However, replenishing the reservoir after the peak demand period and maintaining system pressure should be the main consideration in any operation scheme.

Ground-Level Storage

Since water is stored at ground level, it must be pumped to the point of use. This limits water system effectiveness for fire protection in three ways.

1. There must be excess pumping capacity to deliver the peak demand for normal use and as well as for any firefighting demands.
2. Standby power sources and standby pumping units must be always maintained.
3. Water mains in the system must be oversized to handle peak demand plus fire flow, no matter where a fire occurs.

Elevated Storage

Properly sized elevated water tanks provide dedicated fire storage and are used to maintain constant system pressure. Where elevated tanks are used, ground storage tanks may still be required based on the type of system design. However, the size of storage would be much less. Daily, water from the top 3 to 5 m (10–16 ft) is used for supply and the remaining 70%–75% is held as reserve. When domestic consumption is at its minimum during the day, water in the elevated tank is recycled with fresh water to eliminated the aging of water in the holding tank.

Standpipes

Standpipes are normally constructed using steel or concrete. They are cylindrical in shape with heights exceeding their diameters. Standpipes provide more storage than elevated tanks. But that is useful for equalizing purposes only if the elevation required maintains minimum pressure. The water stored below the minimum elevation can be used for firefighting with fire engines and during other emergencies.

Location of Storage Tanks

Using several small storage tanks or water wells near the major centers of water withdrawal is preferred to using one large tank near the pumping station. Also, it is best to locate tanks on the opposite side of the demand center from the pumping station. This allows more uniform pressure throughout the system and smaller diameter mains and pumps.

Pumping Stations

In most distribution systems the topography of the land requires the use of pumping stations to lift the water to a higher elevation and to maintain a positive pressure within all sections of the system. Several types of pumping arrangements are common. Stations may elect to use a pump large enough to handle the maximum flow rate although this tends to be inefficient during low demand periods. Stations may also use a variable speed motor to decrease the pump's capacity during low demand periods. Most systems have an elevated storage tank or reservoir on high ground floating on the system. For larger municipalities, the pumping station generally consists of two or more pumps of different sizes to match the required rate of flow. Usually, backup pumps will be available in the case of failure.

15.4 PIPELINE LAYOUT

The main distribution pipes within a system are known as the **trunk mains.** Smaller-diameter pipes branch off and distribute water to individual streets. Water mains are generally not less than 150 mm (6-in) in diameter. There are, in general, four different types of networks, commonly used either singly or in combination. A brief description is given in the following paragraphs.

15.4.1 DEAD END SYSTEMS

A dead end system, also called a **tree system**, is shown in **Figure 15.2**. Water from the source goes to a main, which is further divided into many submains. Each submain then divides into several branch pipes called **laterals**. Service connections are provided from laterals to serve users. This type of system may have to be adopted for older towns and cities where the development took place haphazardly. One of the serious drawbacks of the system is the presence of dead ends that prevent recirculation, thus making water stale and vulnerable to contamination. The other disadvantages of the system include the following.

- Since there is only one path, during repairs, a larger part of the system needs to be shut down thus causing inconvenience to consumers and risks being unable to attend to any emergency during this period.
- To maintain the quality of water, dead end hydrants need to be flushed frequently, resulting in wastage of treated water.
- Since there is only one route for water to reach a given point, it is not possible to increase the water supply to meet higher demand during emergencies.

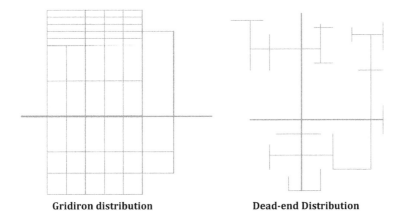

Gridiron distribution   Dead-end Distribution

FIGURE 15.2 Grid iron versus tree distribution

15.4.2 GRIDIRON SYSTEM

The gridiron arrangement of pipes, shown in **Figure 15.2**, is preferred to a layout that has many dead end branches. Dead ends are extremely susceptible to taste and odor problems and require frequent flushing. In a gridiron pattern, valves are used to isolate the broken section of the pipes, and water can still reach consumers from the other side of the loop. The advantages of the grid iron system are as follows.

- Since there is more than one route for water, the flow carried by each pipe is relatively small, thus the size and friction losses are low too.
- In case of repairs, it is easy to shut down or isolate areas without affecting the supply to other users.
- Dead ends are eliminated, and water remains in circulation.
- During emergencies like fires, more water can be diverted to the demand area.
- Pressure variants at points in the system are relatively small.

However, such a system requires more hardware and thus is expensive to construct. The gridiron system is suited to well-planned towns and requires a maximum number of isolation valves and relatively larger diameter pipes. The design of such system is more complex, and the use of computer models is almost a necessity.

15.4.3 RING SYSTEM

Ring system is also called a **circular system** and is shown in **Figure 15.3**. In this case, a ring of pipelines runs around the served area so that flow direction is from the periphery towards the center of the system. This system also eliminates dead ends. The pressure in various parts of the distribution is more uniform. The distribution area is divided into rectangular or circular blocks. This system is very suitable to well-planned and well-laid-out towns and cities.

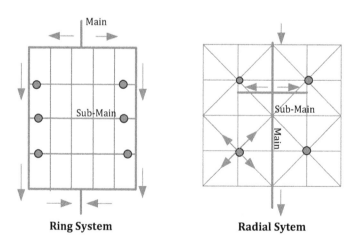

FIGURE 15.3 Ring versus radial distribution

15.4.4 RADIAL SYSTEM

This system is recommended for towns with radial roads emerging from different centers or pressure zones. The general flow direction is from the center towards the periphery, as shown in **Figure 15.3**. Distribution reservoirs are placed at such centers and water is then supplied through radially laid distribution pipes. This method ensures high pressure and different water distribution. The design of such systems is relatively easy and simple.

15.5 PIPE MATERIAL

Water pipes in the distribution system flow under high pressure. Pressure pipes are available in pressure ratings from 100 psi (700 kPa) to 250 psi (1750 kPa). A class 150 (pressure-rating) pipe is the most widely used pressure pipe. American Water Works Association (AWWA) C900 is the most commonly used type of pipe, which adapts easily to existing piping systems. For larger-diameter pipes, reinforced concrete pipes are usually used. Ductile iron is extremely strong, lightly flexible, relatively heavy, and needs corrosion protection. Steel pipes are generally used for high-pressure applications. However, you need to provide corrosion protection for the interior as well as the exterior of the pipe. Pipes used in water distribution systems must have the strength to withstand internal and external pressures and abnormal pressures due to **water hammer**. Pipes should be resistant to corrosion and must not be detrimental to the potability of water. Of lesser consideration is the ease of handling and jointing during installation.

The pipe materials most used in water supply and distribution include polyvinyl chloride (PVC), high density polyethylene (HDPE), asbestos cement (AC), ductile iron (DIP), and reinforced concrete pressure pipe (RCP). Plastic pipes are inexpensive, flexible, lightweight, corrosion free, easily installed, and low in friction. Smoothness of pipes increases the flow-carrying capacity or yields less pressure loss while carrying the same flow.

15.5.1 PLASTIC PIPES

This is currently the most common type of pipe used in distribution systems. It is available in lengths of 3 m (10 ft), 6 m, and 12 m. The main advantage is its lightness, allowing for easy installations. A disadvantage is its ability to withstand shock loads. The National Sanitation Foundation (NSF) in the USA currently lists most brands of PVC pipe as being acceptable for potable use. Plastic pipes have several advantages over metallic pipes, including resistance to rupture from freezing, being corrosion proof, and they can be installed above ground or below ground.

Plastic pipes are made of different strengths. Sometimes, polyvinyl chloride is further chlorinated to obtain greater strength, a higher level of impact resistance, and a greater resistance to extreme temperature. Chlorinated blend pipes are designated as CPVC. High-density polyethylene (HDPE) is flexible and is suitable for household fittings. They are used for minor domestic works. Polythene and PVC pipes can withstand pressures up to 1000 kPa.

15.5.2 CAST IRON PIPES

Cast iron (CI) pipes have been in use for a long time, though they are less common these days, especially in developed countries. However, because of their long life, CI pipes are found in most cities. They are common because these pipes are sufficiently resistant to corrosion, are durable, come at reasonable cost, and are relatively easy to join. They are available in nominal pipe sizes (NPS) of 80 mm to 1200 mm (3 in–48 in) and can withstand pressures up to 2400 kPa (350 psi). Cast iron pipes loses their flow capacity over time due to **tuberculation** and other deposits. They are heavy and uneconomical in larger sections. In addition, they are brittle and likely to break during transportation or handling. Although it is not currently the material of choice there is still a lot of it in the ground.

15.5.3 DUCTILE IRON PIPES

Ductile iron was developed to overcome the breakage problems in CI pipes. Its main advantage is that it can withstand high pressure both internally and externally. It is sometimes protected from highly corrosive soils by wrapping the pipe in plastic sheeting.

15.5.4 STEEL PIPES

Steel pipe is used mostly for transmission mains where internal pressures are usually low. Steel pipes have a smooth interior and withstand pressure up to 1700 kPa (250 psi). Steel pipes need to be protected from internal and external corrosion. To protect the pipe, internal coal tar epoxy linings and cement linings are used. For outside protection, polyethylene linings are used. These are either welded or riveted steel sections. Welded pipes are smoother and stronger than riveted ones and hence are generally preferred. As steel is strong in tension, even pipes of diameters up to 6 m (240 in) can be fabricated out of thin steel shells and they can resist very high internal pressures.

Galvanized iron (GI) pipes with corrugations on their circumferences are stronger than plain steel pipes. They are manufactured in diameters varying from 200 mm (8 in) to 2000 mm. They are comparatively light and can be transported large distances. If they are to be placed above the ground, expansion joints become necessary in order to counteract temperature stress. Steel pipes are liable to quick rusting, they cannot withstand high negative pressures or a vacuum created inside them, or the combined effect of internal vacuum and external stresses due to backfill and traffic loads above.

15.5.5 CEMENT CONCRETE AND RCC PIPES

Plain cement concrete pipes are manufactured up to a maximum diameter of 600 mm (24 in). For greater diameters, reinforced cement concrete (RCC) is used. The diameters of RCC pipes may vary from 1.5 to 4.5 m (5–15 ft). They may be precast or cast in-situ. Plain concrete pipes are used for pressures up to 150 kPa (20 psi), whereas RCC pipes are used for pressures up to 750 kPa (100 psi). For greater pressures, pre-stressed concrete pipes are employed.

They resist corrosion both inside and outside and can resist external compressive loads. They do not fail under traffic loads and nominal vacuum, and are durable, their life being about 75 years. Having a low coefficient of expansion, when placed over ground, no expansion joints are required. If groundwater contains acids, alkalis, or sulphur compounds, RCC pipes may get affected by corrosion. Repairs are difficult and connections are not easy. Shrinkage cracks and porous texture may result in leakage.

15.5.6 ASBESTOS CEMENT PIPES

Asbestos cement (AC) is a homogeneous material prepared out of three ingredients: asbestos, silica, and cement. This has high strength and is quite dense and impervious. Asbestos cement pipes are usually available in diameters varying from 100 mm to 900 mm (4–35 in) and length 4 m. Four grades of pipes are manufactured to suit pressures varying from 350 to 1400 kPa (50–200 psi). To join these pipes, a special coupling called a **simplex joint** is used.

AC pipes are light hence transportation and handling are easy. They are very resistant to corrosion and hydraulically efficient due to their smooth surface. They are suitable for use as small distribution pipes. They do not have much strength and hence are liable to be damaged during transportation and handling or overlying loads or lateral earth pressures. They are quite costly. Asbestos fiber is dangerous to health, so safety precautions are to be taken when handling and joining these pipes.

15.6 PIPE JOINTS

Pipes are manufactured in standard sizes and with various types of joints used to connect them. The most common joints in use are the flanged joint, socket and spigot joint, flexible joint, mechanical joint, expansion joint, and simplex joint.

15.6.1 FLANGED JOINT

Flanged joints are adopted in pumping stations, filter plants, etc. where occasional disjointing is necessary (**Figure 15.4**). Two flanges are brought together with a rubber gasket between them to create watertightness. They are fixed together by nuts and bolts. These joints are rigid, strong, and expensive. They cannot be used where there are vibrations and deflections. They are costly.

15.6.2 SOCKET AND SPIGOT JOINT

Socket and spigot joints are also called **bell and spigot joints**. They are usually used for cast iron pipes. The pipes at the joining ends are made into the shapes of wider

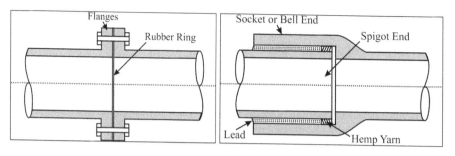

FIGURE 15.4   Flanged joint and socket joint

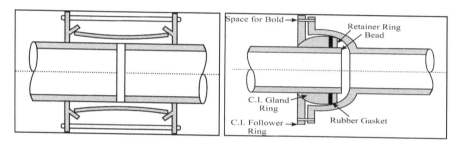

FIGURE 15.5   Flexible joint and dresser coupling

bells or sockets in such a way that the socket and spigot fit properly. The gap between the socket and spigot is filled with jute and molten lead so as to make it watertight. This joint is flexible and can be used for pipes laid on flat curves.

Bell and spigot type joint systems are more common in plastic, ductile iron, and reinforced concrete pipes. The pipes should be installed so that the bell is in the direction of pipe laying. Steel pipes are joined using mechanical joints or **dresser coupling**. Bedding must be a very important consideration as plastic pipes have low resistance to crushing. After laying the pipe, it should be properly backfilled, tested, flushed, and disinfected before being put into operation.

15.6.3   Flexible Joint

Flexible joints are used where pipes are to be laid in rivers with uneven beds and where large-scale settlements are likely to occur or pipes are to be laid on curves (**Figure 15.5**).

Under these circumstances, the joints are likely to break if there is not sufficient flexibility. The socket is made spherical. The spigot is provided with a bead at the end. The special rubber gasket is held in position by a retainer ring. On that is placed a split cast iron "gland ring". They are tightened by bolts and nuts. The spigot end can be moved to tackle the desired deflection and nuts tightened over the gland ring.

15.6.4 MECHANICAL JOINT

A mechanical joint is also called a **dresser coupling**. This joint is adopted when it is necessary to join the plain ends of cast iron pipes. A metallic collar is fitted and tightened over the abutting ends of the pipes, thereby forming a mechanical joint. The usual type adopted is called a dresser coupling. In this joint, an iron ring and a gasket are slipped over the abutting ends of the pipes. An iron sleeve is introduced between the gaskets. These iron rings are tightened by means of bolts. These joints, being rigid and strong, can withstand heavy vibrations. They are used for pipes carried over or below bridges.

15.6.5 EXPANSION JOINT

An expansion joint is necessary to neutralize the effects of temperature stresses (**Figure 15.6**). This is provided at suitable intervals in pipelines. The socket end is cast to have a flange and the spigot end remains plain. The socket end is fixed rigidly to an annular ring which can slide freely over the spigot end. A small gap is allowed between the face of the spigot and the inner face of the socket. A rubber gasket fills up the annular space between the socket and the spigot. The flanges are then tightened by the help of bolts and nuts. When the pipes expand, the socket moves forward, and the gap gets closed. Similarly, when the pipes contracts, the socket contracts, leaving a gap. Meanwhile, the annular ring follows the movements of the socket and maintains the gasket in position, thus maintaining the joint watertight.

15.6.6 SIMPLEX JOINT

A simplex joint is used to connect two asbestos-cement pipes. A sleeve is provided over the abutting ends in such a way that it fits over them. Two rubber rings are compressed between the sleeve and the pipe ends. The joint thus formed is called a simplex joint.

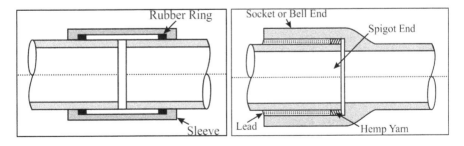

FIGURE 15.6 Expansion joint and simplex joint

15.7 PIPELAYING AND TESTING

The various operations to be undertaken for the laying of water mains are discussed briefly.

PREPARATION OF ROAD MAPS

After a detailed and accurate survey, road maps must be prepared. These must indicate the positions of curbs, sewers, water pipes, and electrical and telephone conduits, etc.

LOCATION OF ALIGNMENT

To mark the trench to be excavated, the central line is marked by driving spikes at 30 m intervals on straight portions and 7.5 m to 15 m intervals on curves.

EXCAVATION OF TRENCHES

The depth and width of the excavated trenches should be adequate to permit the laying and joining of pipes. The width must be about 40 cm more than the outer diameter of the pipe, and there should be a clear ground cover of about 90 cm above the top of the pipe barrel.

PREPARATION OF TRENCH BOTTOM

The bottom of the trench should be carefully prepared so that the barrel can be bedded true to the line and gradient over its entire length on a firm surface. If the excavation is in rock, a 15 cm (6 in) concrete cushioning bed is provided on the bottom of the trench so that adequate protection is given against any possible settlement. For jointing the pipes, joint holes are provided. Pipes must be introduced into the trench carefully so that they do not undergo any damage.

LAYING OF PIPES

Pipes should be laid in an "uphill" direction so that joint making is easy. They are not laid with a flat slope parallel to the hydraulic gradient. They are placed either with a continuous rise to high points or a continuous fall to low points.

JOINTING OF PIPES

For jointing pipes, suitable joints are adopted depending upon the type of pipe.

15.7.1 ANCHORING OF PIPES

Thrust refers to the force exerted due to changes in flow direction as dictated by the law of mechanics. Thus, at all positions where there are bends, toes, valves, branch connections, etc. there is hydraulic thrust. At such positions, to transmit this thrust over a wider area of soil, "thrust blocks" of concrete are provided. In the case of pipes laid on sloping grounds, there will be upward hydraulic pressures. In such cases, "anchor blocks" of concrete are provided at regular intervals, the pipes being rigidly secured to them with steel straps.

One important thing regarding thrust blocks is that they must be placed on the opposite side of the thrust exerted. Another important consideration is that thrust blocks must bear against undisturbed soil since disturbed soil is subject to compression upon loading. The size of the thrust block depends on the water pressure in the pipe, the type of fitting, and the bearing capacity of the soil.

15.7.2 BACKFILLING WITH EARTH

This is the operation of filling up the trench with excavated stuff. The material used around the pipe should be soft and should not contain any rock pieces, lumps, or stone, etc. The filling may be done in layers, each 15 cm to 30 cm thick and well consolidated so that no movements of pipes inside the trench are possible.

The remaining upper portion of the trench may be filled up with excavated debris and the top of the trench brought to a level so that after some time, due to consolidation by traffic moving above, the level of the top of the trench will become flush with the surrounding road.

15.7.3 TESTING OF PIPES

After the pipes are laid and joined, before backfilling, the pipes have to be tested under pressure. The pipe is filled completely with water so that all the air inside is drawn out. It is kept in this condition for some time and a test pressure of about 500 kPa (75 psi) or 50% of the maximum working pressure, whichever is greater, is applied. The pressure may be applied by:

  i. a manually operated test pump; or
  ii. a power-driven test pump in the case of large pipes or longer lengths of mains.

After the test pump registers the test pressure, it is stopped and any fall in pressure is recorded. The pipe is considered to be good if the pipe maintains test pressure without any loss at least for 30 minutes. This test is carried out in sections as the laying of the pipe proceeds. In other cases, test pressure is much greater, typically 1000 kPa and the volume of water pumped into the water main to maintain pressure for one hour is recorded. The volume of water used per joint is worked out and compared with the allowable water loss.

15.7.4 FLOW VELOCITY

Most pipes are sized to carry both fire flows and maximum day demands. Flow velocity usually ranges from 0.5 to 1.5 m/s (1.5–5 ft/s). Whereas low velocity allows sedimentation and stagnation of water, high flow velocity causes a high drop in pressure. A very large flow will cause excessive head losses and hence significant pressure drop. In case of a mains break, or if too many hydrants are open, pressure at critical points may fall below atmospheric due to the larger flow. This can result in contamination of water due to back siphoning. Very low velocities are also not desirable due to the fact that there is a greater risk of contamination and biological growth due to stagnation.

15.8   VALVES

In any water works, valves are used within a distribution system to isolate portions of the system for repair, cleaning, maintenance, or adding additional lines. Valves may also be used to regulate the flow or pressure of water within a pipe, to release or admit air, to prevent the flow of water in the opposite direction, and to drain the lines.

15.8.1   GATE VALVE

Different type of valves may be used for different applications. Most valves are of the gate type, which allow full passage of water without restricting flow. The main purpose of **gate valves**, also called **sluice valves**, is isolation and they should never be used for throttling. For this reason, gate valves are also called **stop valves**. Gate valves are either completely open or closed and should never be operated in a partially open position. Distribution valves generally suffer from lack of use rather than wear. It is recommended practice to **exercise** the valves on a scheduled basis. Based on the movement of the closure element, valves can be categorized as either linear and rotary valves.

15.8.2   GLOBE VALVE

Gate valves and globe valves fall into the category of linear valves. The closure element in this kind of valve has a linear movement when closed or opened. Each turn of the valve handle will move the valve in its seat upwards or downwards by pitch. Thus, the stem of the valve might be turned several times to a fully open or closed position. Usually, a motorized valve operator is installed in large valves. Valves falling in this category include globe, gate, and pressure and flow-control valves.

Compared to a gate valve, the globe valve design allows throttling and pressure regulation. Globe valves are used exclusively on pipes with diameters of 100 mm (4 in) or smaller. In the **diaphragm valve**, a flexible piece inside the valve's body can be moved up or down to adjust the opening size. The diaphragm is made of flexible material, such as rubber or leather.

15.8.3   AIR AND VACUUM RELIEF VALVES

Air and vacuum relief valves consist of a float-operated valve that allows the opening of an orifice to break a vacuum or allow air to escape to the atmosphere when draining or filling a water main. These valves serve three functions: they allow air to escape during filling, permit air to enter the pipeline when draining, and allow entrained air to escape when a line is operating under pressure. Unprotected vacuum conditions can result in the collapsing of the pipe. High points in water systems are more susceptible to air entrapment. Air bubbles in flow stream can cause significant resistance to flow. Reduced buoyancy due to entrapped air causes the float to open a smaller orifice for bubbles to escape.

15.8.4   ROTARY VALVES

As the name indicates, the rotary valve part in the pipe rotates. Usually, this kind of valve can be opened and closed by a one-quarter turn. Thus, they are also known as **quarter turn** valves. The most common valves in this category are butterfly, plug, and ball valves.

> A **butterfly valve** has a movable disk large enough to completely fit in the inside of the pipe. The disk rotates on a spindle or a shaft in only one direction to either a closes or open position. Butterfly valves can be used for throttling as well. However, the valve body being in the pipe, the valve needs to be removed when swabbing the pipe.
>
> **Plug valves** may have a tapered or a cylindrically shaped plug with an opening to the side that can be turned to open, restrict, or close the flow. In smaller sizes, these valves are commonly used as corporation stops on service lines.
>
> A **Ball valve** is very similar in design to the plug valve except the plug is of a ball shape with a cylindrical hole bored through it. When the valve is in the fully open position, the bored hole is parallel to the direction of flow. When rotated through 90 degrees from the fully open position, the opening moves opposite to the direction of flow. Thus, it completely stops the flow. Such valves are not suitable for throttling.

15.8.5   SPECIAL FUNCTION VALVES

Some of the valves are designed to perform a special function. The valve used to allow the flow in flow direction only is called a **check valve**. They are also called reflux valves or non-return valves. There are so many designs of the check valve, including swing check, slanting disk, double door check valve, and foot valve. The **foot valve** is put on the end of a suction pipe of the pump to maintain priming when the pump stops. The discharge side of valves is also equipped with check valves to prevent backflow when the pump is turned off.

Altitude valves are used in reservoirs to prevent overflow. As the water reaches the set altitude or level in the reservoir, the pressure generated activates the valve to close. This valve is also used with tanks where the full system pressure might cause them to overflow. These valves help to maintain a constant water level in the tank or reservoir if pressure in the distribution system is adequate.

Pressure sustaining valves (PSVs) are usually globe valves which are spring loaded to adjust the opening to maintain a set pressure on the downstream side regardless of the fluctuating demand downstream. As the pressure downstream of the valve increases, it compresses the spring and makes the element move to close the opening further and decrease the flow and pressure. When the downstream pressure decreases, the spring will open and allow more flow.

Pressure regulating valves (PRVs) are usually globe valves which control pressure by restricting flow. The pressure downstream of the valve regulates the flow. PRV valves control water pressure and operate by restricting flow. They are used to deliver water from a high-pressure zone to a low-pressure zone and are spring loaded to adjust the opening to maintain a set pressure on the downstream side, regardless of the fluctuating demand downstream. As the pressure downstream of the valve increases, it compresses the spring and makes the element move to close the opening further and decrease the flow and pressure. When the downstream pressure decreases, the spring will open and allow more flow.

In the summit of the mains, it is possible that some suspended impurities may settle down and cause obstruction to the flow of water. At dead ends, water stagnates and over time solids get deposited and may lead to the growth of bacteria. To avoid these difficulties, drain-off valves are provided at all such points. These are also called **scour** or **blown off valves**.

Water mains ending with dead ends are flushed by opening drain valves to wash out the deposits. A similar valve is used at the bottom of elevated tanks to drain and for the cleaning of tanks.

15.8.6 Exercising of Valves

Most of the problems with valves are due to their non-use over a long period. A valve stuck due to corrosion, rust, or freezing becomes a bigger problem when needed to work in an emergency. Hence it is very important to exercise valves. Valve exercising should be done once a year, especially with main lines as part of the preventive maintenance program. A valve inspection should include drawing valve location maps to show distances (ties) to the valves from specific reference points.

15.9 CROSS CONTAMINATION

Cross connection means any actual or potential connection or structural arrangement between a drinking water system and any non-potable water source or system through which it is possible to introduce into the distribution system contaminated water, industrial fluid, gas, or substances other than the intended drinking water with which the system is supplied. Cross connections constitute a serious public health risk. There are numerous well-documented cases of cross connections that have contaminated drinking water and resulted in serious illness.

15.9.1 Back Pressure

Cross contamination in water distribution can occur due to backflow of non-potable water into the water main line. Backflow occurs in two ways: back siphoning and back pressure. Back pressure happens when contaminants enter the drinking water system when the pressure of the contaminant source exceeds the pressure of the water system. The lower the system pressure and/or the increased instances of leakage in the piping network, the greater the probability of contaminant ingress.

15.9.2   BACK SIPHONING

Back siphoning occurs when vacuum conditions occur in the water pipeline and allow water from a contaminated source or system to enter the drinking water supply. Reduction in potable water supply pressure or partial vacuum conditions occur in situations like water line flushing, firefighting, or breaks in water mains. In addition to physical faults in the distribution system, the backflow of contaminants can come from connections to non-potable systems, tanks, receptors, equipment, or plumbing fixtures where inadequate cross connection controls, including **backflow** prevention devices, have been installed or where maintenance has been inadequate.

15.9.3   BACKFLOW PREVENTION

Although backflows through cross connections have caused a broad and varied range of outbreaks of illness associated with drinking water, surveys of water utilities have found that many do not have inspection programmes or have programmes that are insufficient to provide protection against cross connections.

There are many examples of cross connection that speak volumes of the need to protect against cross contaminations. In June 1983, yellow gushy stuff poured from some faucets in the town of Woodsboro, Maryland, USA. This contamination occurred due to backflow of a powerful agricultural herbicide. In 2007–2008, approximately 5000 cases of gastrointestinal illness attributed to campylobacter and salmonella were reported in a population of about 30 000 in Nokia, Finland. The outbreak was caused by cross connection of the sewage system and the drinking water system. **Cross connections** must either be physically disconnected or have an approved backflow prevention device installed to protect the public water system. There are five types of approved devices/methods:

1. Air gap (method).
2. Atmospheric vacuum breaker.
3. Pressure vacuum breaker.
4. Double check valve.
5. Reduced pressure backflow preventer.

The type of device selected for a particular location depends on several factors. First, the degree of risk must be assessed. A high hazard facility is one in which cross contamination could be hazardous to health, such as a chrome plating shop or a sewage treatment plant. Second, the plumbing arrangements must be considered. Third, it must be determined whether protection is needed at the water meter or at a location inside the facility.

15.10   HYDRANTS

Fire Hydrants are used to fight fires and to flush lines and may be used to obtain water for construction purposes. In addition, they provide locations for adding chlorine to disinfect pipes, for the insertion of pigs and swabs for pipe cleaning, and to expel air from the system. Hydraulic performance (friction factor) testing, or fire flow testing can be facilitated at hydrant locations. It is important to be properly trained to operate hydrants. Improper use may result in damage, which could jeopardize the community's firefighting abilities.

There are two basic types of hydrants: dry barrel and wet barrel. Most hydrants are of the **dry barrel** compression type. The operating valve on the dry type is located at the bottom and is provided with a **drain hole**. On closing the valve, water is drained. Thus, the barrel remains dry when not in use. This prevents water from freezing during winter. On **wet barrel** types of hydrants, the operating valve is at the outlet and for this reason these hydrants are limited to climates where freezing is less likely.

Fire hydrants should be inspected and maintained at least twice a year, normally in the spring and fall. Most hydrants have two nozzles and a pumper connection. The typical size of the nozzle is 65 mm (2.5 in). In many cities, hydrants are color-coded to indicate the flow capacity. Blue-coded hydrants have capacity exceeding 100 L/s (1500 gpm), whereas red-coded hydrants are low in capacity – usually less than 30 L/s(500 gpm). This capacity is based on a minimum residual pressure of 150 kPa (20 psi) in the water main.

15.11   SERVICE CONNECTIONS

Water from the distribution main reaches the property line of individual consumers through a service pipe, usually made of copper or plastic, with a minimum diameter of 20 mm. Water connections are made initially when the main is installed (**dry tap**) or later when the main is in service using a **wet tap**. A special fitting called the **corporation stop** is employed to make a service connection. At the property line there is an underground shut-off valve known as a **curb stop**. This valve can be operated from the ground surface.

15.12   WATER METERS

Water meters are used to measure water flows. Most meters read the cumulative flow volume. The difference in readings over a period gives the amount of water used over that period. In developed countries, service connections are all metered. That prevents the wastage of water. In countries like India, there are municipal connections which are not metered. Water meters should be accurate, rugged, durable, and prevent backflow. Water meters used in water distribution systems are usually velocity meters or displacement type meters.

Velocity or **inferential** type water meters measure the flow velocity across a cross section and are suitable for measuring high flows. Rotary and turbine meters fall into this category. **Displacement meters** are more suitable for low flow applications, as in the case of residential services. In this type of meter, the quantity of water passing through is measured by filling and emptying out a chamber of known capacity. The type of displacement meters in use include reciprocating, oscillating, and mutating disc types.

15.13   THRUST CONTROL

Thrust is a force acting on valves and fittings causing them to leak or pull apart entirely. This force is caused by changes in flow velocity (momentum), which may happen due to changes in direction (bends) or changes in the flow area like that in a contraction. There are two types of devices commonly used to control thrust: thrust blocks and thrust anchors. A **thrust block** is a mass of concrete, cast in place between the fitting being restrained and the undisturbed soil at the side or bottom of the pipe. A **thrust anchor** is a massive block of concrete cast in place below the fitting to be anchored.

15.14   DUAL WATER SYSTEMS

A dual water distribution system provides two independent pipeline networks within the same municipal service area. One system is for potable water and the second system carries non-potable water. Non-potable water is usually recycled wastewater that can be used for the irrigation of lawns and gardens, firefighting, and street cleaning.

The potable network of a dual system would require a smaller water mains and pumps, as water requirements are reduced. That also means that less water needs to be treated at the water treatment plant or it may be able to meet future demand without expanding the facility. Dual water systems make more sense where water shortage is a recurring problem. Although dual water systems are uncommon at present, it is a viable option for the future. With the advancement in water treatment technology like nano filtration, it is now possible to reclaim wastewater to drinking water standards.

Discussion Questions and Problems

1. Discuss the advantages and disadvantages of various types of pipe materials used in water supply systems.
2. Define and explain following terms:
    a. tuberculation;
    b. cross connection;
    c. exercising of valves;
    a. thrust;
    e. equalizing demand.
3. Describe with the help of sketches various types of joints used in joining water pipes.
4. Compare the following:
    a. intermittent supply versus continuous supply;
    b. elevated storage versus ground storage;

TABLE 15.2

Storage capacity problem

Hour	1	2	3	4	5	6	7	8	9	10	11	12
Demand, ML/d	2.9	2.9	3.6	4.8	6.0	8.4	18	29	46	52	36	24
Supply, ML/d						48	48	48	48	48	48	48
Hour	13	14	15	16	17	18	19	20	21	22	23	24
Demand, ML/d	1.0	0.8	0.6	1.1	1.5	1.8	1.8	1.6	1.4	0.91	0.35	0.25
Supply, ML/d			48	48	48	48	48	48				

    c. inferential meters versus displacement meters;
    d. dry barrel hydrants, wet barrel hydrants;
    e. back pressure versus back siphoning;
    f. linear valves versus rotary valves;
    a. dry tap versus wet tap;
    b. grid iron versus dead end systems;
    i. reinforced concrete pipes versus prestressed pipes;
    j. curb stop versus corporation stop.
5. Describe various methods of preventing cross contamination.
6. Describe briefly the steps involved in laying down water pipes.
7. Why it is important to install water meters in service connections? Describe desired features of water meters.
8. Describe common types of valves used in a water system.
9. Which kind of valve is the most common in water distribution systems. What is the main reason which causes malfunction of these valves?
10. Compare rotary type valves with linear type valves and give two examples of each.
11. Though a grid iron layout of water pipes is most desirable it is not used everywhere. Explain why.
12. A medium sized city has variable demand, as shown in **Table 15.2**. Determine the storage capacity required to equalize the demand against:
    a. the constant rate of pumping @ 24 ML/d. (9.0 ML).
    b. intermittent rate of pumping (6.0 ML).
13. Briefly discuss variations in water demand over time. Sketch a graph that would illustrate hourly variations.
14. What is a dual water system, and what are its advantages and disadvantages?
15. What is meant by the term "equalizing storage"? What benefits does equalizing storage provide in water distribution systems?

16 Pipeline Systems

In a water distribution system it is important to know if various parts of the system are getting adequate pressure. Pipeline sizes are chosen to avoid very low and high pressures in the system. To achieve this, the flow velocity in water mains usually falls in the range of 0.5 to 1.5 m/s (1.5–5 ft/s). To analyze a given pipeline, flow equations are used. Details of flow equations can be found in any standard textbook of fluid mechanics. Flow equations are discussed in Chapter 3 on hydraulics.

16.1 SERIES AND PARALLEL SYSTEMS

If a pipeline system is arranged so that the fluid flows in a continuous line without branching, it is referred to as a **series system**. Conversely, if the system causes the flow to branch out into more than one path, it is called a **parallel system**.

A pipeline can be connected in series or in parallel. Pipelines connected in series will have the same flow rate to maintain the continuity of flow. However, the head loss in each length of pipeline will be different depending on the diameter, length, and roughness. The total head loss in the system will be the sum of individual head loss components.

In parallel pipeline systems, the head loss in the pipeline is the same. However, flow carrying capacity would vary. Hence the total flow carried is the sum of all the flows carried by individual pipelines. For analysis purposes, different branch lines can be replaced by a single pipe of uniform diameter and of length that will pass the discharge Q, with the head loss h_f. This is called an **equivalent pipe**.

16.2 EQUIVALENT PIPE

In engineering analysis, quite often, if a series of pipes are replaced by a single-diameter pipe, which would have the same head loss and discharge rate, the pipe is called an equivalent pipe and the diameter is called an equivalent diameter. Since head loss in the equivalent pipe is the same as that of all the pipes in series, equating the two, we can express equivalent diameter in terms of individual diameters. Similarly, in the case of pipes in parallel, the friction slope is the same in all the pipes.

Pipes in series and parallel

$$\text{In series,} \; \frac{L_e}{D_e^5} = \frac{L_1}{D_1^5} + \frac{L_2}{D_2^5} + \dots \quad \text{In parallel,} \; \sqrt{\frac{D_e^5}{f}} = \sqrt{\frac{D_1^5}{f_1}} + \sqrt{\frac{D_2^5}{f_2}} + \dots$$

DOI: 10.1201/9781003347941-19

Example Problem 16.1

What is the equivalent pipe size to two pipes each of diameter of 6 in connected in parallel? Assume friction factor f remains the same.

Solution:

Pipes in parallel

$$\sqrt{D_e^5} = \sqrt{D_1^5} + \sqrt{D_2^5} = 2\sqrt{6^5}$$

$$D_e = \left(2\sqrt{6^5}\right)^{2/5} = 7.91 = 8 - in(NPS)$$

Example Problem 16.2

Two water reservoirs are connected by a pipeline system consisting of three pipes connected in series. The three pipes are: 300 m of 30 cm diameter, 150 m of 20 cm diameter, and 250 m of 25 cm diameter respectively, of new cast iron. If the elevation difference between the water levels in the two reservoirs is 10 m, find the rate of flow.

Given:

$$L_1 = 300 \text{ m } L_2 = 150 \text{ m } L_3 = 250 \text{ m}$$
$$D_1 = 30 \text{ mm } D_2 = 200 \text{ mm } D_3 = 250 \text{ mm}$$

$$\Delta Z = 10 \text{ m } C = 120 \quad Q = ?$$

Solution:

Writing the energy equation between water levels in the two reservoirs and noting that points 1 and 2 are in open reservoirs, head losses are equal to the difference in elevation of water levels in the two reservoirs.

$$h_l = Z_1 - Z_2 = 10\,m = 10.7\left(\frac{Q}{C}\right)^{1.85}\left(\frac{L_1}{D_1^{4.87}} + \frac{L_2}{D_2^{4.87}} + \frac{L_3}{D_3^{4.87}}\right)$$

$$Q^{-1.85} = \frac{10.7}{10}\times\left(\frac{1}{C}\right)^{1.85}\left(\frac{L_1}{D_1^{4.87}} + \frac{L_2}{D_2^{4.87}} + \frac{L_3}{D_3^{4.87}}\right)$$

$$Q^{-1.85} = \frac{10.7}{10\times120^{1.85}}\left(\frac{L_1}{D_1^{4.87}} + \frac{L_2}{D_2^{4.87}} + \frac{L_3}{D_3^{4.87}}\right)$$

$$Q^{-1.85} = \frac{10.7}{10\times120^{1.85}}\left(\frac{300}{0.30^{4.87}} + \frac{150}{0.20^{4.87}} + \frac{250}{0.25^{4.87}}\right) = 106.5$$

$$Q = 106.5^{-0.54} = \frac{0.0803\,m^3}{s}\times\frac{1000\,L}{m^3} = 80.3 = 80\,L\,/\,s$$

16.3   SYSTEM CLASSIFICATION

When analyzing a fluid flow system, the primary parameters involved are: energy additions (h_a) or subtractions $(h_r$ or $h_L)$, fluid flow rate (Q) or velocity, diameter (D), length (L) of the pipe, wall roughness, and fluid properties including density (γ) and viscosity (ν). When applying the energy equation only one unknown can be determined at any one time. Basic algebraic principles suggest that we need an equation (relationship) for each unknown. Usually, one of the first three parameters, Q, h_f, and D, is the true unknown, while the remaining items are known or specified by the analyst.

16.3.1   CLASS I SYSTEMS

Knowing the pipeline characteristics (L, D, ε), head losses can be determined for a given flow Q. Knowing h_l, head added by the pump h_a can be determined by applying the energy equation. If the flow is all due to gravity $(h_a = 0)$, h_l can be found applying the energy equation.

Example problem 16.3

A hydraulic gradient test is performed in the field by opening the end hydrant. In the direction of flow, pressure readings at hydrants 1 and 2 respectively were observed to be 59.0 and 56.4 psi. From the map for the area, the elevations of hydrants 1 and 2 are 368.70 ft and 366.60 ft respectively. Given that the two hydrants are connected by 2000 ft of 12 in diameter line, what is the hydraulic gradient for the test conditions?

Given:

Variable	Hydrant 1	Hydrant 2
pressure, psi	59.0	56.4
elevation, ft	368.70	366.60
head loss	?	
length, ft	2000	

Solution:

$$h_l = (Z_1 - Z_2) + \frac{(p_1 - p_2)}{\gamma}$$

$$= (368.70 - 366.60)m + (59.0 - 56.4)psi \times \frac{2.31\,ft}{psi} = 8.106 = 8.1\,ft$$

$$S_f = \frac{h_l}{L} = \frac{8.106\,ft}{2000\,ft} \times 100\% = 0.405 = \underline{0.41\%}$$

Example Problem 16.4

A hydraulic gradient test is performed in the field. Two hydrants 750 m apart were chosen to observe the hydraulic head. For the maximum flow conditions, pressure reading at hydrants 1 and 2 were observed to be 477 kPa and 495 kPa. From the map for the area, the elevations of hydrants 1 and 2 are 112.4 m and 111.9 m respectively. Applying energy equation between the two hydrants find the head loss in the pipeline connecting the two hydrants.

Given:

Variable	Hydrant 1	Hydrant 2
pressure, kPa	477	495
elevation, m	112.4	111.9
head loss, h_L	?	
length, m	750	

Solution:

Friction slope of main line tested

$$h_l = (Z_1 - Z_2) + (p_1 - p_2)/\gamma$$

$$= (112.4 - 111.9)m + (477 - 495)kPa \times \frac{m}{9.81\,kPa} = -1.33\,m$$

$$S_f = \frac{h_f}{L} = \frac{\Delta h}{L} = \frac{1.33\,m}{750\,m} \times 100\% = 0.177 = \underline{0.18\%}$$

16.3.2   CLASS II SYSTEMS

There is added difficulty here due to more than one unknown in solving Class II type problems. If the unknown is Q, then N_R cannot be determined. Hence h_l cannot be estimated. To reduce the number of unknowns to one, you have to guess the other unknown quantities. Based on the assumed value (educated guess), the flow rate is determined. This is called trial or **iteration number one.** Note our answer for Q is based on an assumed value of f, so it needs to be checked. Based on the computed value of Q, calculate N_R and therefore find f. Compare the new value of friction factor f with the assumed value. If the two values are close, your guess was right. Otherwise, start again with the new computed value of f and run another trial. Each trial will lead you closer to the correct value. The number of trials will depend on how much precision is required in the analysis.

Example Problem 16.5

Calculate the water flow rate in a 36 in diameter commercial steel pipe for an allowable head loss of 0.40%.

Given:

$$D = 36 \text{ in} \quad S_f = 0.40\% = 4.0 \times 10^{-3}$$

Solution:

This is essentially a Class II type problem. To solve this by applying Darcy's equation we have to resort to the trial-and-error technique using the Hazen Williams equation; solution is direct. For new pipes C = 120 (old pipes get rougher due to corrosion and incrustation).

$$Q = 0.281 C \, D^{2.63} S_f^{0.54} = 0.281 \times 120 \times (36)^{2.63} \times \left(4.0 \times 10^{-3}\right)^{0.54}$$

$$= 21187 = 21000 \text{ gpm}$$

Example Problem 16.6

At night, water is pumped from a water treatment plant reservoir through a 1.5 km long 250 mm diameter pipeline with a roughness coefficient C of 100 to an elevated storage. Assuming no withdrawal, calculate the discharge pressure required to supply 60 L/s of water to the tank. The water surface in the supply reservoir is at an elevation of 3 m, the pump is at 6 m elevation, and the water level in the elevated tank is 42 m.

Given:

$$L = 1.5 \text{ km} = 1500 \text{ m}, \quad D = 250 \text{ mm} = 0.25 \text{ m}$$
$$C = 100 \quad Q = 60 \text{ L/s} = 0.06 \text{ m}^3/\text{s}$$

$$Z_1 = 3.0 \text{ m (supply)} \quad Z_2 = 6.0 \text{ m (pump)} \quad Z_3 = 42 \text{ m (storage)}$$

Solution:

Head loss due to friction

$$h_f = 10.7 \times \left(\frac{Q}{C}\right)^{1.85} \times \frac{L}{D^{4.87}} = 10.7 \times \left(\frac{0.06}{100}\right)^{1.85} \times \frac{1500 \, m}{0.25^{4.87}} = 15 \, m$$

Flow velocity and velocity head (pipeline)

$$v_2 = \frac{Q}{A_2} = \frac{4Q}{\pi D^2} = \frac{4}{\pi} \times \frac{0.06 \, m^3}{s} \times \frac{1}{(0.25 \, m)^2} = 1.22 \, m/s$$

$$\frac{v_2^2}{2g} = \left(\frac{1.22 \, m}{s}\right)^2 \times \frac{s^2}{2 \times 9.81 \, m} = 0.0759 \, m$$

Noting that velocity head is negligible and $v_3 = 0$

$$p_2 = \gamma \left(Z_3 - Z_2 + h_f - h_{v2}\right)$$

$$= \frac{9.81 \, kPa}{m} \times \left(42 \, m - 6.0 \, m + 15 \, m - 0.076 \, m\right) = 499 = \underline{500 \, kPa}$$

Example Problem 16.7

Let us solve the previous Example Problem using the Darcy flow equation. The corresponding relative roughness is 1.0 mm and the kinematic viscosity of the water is 1.1×10^{-6} m²/s.

Given:

$$L = 1.5 \text{ km} = 1500 \text{ m}, \ D = 250 \text{ mm} = 0.25,$$
$$\varepsilon = 1.0 \text{ mm}, \ Q = 60 \text{ L/s} = 0.06 \text{ m}^3/\text{s}$$

$$Z_1 = 3.0 \text{ m (supply)}, \ Z_2 = 6.0 \text{ m (pump)}, \ Z_3 = 42 \text{ m (storage)}$$

Solution:

Friction factor

$$V_2 = \frac{Q}{A_2} = \frac{4Q}{\pi D^2} = \frac{4}{\pi} \times \frac{0.06 \ m^3}{s} \times \frac{1}{(0.25 \ m)^2} = 1.22 \ m/s$$

$$\frac{v_2^2}{2g} = \left(\frac{1.22 \ m}{s}\right)^2 \times \frac{s^2}{2 \times 9.81 \ m} = 0.0759 \ m$$

$$N_R = \frac{vD}{v} = \frac{1.22 \ m}{s} \times 0.25 \ m \times \frac{s}{1.1 \times 10^{-6} \ m^2} = 2.77 \times 10^5$$

$$f = 0.0055 + 0.0055 \times \sqrt[3]{\left(\frac{20000}{1/250} + \frac{10^6}{2.77 \times 10^5}\right)} = 0.0295$$

Head loss and pressure at section 2

$$h_f = f \times \frac{L}{D} \times \frac{v^2}{2g} = 0.0295 \times \frac{1500 \ m}{0.25 \ m} \times 0.0759 \ m = 13.45 \ m$$

$$p_2 = \gamma (Z_3 - Z_2 + h_f - h_{v2})$$

$$= \frac{9.81 \ kPa}{m} \times (42 \ m - 6.0 \ m + 13.45 \ m - 0.076 \ m) = 484 = \underline{480 \ kPa}$$

Example Problem 16.8

Referring to **Figure 16.1**, water at the filtration plant is pumped into a 350 mm diameter water main (C = 110) at a discharge pressure of 350 kPa. The 1.5 km long water main connects to the load center where the residual pressure during the peak demand period is known to be 170 kPa. The elevations of the pump discharge point (A) and the load center (B) respectively are 102.0 m and 105.0 m.

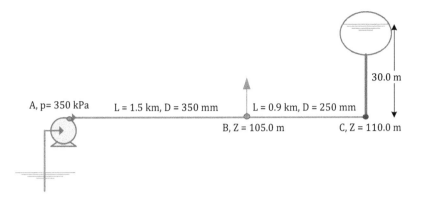

FIGURE 16.1 Simplified water distribution system

i. Calculate the hydraulic gradient in the water main AB.
ii. Applying the Hazen William equation, find how much demand in L/s is served by the clear well at the water plant.
iii. Calculate the flow contribution from the elevated tank.

Given:

	Junction		
Parameter	**A**	**B**	**C**
elevation, Z, m	102.0	105.0	110.0
pressure, kPa	350	170	294
head, m	35.7	17.3	30.0
hydraulic head m	137.7	122.3	140.0

	Pipeline	
Parameter	**AB**	**CB**
length, km	1.5	0.90
diameter, mm	350	250
Coefficient, C	110	110
head loss, m	15.37	17.7
friction slope, S_f, %	1.022	1.96

Pipeline AB

$$h_f = \frac{\Delta p}{\gamma} - \Delta Z = \frac{(350-170)\,kPa.m}{9.81\,kPa} - 3.0\,m = 15.34\,m$$

$$S_f = \frac{h_f}{L} = \frac{15.34\ m}{1500\ m} = 0.0102 = \underline{1.0\%}$$

$$Q_{AB} = 0.278C \times D^{2.63} \times S_f^{0.54} = 0.278 \times 110 \times 0.35^{2.63} \times 0.0102^{0.54}$$

$$= 162.8 = \underline{160\ L/s}$$

Pipeline CB

$$S_f = \frac{\Delta h}{L} = \left(140\ m - \left(\frac{170\ kPa.m}{9.81\ kPa} + 105\ m\right)\right) \times \frac{1}{900\ m} = 0.0196$$

$$Q_{CB} = 0.278C \times D^{2.63} \times S_f^{0.54} = 0.278 \times 110 \times 0.25^{2.63} \times 0.0196^{0.54}$$

$$= 95.4 = \underline{95\ L/s}$$

Demand at the load center

$$Q_B = Q_{AB} + Q_{CB} = 163 + 95 = 258 = \underline{260\ L/s}$$

16.3.3 CLASS III SYSTEMS

This category represents a true design problem because we are required to determine the size of the pipe needed to carry a given flow Q without exceeding the allowable friction slope (head loss per unit length). Without knowing the friction slope D, v, N_R, and f cannot be found. This leaves us with more than one unknown, so a direct solution is not possible. Much like Class II systems, the iteration technique is used to solve problems of this type.

Example Problem 16.9

Determine the size of a concrete main (C = 120) required to carry water at the rate of 60 L/s for an allowable frictional head loss of 0.4%.

Given:

$$Q = 1000\ gpm \quad S_f = 0.4\% = 0.004 \quad C = 120$$

Solution:

Diameter of the pipe

$$D = \left(\frac{Q}{0.281CS_f^{0.54}}\right)^{1/2.63} = \left(\frac{1000}{0.281 \times 120 \times 0.004^{0.54}}\right)^{1/2.63} = 11. = \underline{12\ in\ (NPS)}$$

16.4 COMPLEX PIPE NETWORKS

Modern pipe networks consist of pipe loops, nodes, and junctions. The advantage of this type of system is that a flow direction in each pipe of a loop can change depending on the demand at a given node. Compared to a single pipeline, part of the system can be separated without affecting the supply of water in other sections. Residual pressure at a given node will be determined by the demand and the pipeline characteristics, that is, length, diameter, and roughness. Based on the principles of hydraulics, the following two conditions must be satisfied.

CONTINUITY OF FLOW

Based on the principle of continuity of flow, at a given junction, flow entering must be equal to flow exiting. There must not be any storage at a junction. In mathematical terms, the algebraic sum of all the flows at every junction must be zero.

CONTINUITY OF PRESSURE

This principle applies to a pipe loop in each network. According to this principle, the algebraic sum of all the head losses in a loop must be zero. Put differently, the pressure at a given junction must be the same, whether you do the calculation clockwise or counterclockwise. Applying these two principles, pipe networks can be analyzed for pressures and flows.

16.5   HARDY CROSS METHOD

The Hardy Cross method has no direct solution. It is based on successive iterations such that correction to the flow is almost negligible. Here the correction means the value added or subtracted from the assumed flow for carrying out the next iteration.

Any flow formula can be written in the general form $h_L = KQ^n$. Exponent n is 2 for the Darcy Weisbach equation and 1.85 for the Hazen Williams equation, and constant K is based on the other terms in the flow formula, like length, friction factor, and diameter of the pipe. Expanding the terms, the flow correction at the end of the iteration is:

$$Correction, \Delta Q = \frac{-\Sigma h_L}{n\Sigma\left(h_L / Q\right)}$$

Use of this method is illustrated in Example Problem 16.10

Example Problem 16.10

An elevated tank at location A maintains a pressure head of 42 m, and the minimum desired pressure at remote location D is 210 kPa. Work out flows in all the pipes based on demands as shown in **Figure 16.2**. Check the available pressure at remote point D and other junctions.

Solution:

Flows are assumed such that there is no net flow at a junction. Applying the Hardy Cross method, flows and corresponding head losses are computed as shown in the Excel worksheet (**Table 16.1**). Flows are balanced after three trials.

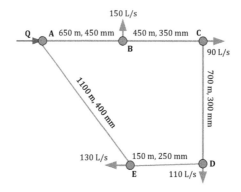

FIGURE 16.2 Pipe network (Ex. Prob. 16.10)

TABLE 16.1
Excel worksheet (Ex. Prob. 16.10)

Pipe	Length	Coefficient	Diameter	Flow	Loss	Ratio
Designation	L, m	C	D, mm	Q, L/s	h_L, m	h_L/Q, m.s/L
Trial 1						
AB	650	110	450	300	6.1	2.0E-02
BC	450	110	350	150	4.0	2.7E-02
CD	700	110	300	60	2.4	4.0E-02
DE	150	110	250	-50	-0.9	1.8E-02
EA	1100	110	400	-180	-7.2	4.0E-02
				$\Sigma =$	4.5	1.5E-01
	$\Delta Q = -\Sigma h_L/(1.85\Sigma(h_L/Q)) =$			-16.75	L/s	
Trial II						
AB	650	110	450	283	5.5	1.9E-02
BC	450	110	350	133	3.2	2.4E-02
CD	700	110	300	43	1.3	3.1E-02
DE	150	110	250	-67	-1.5	2.3E-02
EA	1100	110	400	-197	-8.4	4.3E-02
				$\Sigma =$	0.1	1.4E-01
	$\Delta Q = -\Sigma h_L/(1.85\Sigma(h_L/Q)) =$			-0.3	L/s	
Trial III						
AB	650	110	450	283	5.5	1.9E-02
BC	450	110	350	133	3.2	2.4E-02
CD	700	110	300	43	1.3	3.0E-02
DE	150	110	250	-67	-1.5	2.3E-02
EA	1100	110	400	-197	-8.5	4.3E-02
				$\Sigma =$	0.0	1.4E-01
	$\Delta Q = -\Sigma h_L/(1.85\Sigma(h_L/Q)) =$			0.0	L/s	

16.6 COMPUTER APPLICATIONS

Municipal water distribution systems are, of course, more complex than one or two loops, as illustrated in the previous two example problems. Network analysis is shown by using a computer program spreadsheet. Not very long ago, the same problems were solved doing calculations by hand. One of the very strong points of a computer is that in seconds it can perform repetitive work which took hours by hand.

Mathematical models of distribution networks are now easily created and analyzed using computers. Modern network software combines hydraulic calculations with other software programs like CAD and GIS.

Discussion Questions

1. Starting with the following form of the Darcy Weisbach flow equation, modify it in terms of flow capacity Q.

$$h_f = f \times \frac{L}{D} \times \frac{v^2}{2g}$$

2. Comparing the Darcy Weisbach equation with Manning's flow equation, find the relationship between n and f.
3. Describe the three types of flow equations and their suitability in the hydraulic analysis of pipeline systems.
4. Discuss how the principles of continuity of flow and continuity of pressure are used in pipe network analysis.
5. Describe the steps involved in the application of the Hardy Cross method.
6. When sizing water mains using the Darcy Weisbach flow equation, it might involve number of iterations. However, using the Hazen Williams equation you are able to get a direct answer. Why not then always use the Hazen Williams equation?
7. Search the internet to find commonly used network software, and comment on their usefulness.
8. What is an equivalent pipe? What is the purpose of determining equivalent pipes in water distribution systems?
9. Discuss pipes connected parallel and in series. Using the Hazen Williams flow equation, derive the expressions for equivalent pipes in both cases.
10. Professional bodies recommend the use of the Darcy Weisbach flow equation over the Hazen Williams flow equation. Comment.

Practice Problems

1. A hydraulic gradient test is performed in the field. Two hydrants 750 m apart were chosen to observe the hydraulic head. For the maximum flow conditions, pressure reading at hydrants 1 and 2 were observed to be 477

kPa and 495 kPa. From the map for the area, the elevations of hydrants 1 and 2 are 112.4 m and 111.9 m respectively. Find the hydraulic head at each of the hydrants and hence the head loss in the pipeline connecting the two hydrants. (1.3 m)

2. During water distribution testing, a pressure loss of 1.6 psi was observed over a 3000 ft length of the pipe. What is the friction slope? (0.12%).

3. Calculate the flow-carrying capacity of a 6 in diameter pipe (C = 100) for an allowable friction loss of 0.5 ft per 100 ft length of pipe. (180 gpm).

4. What size pipe (C = 100) should be used to supply 100 L/s so that the head losses do not exceed 2.5 m per km length? (400 mm).

5. A flow test was conducted on part of a 250 mm pipeline. Three hydrants in a row are spaced 300 m apart. When the end hydrant was opened, the pressure difference in the first two was read to be 13 kPa. Based on the Pitot meter reading at the flowing hydrant, flow during the test was found to be 31 L/s. What is the coefficient of roughness C? (80).

6. Two reservoirs are connected by a 300 mm diameter, 1200 m long pipeline. Assuming f = 0.02, find flow rate when the water level in the lower reservoir is 32 m below the water level in the upper reservoir. Assume minor losses are negligible. (200 L/s).

7. In a 200 mm diameter pipeline absolute roughness has increased to 1.1 mm due to tuberculation. What friction factor f value would you use to work out the head losses during the peak demand of 35 L/s? Assume kinematic viscosity at the prevailing temperature is 1.2×10^{-6} m²/s. (0.032).

8. In an 8 in diameter pipeline, due to tuberculation friction factor f, for the peak hour, demand has gone up to 0.032. What slope of the hydraulic grade line would you expect during a peak demand of 530 gpm? (0.86%).

9. In a field test, a pressure drop of 3.5 psi is observed over a length of 1000 ft of a 6 in water main when carrying a flow of 250 gpm. What is the friction coefficient? (110).

10. Estimate the flow capacity of a 12 in diameter main for an allowable friction slope of 0.1%. Assume Hazen Williams coefficient C is 120, a relatively smooth pipe. (560 gpm).

11. A pump installed at an elevation of 170 m delivers 220 L/s of water through a horizontal pipeline to a pressurized tank (**Figure 16.3**). The water level in the

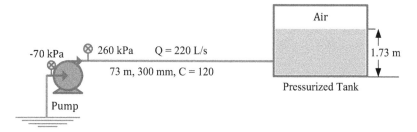

FIGURE 16.3   Water supply to a pressurized tank

tank is 1.73 m. The pressure gauge on the 30 cm diameter suction side of the pump reads 70 kPa vacuum and at the 30 cm diameter discharge side the pressure is 260 kPa. The cast iron discharge pipe is 43 m long. Determine the water pressure in the tank assuming coefficient of friction C is 120? (220 kPa).

12. It has been estimated that due to tuberculation, the effective pipe diameter of a 200 mm water main has been reduced by 3.0 mm and the coefficient of friction C has dropped from 110 to 90. What is the % increase in head losses? (56%).

13. What diameter cast iron pipe (ε = 0.36 mm) is required to carry water at the rate of 250 L/s for an allowable head loss of 2.0%? Assume kinematic viscosity of water at the prevailing temperature of 10°C is 1.3 × 10⁻⁶ m²/s. (350 mm).

14. What diameter cast iron pipe (C = 110) is required to carry water at the rate of 250 L/s for an allowable head loss of 2.0%? (400 mm-NPS).

15. What is the expected head loss/mile length of a 12 in water main with a roughness coefficient of 120 when carrying a flow of 1200 gpm? (22 ft/mile).

16. A 50 mm diameter and 8.0 m long pipeline is connected to a reservoir with a square entry. At the end, it is connected to a 100 mm diameter and 45 m long pipe. What must be the level of water in the reservoir if it is desired to achieve exit velocity of 1.5 m/s. Assume friction factor f = 0.026. (9 m).

17. A 24 in diameter water main is 7.5 miles long. It carries water from the pumping station to the load center. Calculate the pressure loss assuming a coefficient C of 110 when carrying peak flow of 3200 gpm? (17 psi).

18. Two water reservoirs are connected by a pipeline system consisting of three pipes connected in series. The three pipes of cast iron (C = 110) respectively are: 1000 ft of 12 in, 500 ft of 8 in, and 820 ft of 10 in diameter. When the elevation difference between the water levels in the two reservoirs is 32 ft, what is the rate of flow? (1200 gpm).

19. Referring to **Figure 16.4**, water is pumped into a 12 in diameter water main at a discharge pressure of 80 psi. The 5000 ft long water main connects to

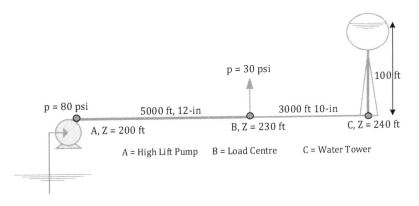

FIGURE 16.4 Simple water distribution system

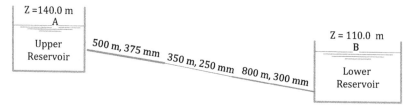

FIGURE 16.5   Two connected reservoirs

the load center where the residual pressure during the peak demand period is 30 psi. During peak demand, the water level in the tower is 100 ft above ground. Assuming C of 110 for all pipes, determine the peak demand. (3700 gpm).

20. Two water reservoirs are connected by three pipelines, as shown in **Figure 16.5**. The length and diameters of the pipelines are indicated in the figure. Assuming friction factor f = 0.02 for all the three pipes, determine the flow rate when the elevation difference for water levels as shown. (155 L/s).

17 Pumps and Pumping

In water distribution, a wide variety of pumps are used to transport water. Though centrifugal types of pumps are common, there are other types of pumps which are also used in special applications. Two broad categories of pumps are positive displacement pumps and kinetic or velocity pumps.

17.1 POSITIVE DISPLACEMENT PUMPS

Positive displacement pumps are designed to deliver a fixed quantity of fluid for each revolution of the pump rotor. Therefore, except for minor slippage, the delivery of the pump is unaffected by changes in the delivery pressure. In general, these pumps will pump against high pressure, but their capacity is low. The pumps are well suited for pumping high viscosity liquids and can be used for metering since output is directly proportional to rotational speed. Reciprocation pumps, piston pumps, and rotatory pumps are all positive displacements type of pumps.

17.2 VELOCITY PUMPS

As the name indicates, this category of pump does add to the kinetic energy of the fluid, which is later converted to pressure energy. A centrifugal pump is the most commonly used velocity pump. Non-positive displacement pumps are generally used to transfer large volumes of liquids at relatively low pressures. However, if pressure is increased, pumping rate drops. This is in strong contrast with positive displacement pumps.

17.2.1 TYPES OF CENTRIFUGAL PUMPS

A centrifugal pump has a radial flow or mixed type impeller. The casing of a centrifugal pump may either be **volute** type or turbine type, also called **diffuser** casing. In a volute type pump, flow velocity is reduced as the water enters the outlet thus increasing the pressure. In turbine pumps, velocity is reduced by stationary guide vanes before the water enters the casing, thus gives better transfer of flow energy or efficiency.

Turbine Pumps

Turbine pumps are usually **multistage** and used for pumping from deep wells. In multistage pumps, each stage adds equal head. The prime mover is kept at the ground surface and impellers are attached to the bottom of a vertical shaft suspended in the borehole.

DOI: 10.1201/9781003347941-20

Submersible Pumps

As the name indicates, both motor and pump are placed under water thus vertical shaft is eliminated. Water rises through a riser pipe to which the whole assembly is attached. This kind of pump can be used for domestic as well as municipal water supplies.

Jet Pumps

A jet pump is a special kind of pump. Basically, it is a combination of a centrifugal pump at the ground surface and a venturi under water. As water is pushed through the venturi, suction is created to lift water from deep wells. These pumps are not very efficient and are common for domestic water supplies where the lift is significantly high.

17.2.2 POSITIVE DISPLACEMENT PUMP CHARACTERISTICS

The capacity of positive displacement pumps is independent of delivery pressure. At high pressures a small decrease in capacity may occur due to internal leakage. Power required to drive the pump varies linearly with pressure. Therefore, it becomes necessary to protect the positive displacement pumps with relief valves to prevent damage by over-pressurization.

17.2.3 PERFORMANCE CURVES OF CENTRIFUGAL PUMPS

Pumping capacity of centrifugal pumps is strongly affected by the operating head. There is an inverse relationship between capacity Q and total dynamic head H (h_a). As the head increases, the capacity decreases.

Performance curves or characteristic curves are relationships of head, efficiency, and power versus capacity at a given speed of the pump. A typical head capacity curve (*H* versus *Q*) is shown in **Figure 17.1**. The pump will never operate at a point which does not lie on this line. In the case of centrifugal pumps, if the operating head changes, capacity will change and correspondingly efficiency and power will change. Keep in mind, though, for small drops in pumping head, flow capacity may increase in greater proportions. In fact, the pump will fail to pump any water if the pump equals or exceeds **shut off head.** The operating point corresponding to peak efficiency is known as **best efficiency point** (bep). To get more out of your pump, it is a good practice to operate a rate of discharge within a range of 60% to 120% of bep.

Example Problem 17.1

Sault city water treatment plant gets its water from the Lake Superior. The intake is below the water surface at an elevation of 230.4 m. The lake water is pumped to the plant influent at elevation 242.4 m. The total head losses are estimated to be 5.5 m when water is drawn at the rate of 340 L/s. Calculate pumping head, water power, and the pump power, assuming pump is 72% efficient?

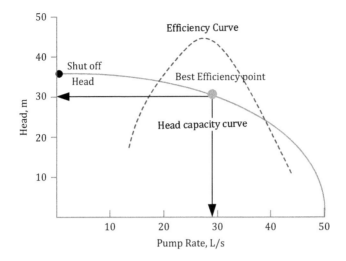

FIGURE 17.1   Centrifugal pump performance curves

Given:

Variable	Suction = 1	Discharge = 2
pressure, p	$p_1 = 0$	$p_2 = 0$
elevation, Z	$Z_1 = 230.4$ m	$Z_2 = 242.4$ m
head loss, h_L	5.5 m	

Solution:

$$h_a = \frac{p_2}{\gamma} - \frac{p_1}{\gamma} + Z_2 - Z_1 + h_l = 0 + (242.4 - 230.4)m + 5.5m = 17.5 = 18m$$

$$P_a = Q \times \gamma \times h_a = \frac{0.340\ m^3}{s} \times \frac{9.81\ kN}{m^3} \times 17.5\ m \times \frac{kW.s}{kN.m} = 58.3 = 58\ kW$$

$$P_P = \frac{P_a}{E_P} = \frac{58.3\ kW}{72\%} \times 100\% = 80.9 = 81\ kW$$

Example Problem 17.2

A city is served partly by a deep well. This well is pumped at a constant rate of 550 gpm to an overhead storage tank though an 8 in diameter and 1500 ft long rising main. The difference in water levels including the drawdown is 100 ft m. Assume coefficient of friction f = 0.022. What power is required by an electric motor, assuming the overall efficiency of pumping unit is 70%?

Given:

D = 8 in = 0.667 ft   L = 850 ft   f = 0.022   Q = 550 gpm
ΔZ = 100 ft P_M = ? E_o = 70%

Solution:
Pump rate

$$Q = \frac{550\ gal}{min} \times \frac{min}{60\ s} \times \frac{min}{7.48\ gal} = 1.225 = 1.23\ ft^3/s$$

Head loss due to friction

$$V_2 = \frac{Q}{A_2} = \frac{4Q}{\pi D^2} = \frac{4}{\pi} \times \frac{1.225\ ft^3}{s} \times \frac{1}{(0.667\ ft)^2} = 3.37\ ft/s$$

$$h_f = f \times \frac{L}{D} \times \frac{v_2^2}{2g} = 0.022 \times \frac{850\ ft}{0.667\ ft} \times \left(\frac{3.37\ ft}{s}\right)^2 \times \frac{s^2}{64.4\ ft} = 8.72\ ft$$

Pumping head and motor power

$$h_a = \frac{p_2}{\gamma} - \frac{p_1}{\gamma} + Z_2 - Z_1 + h_l = 0 + 100\ ft + 8.72\ ft = 108.7\ ft$$

$$P_a = Q \times \gamma \times h_a = \frac{1.225\ ft^3}{s} \times \frac{62.4\ lb}{ft^3} \times 108.7\ ft \times \frac{hp.s}{550\ lb.ft} = 15.1 = 15\ hp$$

$$P_M = \frac{P_a}{E_o} = \frac{15.1\ hp}{0.70} = 21.58 = 22\ hp$$

Example Problem 17.3

Referring to **Figure 17.2**, water is pumped at the rate of 180 L/s from lower reservoir to upper reservoir. Elevation of water surfaces in the lower and upper reservoirs respectively are 110.0 m and 120.0 m. Two reservoirs are connected via pipeline BC of 450 mm diameter and length of 1.2 km. At junction C, it splits in two pipelines, CD and CE. Pipeline CD is 300 mm and 830 m while pipeline CE is 250 mm and 710 m long. For all three pipelines assume friction factor of 0.025.

   i. Calculate head loss due to friction in pipeline BC carrying flow of 180 L/s.
   ii. Flow carried by line CD and CD each.
   iii. Pumping head and pump power, assuming pump efficiency of 80%.

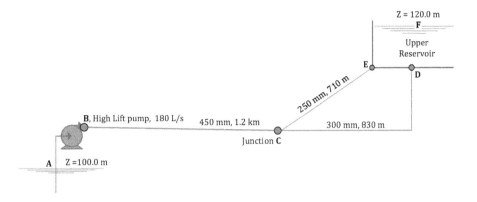

FIGURE 17.2   A simple distribution pumping system

	Point of Interest			
Parameter	**A**	**B**	**C**	**F**
elevation, Z, m	110.0	?	Same as B	120.0
head, m	0	?	?	0
hydraulic head, m	110.0	?	?	120.0

Given:

	Pipeline		
Parameter	**BC**	**CD**	**CE**
length, m	1200	830	710
diameter, mm	450	300	250
friction factor, f	0.025	0.025	0.025
head Loss, m	?	?	same as CD
friction Slope, S_f, %	?	?	same as CD

Solution:

Head loss in pipeline BC

$$h_f = \frac{f \times L}{1.23\,g} \times \frac{Q^2}{D^5} = \frac{0.025\,s^2}{12.1m} \times \frac{1200\,m}{(0.45\,m)^5} \times \left(\frac{0.18\,m^3}{s}\right)^2 = 4.3\,m$$

Head loss in CD = head loss in CE

$$\frac{Q_{CD}}{Q_{CE}} = \sqrt{\frac{L_{CE}}{L_{CD}} \times \left(\frac{D_{CD}}{D_{CE}}\right)^5} = \sqrt{\frac{710}{830} \times \left(\frac{300}{250}\right)^5} = 1.458 = 1.5$$

$$Q_{CD} + Q_{CE} = 180 L/s \ \ or \ 1.458 Q_{CE} + Q_{CE} = 180 L/s$$

$$Q_{CE} = \frac{180 L/s}{2.458} = 73.2 = 73 L/s$$

$$Q_{CD} = 180 - 73.2 = 106.8 = 107 L/s$$

Head loss due to friction in line CD

$$h_f = \frac{fL}{1.23 g} \times \frac{Q^2}{D^5} = \frac{0.0250 s^2}{12.1 m} \times \frac{830 m}{(0.300 m)^5} \times \left(\frac{0.107 m^3}{s}\right)^2 = 8.1 m$$

Check head loss in CE

$$h_f = \frac{f \times L}{1.23 g} \times \frac{Q^2}{D^5} = \frac{0.0250 s^2}{12.1 m} \times \frac{7100 m}{(0.250 m)^5} \times \left(\frac{0.073 m^3}{s}\right)^2 = 8.0 m$$

Pumping head and power

$$h_a = \frac{p_2}{\gamma} - \frac{p_1}{\gamma} + Z_2 - Z_1 + h_l = 0 + 10 \ m + 0 + 4.35 \ m + 8.07 \ m$$

$$= 22.42 = 22.4 \ m$$

$$P_a = Q \times \gamma \times h_a = \frac{0.180 \ m^3}{s} \times \frac{9.81 kN}{m^3} \times 22.42 \ m \times \frac{kW.s}{kN.m} = 39.5 \ kW$$

$$P_p = \frac{P_a}{E_p} = \frac{39.58 \ kW}{80\%} \times 100\% = 49.48 = 49 \ kW$$

17.3   SYSTEM HEAD

The operating point of the pump is determined by the system it is serving. For a given water system, an increase in discharge will cause a greater increase in head loss, since head loss in a straight pipe is proportional to the square of the flow rate. Hence, as shown in **Figure 17.3**, the shape of the system head curve is concave upwards. **System head** consists of two components: fixed or static head and variable component due to head losses.

$$System \ head, h_{sys} = Fixed \ head, h_0 + Variable \ head, h_l$$

When the points in question refer to the open reservoirs, $p_1 = p_2 = 0$ and fixed head equals the total lift. The second term in the system head is the head loss, which can be described by using a flow equation like the Darcy Weisbach equation. System

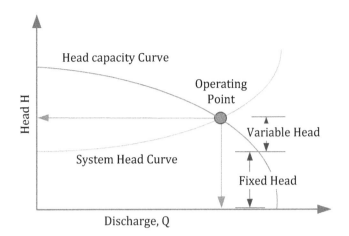

FIGURE 17.3 Pump head and system head

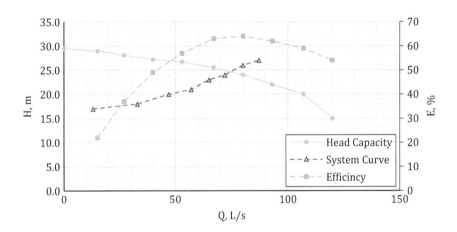

FIGURE 17.4 Pump operating point (Ex. Prob. 17.4)

head is equal to static head plus the head losses. Major head loss is calculated using the flow formula and minor losses are usually neglected or are considered by applying equivalent length technique. The relationship between system head and flow rate is called the **system head curve**. Superimposing the system curve on the pump curve, the point of intersection is the **operating point** of the pump serving the system (**Figure 17.4**). By varying the head (opening or closing valves), the pump operating point will shift along the head capacity curve. For a given system, select the pump which operates in the vicinity of bep. If the system head or demand varies over a wider range, variable speed pump or combination (parallel and series) of pumps may be the answer.

TABLE 17.1
Operating point of a pump

Pump Characteristics			System Head	
Flow, L/s	Head, m	Efficiency, %	Flow, L/s	Head, m
15	29	22	13	17
27	28	37	33	18
40	27	49	47	20
53	27	57	57	21
67	26	63	65	23
80	24	64	72	24
93	22	62	80	26
107	20	59	87	27
120	15	54		

Example Problem 17.4

Pump and system characteristics are given in **Table 17.1**. Plot pump curve and system head curve and read the operating point.

Solution:
Plot is shown in **Figure 17.4** and operating point is the point where both curves meet. Operating point as read from the graph is (74 L/s, 24.5 m, 65%).

17.4   AFFINITY LAWS

The performance of centrifugal pumps can be varied by varying the rotative speed, N. To some degree this can also be accomplished by trimming down the pump impeller. The way pump characteristics change with changes in either speed or diameter is determined by **affinity laws**. For a given pump, impeller diameter is fixed or $D =$ constant. When speed is changed from N_1 to N_2 and ratio $N_2/N_1 = $ Constant, C, pump characteristics change according to following relationships.

Affinity Laws

$$\boxed{\frac{Q_2}{Q_1} = \frac{N_2}{N_1} \quad \frac{Q_2}{Q_1} = \left(\frac{N_2}{N_1}\right)^2 \quad \frac{P_2}{P_1} = \left(\frac{N_2}{N_1}\right)^3}$$

When diameter (D) of the impeller is reduced (trimmed) from D_1 to D_2 such that ratio $D_2/D_1 = $ *Constant, C* and speed remains the same, similar relationships apply except you use new constant value. At various pump speeds the factor H/Q^2 remains constant, thus pump speed could be adjusted to produce desired characteristics.

Example Problem 17.5

A pump delivers 500 gpm at 1000 rpm against a total head of 45 ft. Determine its performance at 1100 rpm.

Given:

$Q_1 = 500$ gpm, $N_1 = 1000$ rpm, $Q_1 = 45$ ft, $N_2 = 1100$ rpm

Solution:

$$Q_2 = Q_1 \times \frac{N_2}{N_1} = 500\,gpm \times \frac{1100}{1000} = 550\,gpm$$

$$H_2 = H_1 \times \left(\frac{N_2}{N_1}\right)^2 = 500\,gpm \times \left(\frac{1100}{1000}\right)^2 = 54\,ft$$

$$P_2 = P_1 \times \left(\frac{N_2}{N_1}\right)^3 = P_1 \times \left(\frac{1100}{1000}\right)^3 = 1.33 \times P_1$$

17.5 SPECIFIC SPEED

To compare the performance of different pumps, it is necessary to have a term which is common to all centrifugal pumps. Specific capacity is the term used for this purpose. Also called **type number**, specific speed is another performance factor that is widely used for both preliminary design and selection of pumps.

Specific speed

$$N_s = \frac{N(rpm)\sqrt{Q(gpm)}}{\left(H(ft)\right)^{3/4}} \qquad N_s\,(metric) = \frac{N(rpm)\sqrt{Q(L/s)}}{\left(H(m)\right)^{3/4}}$$

The values of Q and H correspond to the **best efficiency point** (maximum efficiency) for the shaft speed used. The value in metric units is 61% of the value in USC. Unfortunately, the units shown above do not use a dimensionless form for N_s. In order to write this equation in consistent units, and make it a dimensionless number, the denominator has to contain the value of acceleration due to gravity.

Most published data have been obtained based on the first form of the equation. Multiplying a dimensionless number by a factor of 17 200 gives you the specific speed in the old fashion. For a multistage pump, the value of H should be the same as that for a one-stage pump. A given specific speed refers to a certain combination of head, speed, and capacity, which is typical of a given type of pump. Specific speed can tell us the combination of factors that are both possible and desirable. Low specific speeds refer to centrifugal type pumps, whereas axial flow (propeller type) pumps have high specific speeds. Best efficiencies are obtained when specific speed is in the range 1500–3000.

17.6 HOMOLOGOUS PUMPS

Dynamically similar pumps are called homologous pumps. In addition to geometrical similarity, homologous pumps have the same specific speed, same operating efficiency, and similar flow pattern. When two pumps are dynamically similar, the performance of one can be predicted by knowing the performance of the other. In homologous pumps, discharge varies proportional to cube of diameter, and head varies proportional to square of diameter.

17.7 MULTIPLE PUMPS

When the desired capacity is beyond the range of a single suitable pump, two pumps operating in parallel may be used. Similarly, for high head requirements, two pumps connected in series will be more efficient. When two or more pumps are used in series or in parallel, a combined pump curve should be developed to determine the operating point. When multiple pumps are running in parallel, head remains the same and discharge rates are additive. On the other hand, when pumps are in series, heads are additives and discharge rate remains the same.

In medium to large communities, pumps are usually a combination of small and large pumps connected in parallel to meet the varying demand. Some pumps are running all the time, while other pumps come on when the demand increases.

In smaller communities, sometimes a **variable speed pump,** also called a variable frequency drive (**VFD**) may suffice. As discussed earlier, pump characteristics change with changes in pump speed in accordance with **affinity laws**. In a VFD pump, an adjustable speed motor is used. Using a VFD motor, pump impeller speed is reduced from its design operating point, effectively lowering the position of discharge capacity curve and allowing the pump to operate at low discharge rate without wasting electrical energy. Electronic sensors installed in the system at locations remote from the pump can control VFD pumps so pump speed matches the system requirements. Newer models of VFD pumps are equipped with **microprocessors** to adjust both speed and pressure to meet system requirements without having to place remote sensors.

Example Problem 17.6

Head capacity data of a centrifugal pump is shown in **Table 17.2**. The static head component of system head is 20 m and variable head is $7620Q^2$, where Q is in m³/s. Find the operating head if two pumps are connected in (a) series (b) parallel.

Solution:
Pump and system data are shown in **Table 17.2**. When pumps are connected in series, heads are cumulative. Since pumps are identical for each discharge, values head is doubled as shown in **Table 17.2**, column 2. When pumps are connected in parallel, flows are cumulative. Since pumps are identical for each head, value flow is doubled. System head is the sum of the fixed head (static) and variable head. For example, when Q = 20 L/s, system head is:

$$H_{sys} = 20 + 7620\, Q^2 = 20 + 7620 \left(\frac{20L}{s} \times \frac{m^3}{1000\, L} \right)^2 = 23.04 = 23\, m$$

TABLE 17.2

Pumps in series and parallel

Pump Head, m (Series)		Flow, L/s (Parallel)		System Head
H_1	H_{1+1}	Q_1	Q_{1+1}	H_{sys}
26	52	0	0	20
23	46	10	20	23
21	42	15	30	27
16	32	25	50	39
9	18	35	70	57
5	10	40	80	69

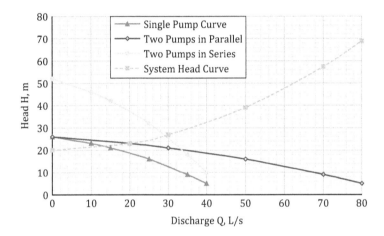

FIGURE 17.5   Two identical pumps in series and parallel

Plot of pump curves and system curve is shown in **Figure 17.5**. Operating point is where the system head curve crosses the pump curve. Operating point for a single pump is (22 L/s, 15 m).

a) In series operation, pump curve moves up, since heads are cumulative, thus operating point moves. In this case operating point as read from Figure 17.5 is (29 L/s, 26 m).

b) For parallel operation, pump curve moves towards right. From Figure 17.5, operating point for pumps running in parallel is (20 L/s, 22 m).

17.8   CAVITATION

The **vapor pressure** of the liquid refers to the pressure at which liquid transforms into vapors. For any liquid there is a definite relationship between the vapor pressure and the temperature of the fluid. Vapor pressure increases with an increase in temperature. At normal atmospheric pressure, water starts boiling at 100°C. In other words, at 100° C, the vapor pressure of water is one atmosphere or about 100 kPa (abs). When operating a pump, pressure at the pump inlet is lower than intake due to losses. if the inlet pressure is allowed to drop close to vapor pressure, bubbles of vapors will start forming. These bubbles collapse as they enter the high-pressure region in the pump body This will create sudden noise and pitting of the metal surface due to the explosion of bubbles This phenomenon is called **cavitation**.

17.8.1 NET POSITIVE SUCTION HEAD

The basic measure to protect the pump against cavitation is to avoid pressures that are close to the vapor pressure of the liquid. Pumps are operated under lift conditions or positive head conditions. Pumps are more susceptible to cavitation when operated under lift conditions. Head available at the pump inlet, called **net positive suction head** (NPSH), is less than at the intake due to losses and lift. To prevent cavitation, pumping systems are designed and operated such that NPSH remains above the vapor pressure head. Using the pump centerline as a reference, net positive suction head available is given by:

$$NPSH_A = h_{atm} - h_{vap} - (Z_2 - Z_1) - h_l$$

h_{atm} = atmospheric pressure as head based on altitude
h_{vap} = vapor pressure head h_1 = head loss due to friction
Z_2 = pump elevation Z_1 = pumping surface elevation
Z_2_Z_1 = suction lift

Note: In suction head conditions $Z_1 > Z_2$ and term $(Z_2 - Z_1)$ is negative.

Example Problem 17.7

Determine the available NPSHA for a pumping system pumping water at 20°C from a well when pumping water level is 2.5 m below the pump. The atmospheric pressure is 101 kPa (abs). Total head losses in the suction line are estimated to be 0.45 m.

Given:

p_{atm} = 101 kPa (abs) $Z_2 - Z_1$ = suction lift = 2.5 m h_l = 0.45 m

T = 20° C p_{vap} = 2.34 kPa (abs) γ = 9.79 kPa/m

Solution

$$NPSH_A = h_{atm} - h_{vap} - (Z_2 - Z_1) - h_l$$

$$= (101 - 2.34) kPa \times \frac{m}{9.79\ kPa} - 2.5\ m - 0.45\ m = 7.05 = \underline{7.1\ m}$$

17.8.2 Permissible Suction Lift

Manufacturers specify the **NPSH required** (NPSHR) for efficient operation of their pumps. If the pump is operated above this value, the operation will be satisfactory. When the pumping level is open to atmosphere and there is a positive suction lift (negative head), it is important to check that available NPSHA > NPSHR. Maximum suction lift which can be allowed for a given pump can be found by knowing the NPSHR and vapor pressure of the liquid being pumped. Suction lift, SL is the height of the pump above the pumping water level.

$$NPSH_A = h_{atm} - h_{vap} - SL - h_l$$

Maximum permissible suction lift (MPSL) without causing any cavitation corresponds to the minimum $NPSHA = NPSHR$. Making these substitutions, equation for calculating maximum permissible suction lift can be found as follows:

$$MPSL = h_{atm} - h_l - NPSH_R - h_{vap}$$

Example Problem 17.8

Find the permissible suction lift for a pump which requires 12 ft of NPSH. Total head losses in the suction line are estimated to be 2.3 ft. The temperature of water is 15°C and the atmospheric pressure head is 33.5 ft.

Given:

$h_l = 0.7$ m, $NPSH_R = 3.5$ m, T = 15°C, $h_{vap} = 0.69$ ft, $h_{ams} = 33.5$ ft

Solution:

$$MPSL = h_{atm} - h_{vap} - h_l - NPSH_R$$

$$= 33.5 \ ft - 0.69 \ ft - 2.3 \ ft - 12 \ ft = 18.51 = 19 \ ft$$

If the pump is placed more than 19 ft above the pumping water level, it will cause cavitation.

17.9 OPERATION AND MAINTENANCE

Proper operation maintenance (O&M) procedures must be followed to obtain satisfactory service from centrifugal pumps. Pumps must be properly lubricated using the recommended lubricant. Remember, both under and over lubrication are damaging.

The rotating part is the only moving part in the casing. A **packing gland** or seal is used where the pump shaft protrudes to stop air from leaking in or water from leaking out. Never overtighten the packing. To make sure the seal is working properly, tighten it to allow some leakage of about 15 to 20 drops per minute. In case of

mechanical seals, no leakage is allowed. Never overtighten. Replace the seal if too much water starts coming out.

A centrifugal pump must be primed or filled with water when it is started. Many pumping units have self-priming units attached. The foot valve is designed to prevent suction line and pump loose prime. If the pump is placed below the pumping water level or suction head conditions, pump remains always primed.

Pump should be started with suction valve open and discharge valve closed. As the motor picks up speed, the discharge valve is opened slowly. In many cases, this is an automatic operation. When shutting off the pump, the discharge valve must be closed slowly to avoid **water hammer**. Water hammer refers to tremendous transient pressures that can damage pipes or pumps. Routine inspection of pumps is important to check for noise, vibrations, alignment, excessive heat, and leakage from gland packing. Vibrations are an indication that the pumping unit is out of alignment, and if not corrected can lead to premature failures.

Discussion Questions

1. Explain why efficiency of diffuser type pumps is better than volute type pumps.
2. Briefly describe the principle on which centrifugal pumps work.
3. Explain why centrifugal type pumps are more common in water systems.
4. Compare reciprocating pumps with centrifugal pumps.
5. How would pump characteristics change if two identical pumps are coupled in series or in parallel?
6. How would you proceed to select the right pump for a given water pumping system?
7. In suction lift conditions, it is recommended to use the minimum of fittings and preferably use a larger suction pipe. Discuss.
8. Describe the relationship between specific speed and type of velocity pump.
9. How does the operation of a pumping system affect operating efficiency?
10. What steps can you take to prevent pump cavitation?
11. Cavitation problems are common when pumping relatively warm liquids and at high altitudes. Comment.
12. What type of pumps are commonly used to withdraw water from water wells?
13. What are the two components of system head? Discuss the shape of system head curve and factors affecting it.
14. Sketch a typical head capacity curve and efficiency curve for a centrifugal pump and discuss these curves to explain shut off head and best efficiency point.
15. Discuss two identical pumps connected in series and in parallel. In water distribution systems, what arrangement is common to meet varying demand?
16. Discuss how change in the impeller speed of a centrifugal pump affects head, capacity, and power requirements.

17. What is VFD? What is the principle on which it works and where it is commonly used?
18. Make a sketch showing the difference between suction head and suction head for a centrifugal pump. Wherever practical, suction head conditions are preferred. Explain why.

Practice Problems

1. Pressure gauges attached to the suction and discharge sides of a centrifugal pump read 12 cm mercury vacuum and 150 kPa respectively. Assuming the gauges are at the same height, determine how far the head is added by the pump. (17 m).
2. A prototype pump has an impeller diameter exactly 3 times that of a model test pump. The model pump delivers 0.1 L/s when operating against a head of 7.5 m at the best efficiency point. Predict the head and capacity of the prototype operating at the same rpm as the test pump. (2.7 L/s, 68 m).
3. Find the NPSHA when pumping water at 80°C from a well. Pumping water level is 2.5 m below the pump and assume atmospheric pressure is at 101 kPa. Head losses in the suction line are estimated to be 0.45 m. (2.7 m).
4. Find the permissible suction lift for a pump that requires 3.5 m of NPSH. Assume total head losses in the suction line are 0.70 m.
   a. The temperature of water is 60°C (4.1 m).
   b. The atmospheric pressure is 80 kPa (abs) and temperature 20 degree (3.7 m).
5. In a pumping system suction gauge reads 6 psi (vacuum) and discharge gauge reads 44 psi. Assuming no elevation difference between the two gauges, how much head is added by the pump? 120 ft
6. The intake in the lake is below the water surface at an elevation of 1230.0 ft. The lake water is pumped to the plant influent at an elevation of 1270.0 ft. The total head losses are estimated to be 18 ft when water is drawn at the rate of 7.8 MGD. Calculate pumping head, water power, and the pump power, assuming pump is 72% efficient? (58 ft, 79 hp, 98 hp).
7. A centrifugal pump with an efficiency of 65% discharges 1600 gpm into a system that includes 3000 ft of 10 in diameter pipe with C of 100. If total static head is 92 ft, compute the required pump power? (100 hp).
8. How many stages of a multistage pump are required to pump water @ 70 L/s against a total head of 185 m. Speed of the pump is 750 rpm and specific speed of the pump is not to exceed 700. (10).
9. In a water pumping system, gauges attached to the suction side (150 mm) and the discharge side (100 mmm) of a pump are read vacuum of 250 mm of mercury and 140 kPa, respectively. The suction side gauge is 0.60 m below pump and the discharge pressure gauge is 0.17 m above the pump. Calculate the power required to pump water @ 30 L/s, assuming pump is 60% efficient? (9.0 kW).

10. A centrifugal pump running at 1400 rpm has the following characteristics:

Q, L/s	13	19	25	31	38	44	50
H, m	28	28	26	25	23	21	18
E, %	65	70	73	74	72	69	63

Plot the performance curves and determine the best efficiency point. At the maximum efficiency, find the power required to drive the pump. (31 L/s, 25 m, 74%, 10 kW).

11. A centrifugal pump is required to discharge water @ 880 gpm against a total static head of 105 ft through a 6 in diameter and 330 ft long pipeline. Assume friction factor f = 0.02 and minor losses equivalent to 30 ft of length. Determine the pump power, assuming pump is 65% efficient. (44 hp).

12. A centrifugal pump is delivering water at the rate of 220 gpm when static suction lift is 12 ft. The pressure gauge on the suction side just before the water enters the pump indicates a reading of 7.2 psia. Find the head losses on the suction side of the pump. (5.43 ft).

13. A centrifugal pump is discharging 7.5 L/s producing a head of 15 m when running at a speed of 1250 rpm and the pump power required to run the pump is 6.0 kW. If the rotational speed of the pump is increased to 1450 rpm, find the new discharge, head, and power. (8.7 L/s, 20 m, 9.4 kW).

14. At a certain location, atmospheric pressure and vapor pressure respectively are 95 kPa and 3.0 kPa expressed as absolute pressure. If the NPSH required to run the pump is 3.0 m, what maximum suction lift can be afforded, assuming head losses on the suction side to be 0.40 m. (6.0 m).

15. In a water pumping system discharge pressure gauge is 0.5 m above the suction gauge. During pumping, suction gauge reads 36 cm of mercury column (vacuum) and discharge gauge reads 370 kPa. What is the pumping head? (43 m).

16. A pumping unit draws 22 kW of power when pumping @ 2.7 m³/min against a pressure of 310 kPa. What is the overall efficiency of the pumping unit? (63%).

17. A water treatment plant gets its water from a reservoir. Water is pumped to the plant influent at elevation 24.2 m above the water level in the reservoir. The total head losses are estimated to be 6.5 m when water is drawn at the rate of 360 L/s. Calculate pumping head, water power, and the pump power, assuming pump is 75% efficient? (31 m, 110 kW, 140 kW).

18. A deep well is pumped at the rate of 35 L/s to overhead storage tank though a 150 mm diameter and 260 m long rising main. The difference in water levels including the drawdown is 32 m. Assume friction factor f = 0.022. Calculate the motor power, assuming the overall efficiency of 65%. (21 kW).

19. In a water pumping system, pressure gauges attached to suction (1) and discharge side (2) of the pump are read at the same elevation. Readings are:

$p_1 = -5.5$ psi and $p_2 = 55$ psi when pumping rate is 750 gpm. Find power required by the pump, assuming pump is 72% efficient. (37 hp).

20. A centrifugal pump is pumping into a 4500 m long water pipe that has an inside diameter of 305 mm and friction coefficient C of 100. The system has total static head of 50 m. The pump characteristics are shown in the table below. Plot the pump head capacity curve and system head curve to determine the operating point. (90 L/s, 83 m).

Q, L/s	0	25	50	75	100
H, m	145	140	126	104	65

18 Water Distribution Operation

The main purpose of this chapter is to illustrate the key hydraulic concepts related to water distribution system operation and maintenance. This includes the use of the energy equation, continuity equation, and working out the pumping head and power. Some of these concepts have already been discussed in previous sections.

Water distribution is an important part of any water system. The water flow takes place under pressure conditions which may be due to an elevated reservoir or directly pumped into the water main by high lift pumps. A properly designed and operated water distribution system is to meet water demand with adequate pressure.

18.1 HEAD LOSSES IN WATER MAIN

In the operation of water works, one important thing to know is head or pressure loss in a water main under various operating conditions. As discussed earlier, an increase in flow rate causes head loss to increase many times. This may result in low pressure in the water main at the downstream end. An **energy equation** coupled with a **flow equation** are used to estimate the head losses or pressure drop for given flow conditions. Under normal operating conditions, the head loss rate as indicated by the hydraulic grade line is typically 0.1% to 0.2%. Head losses at this rate amount to 10–20 kPa for every km length of the water main.

During emergencies, as in cases of firefighting, flows in the water main may be as much as two to three times that of the daily average. This will cause significant head losses. During such events, pressure can drop below the normal of 250–350 kPa (35–50 psi) down to as little as 150 kPa (20 psi). However, under no conditions should pressure be allowed to drop below 150 kPa. This may happen during a main break or after opening too many hydrants on the same water main or from excessive withdrawals for firefighting. Reduction in pressure due to excessive flows may create vacuum conditions and may result in cross contamination due to **back siphonage**. Hence, great caution must be used when operating hydrants.

18.1.1 Flow Capacity

Flow carrying capacity of a given pipe size for a maximum permissible head loss or friction slope can be calculated using the flow equation. Note that in the Hazen Williams equation, the exponent of the diameter D is 2.63 to indicate that a small increase in pipe size results in a much greater increase in flow capacity and vice versa.

DOI: 10.1201/9781003347941-21

18.1.2   PIPE ROUGHNESS, COEFFICIENT C

With age, pipes get rough and cause more head losses. Even in new pipes, this may happen prematurely due to water quality problems including scaling and tubercula- tion. Once a given system shows symptoms of high pressure loss during normal flow conditions, it is important to test the pipe and take corrective actions. In testing the roughness of the pipe in each part of the water distribution system, a hydrant is opened to cause a certain flow through the section of the pipe being studied. Once the flow becomes steady, flow reading and pressure drop between two hydrants is observed. Knowing the flow and the pressure drop, the roughness coefficient of the pipe material can be computed. Friction slope is head loss per unit length of the pipe. Applying the Hazen Williams equation, the roughness coefficient C can be found.

Hazen Williams flow equation

$$C = \frac{Q}{0.278\,D^{2.63} \times S_f^{0.54}} \qquad S_f = 10.7 \times \left(\frac{Q}{C}\right)^{1.85} \times \frac{1}{D^{4.87}}$$

Example Problem 18.1

A field test was performed on a section of newly installed main. A pressure drop of 22 kPa is observed across 400 m length of 250 mm main line when a flow of 61 L/s is carried. What is the friction factor C?

Given:

$$\Delta p = 22 \text{ kPa} \quad Q = 61 \text{ L/s} \quad L = 400 \text{ m} \quad D = 250 \text{ mm} = 0.25 \text{ m} \quad C = ?$$

Solution:

$$S_f = \frac{h_f}{L} = \frac{22\,kPa}{400\,m} \times \frac{m}{9.81\,kPa} = 5.60 \times 10^{-3}$$

$$C = \frac{Q}{0.278\,D^{2.63} S_f^{0.54}} = \frac{0.061}{0.278 \times 0.25^{2.63} \times 0.0056^{0.54}} = 138 = \underline{140}$$

Example Problem 18.2

A water main is sized based assuming a roughness factor of 120. What friction slope is expected when a 300 mm diameter water main carries maximum flow of 75 L/s?

Given:

$$Q = 75 \text{ L/s} = 0.075 \text{ m}^3/\text{s} \quad D = 300 \text{ mm} = 0.30 \text{ m} \quad C = 120 \quad S_f = ?$$

Solution:

$$S_f = 10.7 \times \left(\frac{Q}{C}\right)^{1.85} \times \frac{1}{D^{4.87}} = 10.7 \times \left(\frac{0.075}{120}\right)^{1.85} \times \frac{1}{0.30^{4.87}}$$

$$= 4.44 \times 10^{-3} = 0.44\%$$

18.2 FREE FLOW VELOCITY AND DISCHARGE, Q

Under free flow conditions, the flow velocity through an opening will largely depend on the pressure head exerted. The actual flow velocity, v would be less due to contraction and head loss across the opening.

Flowing hydrant discharge velocity/rate

$$\boxed{v = C\sqrt{2g\Delta h} \qquad Q = 0.785\,D^2 \times C\sqrt{2g \times \Delta h} - Consistant\,units}$$

18.3 HYDRANT TESTING

During hydrant testing, the reduction in pressure at the residual hydrant is being observed by opening the hydrant in the vicinity of the hydrant being studied. This test allows us to estimate the flow capacity of a given hydrant during firefighting or other similar emergencies. Hydrant discharge from a flowing hydrant is found by observing the trajectory of the water jet. For more accurate results, discharge rate is measured by observing velocity pressure of the water jet with a **Pitot gauge**.

Hydrant flow rate

$$\boxed{Q = D^2\sqrt{\Delta p} \qquad (SI) \qquad Q = 25.3D^2\sqrt{\Delta p}\;\; USC}$$

$$p = kPa\,(psi),\; D = m\,(in),\; Q = m^3/s\,(gpm),\; C_d = 0.85$$

Hydrant flow at the residual pressure

$$\boxed{Q_R = Q_F \times \left(\frac{\Delta p_R}{\Delta p_F}\right)^{0.54} = Q_F \times \left(\frac{p_S - p_R}{p_S - p_F}\right)^{0.54}}$$

Q_F = total flow from all the flowing hydrants during testing
Δp_F = drop in pressure at the residual hydrant = $p_S - p_F$
Δp_R = static pressure minus the residual pressure = $p_S - p_R$
p_R = residual pressure = 140–150 kPa (20 psi)

The hydrants are color coded to indicate capacity. Hydrants with maximum flows >100 L/s (1500 gpm) are painted blue, 30–65 L/s (450–1000 gpm) range are orange, 65–100 L/s (1000–500 gpm) range are red, and <30 L/s (500 gpm) are yellow.

Example Problem 18.3

A hydraulic gradient test is performed in the field. Two hydrants, 750 m apart, were chosen to observe the hydraulic head. For the maximum flow conditions, pressure reading at hydrants 1 and 2 were observed to be 477 kPa and 495 kPa. From the map for the area, the elevations of hydrants 1 and 2 are 112.4 m and 111.9 m respectively. Find the hydraulic head at each of the hydrants and hence the hydraulic gradient.

Given:

Variable	Hydrant 1	Hydrant 2
pressure, kPa	477	495
elevation, m	112.4	111.9
head loss, h_L	?	
length, m		750

Solution:

Head loss

$$h_l = (Z_1 - Z_2) + \frac{(p_1 - p_2)}{\gamma}$$

$$= (112.4 - 111.9)m + (477 - 495)kPa \times \frac{m}{9.81\,kPa} = -1.33 = \underline{-1.3\,m}$$

Negative head loss means the water is flowing in the opposite direction.

$$Gradient = \frac{h_l}{L} = \frac{1.33\,m}{750\,m} \times 100\% = 0.177 = \underline{0.18\%}$$

Example Problem 18.4

Applying the principle of flow through an orifice, estimate the leakage rate through a 0.04 in diameter crack in a pipe. The operating pressure is 50 psi and the coefficient of discharge can be assumed to be 60%.

Given:

$$D = 0.04\ in \quad Q = ? \quad C = 0.60 \quad \Delta p = 50\ psi$$

Solution:

Leakage rate through the crack

$$Q = Av = 0.785D^2C\sqrt{2g\Delta h}$$

$$= \frac{\pi}{4}\left(\frac{0.04}{12}ft\right)^2 \times 0.60 \times \sqrt{\frac{2 \times 32.2\ ft}{s^2} \times 50\ psi \times \frac{2.31\ ft}{psi}}$$

$$= 4.515 \times 10^{-4} \frac{ft^3}{s} \times \frac{7.48\ gal}{ft^3} \times \frac{3600\ s}{h} = 90.9 = \underline{91\ gal\,/\,h}$$

Example Problem 18.5

During a fire flow test, the Pitot gauge read a velocity pressure of 55 kPa. Applying the principle of hydraulics, estimate the discharge rate from a 60 mm hydrant nozzle. The coefficient of discharge can be assumed to be 85%.

Given:

$$D = 60 \text{ mm} = 0.060 \text{ m} \quad Q = ? \ C = 0.85 \quad \Delta p = 55 \text{ kPa}$$

Solution:

$$Q = D^2 \sqrt{\Delta p} = (0.060)^2 \sqrt{55} = 2.66 \times 10^{-2} m^3 / s = \underline{27 \ L/s}$$

Example Problem 18.6

In a fire flow test, four hydrants were opened to produce a total flow of 140 L/s. During the test, pressure at the residual hydrant dropped from 480 to 310 kPa. What hydrant flow can be expected at a residual pressure of 150 kPa?

Given:

$$\Delta p_F = 480 - 310 = 170 \text{ kPa} \quad Q_F = 140 \text{ L/s} \quad \Delta p_R = 480 - 150 = 330 \text{ kPa}$$

Solution:

Hydrant flow at the residual pressure

$$Q_R = Q_F \left(\frac{\Delta p_R}{\Delta p_F} \right)^{0.54} = 140 \ L/s \times \left(\frac{330 \ kPa}{170 \ kPa} \right)^{0.54} = 200.3 = \underline{200 \ L/s}$$

Example Problem 18.7

In a fire flow test, four hydrants were opened. The average Pitot tube reading of 15.5 psi was observed through 2.5 in nozzles. Find the total flow. During the test, pressure at the residual hydrant dropped from 90 to 55 psi. What hydrant flow can be expected at a residual pressure of 20 psi?

Given:

$$\Delta p_F = 90 - 55 = 35 \text{ psi} \quad \Delta p_R = 90 - 20 = 70 \text{ psi}$$

Solution:

Total flow

$$Q = 4 \times 25.3 D^2 \sqrt{\Delta p} = 4 \times 25.3 (2.5)^2 \sqrt{15.5} = 2493 = 2490 \, gpm$$

Hydrant flow at the residual pressure

$$Q_R = Q_F \left(\frac{\Delta p_R}{\Delta p_F} \right)^{0.54} = 2493 \, gpm \times \left(\frac{70}{35} \right)^{0.54} = 3624 = \underline{3600 \, gpm}$$

18.4 WATER QUALITY

Contaminants can enter water distribution systems via cross connections, corrosion, biological growth in mains, and dead ends. Water quality monitoring is normally performed on a scheduled basis as dictated by the provincial regulations. When selecting the sampling locations, extreme ends of the system must be included. In addition to compliance sampling, water samples are collected at dead ends, in low flowing pipes, and where corrosion is suspected. Such samples should be analyzed for taste, odor, and color to indicate any changes in water quality.

18.4.1 MONITORING

Water quality monitoring is normally performed on a scheduled basis to detect the presence of contaminants. The frequency of sampling is determined by the size of population and type of water system. In summary, monitoring helps to achieve the following:

- establish potability of the water supply and detect changes, if any;
- determine possible and probable sources of contamination;
- respond to consumer complaints;
- determine the effectiveness of the treatment system.

18.4.2 SECONDARY DISINFECTION

Some chlorine should remain in the water distribution system to continue disinfecting the water. Problems of taste and odor can occur because of excessive chlorine residual or no residual. In drinking water, free chlorine residual of about 0.2 mg/L or about 1.0 mg/L of combined residual is optimal. Operators must be particularly careful to always have chlorine residual in reservoirs and on dead ends. The chlorine residual should be tested regularly.

Taste and odor complaints in part of the distribution system may indicate low levels of chlorine residuals in portions of the distribution systems. In these cases, it is often necessary to dechlorinate the water at various locations throughout the distribution system to ensure the optimal residual level is maintained. The chlorine residual may be in the form of free chlorine or combined chlorine. Combined chlorine has a considerably lower disinfecting power than free chlorine. On the other hand, free chlorine does not last as long. The type of residual should be selected based on the bacteriological and chemical quality of the water and the type of treatment used.

18.4.3 FLUSHING AND CLEANING OF WATER MAINS

Periodically, the mains should be flushed to remove deposits from the pipelines. Opening a hydrant, discharge rate is adjusted to achieve the desired **flushing velocity**. A minimum flushing velocity of 0.8 m/s (3 ft/s) is needed to scour the material from inside the pipe. Failure to flush the distribution pipes may lead to taste, odor, turbidity, and disease problems. In situations where pipe flushing does not help to improve quality or flow capacity of the pipe, the pipe should be cleaned out. Foam swabs, pigs and scrapers are used to remove hardened scale and tuberculation. The deposits and mounds of corrosion products in the water mains are called **tuberculation**.

Water reservoirs are also emptied, cleaned, and disinfected before being put into service. To assure disinfection, a chlorine dosage of 100 mg/L is usually applied to maintain residual of 50 mg/L for 24 hours. Water is sampled for microbiological testing before it is approved to be put back in service. Chlorinated water is dechlorinated before it is put into storm sewers.

18.4.4 REPAIRS AND BREAKS

Leaks in the distribution system are also a concern for operators. Not only do leaks increase the amount of water which must be treated but can also result in contamination of the drinking water during backflow events. Backflow may occur under certain circumstances when the water pressure is turned off, resulting in a negative pressure in the pipe. Leakage water for a given municipality is estimated by doing a **water audit**.

18.4.5 FIELD DISINFECTION

Whenever a pipe is exposed to contamination such as during repairs and installation, the water mains must be disinfected before putting it back in service. This is usually accomplished by applying a high chlorine dosage typically of 50–100 mg/L to obtain a residual of 25 mg/L after 24 hours. Hypochlorite solutions or tablets can be used for applying chlorine. All components of a distribution system operate together as one. All operating personnel must be aware of the system hydraulics, operating limitations, water quality, maintenance requirements, and response times to meet the demands of the community.

Example Problem 18.8
Work out the quantity of 4.5% bleach required to disinfect a 16 in diameter well casing and well screen applying chlorine dosage of 100 mg/L. The depth of water in the well is 80 ft.

Given:

variable	bleach= 1	well water= 2
volume	?	16 in, 80 ft deep
conclusion	4.5% = 45 g/L	100 mg/L

Solution:
 Volume of water standing in well

$$V = \frac{\pi D^2 h}{4} = \frac{\pi}{4} \times \left(\frac{16}{12} ft \right)^2 \times 80\,ft = 111.7 = 110\,ft^3$$

Volume of bleach required to disinfect

$$V_1 = V_2 \times \frac{C_2}{C_1} = 111.7 \ ft^3 \times \frac{0.10 \ g/L}{45 \ g/L} \times \frac{7.48 \ gal}{ft^3} = 1.85 = \underline{1.9 \ gal}$$

Example Problem 18.9

After maintenance and repairs, a 12 m diameter surface reservoir is to be disinfected by dosing it with a chlorine concentration of 100 mg/L. It is thought that a dosage of 100 mg/L should be able to maintain a residual of 50 mg/L during the disinfection period. How many liters of 15% sodium hypochlorite solution should be added to fill the tank to a depth of 3.0 m?

Given:

$$D = 12 \text{ m, } h = 3.0 \text{ m, } C_2 = 100 \text{ mg/L, } C_1 = 15\% = 150 \text{ g/L, } V_1 = ?$$

Solution:

Volume of water dosed

$$V = \frac{\pi D^2 h}{4} = \frac{\pi (12m)^2 \times 3.0m}{4} \times \frac{1000 L}{m^3} = 3.39 \times 10^5 = 3.4 \times 10^5 \ L$$

Volume of hypochlorite required

$$V_1 = V_2 \times \frac{C_2}{C_1} = 3.39 \times 10^5 \ L \times \frac{0.10 g/L}{150 g/L} = 226 = \underline{230 L}$$

Discussion Questions

1. Search the internet and find out legislation pertaining to drinking water quality.
2. Describe the working principle of a Pitot gauge.
3. What is the main purpose of secondary disinfection?
4. When collecting a bacteriological sample from a location you must check for chlorine residual. Discuss.
5. What problems can be expected at dead ends and what should be done to prevent such problems?
6. What is the purpose of flushing a water main and how it is carried out?
7. After repairing a water main, what must be done before putting the water main in service? Describe the procedure.
8. To assure the quality of water, what should the monitoring program include?
9. What is the purpose of hydrant testing? Describe the steps to carry out a hydrant test.
10. From the internet, search the water quality data of the drinking water supply in your area and comment.

Practice Problems

1. A 450 mm main is an old concrete pressure pipe carrying water at the rate of 200 L/s. Calculate friction coefficient C when the pressure difference between two hydrants 300 m apart was recorded as 22 kPa. Assume that hydrants 1 and 2 are at the same elevation and there is no water connection between the two hydrants. (83).

2. Estimate the flow coming out of hydrant connected to a 2-in hose. The velocity pressure of the water jet coming out is observed to be 10.5 psi. What is the rate of flow? (330 gpm)

3. During a hydrant flow test, the water jet coming out of 61 mm nozzle has velocity pressure of 48 kPa. What is the rate of flow in L/s? (26 L/s)

4. Applying the principle of flow through an orifice, estimate the gallons of water leaking in a day from a 0.04 in diameter crack in a pipe. The operating pressure is 60 psi and the coefficient of discharge is 60%. (320 gallons).

5. During a fire flow test, the Pitot gauge reads a velocity pressure of 9.5 psi. Estimate the discharge rate from a 2.5 in hydrant nozzle. The coefficient of discharge can be assumed to be 85%. (490 gpm).

6. In a fire flow test, four hydrants were opened to produce a total flow of 180 L/s. During the test, pressure at the residual hydrant dropped from 460 to 305 kPa. What hydrant flow can be expected at a residual pressure of 150 kPa? (260 L/s).

7. A hydraulic gradient test is performed in the field. Two hydrants 450 m apart were chosen to observe the hydraulic head. For the maximum flow conditions, pressure readings were observed to be 477 kPa and 465 kPa. From the map for the area, the elevations of hydrants 1 and 2 are 112.4 m and 111.9 m respectively. Find the hydraulic gradient and draw HGL for the maximum flow conditions. (0.38%).

8. A field test was performed on a section of newly installed main. A drop in pressure of 25 kPa is observed across a 300 m length of 250 mm main line when a flow of 71 L/s is carried. What is the friction factor C? (130).

9. Due to scaling, the inside pipe diameter has been reduced by 2.0 mm and coefficient of friction C has dropped from 110 to 100. What is the reduction in flow carrying capacity of the 200 mm water main? (12%).

10. A flow test was conducted on part of a 250 mm pipeline. Three hydrants in a row are spaced 300 m apart. When the end hydrant was opened, the pressure difference in the first two was read to be 13 kPa. Based on the Pitot meter reading at the flowing hydrant, flow rate during the test was 31 L/s. Based on this, find the coefficient of roughness C? (80).

11. How many liters of 5.0% sodium hypochlorite will be needed to disinfect a well with a 500 mm diameter casing and well screen. The well is 110 m deep and there is 35 m of water in it. It is recommended to provide an initial dose of 100 mg/L of chlorine. (14 L).

12. A reservoir needs to be disinfected before being placed online. The rectangular tank is 6.0 ft deep, 10 ft wide, and measures 20 ft. An initial dose

of 100 mg/L is expected to maintain the desired residual of 50 mg/L for a period of 24 hours. How many gallons of bleach with 5.0% available chlorine will be needed? (18 gallons).

13. What pressure drop is expected in a 1.4 mile long, 12 in diameter water main with a friction coefficient C of 120 while carrying a flow of 800 gpm? (6.2 psi).

14. After cleaning the 600 m section of 200 mm water main, it was intended to check the improvement in the coefficient by performing a pressure test. During testing, pressure dropped by 28 kPa for a flow of 22 L/s. What is the improved value of roughness coefficient C? (98).

15. A flow test was conducted on an old existing main to determine the roughness coefficient C. A pressure head drop of 3.5 m is observed across a 600 m length of 200 mm main line when a flow of 19 L/s is carried. What is the friction factor C? (76).

16. A Pitot gauge placed in front of a flowing hydrant jet reads 82 kPa of velocity pressure. Assuming the coefficient of 85%, calculate the discharge rate from a 60 mm nozzle. (33 L/s).

17. A fire flow test was conducted on a section of main in a residential area. Two hydrants one on each side of the residual hydrant were opened to produce a flow of 60 L/s. During the test, pressure at the residual hydrant dropped from 350 to 240 kPa. For residual pressure of 150 kPa, what is hydrant flow capacity? (83 L/s).

18. A fire flow test was conducted on a section of a water main. Two hydrants, one on each side of the residual hydrant, were opened to produce a total flow of 250 gpm. During the test, pressure at the residual hydrant dropped from 72 psi to 65 psi. Determine the hydrant flow at the residual pressure of 25 psi. (700 gpm).

19. A 200 mm water main is to be flushed by achieving a flow velocity of 1.0 m/s. What flow rate should be maintained at flowing hydrant? (31 L/s).

20. Maintaining a flow rate of 55 L/s flushes a 250 mm diameter main. What flushing flow velocity is achieved? (1.1 m/s).

Section IV

Wastewater Collection

19 Wastewater Collection System

The sewage collection system is made up of several different components, all working together to carry waste to the wastewater treatment plant. The main components of the sewage collection system are building services (building sewers), sewer mains, lift stations, force mains, and the wastewater treatment plant. Sewage moves from the point of origin to the wastewater treatment plant in the steps shown in **Figure 19.1**.

19.1 SEWER MAINS

Sewage drains from the **building sewer** (**Figure 19.2**) to the **sewer main**. This is a pipe at least 8 in (200 mm) in diameter. It is also graded and slopes to a low point in the town where the sewage is collected in a **lift station**. Depending on the size of the city, sewer lines may consist of laterals, sub-mains, mains, trunk sewers, and intercepting sewers. A **lateral** is the first line to no other joins except the building sewer. **Sub-mains** can receive flow from more than one lateral or be directly connected to building sewer. **Trunk sewers** receive wastewater from a large area. A lateral being the smallest in size is subjected to greater variations of flow, and as we go up in the hierarchy pipe size increases and flows also get smoothed out. For this reason, laterals are designed to carry peak flow flowing half full and mains and trunk sewers are designed to flow more than half full.

Manholes for servicing are located along the sewer mains at distances of 100–150 m (250–500 ft). At low points in the collection system, a lift station will be installed. All the nearby sewers drain into these low areas. Pumps located in the station then move the collected sewage via a force main to the wastewater treatment plant or main sewer. A **force main** is an isolated line linking the lift station and the disposal area. It is called a forced main because it flows under pressure, while other sewer mains are designed to flow partially full under gravity or open channel flow conditions.

19.1.1 COMBINED SEWERS

Most communities have two sewer systems, one for carrying sewage and one for carrying storm water. However, some older communities have only one sewer system, which collects wastewater and storm water together in one sewer. This is called a **combined sewer system**. Combined sewers save cost on sewer installation; however, the sewers must be much larger in diameter to accommodate storm flow, and the wastewater treatment plant is also larger than necessary to treat just sewage. Most communities are now switching to separate sewer systems since during wet weather,

DOI: 10.1201/9781003347941-23

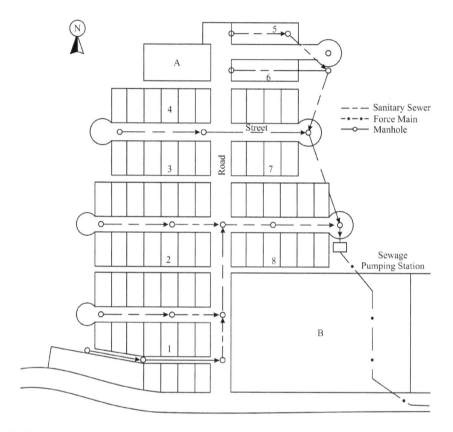

FIGURE 19.1   Sewage collection system

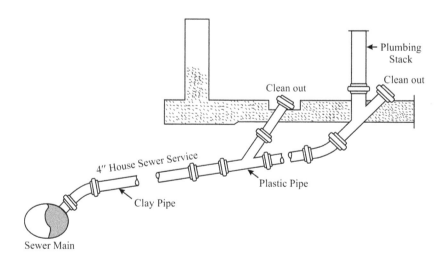

FIGURE 19.2   Building sewer connection

TABLE 19.1

Storm sewers versus sanitary sewers

Parameter	Sanitary Sewers	Storm Sewers
size	relatively smaller	much larger
depth	much deeper	shallow
flow and season	flow year around	during runoff season
design	based on 3–4× average daily	peak run off rate
flow depth	partial	full
flow velocity	<0.5 m/s	>1.0 m/s
surcharging	not permitted	in extreme events

---

combined discharges exceed the capacity of the plant and so flows are bypassed and can result in pollution of the receiving water body.

19.1.2 STORM SEWERS

The purpose of the **storm sewers** is to carry drainage water or storm runoff. Since a combined sewer system leads to pollution and contamination, a separate sewer system is the norm in all developed areas. Storm sewers are different in many aspects compared to sanitary sewers. Since storm sewers are supposed to carry peak storm flows, they are larger in size and are not as deep as sanitary sewers. Storm sewers are designed to flow full and are allowed to surcharge during extreme events. Some of the striking differences between the two types of sewer systems are listed in the **Table 19.1**.

19.2   INFILTRATION & INFLOW

Infiltration & Inflow, or **I&I**, is a term used to describe the flow of non-wastewater sources into the wastewater collection system. **Infiltration** refers to groundwater that enters the collection system through building connections, defective pipes, pipe joints and seals, manhole walls, and other parts of the collection system that are not well sealed. **Inflow** refers to water that enters the system from illegal connections such as sump pumps, roof, and eave trough drains, catch basin connections, and storm water runoff over improperly sealed manhole covers or other access points. **Exfiltration** is the term used to indicate the wastewater leaking out in the ground during dry weather when the ground water level is below that in the sewer pipe. If uncontrolled, exfiltration can lead to contamination.

Infiltration & inflow can cause significant problems to the wastewater collection system if it has not been sized in such a way to handle the extra flow. Under normal conditions, not more than 10% of the daily flow should come from I&I. In a newer system, this should be less than 10%. It may also cause a problem at the wastewater treatment plant if too much water is coming through the plant to be treated efficiently. Infiltration & inflow may also dilute the wastewater. Part of the wastewater collection system personnels' job will be to monitor the wastewater collection system at different times of year for infiltration & inflow and to mitigate it where possible.

19.3 WASTEWATER FLOWS

The **rated capacity** of a wastewater treatment plant is usually based on the average annual daily flow rate. However, from a practical point of view, it makes sense to understand how flow hydrographs change over the course of a day, week, or even a year, to maximize treatment efficiency. Wastewater flow rates will vary depending on the use, the source of the wastewater, and the time of day. A properly designed and operated collection system must be able to handle all these variables. If a wastewater collection system cannot handle all the volume generated, it may overflow to the surface, or back up the building sewers into basements or floor drains.

Typical residential flow patterns can be described as **bimodal**, which means the pattern has two main peaks over the course of a day. The first peak usually occurs in the morning, as people are getting up and getting ready to go to work. The second peak usually occurs in the evening, between 7 and 9 p.m., when people return to their homes. Since a wastewater collection system is not only collecting and transporting residential wastewater, but industrial, commercial, and institutional wastewater as well, it is important to recognize that different design factors will have to be considered for different parts of the collection system.

A flow pattern like a residential flow pattern is experienced at the plant as shown in **Figure 19.3**. However, peaks are bit smaller compared to a collection system and the timings are lagged depending on the extent of the collection network. In a typical **dry weather flow** hydrograph, minimum flows at the plant are experienced during the early morning hours and peak flow just in the afternoon hours. Flow peaks are

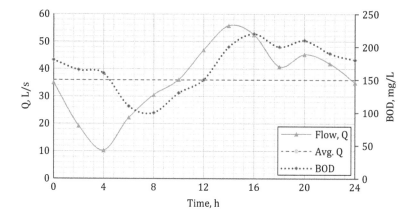

FIGURE 19.3   Dry weather wastewater flow hydrograph

smaller due to the levelling out effect as wastewater flows through various sections of the collection system.

Excessive **infiltration & inflow** as experienced in poorly maintained and older systems can run havoc both on the collection system and plant operations. The difference in volume as indicated by the difference in the **wet weather** and **dry weather** flow hydrograph is indicative of the degree of infiltration & inflow. When describing the variable flow conditions in a collection system, several different terms are used. Some of these terms are described below.

19.4   SEWER MAINS

Sewer pipes, also called sewer mains, are installed in one of two ways: gravity mains or force mains. In gravity mains, the sewage moves through the pipe by gravity alone. When **gravity sewers** are installed, they must be **graded** or sloped so that the sewage in them can be moved by gravity alone, with no other force or pressure required. In a **force** mains, the sewage is pushed, or forced, through the sewer by means of a pump.

19.4.1   PIPE SIZE

Minimum pipe sizes (diameters) have also been established for different parts of the collection system. Sewer mains must be a minimum of 200 mm (8 in), house connections or **building services** are a minimum of 100 mm (4 in), and commercial buildings (such as office buildings or apartments) are 150 mm (6 in) or larger.

Circular sewer pipes are manufactured with an inside diameter of 4 in to as much as 12 ft. In the range from 4 in (100 mm) to 12 in (300 mm), available pipe sizes are at an interval of 2 in in diameter. In the middle range, a 12 in and 36 in (900 mm), 3 in or 75 mm interval is used. In the larger pipe sizes in the diameter range of 36 in to 144 in, pipe sizes are available at 6 in intervals. Maximum pipe sizes vary with the material of the pipe. For example, the maximum pipe size for a concrete pipe is 42 in in diameter whereas a reinforced cement concrete (RCC) pipe is available in larger sizes too.

19.4.2   SEWER GRADE

Sewer lines are usually graded so that sewage will flow at a rate of at least 0.6 m/s (2 ft/s) when the sewer pipe is half full or full. This is called the minimum **scour velocity** for gravity sewer pipes. To achieve a self-cleansing velocity, the pipe must be laid on a minimum slope or grade. If the sewer or building service line has less slope than is indicated in the sewer pipe grading (**Table 19.2**), solids will sink to the bottom of the line and build up there. This process will eventually block the sewer line.

Too much of a slope will cause the solids and the liquids in the sewage to separate, resulting in loss of scrubbing action in the pipes. Paper and other debris will be left behind in the pipe, causing eventual blockage of the sewer line. To prevent erosion of the pipe surface, velocities of flow of more than 2.5 m/s (8 ft/s) should be avoided.

TABLE 19.2

Gravity sewer pipe grading and size

Pipe Size mm (in)	mm/ 30 m	%	In/100 ft
100 (4″)	312.5	1.0	12
150 (6″)	180	0.6	7.2
200 (8″)	120	0.4	4.8
250 (10″)	85	0.3	3.6
300 (12″)	65	0.2	2.4

19.4.3   PIPE FLOW VELOCITY AND CAPACITY

The sewer pipe is graded to provide the scouring velocity of 0.6 m/s when flowing full. The minimum slopes indicated in Table 19.2 define the minimum slope needed to achieve this velocity. In circular pipes flowing full, the hydraulic radius is one-fourth of the diameter of the pipe. Thus, **Manning's flow equation** for full flow becomes:

$$v_F = \frac{0.4}{n} \times D^{2/3} \times \sqrt{S} \qquad Q_F = \frac{0.312}{n} \times D^{8/3} \times \sqrt{S} \; (SI)$$

$$v_F = \frac{0.6}{n} \times D^{2/3} \times \sqrt{S} \quad Q_F = \frac{0.464}{n} \times D^{8/3} \times \sqrt{S} \; (USC)$$

This formula yields the flow velocity for **full flow** conditions. A pipe flowing half full would carry exactly half of this full flow capacity. The factor n is the **roughness factor.** The value of n is typically 0.013.

Example Problem 19.1

Work out the full flow velocity and flow capacity of a 300 mm sewer pipe laid on a slope of 0.20%.

Given:

$$v =? \quad D = 300 \text{ mm} \quad S = 0.20\%$$

Solution:

$$v_F = \frac{0.4}{n} \times D^{2/3} \times \sqrt{S} = \frac{0.4}{0.013} \times (0.30)^{\frac{2}{3}} \times \sqrt{\frac{0.20\%}{100\%}} = 0.643 = \underline{0.64\,m/s}$$

$$Q_F = A \times v = \frac{\pi(0.30\ m)^2}{4} \times \frac{0.643\ m}{s} \times \frac{1000\ L}{m^3} = 43.5 = \underline{44\,L/s}$$

Example Problem 19.2

What is the maximum population that can be served by a 200 mm sanitary sewer laid on a grade to provide the full flow velocity of 0.6 m/s. Assume the design flow to be 1600 L/c.d.

Given:

$$v = 0.60 \text{ m/s} \quad D = 200 \text{ mm} \quad PE =?$$

Solution:

$$Q = A \times v = \frac{\pi(0.60\,m)^2}{4} \times \frac{0.60\,m}{s} \times \frac{1000\,L}{m^3} = 18.8 = 19\,L/s$$

$$Pop. = \frac{18.8\,L}{s} \times \frac{60\,s}{min} \times \frac{1440\,min}{d} \times \frac{p.d}{1600\,L} = 1017 = \underline{1000\ people}$$

Example Problem 19.3

Determine the full flow carrying capacity of an 8 in sewer pipe (n = 0.012) laid on a 0.33% slope.

Given:

$$D = 8 \text{ in} \qquad n = 0.012 \qquad S = 0.33\% = 0.0033$$

Solution:

$$Q_F = \frac{0.464}{n} \times D^{8/3} \times \sqrt{S} = \frac{0.464}{0.012} \times (8/12)^{8/3} \times \sqrt{\frac{0.33\%}{100\%}} = 1.695 = 1.7\,ft^3/s$$

19.4.4   GRAVITY SEWER MAINS

Gravity sewer mains are large-diameter pipes that carry sewage from the building service to a lift station. These are called gravity sewers since the flow in these sewers is due to gravity. Gravity sewer mains have a minimum slope and diameter that must be maintained along the length of the pipe to keep the sewage flowing by gravity (i.e., there are no pumps in these sections to move sewage along).

Gravity sewers may be laterals, branch lines, and main sewer lines and have a minimum diameter of 200 mm. **Laterals** are the sewer lines at the top of the system, such that there is no tributary to these lines. A **building sewer** is the sewer that connects the house or business to the main sewer in the street in front of the building. A building service sewer line is usually a gravity line and has a minimum diameter of 100 mm for residential applications and 150 mm for commercial applications.

19.4.5 Force Mains

A **force main** is any part of the sewage collection system that is under positive pressure. Pumps are employed to create the pressure in the pipe. A force main is always downstream of a lift station, and sometimes connects the lift station and the wastewater treatment plant. A **hydraulic grade line** (HGL) in a force mains would be higher than the pipe. However, the pressure is much lower compared to water distribution systems.

All the sections of the collection system that are under pressure will be indicated on the as-built drawings as force mains. Force mains also have additional features that gravity mains do not have, such as vents and shut-off valves. Since a force main is under pressure, it always flows full and the pipe material is usually ductile iron or steel. Force mains are sized such that during low flow pumping, a minimum flow velocity of 0.60 m/s is achieved. This flow velocity will allow the flow stream to keep solids in suspension and prevent deposition of solids in the force main.

The **shut-off valves** will be located at each end of the force main – at the lift station and just before the discharge end of the force main – and are normally open. These valves can be used to shut off flow through the force main when maintenance is needed. The valves should be opened and closed (exercised) twice per year to ensure that they will operate when needed. The force main route should be inspected periodically to signs of digging or construction activity, which might endanger the line.

19.5   OPERATION AND MAINTENANCE

As part of wastewater collection system maintenance, the main activities include: detecting an obstruction in a sewer main through your inspection of the manholes in your sewage collection system, removal of obstructions in sewer mains, repairs undertaken to collapsed or separated sewer mains, and the repair of frozen or blocked building service sewer lines.

19.5.1 Detecting and Repairing an Obstruction

The wastewater collection system operator's daily maintenance schedule will include an inspection of the manholes in the system. An operator may sometimes find that one manhole has very little flow coming through it, while the next manhole upstream has significant depth of water in it. This condition indicates that the section of sewage main between the two manholes is plugged by some obstruction, which is blocking the flow.

To remove a blockage in a plugged sewer main an operator must obtain a **sewer tape** or **sewer rod**. The blockage will be removed from the manhole below the flooded one; therefore, this manhole should be thoroughly ventilated using a forced air blower. Remember to follow all the confined space entry rules, including documentation, before entering the manhole. To clear obstructions from sewer mains, jointed **sewer rods** can be used. They are made of wood in lengths of about 1 to 1.5 m (3–5 ft) or of metal in lengths of 0.6 m (2 ft) or more. The rods are designed so that sections are jointed at an angle of nearly 90 degrees. There is no chance of the sections coming apart while the string of rods is in the sewer. The last section is a crank or T-handle, which allows the operator to push or turn the rods against the obstruction.

Sewer tapes are made of spring steel. They are 2 cm to 2.5 cm wide, and 30 m to 60 m long. They are wound onto a cradle or holder, which is like a reel. Two lengths of sewer tape can be bolted together to increase the length when required. As well, there are a variety of points (such as a ball or a spear point), which can be attached to the leading end. Work is started from the downstream manhole since it will be dry. After the downstream manhole has been ventilated, an operator may enter the manhole and insert the sewer tape or sewer rod into the plugged main to try to bump any obstruction in the sewer main loose. If the blockage still cannot be loosened, attach a fire hose to a hydrant and direct a jet of water against the obstruction. Sewer cleaning using a **high-pressure jet** is a very effective way of removing stoppages and cleaning sewers. The cleaning truck is parked such that the hose reel is over the downstream manhole. A high-pressure nozzle is attached to the hose and inserted through the manhole and into the sewer pipe. Water pumped through the hose is ejected out through the orifices at the rear of the nozzle at a very high pressure of 12–15 MPa. The jet coming out in the downstream direction loosens up the broken solids and thrusts the nozzle forward or upstream.

A dam may be placed in the manhole to retain solids and vacuumed up in the tank in the truck. Solids are separated with a cyclone separator and water is recycled for further cleaning. Chemicals may also be used to loosen up the blockages and make cleaning more efficient.

19.5.2   CROWN CORROSION

If anaerobic conditions exist in sewer pipes, sulphates in water will be reduced by bacteria to produce hydrogen sulphide gas (H_2S). As gas rises to the top, it mixes with the moisture or condensation to produce sulphuric acid. This acid corrodes the crown section of the pipe and result in weakening of the pipe. To control **crown corrosion**, pipes should be ventilated, and septic conditions avoided.

The accumulation of hydrogen sulphide is known to cause at least three detrimental effects. First, it is malodorous and is a hazard to people who work in the vicinity – and can be fatal at high concentrations. Second, it is flammable and explosive. Finally, it is acidic and can cause corrosion problem in sewers and sewage treatment works. Crown corrosion can be controlled by lining of the pipes, ventilating the sewers, making the sewer run full, and aerating and chlorinating septic sewage.

19.5.3   REPAIRING BROKEN SECTIONS

Sewer main collapses or breaks can also be detected by diligent manhole inspections. When a sewer main has collapsed or separated, the operator will see stones, sand, and gravel in the manholes downstream of the problem area. Collapsed mains can be caused by freezing, which shatters the pipe, or by poor bedding or rock under the pipe. Separated mains are usually caused by water washing the support bedding away from under the pipe, allowing several sections of the main to drop. This separates the pipe sections by pulling them apart at the joints.

To repair the main, an operator must locate the problem spot by using a sewer tape or rod to measure the distance from the manhole to the problem area. A backhoe may then be used to excavate down to the broken or collapsed section of the pipe, and new pipe may be laid as required. All safety precautions, including appropriate signage and barricading on the street and shoring of the excavation must be followed when conducting a pipe repair.

19.5.4 BUILDING SERVICES

When unplugging a building sewer (**Figure 19.2**), the entry will be through a **clean out** placed next to the wall facing the sewer, or at the heel of the plumbing stack. The building service line is usually blocked for one of two reasons: it is frozen or it is plugged with solids. To thaw a frozen building service sewer, remove the clean-out cover and thaw the line with a steam hose from the basement. To unplug a blocked building service sewer, remove the clean-out cover and prod the obstruction with a metal sewer tape or sewer rod until it is dislodged.

19.5.5 FORCE MAIN MAINTENANCE

Materials such as sand or grit will build up in the forced main over time. Forced mains should be flushed out in the spring and fall of each year to clean out this accumulation. When using a fire truck, be sure you do not exceed the pressure rating of the pipe in the force main. Piping in some force mains will not withstand pressures of more than 400–450 kPa (60–70 psi).

19.6 INSPECTION

As part of inspection of sewer systems, manholes and illegal connections are checked frequently. As part of this inspection, two tests are usually common: smoke test and dye test.

19.6.1 SMOKE TEST

Smoke testing is done to determine illegal connections, the source of surface flow, to confirm the buildings is connected to the sewer system, and to locate broken manholes and lost sewers.

A smoke test is usually performed when the ground water table is low. This allows the smoke to rise through openings and soil. Smoke is introduced in the blocked section of the sewer to be tested. The two end manholes of the test section are plugged, and a smoke bomb is fired in the middle. An important consideration is to establish air flow before the smoke bomb is fired. To achieve this, a blower is run for 10 to 15 minutes before firing. Residents in the area must be notified to avoid panic. Smoke emanating from the ground surface is indicative of the infiltration or exfiltration and confirms the connections, and legal or illegal discharging into the sewer system.

19.6.2 DYE TEST

A dye test is carried out by introducing a dye at the point of discharge. This test allows us to check if the said point is connected to the sewer collection system. This test can help to detect if non-sanitary water is being discharged into the collection system. If the point of entry is flowing, dye is directly introduced and if it is running dry, then the dye is first diluted and then poured. Confirmation is made if the dye appears in the downstream sewer pipe. When performing multiple dye tests in each area of the system, start at the downstream end and progress upstream. The time for the dye to travel from the application point to the observation point must be estimated. This calculation helps to figure out the approximate time the dye appearance can be expected at the observation point.

19.6.3 CLOSED CIRCUIT TELEVISION

Closed circuit television is the best method for inspecting sewers, especially small-diameter pipes. It is becoming more common recently since prices of the equipment have come down. After a sewer is cleaned, a skid-mounted video camera is pulled through the pipe and a continuous picture is transmitted to the receiver in the service truck. This technique allows visual inspection and a determination of the location of structural defects, infiltration inflow, illegal connections, root growth, and grease build-up. In addition, video can be replayed for further examination and for record keeping.

19.7 INVERTED SIPHON

An **inverted siphon** is a **depressed sewer** that drops below the hydraulic grade line. Pressure conditions in these sewer lines are those of positive pressure. It is since atmospheric conditions exist at both inlet and outlet and water level at the outlet is below inlet water level. It is called inverted since the pipe is below the inlet and outlet. Flow in the siphon occurs due to the difference in water levels at the inlet and outlet.

Inverted siphons are needed to cross obstructions such as a streams, railway lines, or a depressed highway. In inverted siphons, flow velocity in the pipes should be high to prevent the solids from depositing. Inverted siphons are designed to provide flow velocities of 1.0 m/s (3.0 ft/s) or more. This is accomplished by constructing an **inlet splitter** box that directs flow to two or more siphon pipes placed in parallel. Depending on the flow, one or more pipes are in operation to maintain the minimum flow velocity. In addition to control of flow, the inlet and outlet structures provide access for cleaning sewer lines. The difference in water level between the inlet and outlet is the operating head and equals the head loss in the siphon.

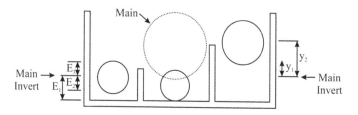

FIGURE 19.4 Inverted siphon

19.7.1 DESIGN OF INVERTED SIPHON

If an inverted siphon fails to achieve self-cleansing velocity of 1.0 m/s (3.0 ft/s) it will clog and will not function. Moreover, it is very difficult to clean or unclog siphon pipes. For this reason, a siphon usually consists of three pipes laid side by side. One of these pipes is for carrying minimum flow, one for maximum flow, and the third one for carrying combined flow (**Fig. 19.4**). If siphon is to carry only sanitary flow, two pipes may suffice. If the length of the pipe is more, **hatched boxes** should be provided every 100 m (300 ft) to facilitate rodding. Minimum pipe size is taken to be 150–200 mm (6–8 in) in diameter. It is recommended to provide a diversion for the sewage flow in case the siphon gets fully clogged. The inlet chamber should be provided with screens to prevent the entry of any debris.

Example Problem 19.4

Design an inverted siphon to cross a stream. The total length of the siphon including the slopes is 75 m. For minor head loss assume an equivalent length of 12 m. Available head is 0.50 m and average flow is 0.35 m³/s. Assume maximum and minimum flow to be 250% and 50% of average flow respectively.

Given:

$$h_f = 0.50 \text{ m}, \; n = 0.013, \; L = 75 \text{ m} + 12 \text{ m} = 87 \text{ m}, \; Q_{avg} = 0.35 \text{ m}^3/s$$

$$Q_{min} = 0.5 \times 0.35 = 0.175 \text{ m}^3/s, \; Q_{max} = 2.5 \times 0.35 = 0.875 \text{ m}^3/s, \; v = 1.0 \text{ m/s}$$

Solution:
Minimum flow conditions

$$D_l = \sqrt{\frac{1.27\,Q}{v}} = \sqrt{1.27 \times \frac{0.175\,m^3}{s} \times \frac{s}{1.0\,m}} = 0.471\,m = 450\,mm$$

Applying Manning's equation

$$h_f = \left[\frac{Qn}{0.312 \times D^{8/3}}\right]^2 \times L = \left[\frac{0.175 \times 0.013}{0.312} \times \frac{1}{0.45^{8/3}}\right]^2 \times 87\,m = 0.33\,m$$

This is less than 0.5 m, so okay

$$Q_{II}\,(Excess) = Q_{max} - Q_{min} = 2 \times Q_{avg} = 2 \times 0.35\,m^3\,/\,s = 0.70\,m^3\,/\,s$$

Maximum flow conditions

$$D_{II} = \sqrt{\frac{1.27\,Q}{v}} = \sqrt{1.27 \times \frac{0.70\,m^3}{s} \times \frac{s}{1.0\,m}} = 0.942\,m = 900\,mm$$

$$h_f = \left[\frac{Qn}{0.312 \times D^{8/3}}\right]^2 \times L = \left[\frac{0.7 \times 0.013}{0.312} \times \frac{1}{0.90^{8/3}}\right]^2 \times 87\,m = 0.13\,m$$

19.8 MANHOLES

Sewer **manholes** are one of the important appurtenances of sewer systems. They consist of an opening constructed on alignment of a sewer for facilitating a person access to the sewer for the purpose of inspection, testing, cleaning, and removal of obstruction from the sewer line.

Manholes, more appropriately called **maintenance access points** (MAPs), are built into a sewage collection system at every change of alignment, gradient, or diameter, at the head of all sewers and branches, and at every junction of two or more sewers. **Figure 19.5** shows some of these different configurations.

19.8.1 ORDINARY MANHOLE

Manholes are generally constructed of concrete rings placed one on top of the other. The rings are usually wide enough in diameter to allow an operator to comfortably descend into the manhole. It is equipped with steel ladder rungs on one side for accessibility. Manholes are spaced along the sewer lines at distances of not more than 150 m (500 ft). This spacing allows the operator to reach obstructions in the sewer main with the tools used to clear them.

19.8.2 CONSTRUCTIONAL DETAILS

Manholes are generally constructed directly over the center line of the sewer. They are usually circular in shape but in some cases a rectangular shape may be used. Circular sewers are stronger, so they are preferred.

Manholes are straight down in lower portion so as to narrow down to top cover opening equal to internal diameter of the manhole cover. Depending on the depth, the diameter of the manhole changes. The opening of entry to the manhole should be of such minimum size as to allow a workman to get access to the interior of the manhole without difficulty. A minimum circular opening of 600 mm (2.0 ft) is usually provided. Manhole covers and frames are usually cast iron with a minimum clear opening of 540 mm (21 in). Solid covers are used on sanitary sewers and open type covers are used on storm sewers. Steps or ladder rings are placed for access. The material of the rings needs to be corrosion proof. Walls may be constructed of precast rings, concrete block, brick, or poured concrete.

Wastewater is conveyed though the manhole in a smooth U-shaped channel formed in the concrete base. In junction manholes, where more than one sewer enters a manhole, the flowing through channels should be curved to merge the flow streams. If sewer changes direction without change in size, a drop of 50–75 mm (2–3 in) is provided to account for head loss. When a smaller sewer joins a larger sewer, the bottom of the larger sewer should be lowered sufficiently to maintain uniform flow transition and not cause any back up in the smaller sewer. To achieve this, either the 80% of depth of both the sewers should be at the same elevation or the crowns of both sewers are matched in elevation.

19.8.3   DROP MANHOLES AND DEAD-END MANHOLES

Drop manholes are used where laterals join a deep main line, or where the slope of the ground makes the sewer grade too steep. Drop manholes may be used to bring a sewer main down a hill in steps to avoid steep sewer grades for proper sewage flow. **Figure 19.5** illustrates this concept.

Poisonous or explosive gases can build up in these manholes. Because of its location, the dead-end manhole is seldom inspected. The cover may have several inches of dirt sealing the vent holes, thus creating the ideal gas trap.

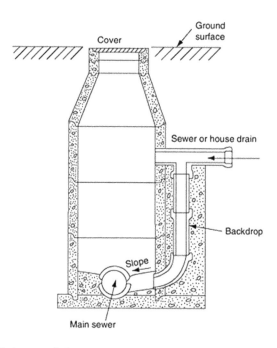

FIGURE 19.5   A drop manhole

19.8.4 MANHOLE SAFETY

A manhole is a very hazardous place to work in, so very great care must be taken to prevent injury and loss of life. The hazards include: oxygen deficiency, presence of toxic gases likes hydrogen sulphide or carbon monoxide, and combustible material. Before entering a manhole, one must check for the presence the obnoxious gases or the deficiency of oxygen with a gas meter. If needed, allow natural ventilation or use forced air to help make the environment in the manhole safer to work. Since it is a case of confined safe entry, all regulations and approved procedures must be followed. The worker is the first one to be responsible for his or her safety. If the working conditions are unsafe, workers have the right to refuse to work. Before starting any work, place proper traffic signs. A team of two people is required, with one waiting on the ground to pull the harness in case it becomes dangerous to work in the manhole. Persons entering the manhole need to have a self-contained breathing apparatus (SCBA) if there is danger of oxygen deficiency. No smoking or open flames are permitted near the work area.

19.8.5 SEWER VENTILATION

As discussed above, sewer lines contain acidic vapors, combustibles and noxious gases, and strong odors. The stack connected to the building sewer usually provides sewer ventilation. Where there is need, venting shafts connected to sewer manhole may be provided. In some cases, forced ventilation may also be done.

Example Problem 19.5

You need to enter a 2.5 diameter manhole. Before entering, you need to make sure that for safety reasons, you ventilate the manhole such that the whole air is replaced by fresh air by using an air blower. If the air blower capacity is 2.0 m³/min and the depth to water in the manhole is 3.0 m, calculate the minimum time for which the blower should be run.

$$t = \frac{V}{Q} = \frac{\pi(2.5\,m)^2}{4} \times 3.0\,m \times \frac{min}{2.0\,m^3} = 7.35 = 7.4\,min\,(say\,8\,min)$$

19.7.4 MANHOLE INSPECTION AND MAINTENANCE

The operators of the wastewater collection system should establish a daily maintenance inspection schedule for the manholes in your system. In order to establish a maintenance schedule, the **as-constructed** drawings for the sewage collection system will need to be consulted. These drawings show the location and number of each manhole in the collection network. The total number of manholes should be divided into groups for daily inspections (a different group will be inspected each day). If properly scheduled, most of the manholes in a system can usually be inspected in the course of a week.

Careful written records should be kept of all manhole inspections. Any build-up of solids in the manholes must be cleaned out so that sewage flow through the manhole is not slowed down. Experience and keeping good records will soon identify which manholes need regular attention and which are trouble-free. In severe winter conditions, ice build-up in some manholes can completely fill the upper part of the structure.

19.9   SAMPLING AND FLOW MEASUREMENT

Monitoring sewer use requires **composite sampling** and flow measurements at selected points in the sewage collection systems. Flow measurement and sampling of wastewater is important to monitor industrial discharges and inflow and infiltration studies. For medium to large industries, sampling stations are built to accommodate automatic sampler and flow measuring devices.

19.9.1   Flow Measurement in Sewers

First estimates of flow in the sewer can be achieved by observing the depth of flow. Applying Manning's formula, flow calculation can be made. An alternative is to observe flow velocity with a float, **current meter**, or a dye float and multiply it by the wetted area.

Installation of a flume is necessary for accurate flow measurement and automatic composite sampling. A **Palmer-Bowlus flume** with a trapezoidal section is most common. By observing the head upstream of the flume, flow is read from the rating equation or table for the flume provided by the manufacturers. Palmer-Bowlus flumes are constructed for temporary installation in the half section of the sewer in a manhole. These flumes are prefabricated of materials like fiberglass or plastic and are available in various sizes to match the diameter of the sewer pipe. Weirs can also be used but are not preferred because of silting problems.

Various types of **depth sensors**, including ultrasonic, submerged probes and bubblers are common. An ultrasonic sensor is based on the delay in the ultrasonic pulse as it is reflected from the water surface. This sensor is installed above the water surface and is not affected by the grease or other floating stuff, so far flow is smooth. A submerged probe works on the pressure differential principle. Floating debris, grease, or foaming does not affect the accuracy of the meter. A bubbler gauge measures the depth by sensing the pressure required, forcing air to bubble through the water column. Many of the flow meters are programmable so as to record depth of water and display flow rate and totalized flow.

19.9.2   Sample Collection

Samples can be collected manually or using automatic samplers. Irrespective of technique used, samples must be representative. Sampling locations should be selected where the flow stream is smooth and well mixed. Scoping the sample from the surface or from the bottom would be a biased sample. **Automatic sampling** is becoming very common. Manual sampling is usually done for spot-checking and similar uses.

Though composite samples are more common, for inspection and auditing an industrial user, sometimes a **grab sample**, is collected. However, an analysis of a grab sample represents the point value with respect to time and place. A composite sample test results would give an average value over the sampling period. In some cases, like pH and dissolved oxygen measurements, only grab samples are not composited, since it would change the accuracy of results due to chemical changes.

19.10 WASTEWATER PUMPING

Wastewater pumping is in many ways different from freshwater pumping. Wastewater flow varies hourly, seasonally, and yearly depending on growth and climatic conditions. The pumping system therefore should be capable of handling peak and minimum flows. Retaining the wastewater too long in the **wet well** can lead to septicity and odors. Therefore, the detention time in the sump should not exceed 30 minutes and wells should be properly ventilated to release gases such as produced due to decomposition of organics in wastewater. Since wastewater carries solids, the diameter of the rising main should be such as to provide minimum flow velocity of 0.6 m/s (2 ft/s).

To minimize premature wear out of pump and motor, frequency of pump starts is usually restricted to 5/h. More frequent starts will shorten the useful life of a pump and motor. If the minimum flows are less than 20% of the average flows, odors can develop, so in such situations providing proper ventilation is very important.

A wet well pumping system usually consists of two pumps of at least capacity equal to peak flow. Two pumps ensure 100% standby capacity. In situations when flows exceed peak flows, both pumps operate. Operation is such that both pumps take turns to operate as lead pump. This keeps both the pumps in good condition and does not cause overwearing of one.

19.10.1 WET WELL LIFT STATIONS

A **wet well lift** station consists of a holding reservoir which receives and holds the sewage as it pours in from the sewer mains, pumps to lift the sewage to the disposal area, electric motors to power the pumps, the necessary pipes, valves, and fittings to move the sewage to the pump and then to the point of discharge, and a control system which starts and stops the pumps automatically. In a wet well station, the pumps are situated in the holding reservoir.

The control system maintains maximum and minimum liquid levels in the wet well and may consist of three or more floats, or a combination of floats and timers. In small installations, usually a submersible pump with controls is employed to pump wastewater. In case a pump needs repair or replacement, you need to pull the pump out of the well. In larger installations, however, a dry well is used to house the pump, motor, and electrical connections so as to have easy access for control, inspection, and maintenance.

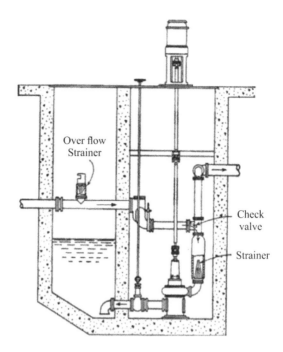

Over flow
Strainer

Check
valve

Strainer

FIGURE 19.6   Dry well lift station

19.10.2   DRY WELL LIFT STATIONS

Dry well lift stations consist of two pits or wells – a wet well and a dry well (**Fig. 19.6**). The wet well receives and holds the sewage as it comes from the sewer main. The **dry well** houses the pumps, motors, piping, valves, fittings, and electrical controls necessary to move the sewage from the wet pit to the disposal area. A dry well allows easy access to pump, motor, and electrical controls. The standby motor (if one is present) is mounted on the floor of the lift station above the dry pit. This motor supplies power to one of the pumps in the event of power failure. **Figure 19.6** shows a typical dry well lift station with separate pits for wastewater and pumps.

19.11   WASTEWATER FLOW PUMPS

One main difference compared to potable water pumps is that sewage pumps must be capable of passing solids. For this reason, non-clog or pumps with large impellers are used. Screens are placed to retain large-size debris, and in some installations a shredder may also be used. The centrifugal pump is the most common used in lift conditions.

In the case of submersible pumps, a rail system is used to remove the pump. Self-priming sewage pumps are suitable when the suction lift is less than 5.5 m (18 ft). Compared to submersible pumps, access is direct. Direct access allows easy maintenance and the operation of the pump can be observed readily. In dry well applications, non-clog vertical-mounted centrifugal pumps are used. Wastewater from the wet well side is conveyed to the pump suction by piping. Since the pump is sitting below the wastewater level in the dry well, the pump always operates under positive head conditions.

19.11.1   WET WELLS

The main purpose of the wet well is to provide short-term storage. This reduces stress on the pumping equipment by keeping the pump cycle time more than 8 min. Frequent start time results in overheating and shortens the useful life of motors.

19.11.2   SCREENS

As the name indicates, screens are installed to retain large debris from the wastewater. To prevent large head built up and causing restrictions to flow, screens must be cleaned up frequently. **Comminutors** and **barminutors** are used where too much debris is expected.

19.11.3   ELECTRICAL AND CONTROLS

The electrical system in the lift station consists of the main breaker that feeds the motor control panel and auxiliary electrical systems. The motor control panel houses the control system for the motors and other electrical systems in the lift station. The controls include starters, fuses, heater strips, coils, relays, and switches.

The level sensing devices used in controlling the pump operation can be floats, electrodes, bubblers, mercury tilt and sonic. In the case of a bubbler tube, compressed air is supplied. The pressure required to bubble the air through the wastewater column is proportional to the height of wastewater above the tube end. In sonic units, the time to return for the high frequency sound indicates the depth of wastewater.

19.12   LIFT STATION MAINTENANCE

In both wet pit and dry pit type lift stations, maintenance involves mainly keeping the station and the wet pits clean and keeping the pump(s) unclogged and operating at the proper capacity. The main items requiring regular attention are screen baskets and bar screens, sand, grease, or debris build-ups on the walls or floor of wet pits and sump pumps in dry pits.

19.12.1   SCREENING BASKETS AND BAR SCREENS

To empty a **screening basket**, the basket is pulled up to the surface and the accumulated material is emptied into a bucket or similar receptacle. Material must be appropriately disposed of (e.g., burning, burying, or sent to land fill).

A **bar screen** serves the same purpose as the screening basket. It collects the larger pieces of debris and sewage before they reach the lift station pumps and clog the impellers. A bar screen is usually installed in the nearest manhole upstream from the lift station. The manhole will be no more than 30 m from the lift station and in most cases much closer. Screening baskets and bar screens must be checked regularly and cleaned out before they are full. Once the screens are full, the material starts to wash over and may go into the pumps. If this happens, you will have to dismantle and clean or repair the lift station pumps.

19.12.2 WET WELL FLOOR MAINTENANCE

Grease, grit, sand, and other material present in sewage will build up over time on the walls, floor, piping, and pumps in a wet pit. If the sand and grit are allowed to build up, they will work into the pumps, causing rapid wear of the impellers and impeller chambers. The wet well should be cleaned on a regular basis to avoid excessive wear or damage to the pumps. In order to clean the wet pit an operator may follow the steps outlined below.

1. Remember, a wet well is a confined space, so you must work in pairs and have all the appropriate safety equipment and documentation.
2. Obtain the required clothing and safety equipment for working in manholes and wet pits.
3. Obtain a suitable length of fire hose from your water treatment plant and attach it to the nearest fire hydrant. Also obtain a stiff-bristle push broom or other tool suitable for use as a scrubber on walls, pipes, etc.
4. Thoroughly hose down the walls and piping in the wet pit. Use the scrubber where necessary to remove dirt and grime.
5. To clean out the bottom or floor of the wet pit, it will be necessary to close off the sewage flow into the wet pit. Flow into the wet pit may be stopped in two ways. If an inlet valve has been provided, closing this valve may stop flow into the wet pit. If no valve is present, an inflatable bladder must be used. The bladder is installed in the first manhole upstream from the lift station. The bladder serves as a temporary plug in the inlet line. Observing proper safety procedures, an operator descends into the nearest manhole upstream of the lift station. The bladder is inserted into the line leading from the manhole to the lift station. Using a hand pump or portable air tank, the bladder is inflated. An anchor rope is attached to the D-ring on the rubber plug and the air release line is attached to the air valve on the rubber bladder. The air release line and anchor line are carried out of the manhole.
6. When the flow into the lift station has stopped the wet pit floor may be cleaned. An operator (following all appropriate safety precautions) may go down to the floor of the wet pit and shovel out the sand and grit built up on the floor of the wet pit. Bucket loads will be lifted out of the pit and emptied by your topside helpers. When the wet pit floor has been cleaned, and the operator has exited the wet pit, the air may be let out of the bladder through the air release valve and the bladder may be removed from the manhole, restarting the flow to the wet well.

19.12.3 Sump Pump Operation and Maintenance

Sump pumps are located in the dry pit of a lift station. The sump pump removes water caused by seepage of ground water into the dry pit, excess from water-lubricated bearings, and spills left from servicing or washing out sewage pumps. The sump pit must be regularly cleaned to prevent clogging of the sump pump. The sump pit should be checked for clogging material and any material found should be promptly removed and disposed of. The sump pump should also be checked to make sure that it is operable. Lifting the float that triggers the pump does this. If a hum is heard, then the pump is operable. If the pump does not turn on when the float is lifted, the pump is not working and should be repaired or replaced.

19.13 PUMP OPERATING SEQUENCE

Most lift stations are controlled automatically. Level controllers in the wet well turn pumps on and off depending on the level in the well. On a dual pump system, the sequence of operation is described below:

1. Lead pump comes when the water level reaches pump start level. Below this level, no pump is running, and the incoming wastewater is being stored.
2. At some stations the lead pump automatically alternates between the two pumps each pumping cycle. As discussed before, each pump receives equal wear. For stations not equipped with an alternating device, the lead pump is alternated manually every week.
3. During peak flows, water level will continuously be rising when the lead pump is running. As water level reaches the higher start level, pump number 2 comes on.
4. If water flows are exceptionally high, such as during storm periods, the water level may still be rising. In such cases, as the water level reaches the flooding level, alarms are signaled
5. As the water level starts dropping, the first pump number will stop. During low flows, water level drops below the lead pump start level and the lead pump stops to complete on the pumping cycle.

19.13.1 Level Setting

Too many frequent starts is damaging to the pumping equipment. Experience has shown that a pumping cycle should be preferably more than ten minutes or the pump cycled should be in the range of 6–8/h. Frequent starts will not allow the motor and starters to cool down properly and may result in overheating. The lead pump stop level is normally set at the minimum submergence level to prevent pump cavitation. The start level is determined using the following formula.

$$h_{start} = \frac{V_S}{A} = \frac{t_C}{A} \times \frac{Q_i}{Q_p} \times \left(Q_p - Q_i\right)$$

V_S = volume of storage between stop and start level
A = area of cross section of wet well, t_c = desired pump cycle time
Q_p = pumping rate, Q_i = wastewater influent rate

Example Problem 19.6

Work out the pump start level in the case of a 2.5 m diameter wet well if the desired pumping cycle time is 10 min (6/h). Wastewater average inflow rate is 12 L/s and pump rate is 15 L/s.

Given:

$$D = 2.5 \text{ m} \quad t_c = 10 \text{ min} \quad Q_i = 12 \text{ L/s} \quad Q_p = 15 \text{ L/s}$$

Solution:

$$h_{start} = \frac{t_C}{A} \times \frac{Q_i}{Q_p} \times (Q_p - Q_i)$$

$$= \frac{10\,min}{0.785(2.5m)^2} \times \frac{12}{15}(15-12)\frac{L}{s} \times \frac{60\,s}{min} \times \frac{m^3}{1000\,L} = 0.293 = \underline{0.29\,m}$$

Pump start must be set at 30 cm above the submergence level (stop level).

19.13.2 PUMPING RATE IN LIFT STATIONS

Even with good maintenance of screening baskets and lift station walls and floors, some debris may enter pumps or force the mains and begin to clog them up. Clogging in pumps or force mains will often show up first in a drop in the lift station pump's capacity. If the pump is not moving sewage at the rate that is indicated by its capacity rating, debris clogging the pump or force main may be the problem. This situation must be corrected before it causes serious problems leading to major pump repairs and to complete blockage of the sewer lines.

The actual pumping rate of a lift station can be checked against its rated capacity by observing and timing the pump during its operation. Steps are as follows.

1. Find out and record the capacity rating for the pump as indicated by the manufacturer's information sheets or pump table.
2. It will be preferred to check the pumping rate for the operating head from the performance curves. A decrease in pumping rate might indicate a problem with the pump that needs to be investigated.
3. Select and tag two ladder rungs close to the bottom of the wet pit. These will be used as benchmarks for high and low wastewater levels in the wet pit.

4. Measure the distance in meters between the two rungs. This measurement will give the depth between the high and low water levels.
5. Calculate the number of liters of wastewater contained in the wet pit between these two levels by applying formulae in the following section.
6. Shut off the pump and wait for the wastewater level to rise over the top rung selected.
7. Restart the pump when the water has risen over the top rung. As the water level drops, start timing on your wristwatch when the top rung reappears. Continue timing until the lower rung is out of the water.
8. Write down the time it took for the water level to drop from the top to the bottom ladder rung in minutes.

Example Problem 19.7

A wet well measures 10 ft × 8.0 ft. To check the pumping rate, close down the influent valve and observe the drop in water level. Over a duration of 5.0 min, the water level in the well dropped by 2.2 ft. Work out the pumping rate.

Given:

$$A = 10 \text{ ft} \times 8.0 \text{ ft} \quad \Delta d = 2.2 \text{ ft} \quad \Delta t = 5.0 \text{ min} \quad Q = ?$$

Solution:

$$Q = \frac{\Delta V}{\Delta t} = \frac{10 \text{ ft} \times 8.0 \text{ ft} \times 2.2 \text{ ft}}{5.0 \text{ min}} \times \frac{7.48 \text{ gal}}{ft^3} = 263.3 = \underline{260 \text{ gpm}}$$

Example Problem 19.8

A wet well measures 2.5 m × 3.0 m. Influent to the well measured before the test is 22 L/s. If the water level drops by 6.0 cm in 5.0 min, what is the pumping rate in L/s?

Given:

$$A = 3.0 \text{ m} \times 2.5 \text{ m} \quad \Delta d = 6.0 \text{ cm} \quad \Delta t = 5.0 \text{ min} \quad Q_i = 22 \text{ L/s} \quad Q_p = ?$$

Solution:

$$\Delta Q_S = \frac{\Delta V}{\Delta t} = \frac{2.5 \text{ m} \times 3.0 \text{ m} \times 0.06 \text{ m}}{5.0 \text{ min}} \times \frac{1000 \text{ L}}{m^3} \times \frac{min}{60 \text{ s}} = 1.5 \text{ L/s}$$

$$Q_p = Q_i - -\Delta Q_S = 22 \text{ L/s} + 1.5 \text{ L/s} = 23.5 = \underline{24 \text{ L/s}}$$

Change in storage is negative since water level is dropping. Change will be positive if the water level rises during pumping.

Example Problem 19.9

A pumping station pumps the wastewater from an intercepting sewer to the treatment plant. The wet well is 3.8 m by 3.2 m. After the influent valve is closed, the water level in the wet well drops by 0.6 m in 5.0 min. after the valve is opened, the water level rises by 5.0 cm in 10.0 min time. Calculate the influent rate.

Given:

$$\text{valve closed: } A = 3.8 \text{ m} \times 3.2 \text{ m} \quad \Delta d = 0.6 \text{ m} \quad \Delta t = 5.0 \text{ min } Q_p =?$$

$$\text{valve open: } A = 3.8 \text{ m} \times 3.2 \text{ m} \quad \Delta d = 5.0 \text{ cm} \quad \Delta t = 10 \text{ min } Q_i =?$$

Solution:

$$Q_p = \frac{\Delta V}{\Delta t} = \frac{3.8 \, m \times 3.2 \, m \times 0.60 \, m}{5.0 \, min} \times \frac{min}{60 \, s} \times \frac{1000 \, L}{m^3} = 24.3 = \underline{24 \, L/s}$$

$$\Delta Q_S = \frac{3.8 \, m \times 3.2 \, m \times 0.050 \, m}{10 \, min} \times \frac{min}{60 \, s} \times \frac{1000 \, L}{m^3} = 1.01 = 1.0 \, L/s$$

$$Q_i = Q_p + \Delta Q_S = 24.3 \, L/s + 1.01 \, L/s = 25.4 = \underline{25 \, L/s}$$

Example Problem 19.10

At a lift station, wastewater is lifted by 35 ft via a 280 ft long rising main. Assuming pump efficiency of 65% and head loss of 1.2%, work out the pump power required to lift wastewater @ 1200 gpm during peak flow conditions.

Given:

$$\Delta Z = 35 \text{ ft} \quad h_l = 1.2\% \quad Q_p = 1200 \text{ gpm} \quad P_P =?$$

Solution:

$$h_a = lift + h_l = 35 \, ft + 280 \, ft \times 0.012 = 38.36 = 38 \, ft$$

$$P_a = Q \times \gamma \times h_a = \frac{1200 \, gal}{min} \times \frac{8.34 \, lb}{gal} \times 38.36 \, ft \times \frac{hp.min}{33000 \, ft.lb} = 11.63 = 12 \, hp$$

Assuming pump efficiency of 65%, pump power is worked out

$$P_p = \frac{P_a}{E_p} = \frac{11.63 \, hp}{65\%} \times 100\% = 17.89 = 18 \, hp$$

Example Problem 19.11

A town has a population of 45 000. Average per capita sewage production is 400 L/c.d. Determine the power required to pump the sewage during peak flow conditions against a static head of 8.5 m. Make suitable assumptions and determine the size of rising main. The length of the rising main is 220 m (f = 0.02).

Given:

$$\Delta Z = 8.5\ m \quad Q = 400\ L/c{\cdot}d \quad L = 220\ m \quad f = 0.025 \quad P_1 =?$$

Assumptions:

1. Maximum flow is 2× average.
2. Velocity in the rising main = 1.0 m/s.
3. Minor losses = 0.50 m.
4. Pumping system efficiency = 65%.
5. Three pumps with one as standby unit.

Solution:
Rising main

$$Q_p = 2 \times \frac{400\,L}{p.d} \times 45000\,p \times \frac{m^3}{1000\,L} \times \frac{d}{24 \times 3600\,s} = 0.416\,m^3 / s$$

$$D = \sqrt{\frac{4Q}{\pi v}} = \sqrt{\frac{4}{\pi} \times \frac{0.416\,m^3}{s} \times \frac{s}{1.0\,m}} = 0.728 = 750\,mm\,(30\,in)$$

$$h_f = f \times \frac{L}{D} \times \frac{v^2}{2g} = 0.025 \times \frac{220\,m}{0.75\,m} \times \left(\frac{1.0\,m}{s}\right)^2 \times \frac{s^2}{2 \times 9.81\,m} = 0.373 = 0.37\,m$$

Capacity of sump

$$V_{20\,min} = \frac{0.416\,m^3}{s} \times 20\,min \times \frac{60\,s}{min} = 499.2 = 500\,m^3$$

$$V_{main} = \frac{\pi D^2}{4} \times L = \frac{\pi(0.75\,m)^2}{4} \times 220\,m = 97.19 = 97.2\,m^3$$

$$V_{tot} = 499.2 + 97.2 = 596.3 = 596\,m^3$$

Providing 3# of sump wells, when 2 will store wastewater flow for 20 min

assuming depth of 3.5 m diameter is found

$$D = \sqrt{\frac{1}{2} \times \frac{4V}{\pi d}} = \sqrt{\frac{2}{\pi} \times \frac{596.3\,m^3}{3.5\,m}} = 10.41 = 10.5\,m$$

Pumping system

$$h_a = \Delta Z + h_f + h_m = 8.5\,m + 0.37\,m + 0.5\,m = 9.373 = 9.4\,m$$

$$P_i = \frac{Q\gamma h_a}{E} = \frac{1}{2} \times \frac{0.416\,m^3}{s} \times \frac{9.81\,kN}{m^3} \times \frac{9.373\,m}{0.65} \times \frac{kW.s}{kN.m} = 29.42 = 29\,kW$$

Discussion Questions

1. Modify Manning's equation for flow capacity of a circular sewer of diameter D in SI units.
2. Prove that for a circular sewer flowing full or half full, hydraulic radius is 1/4th of the diameter of the sewer pipe.
3. Define crown corrosion and describe measures to prevent it.
4. What are the functions of a drop manhole?
5. Differentiate between sewage and sewerage.
6. State the merits and demerits of: (i) a separate system of sewerage; (ii) a combined system of sewerage.
7. Discuss diurnal flow variation in sewage flow. How is this variation considered in the design of sewers?
8. What are the various types of storm water regulators used in a sewerage system?
9. Why is ventilation important in sewers? How is it done?
10. Explain the construction details of a manhole.
11. Describe the different methods of sampling of wastewater.
12. Name common flow measurement devices used in sewer collection system.
13. Compare storm sewer with storm sewers.
14. How is pumping wastewater different than pumping potable water?
15. During inspection of pumping stations, what important tasks should be performed?
16. Describe the procedure to verify pumping rate at a wastewater pumping station.
17. Describe pump-operating sequence at a lift station.
18. Compare dry well pumping station with a wet well pumping station.
19. Discuss the type of pumps used for wastewater pumping.
20. What are the main considerations when selecting the pumping capacity?
21. Explain, with the help of a sketch, the main components of a lift station with a separate dry well and wet well?
22. Explain the importance of preliminary treatment before water enters the wet well? What devices are commonly employed for this purpose?

Practice Problems

1. In a leak test, water level in a 4.0 ft diameter manhole is dropping at the rate 3.0 in/h. What is the leakage rate? (0.39 gpm)
2. Design an inverted siphon to cross a river. The total length of the siphon, including the slopes, is 85 m. Assume 0.30 m minor head losses and 0.012 for roughness coefficient. Available head is 0.75 m and average flow is 450 L/s. Assume maximum and minimum flows to be 250% and 50% of average flow respectively. (500 mm, 1000 mm, head loss = 0.55 m and 0.40 < 0.75 m).
3. You need to enter a 3.0 diameter manhole. Before entering, you need to make sure that, for safety reasons, you ventilate the manhole such that the whole air is replaced by fresh air using an air blower. If the air blower capacity is 2.0 m³/min and the depth to water in the manhole is 3.0 m, calculate the minimum time the blower should be run. (11 min).
4. Work out the full flow velocity and flow capacity of a 16 in sewer pipe laid on a slope of 0.20%. (2.5 ft/s, 1600 gpm).
5. Design an inverted siphon to cross a canal. The length of the sewer pipe including slopes is 60 m. Consider an equivalent length of 9.0 m to account for minor losses. The available head is 0.90 m, and the average flow is 350 L/s. The maximum and minimum rates of flow may be taken as 300% and 50% of average flow respectively. (450 mm, 0.26 m, 1000 mm, 0.1 m).
6. Work out the full flow velocity and flow capacity of a 500 mm sewer pipe laid on a slope of 0.30%. (1.1 m/s, 210 L/s).
7. A 12 in stoneware sewer is laid on a grade of 1.0%. Assuming n = 0.012, what flow velocity is achieved when flowing full. (5.0 ft/s).
8. What diameter sewer pipe (n = 0.012) is required to carry wastewater with a flow velocity of 2.0 ft/s flowing full. The sewer pipe is laid on a gradient of 1 in 500. (10 in).
9. To carry solids, it is required to generate a flow velocity of 1.4 m/s in a 300 mm concrete sewer pipe flowing full. Assuming n = 0.013, what is the minimum grade on which the pipe should be laid? (1 in 100).
10. On a city map, the distance between two manholes is measured to be 21 cm. If the map scale is 1:500, what is the linear distance between the manholes? (110 m).
11. The average time taken by a dye to travel a distance of 160 m between two manholes is 4 min 15 seconds. What is the average sewage flow velocity? (0.63 m/s).
12. It took 2 min 20 seconds for a float to travel a distance of 125 m between two manholes. Estimate the velocity of the wastewater in the sewer line assuming flow velocity to be 88% of the float velocity. (0.79 m/s).
13. A rectangular wet well is 2.0 m × 3.0 m. With no pump running, water level in the well was rising at the rate of 0.5 m in 2.5 min. What is the rate of inflow? (20 L/s).

14. A wet well has water surface of 6.0 m². With the lead pump on, drawdown rate is observed to be 30 cm in 3.0 min. What is the pumping rate given that influent rate is 23 L/s? (33 L/s).

15. A flow of 720 gpm is passing through a square wet well with each side of 30 ft and 11 ft deep. What is the average detention time in the wet well? (100 min).

16. The diameter of a wet well is 10 ft. How much wastewater does it hold when filled to a depth of 8.0 ft? (4700 gal).

17. A pumping station wet well is 3 m × 2 m. To check pumping rate, influent valve to the wet well is closed for a 5 min drawdown test. During the test, the water level dropped by 0.6 m. What is the pumping rate in m³/min? (0.7 m³/min).

18. A pumping station pumps the wastewater from an intercepting sewer to the treatment plant. The wet well measures 13 ft × 10 ft. After the influent valve is closed, the water level in the wet well drops by 34 inches in 5.0 min. What is the pumping rate? (550 gpm).

19. Checking the pumping rate at a lift station it is important to know whether or not the pump is pumping the right volume of wastewater. A circular lift station has a diameter of 2.5 m and the distance between the two rungs of the ladder used in the test is 0.5 m. During the test, it takes 30 minutes to lower the water level by 0.50 m. The measured inflow into the chamber for one minute after the pumps were shut off was an additional 19 liters. What is the actual pumping rate of the pump in liters per minute? (100 L/min).

20. Select pumping units to pump average wastewater flow of 800 m³/d, assuming peak flow is 2.5× and both the pumps operate during peak flow period. Rising main is 200 mm in diameter and 550 m long and wastewater must be lifted through a height of 11 m. Assume friction factor f = 0.03 and 50 m minor losses equivalent length. Determine operating head and power added to wastewater. (14 m, 1.5 kW).

21. Work out the pump start level in case of a 3.0 m diameter wet well if the desired pumping cycle should not exceed 5/h. Wastewater average inflow rate is 8.5 L/s and pump rate is 13 L/s. (0.30 m).

22. At a lift station, wastewater is lifted by 31 ft via a 250 ft long rising main. Assuming pump efficiency of 65% and head loss are 2%, work out the power required to pump wastewater at the rate of 950 gpm during peak flow conditions. (13 hp).

23. A flow of 50 L/s is passing through a wet well measuring 15 m long by 5.5 m wide by 3.4 m deep. What is the average theoretical detention time? (94 min)

24. If the lead pump is pumping at the rate of 10 L/s and the water in the 2.5 m × 1.8 m wet well is rising at a rate of 0.1 m in 40 s, calculate the rate of inflow? (21 L/s).

25. The storage in a well between the high (pump starts) and low (pump stops) levels is 4.8 m³. If the pumping rate is 4.2 L/s and the inflow to the wet well is 2.3 L/s, what is the pumping operating time? (36 min)

26. The water level in a 16 ft × 16 ft wet well is sufficiently low to allow shutting off all pumps. Before the pumps are restarted, the rise rate is observed to be 20 inches in 2 min and 46 seconds. Estimate the inflow rate. (1200 gpm).

27. Work out the pump start level in the case of an 8.0 ft diameter wet well if the desired pumping cycle time is 12 min (5/h). Wastewater average inflow rate is 180 gpm and the pump rate is 230 gpm. (2.0 ft above the stop level).

20 Design of Sewers

Sewer pipes, also called sewer mains, are installed in one of two ways: **gravity mains**, or **force mains**. In gravity mains, the sewage moves through the pipe by gravity alone. No other pressure or force is required. In force mains, the sewage is pushed, or forced, through the sewer by means of a pump. Before we talk about design, a refresher about flow hydraulics is necessary

20.1 OPEN CHANNEL FLOW

In comparison to closed conduit flow, open channel flow has one surface, which is free and open to the atmosphere. Pipe flow conditions exist in water distribution mains and in sewer mains flow is primarily due to gravity. In closed conduits (pipe flow) the flow is primarily due to pressure, whereas in open conduits it is all due to gravity, therefore open channel flow is also called **gravity flow**.

20.1.1 Flow Classification

Flow may be classified as being steady (independent of time) or unsteady (time dependent) and based on the depth of the flow. Therefore flow velocity is uniform and non-uniform.

   Uniform steady flow occurs when the flow rate (discharge) remains constant with time (steady) and the depth of the liquid does not vary along the length of the channel. Steady uniform flow occurs in channels that are long and straight and whose depth and slope are constant (prismatic) along the length of the channel. In this type of flow, the slope of the free water surface (S_w) is parallel to the slope of the bed of the channel (S_o), as shown in **Figure 20.1**.

   Steady non-uniform (varied) flow occurs where the depth of flow varies along the length of the channel. This will occur if the channel is non-prismatic. Varied flow can be further classified as **rapidly varied flow** (RVF) or **gradually varied flow** (GVF) depending on the rate of change in depth along the channel. Unsteady non-uniform flow occurs when the depth of the fluid varies both with time and along the channel bed. It is not practicable to achieve unsteady uniform flow.

20.1.2 Hydraulic Slope

In open channel flow, the common slope terms used include: bed slope, slope of hydraulic grade line or hydraulic gradient, energy slope, and slope of free liquid surface.

   Channel bed slope, S_o, refers to the change in elevation of the channel bed per unit horizontal length of the channel. This could be expressed as a fraction, a percent, or the angle the channel bed makes with the horizontal plane.

DOI: 10.1201/9781003347941-24   **327**

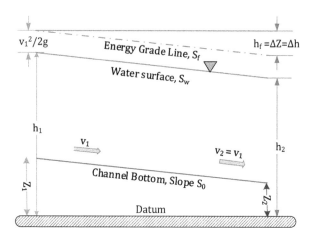

FIGURE 20.1   Open channel flow energy

Water (liquid) surface, S_w is the slope of the water surface. As in the case of open channel flow, flow is all due to gravity. Therefore the hydraulic head at a given section is equal to the elevation of the water surface. It can also be said that water surface represents **hydraulic grade line** (HGL) in open channel flow conditions. Therefore, **hydraulic slope**, S_w is a fall in water surface elevation per unit horizontal length of the channel.

20.2   MANNING'S FLOW EQUATION

As discussed previously, Manning's equation is empirical in nature; units must be consistent.

Manning's equation for flow velocity

$$v = \frac{1}{n} \times R_h^{2/3} \times \sqrt{S} - SI \qquad v = \frac{1.49}{n} \times R_h^{2/3} \times \sqrt{S} - USC$$

The average velocity of flow v will be in m/s when the hydraulic radius, R_h, is in m. The term n is a roughness factor, commonly called Manning's n. Friction slope, S, is dimensionless and is equal to slope of the channel bed for uniform flow conditions.

Manning's equation for discharge rate

$$Q = A \times v = \frac{A}{n} \times R_h^{2/3} \times \sqrt{S} - SI = \frac{1.49\,A}{n} \times R_h^{2/3} \times \sqrt{S} - USC$$

Flow rate (discharge) for which uniform flow will occur is referred to as **normal discharge**. The corresponding depth of flow is called **normal depth**.

20.2.1 Hydraulic Radius

The term R in Manning's flow equation is the characteristic dimension of the channel. **Hydraulic radius** is defined as the ratio of the liquid section area to the wetted perimeter, P_w.

Wetted perimeter, as the name implies, is the length of the wetted surface of the channel. It is important to note that the wetted perimeter is equal to the total perimeter of the liquid section minus the length of the free liquid surface.

20.2.2 Uniform Flow Problems

As indicated earlier, under uniform flow conditions friction slope is equal to the channel bed slope. By applying Manning's formula, the flow-carrying capacity of a channel of given shape, size, and roughness can be calculated for a specified depth (normal depth) of flow. In varied flow conditions, friction slope will not be equal to the channel bed slope. However, Manning's equation is still applicable provided the slope term in the equation refers to friction slope.

20.2.3 Circular Pipes Flowing Full

For circular pipes flowing full, $R_h = 0.25D$ and $A = 0.785D^2$. Making these substitutions, Manning's flow equation becomes:

Full flow velocity

$$v_F = \frac{0.4}{n} \times D^{2/3} \times \sqrt{S} \qquad Q_F = \frac{0.312}{n} \times D^{8/3} \times \sqrt{S} - SI$$

$$v_F = \frac{0.6}{n} \times D^{2/3} \times \sqrt{S} \qquad Q_F = \frac{0.464}{n} \times D^{8/3} \times \sqrt{S} - USC$$

Example Problem 20.1

Desired full flow capacity of a sewer main 30 L/s and available grade is 0.10%. Select the size of the sewer main?

Given:

$$D = ? \quad n = 0.013 \quad Q = 30 \text{ L/s} = 0.03 \text{ m}^3/\text{s} \quad S = 0.10\% = 0.001$$

Solution:

$$D = \left[\frac{Q_F \times n}{0.312\sqrt{S}} \right]^{3/8} = \left[\frac{0.03 \times 0.013}{0.312 \times \sqrt{0.001}} \right]^{3/8} = 0.297 \text{ m} = 300 \text{mm} \left(12 \text{ in NPS} \right)$$

Example Problem 20.2

Determine the flow carrying capacity of a 12 in sewer pipe (n = 0.013) running half full and laid on a slope of 0.11%.

Given:

$$D = 12 \text{ in} = 1.0 \text{ ft} \quad n = 0.013 \quad S = 0.10\% = 0.001$$

Solution:

$$Q_F = \frac{0.464}{n} \times D^{8/3} \times \sqrt{S} = \frac{0.464}{0.013} \times 1.0^{8/3} \times \sqrt{0.0011} = 1.183 \text{ ft}^3 / s$$

$$Q_{0.5F} = 0.5 \times 1.1838 \text{ ft}^3 \times \frac{7.48 \text{ gal}}{\text{ft}^3} \times \frac{60 \text{ s}}{\text{min}} = 4.42 = 4.4 \text{ gpm}$$

Example Problem 20.3

Calculate the minimum slope on which a rectangular channel with 1.2 m of width should be laid to maintain a velocity of 1.0 m/s while flowing at a depth of 0.60 m. The channel is made of unfinished concrete (n = 0.017).

Given:

$$B = 1.2 \text{ m} \quad d = 0.60 \text{ m} \quad v = 1.0 \text{ m/s} \quad n = 0.017 \quad S = ?$$

Solution:

$$R_h = \frac{A}{P_w} = \frac{1.2 \text{ m} \times 0.60 \text{ m}}{1.2 \text{ m} + 2 \times 0.6 \text{ m}} = 0.30 \text{ m}$$

$$S = \frac{n^2 v^2}{R_h^{4/3}} = 0.017^2 \times \frac{1.0^2}{0.30^{4/3}} = 1.44 \times 10^{-3} = \underline{0.14\%}$$

Example Problem 20.4

A finished concrete rectangular trench is to be built to carry a discharge of 9.0 m³/s. The available slope is 1.0%. Determine the size of trench, assuming depth is half of the width.

Given:

$$d = b/2 \quad S = 1.0\% = 0.01 \quad Q = 9.0 \text{ m}^3/s \quad n = 0.013 \quad b = ?$$

Solution:

$$R_h = \frac{bd}{b + 2d} = \frac{b \times 0.5b}{b + b} = 0.25b$$

$$A \times R_h^{2/3} = 0.5b^2 \times (0.25b)^{2/3} = 0.198b^{8/3}$$

$$\frac{Qn}{\sqrt{S}} = \frac{9.0 \times 0.013}{\sqrt{0.01}} = 1.17$$

$$0.198b^{8/3} = 1.17, \quad b = (65/11)^{3/8} = 1.946 = 1.95 \text{ m}$$

$$d = 0.5b = 0.5 \times 1.95 = 0.98 = \underline{1.0 \text{ m}}$$

The width of the channel should be 2.0 m. Design depth of the channel will be more than 1.0 m to provide freeboard.

20.3   EFFICIENT CONVEYANCE SECTION

Based on Manning's equation, it can be said that for a given slope and roughness, the velocity increases with the hydraulic radius. Therefore, for a given area of cross section, flow velocity will be maximum when R is maximum, since for a given section area hydraulics radius solely depends on the shape of the channel. Hydraulic radius will be greatest when the wetted perimeter is minimum. Such a section is called **most-efficient section.**

Of all geometric shapes, the circle has the least perimeter for a given area. However, it is not practicable to construct circular shapes for all kinds of materials Therefore there are other factors besides hydraulic efficiency which determine the best cross section. For rectangular and trapezoidal sections, it can be shown that the hydraulic radius of the most efficient section is one for which R = y/2. This corresponds to a rectangle whose depth is one half the width, or b = 2y, as was the case in the previous example problem.

20.4   MAXIMUM AND MINIMUM FLOW VELOCITIES

Flow velocities in sewers should be such that they are able to carry solids without scouring the surface of the pipe. The first condition limits minimum velocity and the second limits the maximum velocity.

20.4.1   MINIMUM FLOW VELOCITY

Minimum velocity should be such that at least once a day it allows the self-cleansing of the sewer pipe. If such velocity is not achieved, it would result in deposition of solids causing blockage and the production of toxic gases due to anaerobic decomposition. The velocity that would not permit the solids to settle down and even scour the deposited particles of a given size is called **self-cleansing velocity**. This minimum velocity should at least develop once in a day so as not to allow any deposition in the sewers. Otherwise, if such deposition takes place, it will obstruct free flow causing further deposition and finally leading to the complete blocking of the sewers. The **Shields formula** suggested the following expression for self-cleansing velocity based on particle size, specific gravity, grade, roughness coefficient, and hydraulic radius of the sewer pipe.

Self-cleansing velocity

$$v_s = \frac{R_h^{1/6}}{n} \times \sqrt{kD(G_s - 1)}$$

Where k is a factor depends on particle properties. Typical values of k for clean inorganic and organic matter present in sewage are 0.04 and 0.60 respectively. From the above equation, the self-cleansing velocity for a sand particle of diameter 1.0 mm, specific gravity 2.65, and sewer pipe of size 200 mm is shown below.

$$v_s = \frac{R_h^{1/6}}{n} \times \sqrt{kD(G_s - 1)}$$

$$= \left(\frac{0.2}{4}\right)^{1/6} \times \frac{1}{0.013} \times \sqrt{0.04 \times 0.001 \times 1.65} = 0.39 \text{ m/s}$$

Based on this it can be said that to achieve self-cleaning, flow velocity during minimum flow conditions (usually d/D = 0.33) is kept around 0.45 m/s (1.5 ft/s) or flowing full velocity of about 1.0 m/s (3.3 ft/s). Hence, for removing the impurities present in sewage (i.e., sand up to 1 mm diameter with specific gravity 2.65 and organic particles up to 5 mm diameter with specific gravity of 1.2), it is necessary that a minimum velocity of about 0.45 m/s and an average velocity of about 0.9 m/s should be developed in sewers. Hence, while finalizing the sizes and gradients of the sewers, they must be checked for the minimum velocity that would be generated at minimum discharge, i.e., about 1/3 of the average discharge. While designing the sewers the flow velocity at full depth is generally kept at about 0.8–1.0 m/s. (3.0–3.2 ft/s). Since sewers are generally designed to be ½ to ¾ full, the velocity at "designed discharge" (i.e., ½ to ¾ full) will even be more than full-depth velocity. Thus, the minimum velocity generated in sewers will transport suspended solids and prevent the sewage from decomposition by moving it faster, thereby preventing the evolution of foul gases.

20.4.2   MAXIMUM VELOCITY OR NON-SCOURING VELOCITY

The interior surface of the sewer pipe gets scored due to the continuous abrasion caused by suspended solids present in sewage. The scoring is pronounced at higher velocity than what can be tolerated by the pipe materials. This limiting or non-scouring velocity mainly depends upon the material of the sewer. Thus, to prevent abrasion and scouring of the pipe surface, maximum flow velocities should not be allowed to exceed 4.0 m/s (13 ft/s) for vitrified clay, PVC, and cast iron, 3.0 m/s or 10 ft/s for cement concrete pipes, and 1.5–2.5 m/s for brick-lined sewers. The following points should be kept in mind while designing the sewers in connection with the self-cleaning velocity and non-scouring velocities.

- During the design of sewer, the discharge to be carried out by the sewer should be determined first.
- While designating sewer lines in flat topography, sewer grades and sizes are provided so that minimum flow velocity is developed at least during the peak flow conditions.
- In hilly areas, sewers should be designed with the maximum grades and sizes permissible (no-scouring) at maximum flow and with the lowest discharge. To reduce the fall, if needed, drop manholes should be provided.
- Manholes should be provided at all transitions and drop should be provided to take care of head loss during transition.

20.5 PARTIAL FULL PIPES

In sewer pipes, flow frequently occurs at partial depth. Manning's flow equation for sewer pipes flowing full or half full has already been discussed. For other depths of flow this is a complex relationship. The simplest way to handle partly full pipes is to compute the velocity or flow rate for the pipe's full condition and adjust to partly full conditions by using a chart as seen in **Figure 20.2**. or **Table 20.1**. The mathematics of partial full pipes is illustrated in the following example problems.

Example Problem 20.5

A 18 in sewer pipe (n = 0.013) is laid on a slope of 0.25%. At what flow depth is the flow velocity 2.0 ft/s?

Given:

$$D = 18 \text{ in} = 1.5 \text{ ft} \quad n = 0.013 \quad S = 0.25\% = 0.0025 \quad v = 2.0 \text{ ft/s}$$

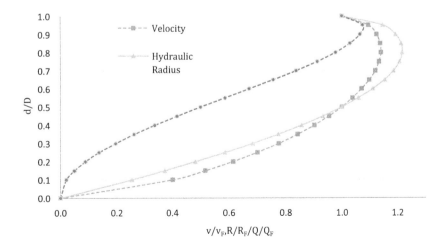

FIGURE 20.2 Standard chart for proportionate elements

TABLE 20.1
Partial flow and equivalent hydraulic elements

d/D	rad	deg	P/P_F	A/A_F	v/v_F	Q/Q_F	S_s/S_F	v_s/v_F	Q_s/Q_F
0.10	1.29	73.7	0.25	0.05	0.40	0.02	3.94	0.80	0.04
0.15	1.59	91.1	0.37	0.09	0.52	0.05	2.69	0.85	0.08
0.20	1.86	106.3	0.48	0.14	0.62	0.09	2.07	0.89	0.13
0.25	2.09	120	0.59	0.20	0.70	0.14	1.71	0.92	0.18
0.30	2.32	132.8	0.68	0.25	0.78	0.20	1.46	0.94	0.24
0.35	2.53	145.1	0.77	0.31	0.84	0.26	1.29	0.96	0.30
0.40	2.74	156.9	0.86	0.37	0.90	0.34	1.17	0.98	0.36
0.45	2.94	168.5	0.93	0.44	0.95	0.42	1.07	0.99	0.43
0.50	3.14	180	1.00	0.50	1.00	0.50	1.00	1.00	0.50
0.55	3.34	191.5	1.06	0.56	1.04	0.59	0.94	1.01	0.57
0.60	3.54	203.1	1.11	0.63	1.07	0.67	0.90	1.02	0.64
0.65	3.75	214.9	1.15	0.69	1.10	0.76	0.87	1.02	0.71
0.70	3.97	227.2	1.19	0.75	1.12	0.84	0.84	1.03	0.77
0.75	4.19	240	1.21	0.80	1.13	0.91	0.83	1.03	0.83
0.80	4.43	253.7	1.22	0.86	1.14	0.98	0.82	1.03	0.89
0.85	4.69	268.9	1.21	0.91	1.14	1.03	0.82	1.03	0.94
0.90	5.00	286.3	1.19	0.95	1.12	1.07	0.84	1.03	0.98
0.95	5.38	308.3	1.15	0.98	1.10	1.08	0.87	1.02	1.00
1.00	6.28	360	1.00	1.00	1.00	1.00	1.00	1.00	1.00

Solution:

$$V_F = \frac{0.60}{n} \times D^{2/3} \times \sqrt{S} = \frac{0.60}{0.013} \times 1.5^{2/3} \times \sqrt{0.0025} = 3.02 = 3.0 \text{ ft}/s$$

$$\frac{v}{V_F} = \frac{2.0}{3.0} = 0.6, \quad \frac{d}{D} = 0.23 \text{ from Table 20.1}$$

$$d = 0.23\,D = 0.23 \times 1.5 \text{ ft} = 0.345 \text{ ft} = \underline{0.35 \text{ ft}}$$

Example Problem 20.6

A 24 in diameter sanitary sewer ($n = 0.020$) is laid on a grade of 0.15%. When the sewer is carrying flow of 600 gpm (minimum), what is the unused capacity of the pipe as a percentage of full flow?

Given:

$$n = 0.02 \quad D = 24 \text{ in} = 2.0 \text{ ft} \quad S = 0.15\% \quad Q = 600 \text{ gpm}$$

Solution:

$$Q_F = \frac{0.464}{n} \times D^{8/3} \times \sqrt{S} = \frac{0.464}{0.02} \times 2.0^{8/3} \times \sqrt{0.0015}$$

$$= \frac{5.705 \text{ ft}^3}{s} \times \frac{7.48 \text{ gal}}{\text{ft}^3} \times \frac{60 \text{ s}}{\text{min}} = 2560.5 = 2600 \text{ gpm}$$

$$\frac{Q}{Q_F} = \frac{600}{2560.5} = 0.234 = 23.4\% \text{ or } 77\% \text{ unused}$$

Example Problem 20.7

Select a sewer main for carrying a flow of 280 L/s when running half full. Allowable slope is 0.1% and n = 0.013.

Given:

$$n = 0.013 \quad D = ? \quad S = 0.1\% = 0.01 \quad Q \text{ (half full)} = 280 \text{ L /s} = 0.28 \text{ m}^3/\text{s}$$

Solution:

$$Q_F = 2 \times Q = 2 \times 0.28 = 0.56 \text{ m}^3 / s$$

$$D = \left(\frac{Qn}{0.312\sqrt{S}} \right)^{0.375} = \left(\frac{0.56 \times 0.013}{0.312 \times \sqrt{0.001}} \right)^{0.375} = 0.892 \text{ m} = 900 \text{ mm}$$

$$V_F = \frac{0.40}{n} \times D^{2/3} \times \sqrt{S} = \frac{0.40}{0.013} \times 0.90^{2/3} \times \sqrt{0.001} = 0.9 \text{ m/s} > 0.6 \text{ m/s}$$

20.6   STORM DRAINAGE

Providing adequate drainage of storm water is one of the many jobs of an environmental engineer. This job has gained new dimensions since drainage works have changed from primitive ditches to complex networks of curbs, gutters, and underground conduits. Simple rules of thumb and crude empirical formulas are generally inadequate. In addition, demands by society for better environmental control require that water quality considerations be superimposed on estimates of quantity so that management of the total water resource can be affected. To design a system with sufficient capacity to yield adequate drainage of water, for most of the time, a complete understanding of the rainfall–runoff relationship for a given watershed system is very important.

A number of design methods for estimating peak discharges are available and vary from a rule of thumb approach, macroscopic approach, microscopic approach, and continuous simulation. The most popular methods for estimating peak flows in urban catchments is the Rational method.

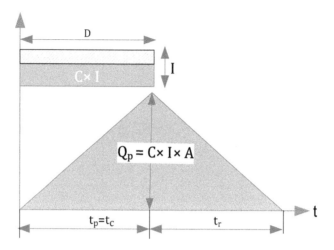

FIGURE 20.3 Rational method

20.7 RATIONAL METHOD

The most commonly used peak discharge design method is the **Rational method**, derived by Emil Kuichling in 1889. A peak flow is associated with certain watershed parameters, such as area, topography, soil texture, vegetation, and surface storage, and with the storm characteristics of duration, intensity, and frequency (**Figure 20.3**).

Peak flow rate

$$Q_p = C \times I \times A$$

C = runoff coefficient, proportion of the total rainfall which runs off
I = the maximum rainfall intensity for a duration equal to the time of concentration of the basin (t_c) and for a storm of design return period, T
A = basin area contributing.

20.7.1 RUNOFF COEFFICIENT

Estimate the runoff coefficient, C, from knowledge of the watershed characteristics. From **Tables 20.2 and 20.3**, select a value of runoff coefficient. When not sure, it is recommended to select the mid-range.

20.7.2 TIME OF CONCENTRATION

Time of concentration, t_c, is the time for water to travel from hydraulically remote areas of the catchment to the outlet or drain inlet. The estimation of time of

TABLE 20.2

Runoff coefficients for rural areas

Topography and Vegetation	Sandy Loam	Clay and Silt Loam	Tight Clay
Woodland			
(flat 0%–5% slope)	0.10	0.30	0.40
rolling (5%–10% slope)	0.25	0.35	0.50
hilly (10%–30% slope)	0.30	0.50	0.60
Pasture			
flat	0.10	0.30	0.40
rolling	0.16	0.36	0.55
hilly	0.22	0.42	0.60
Cultivated			
flat	0.30	0.50	0.60
rolling	0.40	0.60	0.70
hilly	0.52	0.72	0.82
Developed Areas			
impervious	30%	50%	70%
flat	0.40	0.55	0.65
rolling	0.50	0.65	0.80

concentration can be calculated by dividing distance of travel by the average flow velocity, as found by Manning's flow equation. For rural catchments, time of concentration can be estimated by the empirical formula. One such formula is shown below.

Time of concentration (bare earth)

$$t_c = \frac{0.02\,L^{0.77}}{S^{0.385}} = \frac{0.02\,L^{1.2}}{H^{0.385}}$$

t = time in min, L = length in m, H = drop in m, S = slope as decimal fraction

For covered surfaces, this value can be modified. For example, if fully covered with short grass, time of concentration will be twice as much.

In an urban catchment, time of concentration would consist of two components: overland flow time, usually called inlet time to reach inlet, and travel time through the drain or gutter, called flow time. Flow time, t_f, is determined by estimating flow velocity from Manning's equation, as discussed earlier. Inlet time is found by using empirical formula, as discussed earlier.

$$t_c = t_i + t_f = \frac{0.02\,L^{0.77}}{S^{0.385}} + \frac{L}{v}$$

v = flow velocity in m/min t_i = inlet time, t_f = flow time

TABLE 20.3

Typical C values (urban areas)

Description of Area	Runoff Coefficients
Business	
downtown areas	0.70 - 0.95
neighborhood areas	0.50 - 0.70
Residential	
single-family areas	0.30 - 0.50
multi-family units, detached	0.40 - 0.60
multi-family units, attached	0.60 - 0.75
Residential (suburban)	0.25 - 0.40
Apartment dwelling areas	0.50 - 0.70
Industrial	
light areas	0.50 - 0.80
heavy areas	0.60 - 0.90
parks, cemeteries	0.10 - 0.25
Playgrounds	0.20 - 0.35
Railroad yard areas	0.20 - 0.40
Unimproved areas	0.10 - 0.30
Streets	
asphaltic	0.70 - 0.95
concrete	0.80 - 0.95
brick	0.70 - 0.85
drives and walks	0.75 - 0.85
roofs	0.70 - 0.95
Lawns; sandy soil:	
flat, 2%	0.05 - 0.10
average, 2%–7%	0.10 - 0.15
steep, 7%	0.15 - 0.20
Lawns; heavy soil:	
flat, 2%	0.13 - 0.17
average, 2%–7%	0.18 - 0.22
steep, 7%	0.25 - 0.35

20.7.3 RAINFALL INTENSITY

For the estimated time of concentration and a selected rainfall frequency the design rainfall intensity *I* is obtained from a plot of rainfall intensity–duration–frequency (IDF) curves or equations. Standard drainage systems and small spillways are normally designed for rainfall events of 5-year to 10-year frequency.

Sample IDF function

$$I = \frac{850\,T^{0.2}}{\left(D+15\right)^{0.75}}$$

$$T = \text{return period} \quad D = \text{duration in min} = t_c \quad I = \text{Rainfall intensity, mm/h}$$

20.7.4 AREAL WEIGHING OF RUNOFF COEFFICIENTS

When a clear pattern of variation in the catchment surface is apparent, an area-weighted runoff coefficient should be used.

Weighted Coefficient

$$\boxed{\bar{C} = \frac{\sum C_i A_i}{\sum A_i} = \frac{\sum C_i A_i}{A}}$$

20.7.5 LIMITATIONS OF RATIONAL METHOD

The following limitations should be recognized for application of the Rational method.

1. The runoff coefficient is difficult to establish consistently and objectively.
2. The time of concentration t_c is not as constant as the formula suggests and may vary with the storm characteristics.
5. The method is not recommended for use on catchments larger than 500 ha or 5.0 km².
6. Runoff frequency is not always equal to rainfall frequency.
7. This method yields a point value peak runoff *rate* and says nothing about the nature of the rest of the hydrograph.

In urban watersheds, where most of the area is impervious, infiltration does not play a major role in the production of runoff. This is the main reason that the Rational method has been used more successfully in its application to small urban watersheds.

Example Problem 20.8

Calculate the peak runoff rate from a 8.7 acre area parking lot. The time of concentration and runoff coefficient respectively are 20 min and 0.75. The rainfall intensity for 20 min duration is 3.0 in/h.

Given:

$$D = 20 \text{ min} \quad I = 3.5 \text{ in/h} \quad A = 8.7 \text{ acres}$$

Solution:

Peak runoff rate

$$Q_p = CIA = 0.75 \times \frac{3.0\,\text{in}}{h} \times 8.7\,\text{acre} \times \frac{ft^3 / s}{acre.in / h} = 19.5 = \underline{20\ ft^3 / s}$$

Example Problem 20.9

It is being proposed to develop a 1530 acre drainage area. Pre-development and post-development conditions suggest the following hydrologic parameters.

Development	Surface	C	D = t_c, h
pre-development	natural	0.3	3.0
post-development	partially paved	0.6	2.0

Estimate the peak discharge for pre-development and post-development conditions using 5-year storm.

Given:

IDF function for a given area is I (in/h) = 35 T $^{0.2}$/ (D + 10)$^{0.75}$

T = return period in y   D = rainfall duration in min

Solution:
Pre-conditions

$$I = \frac{90\, T^{0.2}}{(D+10)^{0.75}} = \frac{35\times(5.0)^{0.2}}{(180+10)^{0.75}} = 0.9436 = 0.94\, in/h$$

$$Q_p = CIA = 0.3\times\frac{0.9436\, in}{h}\times150\, acre\times\frac{h}{acre.in}\times\frac{ft^3}{s} = 42.46 = \underline{42\, ft^3\,/s}$$

Post-conditions

$$I = \frac{90\, T^{0.2}}{(D+10)^{0.75}} = \frac{90\times(5.0)^{0.2}}{(120+10)^{0.75}} = 1.25 = 1.3\, in/h$$

$$Q_p = CIA = 0.6\times\frac{1.254\, in}{h}\times150\, acre\times\frac{ft^3\,/s}{acre.in\,/h} = 84.9 = \underline{85\, ft^3\,/s}$$

Example Problem 20.10

Two adjacent fields, A and B, contribute runoff to a collector whose capacity is to be determined. It is estimated that in 25 min areas from both fields start contributing. Field A has an area of 2.5 acres and runoff coefficient of 0.35. Field B has an area of 5.2 acres and runoff coefficient of 0.65. Rainfall intensity in in/h for the 15-year storm is given by the following IDF function, where D is duration of the storm in minutes. $I = \dfrac{60}{(D+15)^{0.8}}$

Given:

$$C_1 = 0.35, C_2 = 0.65, A_1 = 2.5 \text{ acres}, A_2 = 5.0 \text{ acres}, D = t_c = 25 \text{ min}$$

Solution:

$$I = \frac{60}{(D+15)^{0.8}} = \frac{60}{(25+15)^{0.8}} = 3.13 = 3.1 \text{ in/h}$$

$$C = \frac{\Sigma C_i A_i}{\Sigma A_i} = \frac{0.35 \times 25 + 0.65 \times 5.2}{2.5 + 5.2} = 0.5526$$

$$Q_p = CIA = 0.5526 \times \frac{3.13 \text{ in}}{h} \times 7.7 \text{ acres} \times \frac{ft^3/s}{acre.in/h} = 13.3 = 13 \text{ ft}^3/s$$

20.7.6 URBAN CATCHMENTS

Application of the Rational method to urban catchments ($1 < A < 2.5 \text{ km}^2$) requires special techniques. Flow varies widely along the length of the main channel (sewer) – generally small at the upstream and larger at the downstream. It may be difficult to determine average t_c. An alternative is to apply the Rational method incrementally. The method requires the subdivision of the catchment into several sub-areas, as shown in **Figure 20.4**. Water moves through a watershed as sheet flow, shallow concentrated flow, open channel flow, or some combination of these before it exits the sewer line. This is called inlet time. Referring to **Figure 20.4**, the time of concentration for inlet 1 is inlet time for runoff from catchment area A to reach the inlet. This is the time taken by the overland flow to reach point B.

Based on flow in reach AB of sewer line, travel time from A to B, t_{AB} is calculated using Manning's equation. Time of concentration for inlet 2 is equal to inlet time for area A plus travel time from 1 to 2, or inlet time for catchment area B, whichever is greater. This process is continued until the peak flow at the outlet is worked out. Except for the first inlet, more than one area contributes to runoff. Thus, it becomes a case of composite catchment. Various flow paths are tried and the one producing maximum flow is selected, as shown in the following example problem.

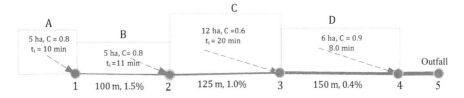

FIGURE 20.4 Urban draining system (Ex. Prob. 20.11)

Example Problem 20.11

A storm drainage system comprises the four areas as shown in **Figure 20.4**. Determine the 5-year design flow for each section of sewer line.

Given:

Catchment	Area A, ha	Coefficient C	Inlet time, min
A	5	0.8	10
B	5	0.8	11
C	12	0.6	20
D	6	0.9	8

Solution:

Complete solution is shown in **Table 20.4**. For a given outlet, the longest path is used to calculate time of concentration and to calculate rainfall intensity.

Sample of Calculations (inlet #2)

$$\Sigma C_i A_i = 0.8 \times 5 + 0.8 \times 5 = 8.0\,\text{ha}$$

$$t_{A-2} = 10\text{ min} + 100\text{ m} \times \frac{\text{s}}{1.5\text{ m}} \times \frac{\text{min}}{60\text{ s}} = 11.1\,\text{min}$$

$$I = \frac{2500}{D+15} = \frac{2500}{11.1+15} = 95.7\,\text{mm/h}$$

$$Q_p = I \times \Sigma C_i A_i = \frac{95.7\text{ mm}}{\text{h}} \times 8.0\text{ ha} \times \frac{10\text{ m}^3}{\text{ha.mm}} \times \frac{\text{h}}{3600\text{ s}} = 2.10 = \underline{2.1\,\text{m}^3/\text{s}}$$

TABLE 20.4
Table of computations

Inlet	C	A (ha)	CA (ha)	ΣCA (ha)	Path	Time of Concentration, minutes: Inlet Time, t_i	L, m	v (m/s)	Flow Time (t_f)	Totals	I (mm/h)	Q_p (m^3/s)
1	0.8	5	4	4	A-1	10				10.0	100.0	1.1
2	0.8	5	4	8	A-2	10	100	1.5	1.1	11.1	95.7	2.1
3	0.6	12	7.2	15.2	C-3	20			0.0	20.0	71.4	3.0
4	0.9	6	5.4	20.6	C-4	20	150	1.5	1.7	21.7	68.2	3.9

Example Problem 20.12

Find the size of a combined sewer pipe to carry the combined flow from an area of 110 ha with population density of 250 p/ha without exceeding flow velocity of 3.0 m/s. Assume water supply rate of 220 L/p·d of which 80% becomes sewage water. Weighted runoff coefficient is 0.45 and inlet time and flow time respectively are 13 min and 22 min. The rainfall intensity for the design storm is given by the following IDF function: $I = \dfrac{1500}{(D+15)^{0.8}}$

Given:

$$C = 0.45 \quad A = 110 \text{ ha} \quad Q = 220 \text{ L/p·d} \quad \text{Population} = 250 \text{ p/ha}$$

Solution:

Average wastewater flow

$$Q_{ww} = 0.80 \times \frac{220L}{p.d} \times 110 ha \times \frac{250p}{ha} \times \frac{m^3}{1000L} \times \frac{d}{24h} \times \frac{h}{3600s}$$

$$= 0.0560 = 0.056 \, m^3/s$$

$$t_c = t_i + t_f = 13\,min + 22\,min = 35 \text{ min (duration)}$$

$$I = \frac{1500}{(D+15)^{0.8}} = \frac{1500}{(35+15)^{0.8}} = 65.6 \text{ mm/h}$$

Peak runoff rate

$$Q_p = CIA = 0.45 \times \frac{65.6 \text{ mm}}{h} \times 110 \text{ ha} \times \frac{10 \, m^3}{ha.mm} \times \frac{h}{3600 \, s} = 9.02 = 9.0 \, m^3/s$$

$$Q_c = Q_{ww} + Q_{sw} = 0.056 + 9.02 = 9.076 = 9.1 \, m^3/s$$

These calculations indicate that wastewater flow is relatively negligible compared to storm flow. Thus, during dry weather, the sewer may not be able to achieve self-cleansing velocities.

$$D_{min} = \sqrt{\frac{4}{\pi} \times \frac{Q_p}{V_{max}}} = \sqrt{\frac{4}{\pi} \times \frac{9.076}{s} \times \frac{s}{3.0 \, m}} = 1.96 \, m = 2.0 \, m$$

Discussion Questions

1. Comparing the Darcy Weisbach flow equation with Manning's flow equation, develop the relationship between f and n.
2. Show that for circular pipes flowing full or half full, hydraulic radius is one-fourth of the diameter of the pipe.

3. Explain why a rational formula is more successful for small urban catchments.
4. What are the limitations of the rational formula?
5. Define time of concentration and the factors affecting.
6. Describe the methodology to size storm sewers in a given catchment area.
7. Compare storm sewer with sanitary sewer.
8. Modify Manning's equation (SI) for circular sewer pipes flowing full.
9. What is normal depth? Compare gradually varied flow with rapidly varied flow, and give examples.
10. As it is, Manning's equation is applicable for uniform flow conditions. If you are to apply this equation to GVF, what modifications would you need to make?
11. Define self-cleansing velocity and non-scouring velocity and the accepted range of values for this in sewer pipes.
12. Explain why it is necessary to achieve this flow velocity at least for some time during the 24 h period?
13. Describe the relationship for equivalent self-cleansing slope and velocity at partial depth.
14. What considerations should you have when designing sanitary sewer pipes?
15. What are the factors that affect the hydraulics of sewer lines?
16. Wastewater flows have diurnal variation. How does this affect the flow velocity in circular sewer pipes?

Practice Problems

1. A storm sewer pipe drains an area of 5.0 acres with runoff coefficient of 0.55. Assuming rainfall intensity of 1.3 in/h, what is the required flow capacity of the drain? (3.6 ft^3/s).
2. What flow velocity is achieved in a 12 in diameter sewer pipe (n = 0.012) flowing half full laid on a slope of 0.2%? (2.2 ft/s).
3. What is the peak runoff rate from a 1.5 ha parking lot with runoff coefficient of 0.85 and rainfall intensity of 80 mm/h? (0.28 m^3/s).
4. Determine the size of a sanitary sewer to serve a population of 50 000 when flowing at a depth of 70% of the diameter. Peak hourly flow is 2.5 times that of average flow and the available grade is 1/1200. Assume roughness coefficient n of 0.013. (1.0 m).
5. What is the full flow capacity of a 10 in sewer pipe with n = 0.015 and laid on a grade of 1:400? (430 gpm).
6. Determine the size of a combined sewer pipe to carry combined flow from an area of 90 ha with a population density of 250 p/ha without exceeding flow velocity of 3.0 m/s. Assume water supply rate of 250 L/p·d of which 80% becomes sewage water. Weighted runoff coefficient is 0.45 and inlet time and flow time respectively are 11 min and 18 min. The rainfall intensity for the area is $I(mm/h) = 1500/(D(min)+12)^{0.8}$ answer (2.0 m)

7. A town with a population of 40 000 is spread over a drainage area of 190 acres with a runoff coefficient of 0.70 and time of concentration of 40 min. Average daily water consumption is 100 gal/c.d, 75% of which reaches the sewer. Calculate the peak storm flow and peak sanitary flow rate. Rainfall intensity function: $I(\text{in}/h) = 3.2/(1+D(h))$ (260 ft³/s, 12 ft³/s).

8. Design a sanitary sewer flowing 70% full to carry the wastewater from a community with a population of 55 000 people. Assume daily per capita wastewater production is 150 L/c.d. Sewer main is lined, n = 0.012, and available gradient is 0.20%. Check for self-cleansing velocity at peak flow (250%) conditions. (650 mm, 1.2 m/s)

9. You are asked to figure out the required flow capacity of a drainage sewer pipe which drains two adjacent fields. It is estimated that in 28 minutes areas from both the fields start contributing. Assume that IDF function is the same as in Practice Problem 21.6 above. (250 L/s).

Area, ha	Coefficient, C
1.5	0.45
0.8	0.60

10. For a storm drainage system shown in **Figure 20.5**, catchment area, runoff coefficient, and inlet time for each inlet are given. Also, for each reach length and slope are indicated. Using Rational method, compute the design flow and required pipe size for each reach. For a 10-year return period, maximum rainfall intensity for various durations is shown below:

duration, min	5	10	15	20
intensity, mm/h	150	138	125	110

11. A 450 mm sewer pipe of n = 0.013 is laid on a slope of 0.0025. At what depth of flow, is the flow velocity 0.60 m/s? (100 mm).

12. A 1.5 ha residential area is comprised of 20% roof area (C = 0.90), 25% paved area (C = 0.85), 50% area is open ground and lawns (C = 0.10), and the remaining 5% is wooded (C = 0.05). Assuming maximum rainfall intensity of 65 mm/h, what is the peak runoff rate for the area? (120 L/s).

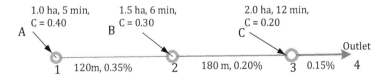

FIGURE 20.5 Storm drainage system

13. A sewer main is to be designed to carry peak flow while flowing 75% full at a velocity of 3.0 ft/s. If the ratio of the maximum/average flow is 3.0 and average/minimum flow is 2.5, determine proportionate depth of flow and flow velocity at average flow and minimum flow.
    (37%, 2.3 ft/s, 22%, 1.7 ft/s).
14. A town has a population of 82 000 with water supply demand of 250 L/c.d. Design a sewer main running at partial depth of 75% of diameter. Available slope is 0.18%. Assume peaking factor of 2.5 and n = 0.013. (825 mm NPS).
15. A 12 in sewer pipe of n = 0.013 is laid on a slope of 0.1%. At what depth of flow is the flow velocity 1 ft/s? (0.25 ft).

21 Construction of Sewers

Sewer pipes are as small as building sewers and as big as intercepting and trunk sewers. There is no one material which suits all kinds of sewer pipes. In addition to flow characteristics, properties of the material of the pipe like resistance to corrosion, available sizes, structural strength, ease of handling, and installation and cost should be considered.

21.1 MATERIALS FOR SEWERS

The most common materials used to convey sewage include vitrified clay pipe (VCP), asbestos-cement pipe (ACP), plastic pipe, and concrete pipe. There are two main types of plastic pipes available. Rigid plastic pipe can be either polyvinyl chloride (PVC) or ABS (Acrylonitrile butadiene styrene). PVC pipes are usually gray, but sometimes are green or blue in color. ABS pipes are black. Soft plastic pipe called high-density polyethylene (HDPE) and is often used for force mains.

21.1.1 VITRIFIED CLAY PIPE (VCP)

Vitrified clay is the most common material used in sanitary sewers. They are made in standard sizes up to 900 mm (36 in) in diameter. They are popular because of their resistance to corrosion and relatively smooth surface. Because they are brittle and break easily, they are not manufactured in large sizes. For such cases, **reinforced concrete pipe** (RCP) coated with protective lining is commonly used. VCP pipe is available in various lengths ranging from 600 mm (24 in) to 2.0 m (80 in) depending on diameter.

21.1.2 PLASTIC PIPE

The plastic pipe used most frequently in sewer systems is **polyvinyl chloride** (PVC) or **polyethylene** (PE). PVC pipe is usually made in sizes 100 mm to 300 mm (4–12 in), with certain manufacturers making them up to 750 mm in diameter. These pipes are light and are usually made in 6.0 m (20 ft) lengths, though other lengths are also available. PVC pipe is used for building connections and branch sewers. The use of PE pipes is more common in the case of long pipelines, often laid under adverse conditions, like swamps or underwater crossings.

21.1.3 FIBERGLASS POLYMER PIPE (FRPP)

The fiber-reinforced polymer pipe (FRPP) and glass fiber reinforced pipe (GFRP) is available in sizes from 150 mm to 3000 mm and in lengths up to 6.0 m. Bell and spigot O-ring compression joints are generally used to join the pipe sections.

DOI: 10.1201/9781003347941-25

21.1.4 Concrete Pipe (CP)

Concrete pipe (CP) is used extensively in storm sewer systems since it is available in large sizes and is resistant to abrasion. Concrete pipe is not used in small-diameter sanitary sewers because of its vulnerability to **crown corrosion**. Corrosion is caused by the formation of hydrogen sulphide gas under anaerobic conditions. Oxidation of H_2S to sulphuric acid collects in droplet form at the crown of sewer pipe and results in corrosion, hence the term crown corrosion.

Non-reinforced concrete pipe is available in sizes up to 600 mm (24 in) and usually in shorter lengths. Unreinforced concrete pipe in sizes 100 to 600 mm in diameter and reinforced concrete pipe in sizes 200 mm to 3.0 m in diameter are generally available for gravity sewers. Several joint designs are available depending on the degree of water tightness required. Casketed tongue-and-groove joints can be used when infiltration is a problem. Protective linings should be used where excessive corrosion is likely to occur.

In case of large sizes, **reinforced cement concrete** (RCC) pipe is used for strength. Reinforced cement concrete pipe is used in large sanitary sewers like trunk sewers where VCP sizes are not available. In such cases, the pipe is protected against corrosion by applying epoxy, providing ventilation, and laying pipes on a relatively steeper grade to achieve high flushing velocities. The advantages of concrete pipe are the relative ease with which the required strength may be provided, the wide range of pipe sizes, the long laying lengths, and the rapidity with which the trench can be opened and backfilled. *Concrete pipes are preferred for medium and large sewer pipes such as main and branch sewers.* When specifying concrete pipe, it is necessary to give the pipe diameter, class or strength, the method of jointing, the type of protective coating and lining if any, and any other special requirements for concrete.

21.1.5 Asbestos Cement Pipe (ACP)

Pipe of asbestos fiber and cement, in sizes ranging from 100 mm to 900 mm in diameter, is available and is used in sewerage systems. Jointing is accomplished by compressing rubber rings between pipe ends and sleeves. Asbestos cement or cast-iron fittings are used. The advantages of using this pipe are its light weight and ease of handling, with long laying lengths, tight joints, rapidity of installation, and corrosion resistance to most natural soil conditions. However, the pipes are prone to sulphide corrosion. When specifying the pipe, the diameter, class or strength, and type of joint should be specified. ACP have a smooth surface, so they are hydraulically more efficient. They are brittle and structurally weak, so more commonly are used as verticals.

21.1.6 Brick Masonry

Brick masonry has been used for large-diameter sewers in the past. Due to high cost, lack of durability, and other factors, brick is now used only in special applications.

21.1.7 CAST-IRON PIPE

Cast iron pipe is available from 100 mm to 1200 mm in diameter with a variety of jointing methods. They are used in gravity sewers where tight **joints** are essential. The advantages of cast iron pipes are their long laying lengths with tight joints, their ability to withstand high internal pressure and external load, and corrosion resistance in most natural soils. They are very vulnerable to corrosion unless protected by **lining** with paint or cement concrete. They are relatively costlier and are used under special circumstances such as heavy external loads, protection against contamination, crossing low-level areas, the prevalence of wet conditions, and temperature variations and vibrations. When specifying cast iron pipe, it is necessary to give the pipe class, the joint type, the type of lining, and the type of exterior coating.

21.1.8 STEEL PIPE

Steel pipe is used for force mains and inverted siphons. Other pipe materials for special applications in wastewater collection system are smooth-wall and corrugated steel pipe bituminized fiber and reinforced resin pipe. The flexibility of galvanized corrugated steel permits the fabrication of a variety of conduit shapes with a choice of protective coatings. Available circular sizes are in the range of 200 mm to 2400 mm in diameter. Pipe sections are generally furnished up to 6.0 m in length in multiples of 60 cm. Coupling bands that may be single piece, two pieces, or an internal expanding type used in lining work, joint the sections. To increase durability and to resist corrosion, galvanized pipe can be coated with bituminous material. The advantages of steel pipe are lightweight, long laying lengths, ease of shipping, adaptable to stacking methods, flexibility, strong mechanical joints, and ability to adjust to trench-loading conditions.

21.1.9 CAST-IN-PLACE REINFORCED CONCRETE

Sewers are constructed of cast-in-place reinforced concrete when the required size is more economical to construct in situ, when a special shape is required, and when headroom and working space are limited. Forms for concrete sewers should be unyielding and tight and should produce a smooth sewer interior. Methods for resisting corrosion are the same as those for concrete pipe.

21.2 LAYOUT AND INSTALLATION

The various steps involved in the layout and construction of sewer mains are described in the following sections.

21.2.1 SETTING OUT

This is the first step in the construction of any work. In the sewerage work, this is carried out starting from the tail end or the outfall end and proceeding upwards, marking sewer lines along proposed routes on the ground, fixing pegs at intervals of 15 m or 50 ft and establishing temporary benchmarks with respect to the fixed **benchmarks**.

The advantage gained by starting from the tail end is the utilization of the sewers from the beginning, thus ensuring that the functioning of the sewerage scheme has not to wait until the completion of the entire project work.

21.2.2 ALIGNMENT AND GRADIENT

The sewers are laid to the correct alignment and gradient setting the positions and levels of sewers so as to ensure a smooth gravity flow. This is done with the help of suitable boning rods and sight rails and with accurate levelling instruments. The sight rail is a horizontal wooden board secured to two vertical posts called uprights by means of heavy steel clamps, fixing it immovably to the correct line and level. A boning rod or traveler is a vertical wooden post suitably shod with iron and fitted with a crosshead or tee and is of length equal to the height of the sight rail above the invert line of the sewer. This can move to and fro in the trench to give the invert line on the prepared bed of the sewer. Both the boning rod and the sight rail have their centerline accurately marked with thin saw-cut and painted black and white to aid in proper visibility. At least three to four sight rails are always maintained in correct level and gradient along the line of the sewer at the sight of sewer construction.

21.2.3 EXCAVATION OF TRENCHES

This work is usually carried out in open cutting. Tunnelling is adopted only when large-sized sewers are to be laid at considerable depth below the ground level. The excavation is made to have trenches of such lengths, widths, and depths as enable the sewers to be properly constructed. Usually, not more than about 18 meter (60 ft) length of any trench in advance of the end of the constructed sewer is to be left open at any time in busy streets and localities.

The width is chosen to permit thorough ramming of the back-fill material around the pipe and to enable the construction of tight joints. At least 20 cm (3/4 in) of clear space should be left on each side of the barrel of the pipe so that the minimum clear width of the trench is equal to the external diameter of the pipe-barrel plus 40 cm (16 in). For other types of sewers, the minimum clear width should be the greatest external width of the structures to be built therein. The depth of excavation should be such as to enable the sewers to be laid at their proper grades on the bed of the trenches. Except where the soil is very firm, it is usual to provide a bed of concrete on the bottom of the trench and to rest the sewer thereon. Suitable recesses are left on the bed to accommodate the socket-end of the pipe sewer.

Timbering of Trenches

For depths of sewers exceeding 2 meters, it becomes necessary to adopt timbering of trenches. The object is to prevent the sides of trenches from caving in that may cause possible damage to the foundations of the adjoining property. The type of timbering depends chiefly upon the type of soil met with. It consists of vertical pol- ing boards with horizontal whaling suitably strutted through hardwood wedging, having adequate dimensions and strength. For a soil which is loose or sandy, close timbering formed by additional vertical boards called "runners" is used. Steel sheet piling is resorted to sometimes in place of timbering in case of badly waterlogged areas or in other situations where timber is not easily available. Steel sheeting is more watertight, stronger, and more durable. Though costlier than timber, it can be used many times without disintegration and hence is more economical in works of larger scale. Timbering or sheeting is usually withdrawn after the sewer has been laid, though sometimes it is necessary to leave it off as such, particularly in the case of wet trenches, which may otherwise be damaged.

Dewatering in Trenches

Where the sub-soil water level is very near the ground level, the trench becomes wet and muddy, and it becomes difficult for sewers to be constructed. Methods suggested to dewater the trenches include direct drainage, drainage by an under-drain, sump pumping, and well point drainage.

Direct drainage is possible by giving a uniform slope to the bottom of the trench and taking out water at some forward point. This method is not satisfactory as some water always remains in the trench, which, therefore, becomes muddy. For drainage by under-drain, an open-jointed tile drain is laid in a small trench 30 cm x 30 cm (1ft x 1ft) constructed below the usual trench grade. Discharge from the underdrain goes to a natural watercourse or into a sump from where water is pumped. The method is useful provided the trench is not very deep and the underdrain can withstand the load without giving way.

In sump pumping, water is collected in a sump made out in the trench from where water is pumped. The pump is required to be worked continuously day and night, otherwise water keeps on flowing into the trench. This method can be used on small jobs and where the sub-soil strata is not very sandy, otherwise sides of the trench are likely to cave in due to continuous pumping. Well point drainage is particularly suit- able for large jobs and where the sub-soil strata consists of quicksand or "running sand".

21.2.4 BEDDING

When installing a new or repaired section of sewer main, careful attention must be paid to the pipe bedding. **Bedding** refers to the type of material and the preparation of the material that the pipe is placed in when it is being installed and repaired. The preferred method of bedding pipe is to install it in bedding that is supported along its whole length by a bed of sand or gravel spread to an even depth. The sand or gravel should be slightly compacted. In some cases, it may be preferable to lay the pipe directly on the bottom of the trench. This is acceptable only if the trench has been smoothed out to support the pipe along its entire length, and the areas under the couplings have been slightly scooped out. The couplings should never rest directly on the trench bottom because the coupling is slightly flexible and will break if not allowed to move. Bed preparation in the case of plastic pipes need to be done more carefully as pipes deform when backfilled. Bedding is classified into four types.

Class D bedding is the weakest and least desirable type and is rarely recommended for sewer construction. In this type of bedding, the bottom of the trench is left flat and the barrel of the pipe is not fully supported because of the protruding bell ends. Backfill is placed loosely without proper compaction.

Class C, or **ordinary bedding**, has compacted granular material placed under the pipe and partially extending up the pipe barrel. This provides reasonable support, with a load factor of 1.5. Thus, field strength is increased to 150% of the crushing strength.

Class B, or **first-class bedding**, has compacted granular material extending halfway up the barrel of the pipe. In addition, backfill is compacted carefully over the top of the pipe, thus achieving a load factor of 1.9.

Class A bedding is the superior type of bedding, with a load factor as high as 2.8. In class A bedding, the pipe barrel is cradled in concrete and the backfill is carefully compacted.

21.2.5 LAYING

After setting the sight rails over the trench, small pegs at intervals of 3 m (10 ft) or so are driven such that the tops of the pegs are level with the invert line of the pipe sewer. As shown in Figure 21.1, this is achieved by adjusting the uprights of sight rails, stretching a string line "bb" across the center marks on the sight rails and moving the boning rod with its crosshead nearly touching the string line, its base-shoe resting on the trench bed.

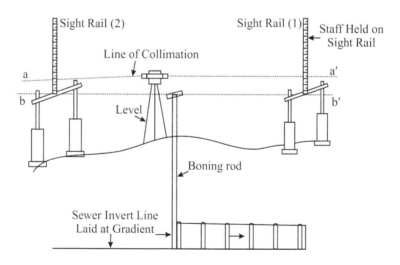

FIGURE 21.1   Laying of sewer

If the horizontal distance between consecutive sight rails is 15 m (50 ft), and the desired gradient of sewer is 1:60, the drop in elevation must be 0.25 m or 25 cm (10 in).

After transferring the centerline of the sewer to the bottom of the trench, the latter is trimmed off slightly to enable the inside of the pipe barrel to conform to the invert line. The pipe can now be laid on the prepared bed. In order to enable the pipes to be laid in straight lengths, the method is to measure, from the center line already marked, a distance equal to half the external diameter of the socket. At this distance, stretch a line at half height of pipe, so that when the pipes are laid just not touching this line they will be in a straight line.

The pipes are usually laid uphill with their sockets facing the direction of flow. In this way, the spigot of each pipe can be easily inserted in the socket of the pipe already laid. The pipe when laid and pressed in position is tested for level by passing a straight edge through it and seeing that the edge rests squarely upon the level-pegs as well as on the invert of the pipe throughout its length. Any departure from there would mean that the pipe is laid either too high or too low. This should accordingly be corrected.

21.2.6 LASERS

In modern sewer construction, a laser beam is used to establish the specified slope of the pipe. A laser can project an intense but narrow beam of light for a long distance. The laser is securely mounted in the manhole, and the slope of the light beam is accurately set to match the required slope of the pipe. A transit mounted on the manhole is used to establish pipe alignment from the field reference points and to transfer the alignment down to the laser. A laser can maintain accuracies of 0.01% over a length up to 300 m (1000 ft). In other words, the invert elevations can be set accurately to within 30 mm in a kilometer length of the sewer line.

Example Problem 21.1

A 120 m reach of sewer is to be designed with a flow capacity of 100 L/s, flowing full. The street elevation at the upper manhole is 80.00 m and at the lower manhole is 77.60 m. Select appropriate size and slope of the pipe invert elevations. Assume minimum cover of 2.0 m above the crown of pipe.

Given:

$$n = 0.013 \quad D = ? \quad Z_1 = 80.00 \text{ m} \quad Z_2 = 77.60 \text{ m} \quad L = 120 \text{ m} \quad Q = 100 \text{ L/s}$$

Solution:

$$Available, S = \frac{(80.0 - 77.6)m}{120\ m} = 0.0193 = 0.02\ m/m$$

$$D = \left(\frac{Qn}{0.312\sqrt{S}}\right)^{0.375} = \left(\frac{0.1 \times 0.013}{0.312 \times \sqrt{0.02}}\right)^{0.375} = 0.266\ m = 300\ mm$$

$$Q_F = \frac{0.312}{n} \times D^{8/3} \times \sqrt{S} = \frac{0.312}{0.013} \times 0.30^{8/3} \times \sqrt{0.02} = 0.136 \ m^3/s = \underline{136 \ L/s}$$

$$\frac{Q}{Q_F} = \frac{100}{136} = 0.74 \quad \text{From table} \quad \frac{d}{D} = 0.64, \ \frac{v}{v_F} = 1.1$$

Actual depth, $d = D \times 0.64 = 300 \ mm \times 0.64 = 190 \ mm$

$$v_F = \frac{0.40}{n} \times D^{2/3} \times \sqrt{S} = \frac{0.40}{0.013} \times 0.30^{2/3} \times \sqrt{0.02} = 1.94 = 1.9 \ m/s$$

Upper invert $= 80.00 \ m - 2.0 \ m - 0.30 \ m = 77.70 \ m$

Lower invert $=$ *upper* $-$ *drop* $= 77.70 \ m - 0.02 \times 120 \ m = 75.30 \ m$

21.2.7 JOINTING

The characteristics of a good joint include water tightness, resistance to root penetration, resistance to corrosion, a reasonable degree of flexibility, and durability. There are several types of joints. For small size concrete pipes made with **bell and spigot** ends, the joints may be like those used on vitrified clay pipes. A bell and spigot type of joint is shown in **Figure 21.2**. For a larger-size concrete pipe with **tongue-and-groove** ends, joints can be made with mortar or bituminous compounds or with rubber gaskets. For an asbestos-cement pipe, the joint consists of a **collar or coupling** and a pair of rubber rings. For a bell and spigot concrete pipe, cement mortar is used to pack against the hemp or jute caulked into the annular space of the pipe and to butter around the pipe joint forming a 45 degree bevel.

After truly bedding the first pipe, the second pipe is laid. A ring of tarred yarn soaked in cement slurry is passed around the spigot of the second pipe so that when driven home it is supported in the spigot of the first pipe. The yarn ring is caulked tightly to fill about one-fourth of the depth of the socket. Caulking helps in making the joints watertight.

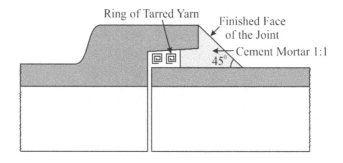

FIGURE 21.2   Bell and spigot joint

The two pipes can be tested for gradient by passing the straight edge along the inverts of the two pipes and on to the nearest level-peg up the gradient. The level-pegs are removed as the work proceeds. When the second pipe has been truly bedded, the socket of the first is filled with cement mortar of a stiff consistency (1:1), worked well, and finally finished off with a splayed joint or fillet formed with a trowel making an angle of 45° with the barrel of the pipe. Any cement or other extraneous matter sticking inside the pipe or joint is removed with a scraper before the next pipe is laid.

Jointing VCP Pipes

Bell and spigot VCPs are connected by compression joints with seals of plastic to prevent leakage, infiltration, and penetration of roots. The bell end has a polyester liner and the spigot end has a compression ring in the annular space. In the bell section of the pipe already laid, the spigot end of the new section is pushed to form a tight seal. In other cases, both bell and spigot have polyurethane elastomer seals. Before joining, seals are lubricated, and the spigot end is pushed into the bell section of an already laid pipe using a pipe puller. In the case of plain end pipes, a PVC sleeve or collar is attached in the factory. The seal on the plain end is lubricated and pushed into the collar to form a compression joint.

Jointing an Asbestos-Cement Pipe

An asbestos-cement pipe (ACP) has tapered ends. To join two ACPs together, an **asbestos-cement collar** is used. This collar has rubber gaskets at each end. To install the collar, the gaskets must be free of all grit and dirt. Grease the gaskets with manufacturer-recommended lubricant. Make a joint with two asbestos-cement pipes by carefully forcing the tapered end of each pipe into the coupling. Some makes of couplings have a wedged type of gasket that must be placed in the correct position to make a good seal. A-C pipe must be very carefully bedded or supported, as it has limited resistance to shearing. A-C pipe comes in longer lengths than clay pipe and therefore requires fewer joints.

Jointing Plastic Pipe

As mentioned above, there are three types of plastic pipe: ABS, PVC, and HDPE. Plastic pipe has been used more frequently in sewage installation in recent years, especially for building service connections. Plastic pipe is very smooth and has excellent flow characteristics. Plastic pipes may be manufactured with either plain ends or with a bell and spigot arrangement. Both types use a special jointing solvent (glue) which cements the sections together. When jointing a bell and spigot pipe, the solvent is spread either on the spigot or in the bell and the ends are quickly joined.

Plain end pipes are joined by means of a collar. Solvent is applied to the end of one pipe and the collar is immediately fitted. Solvent is then applied to the end of the other pipe to be joined and it is fitted into the collar. Plastic jointing solvents have a very quick drying time. The joint must be made within seconds of the application of the solvent. Solvents are specific to each pipe. Use only ABS solvent with ABS pipe and PVC solvent with PVC pipe. High-density plastic pipe can only be jointed in two ways: by a special fusion-welding machine used by the contractor at the time of installation or using a repair clamp.

21.3   TESTING

Sewers are normally tested for leakage and straightness before they are put into service.

After installation of pipes, the integrity of joints and leakage needs to be checked. A water test or low-pressure air test is used for this purpose. The air test is simple but in some situations a water test is the only option.

21.3.1   WATER TEST

After sufficient time has been given for the joints to set, a section of pipe usually between two manholes is subjected to a test pressure head of 1.5 m (5 ft) of water in case of stoneware pipes and concrete pipes and 9 m (30 ft) of water in case of cast iron pipes. See **Figure 21.3**. The test is carried out by plugging the lower end of the pipe sewer with a rubber bag equipped with a canvas cover inflated by blowing air, and then filling the section with water, lowered until the required head is maintained for observation.

The upper end is plugged with a connection to a hose ending in a funnel, which could be raised. The tolerance of 1 to 2 L/cm.km (1-2 gal/in.mile) length of sewer may be allowed for a period of 10 minutes. Any subsidence of the test water may be attributed to leakage at pipe-joints or in the defective lengths of pipes, which should, therefore, be cut out and made good. Subsidence may be also due to some absorption or sweating of pipes or joints.

21.3.2   AIR TESTING

Leakage and integrity of joints can also be done by low pressure testing. To do this, plug both ends of a section of sewer to be tested before subjecting that section of pipe to low pressure air. The air must be maintained at a minimum pressure of 25 kPa for the specified time for each diameter. Duration times vary with changes in pipe size and length of test sections.

A maximum drop in pressure of 3.5 kPa is permitted within the specified time duration. Should the pressure drop be greater than 3.5 kPa (0.5 psi) within the

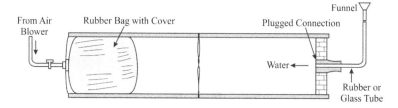

FIGURE 21.3   Water test

specified time, the repair of any deficiencies and retesting must be performed until a successful test is achieved. Sources of leaks may be dirt in an assembled gasket joint, incorrectly tightened service saddles, or improper plugging or capping of sewer lateral piping. If there is no leakage (i.e., zero pressure drop) after one hour of testing, the section should be passed and presumed free of defects. If there is groundwater present at a level higher than the pipe invert during the air-test, the test pressure should be increased to a value of 25 kPa (3.5 psi) greater than the water head at the bottom of the pipe.

Tests for straightness and obstruction are carried out to check pipe alignment. Two common tests for this purpose are the ball test and mirror test.

21.3.3 BALL TEST

A ball test is carried out by inserting at the high end of the sewer or drain a smooth ball of diameter 13 mm (1/2 in) less than the pipe bore. In the absence of any obstruction, such as yarn or mortar projecting through the joints, the ball will roll down the invert of the pipe and emerge at the lower end.

21.3.4 MIRROR TEST

By placing a mirror at one end of the sewer line and a lamp at the other end, if the pipeline is straight, the full circle of light can be observed. If the pipeline is not straight, this would be apparent. The mirror will also indicate any obstruction in the pipe barrel.

21.3.5 SMOKE TEST

This is carried out for drainage pipes located in buildings. All soil pipes, waste pipes, vent pipes, and all other pipes when above ground should be approved gas-tight by a smoke test conducted under a pressure of 25 mm(1 inch) of water, maintained for 15 minutes after all trap seals have been filled with water. The smoke is produced by burning oil waste or tarpaper or similar material in the combustion chamber of a smoke machine (chemical smokes are not considered satisfactory).

21.3.6 BACK FILLING

After a sewer has been constructed and tested, the trenches are required to be refilled. The work involved should be carried out with due care, particularly the selection of the soil used for back filling around the sewer so as to ensure the further safety of the sewer.

The filling in the trenches and up to about 0.75 m (2.5 ft) above the crown or soffit of the sewer should be made in the finest selected material placed carefully in layers of 15 cm (6 inches) thickness, watered and evenly rammed. After this, the excavated topsoil, turf, pavement, or road metal are replaced as the top filling material, rammed and satisfactorily maintained until the surface has been restored.

21.4 STRUCTURAL REQUIREMENTS

The structural design of a sewer requires that the supporting strength of the conduit as installed, divided by a suitable factor of safety, must equal or exceed the loads imposed on it by the weight of earth and any superimposed loads. The supporting strength of buried conduits is a function of installation conditions as well as the inherent strength of the pipe itself. Since installation conditions have such an important effect on both load and supporting strength, a satisfactory sewer construction project requires attainment of design conditions in the field.

a. Rigid pipes support loads in the ground by virtue of resistance of the pipe wall as a ring in bending.
b. Flexible pipes rely on the horizontal thrust from the surrounding soil to enable them to resist vertical load without excessive deformation.
c. Intermediate pipes are those pipes which exhibit behavior between those in (a) and (b). They are also called semi-rigid pipes.

21.5 LOADING CONDITIONS

The three general types of loading conditions are trench, embankment, and tunnel. Out of these, trench conditions are most common.

Trench Conditions

Trench conditions exists when a conduit is laid in a relatively narrow trench or ditch of width W_d cut out of the undisturbed soil and then backfilled with original soil level with the ground surface. If the trench has sloping sides, the width, W_d, is taken equal to a horizontal plane tangential to the top of the sewer pipe.

Embankment Conditions

Embankment conditions are defined as those in which the conduit is covered with fill above the original ground surface or when a trench in undisturbed ground is so wide that trench wall friction does not affect the load on the pipe.

Tunnel Conditions

Tunnel conditions are defined as those in which the conduit is installed in a relatively narrow trench cut in undisturbed ground and covered with earth backfill to the original ground surface.

21.6 DEAD LOADS

The vertical load on a sewer pipe is the resultant of the weight of the prism of soil within the trench and above the top of the pipe, and the friction or shearing forces generated between the prism of soil in the trench and the sides of the trench. The backfill tends to settle in relation to the undisturbed soil in which the trench is excavated. This tendency for movement induces upward shearing forces which support a part of the weight of the backfill. Hence, the resultant load on the horizontal plane at the top of the pipe within the trench is equal to the weight of the backfill minus the upward shearing forces, as indicated in **Figure 21.4**.

The width of the trench and the unit weight of the backfill soil have a direct influence on the load on the pipe. So, the width should be kept to an absolute minimum with the provision of sufficient working space at the sides of the pipe to calk joints properly, to insert and strip form, and to compact backfill. The load is also influenced by the coefficient friction between the backfill and the sides of the trench and by the internal friction of the backfill soil.

Anson Marston developed methods for determining the vertical load on buried conduits due to gravity earth forces in all of the most commonly encountered construction conditions. His methods are based on a theory which states that the load on a buried conduit is equal to the weight of the prism of earth directly over the conduit, called the **interior prism**, plus or minus the frictional shearing forces transferred to that prism by the adjacent prisms of earth. The magnitude and direction of these frictional forces depend upon the relative settlement between the interior and adjacent earth prisms.

Marston's equation

$$W = C_d \times \gamma \times B_d^2$$

W = Vertical load per unit length, kN/m (lb/ft)
γ = Specific or unit weight of earth, kN/m³ (lb/ft³)
B_d = Trench width at the crown of pipe, m (ft)
C_d = A dimensionless coefficient that measures the effect of
- ratio of the height of fill to the width of the trench;
- shearing forces between interior and adjacent earth prisms;
- direction and amount of relative settlement between interior and adjacent earth prisms for embankment conditions.

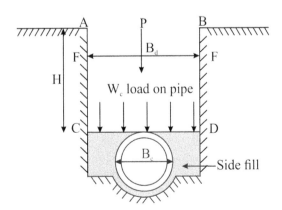

FIGURE 21.4 Load on pipe (trench conditions)

EMPIRICAL RELATIONSHIP

The value of C_d can be computed using an empirical formula. The value of the coefficient for a given back fill soil decreases exponentially as the ratio of soil cover to width increases. An empirical relationship is developed for computing the value of the coefficient.

$$Coefficient, C_d = \left[\frac{1-e^{-2K\mu'H/B_d}}{2K\mu'} \right]$$

Where μ' is the friction coefficient and K is the ratio of lateral unit pressure to vertical unit pressure. Maximum values of $K\mu'$ for fill materials are shown in **Table 21.1**.

Example Problem 21.2

A 12-in diameter sewer pipe is placed in an 8.0 ft deep rectangular trench that is 2.0 ft wide. The trench is backfilled with clay that has unit weight of 120 lb/ft³. Compute the dead load.

Given:

$$D = 12 \text{ in} = 1 \text{ ft} \quad H = 8.0 \text{ ft} \quad \gamma = 120 \text{ lb/ft}^3 \quad B = 2.0 \text{ ft} \quad W = ?$$

Solution:

$$\frac{H}{B} = \frac{8.0 ft}{2.0 ft} = 4.0$$

From Table 21.1, $K\mu' = 0.13$

$$C_d = \left[\frac{1-e^{-2K\mu'H/B_d}}{2K\mu'} \right] = \frac{\left(1-e^{-2\times0.13\times4.0}\right)}{2\times0.13} = 2.486 = 2.5$$

$$W = C_d \times \gamma \times B_d^2 = 2.486 \times \frac{120 \ lb}{ft^3} \times (2.0 \ ft)^2 = 1193.6 = \underline{1200 \ lb / ft^3}$$

TABLE 21.1
Factor $K\mu'$ for various fill materials

Type of Fill	$K\mu'$	SG
granular material	0.192	1.75
sand and gravel	0.165	1.80
saturated topsoil	0.150	1.70
ordinary clay	0.130	1.90
Saturated clay	0.110	2.10

Example Problem 21.3

Determine the load on a 600 mm diameter rigid pipe under 3.0 m of saturated topsoil (16 kN/m³) in trench conditions. The side fill is 30 cm wide on each side. Assume the pipe wall thickness as 50 mm.

Given:

$$D = 600 \text{ mm} = 0.60 \text{ m} \quad H = 3.0 \text{ m} \quad \gamma = 16 \text{ kN/m}^3 \quad W = ?$$

Solution:

$$B_d = 0.60\,m + 0.1 + 2 \times 0.30\,m = 1.3\,m, \quad \frac{H}{B} = \frac{3.0\,m}{1.3\,m} = 2.31$$

From Table 22.1, the value of the factor Kμ' for saturated topsoil is 0.15.

$$C_d = \left[\frac{1 - e^{-2K\mu'H/B}}{2K\mu'}\right] = \left[\frac{1 - e^{-2 \times 0.15 \times 2.31}}{2 \times 0.15}\right] = 1.666 = 1.67$$

$$W_c = C_d \times \gamma \times B_d^2 = 1.67 \times \frac{16\ kN}{m^3} \times (1.3\ m)^2 = 45.06 = \underline{45\ kN/m}$$

If the pipe is **flexible** and the soil at the sides is compacted to the extent that it will deform under vertical load the same amount as the pipe itself. The side fills may be expected to carry their proportional share of the total load. For a situation of this kind, the trench load formula is modified as follows:

Marston's equation (flexible pipes)

$$\boxed{W = C_d \times \gamma \times B_d \times B_c}$$

B$_d$ = width of trench or ditch at the crown of the pipe, m
B$_c$ = outside width of the conduit or pipe, m

Example Problem 21.4

Determine the load on a 750 mm diameter flexible conduit, with a wall thickness of 50 mm, installed in a trench 1.4 m wide at a depth of 3.4 m. The trench is cut in clay soil with specific gravity of 1.90.

Given:

$$D = 750 \text{ mm} = 0.75 \text{ m} \quad H = 3.4 \text{ m} \quad B_d = 1.4 \text{ m} \quad SG = 1.90 \quad W = ?$$

Solution:

$$B_c = 0.75\,m + 2 \times 0.050\,m = 0.85\,m \qquad H = 3.4\,m - 0.85\,m = 2.55\,m$$

$$\frac{H}{B_d} = \frac{2.55\,m}{1.4\,m} = 1.82\,m \qquad K\mu' = 0.13$$

$$C_d = \left[\frac{1-e^{-\frac{2K\mu'H}{B}}}{2K\mu'}\right] = \left[\frac{1-e^{-2\times0.13\times1.82}}{2\times0.13}\right] = 1.449 = 1.45$$

$$W_c = C_d \times \gamma \times B_c \times B_d = 1.449 \times \frac{1.9\times9.81\,kN}{m^3} \times 0.85\,m \times 1.4\,m = \underline{32\;kN\,/\,m}$$

21.7 FIELD SUPPORTING STRENGTH

Sewers must be supporting both lead loads and live loads. Dead loads refer to loads due to backfill and live loads refer to loads due to vehicular traffic. Sanitary sewers are placed very deep, so traffic loads or live loads are negligible compared to dead loads. After knowing the load on the sewer pipe, it is important to compare with the load-carrying capacity of the sewer pipe.

21.7.1 LOAD-CARRYING CAPACITY

The **load-carrying capacity** of a sewer will depend on two key factors, viz., pipe crushing strength and the type of class of pipe bedding. The crushing strength of a pipe is determined by a standard laboratory procedure, and it is specified in terms of load per unit length, kN/m (lb/ft). Minimum crushing strengths of pipe materials and sizes are published by pipe manufacturers. For example, typical values of crushing strength of vinyl chloride pipe (VCP) are presented in **Table 21.2**.

21.7.2 LOAD FACTOR

Bedding refers to the way in which the pipe is placed on the bottom of trench. Proper bedding always increases the actual supporting strength of the installed pipe by

TABLE 21.2
Strength of vinyl chloride pipe of various sizes

Nominal Size mm (in)	Strength, kN/m*	
	Standard	Extra
200(8)	20.4	32.0
250(10)	23.2	35.0
300(12)	26.3	37.9
380(15)	29.2	42.3
460(18)	32.0	48.1

* kN/m = 68.5 lb/ft

distributing the load over the pipe circumference. The ratio of the actual field-supporting strength to the crushing strength is called the load factor **Table 21.3**).

$$Load\ Factor, LF = \frac{Field\ supporting\ Strength}{Crushing\ Strength}$$

$$Field\ supporting\ Strength = LF \times Crushing\ Strength$$

In addition to the load factor provided by the pipe bedding, a **safety factor** (SF) is applied to the computations to arrive at the safe supporting strength as follows.

$$Safe\ supporting\ Strength = \frac{LF}{SF} \times Crushing\ Strength$$

A safety factor of 1.5 is commonly used for clay and plain concrete sewers to compensate for the poor quality of the material or for faulty construction.

Example Problem 21.5

Assume the sewer pipe in Example Problem 21.2 is of standard strength VCP. What class of bedding should be specified for construction, using a safety factor of 1.5?

Given:

W = 1193.6 lb/ft   SF = 1.5   crushing strength = 26.3 kN/m from Table 21.2
26.3 kN/m × 68.5 = 1801.9 lb/ft   LF = ?

Solution:

$$LF = \frac{Safe\ supporting\ Strength \times SF}{Crushing\ Strength}$$

$$= \frac{1193.6\,lb}{ft} \times 1.5 \times \frac{ft}{1801.9\,lb} = 0.9936 = 1.0$$

As per calculations, the lowest class of bedding with a LF of 1.1 should be adequate. However, since class D bedding is not recommended for good construction, class C bedding is selected.

TABLE 21.3
Load factor for different classes of bedding

class of bedding	A	B	C	D
load factor	2.8	1.9	1.5	1.1

Discussion Questions

1. Explain the steps involved in sewer construction.
2. Describe the procedure of laying sewers in correct alignment and slope with the use of lasers.
3. Why is pipe bedding important in sewer construction? Discuss the different classes of bedding.
4. Explain the tests on sewers for straightness and obstruction.
5. Why are pipe joints important in sewer construction? What are the requirements of a good pipe joint?
6. Explain the terms:
   a. timbering of trenches;
   b. Backfilling of trenches;
   c. well-point drainage.
7. Explain how the load on the sewers are taken into consideration while designing sewers.
8. What are the different materials from which sewers are made? Discuss their suitability with respect to size and type of sewer pipe.
9. List five points that should be important considerations in the selection of materials for sewer pipes.
10. Explain the procedure of jointing plastic pipes.
11. What are the tests used for finding illegal connections to the sewer system?
12. What is the purpose of a water test? And discuss how it is carried out in the field.
13. Describe different types of loads on sewer pipes laid underground?
14. Preparation of bedding for flexible pipes needs extra care compared to rigid pipes. Comment.
15. Define the loading factor and how it is used to select proper bedding for sewers.
16. Is it good practice to overdesign a sewer pipeline? Why?

Practice Problems

1. Work out fill load on 1200 mm inside diameter concrete sewer pipe installed in a trench of 2.2 m width and depth of 3.2 m. Sewer pipe is 65 mm thick and the unit weight of the fill ($K\mu' = 0.15$) is known to be 19 kN/m^3. (69 kN/m).
2. Determine the load on a 450 mm diameter concrete pipe with a wall thickness of 35 mm, installed in a trench 0.85 m wide at a depth of 3.0 m. The trench is cut in clay soil with specific gravity of 1.8. (26 kN/m).
3. Determine the dead load on a 600 mm diameter rigid pipe under 2.5 m of saturated topsoil in trench conditions. The side fill is 30 cm wide on each side. Assume the pipe wall thickness as 50 mm and weight density of the wet fill is 17 kN/m^3. (42 kN/m).

4. Determine the dead load on a 24 in diameter rigid pipe under 8.0 ft of top-soil in trench conditions. The side fill is 1.0 ft wide on each side. Assume the pipe wall thickness as 2 in and specific gravity of topsoil is 1.6. (2700 lb/ft).
5. Determine the load on a 750 mm diameter flexible conduit, with a wall thickness of 50 mm, installed in a trench 1.5 m wide at a depth of 3.5 m. The trench is cut in topsoil with specific gravity of 1.60. (28 kN/m).
6. A 100 m reach of sewer is to have minimum capacity of 200 L/s. The street elevation at the upper manhole is 105.55 m and the lower manhole is 103.05 m. Select appropriate size and slope for this reach and establish the pipe invert elevations. Assume minimum cover of 2.0 m above the crown of the pipe. (350 mm, 103.20 m, 100.70 m).
7. A 300 ft reach of sewer is to have minimum capacity of 2200 gpm. The street elevation at the upper manhole is 508.65 ft and the lower manhole is 502.05 ft. Select appropriate size and slope for this reach and establish the pipe invert elevations. Assume minimum cover of 7.0 ft above the crown of the pipe. (12 in, 500.65 ft, 494.05 ft).
8. A sewer reach of length 100 m is to carry 40 L/s flowing full (**Figure 21.5**). Select the required size of sewer and invert elevations. Assume cover of 2.5 m.
   (200 mm, 347.30 m, 345.17 m).
9. Second reach of the sewer line in **Figure 21.5** is of length 120 m and required to carry wastewater at the peak rate of 80 L/s. Select the required size of sewer and invert elevations. (300 mm, 345.07 m, 344.20 m).
10. A 200 mm diameter sewer pipe is placed in a 3.0 m deep, and 0.90 m wide trench and backfilled with sand (17.2 kN/m^3). Using a safety factor of 1.5, select appropriate bedding for the standard strength VCP sewer.
    (LF = 1.85, Class B bedding).
11. Repeat Practice Problem 10 where the VCP sewer is of extra strength.
    (LF = 1.2, Class C bedding).
12. A sewer pipe with a diameter of 300 mm is placed in a 3.2 m deep, and 0.75 m wide trench that is filled with saturated clay of SG = 2.1. Work out the dead load on the pipe. (30 kN/m).
13. A 12 in sewer pipe is laid in a 10 ft deep and 3.0 ft wide trench that is filled with saturated clay (125 lb/ft^3). Compute the dead load on the pipe. (2500 lb/ft).
14. A 15 in VCP sewer pipe is laid in a 9.0 ft deep and 3.5 ft wide trench that is filled with saturated topsoil (SG = 1.7). Compute the dead load on the pipe. Using a safety factor of 1.5, select appropriate bedding for the standard strength VCP sewer (2100 lb/ft, LF = 1.6, Class B bedding).

FIGURE 21.5   Sewer line reaches

15. A 12 in extra-strength sewer pipe is laid in a 10 ft deep, and 3.0 ft wide trench that is filled with saturated clay (125 lb/ft³). Compute the loading factor, assuming a safety factor of 1.5. (2500 lb/ft).

16. A 1.2 m × 2.4 m × 20 m trench is to be backfilled with sand. How many 9 m³ capacity truck loads are needed? (7).

17. The upstream invert elevation for a 250 mm sewer is fixed at 130.58 m. You are required to lay the 140 m long sewer at a slope of 0.40%. Find the elevation of downstream invert. (130.02 m).

18. An 8 in diameter inflated plug is holding water in a sewer pipe tested for leakage. Find the pressure force against the plug when the water depth in the upstream manhole is 3.0 ft above the center of the plug. (65 lb)

19. How many tonnes (2000 lb) of soil (SG = 2.4) are to be dug to excavate a 4.0 ft × 8.0 ft × 65 ft trench? (160).

20. A sewer main is laid on a grade of 0.09%. What is the drop in elevation over 1 mile length of the sewer pipe? (4.75 ft).

Section V

Wastewater Treatment

22 Natural Purification

To prevent pollution of the receiving water bodies, it is necessary to treat wastewater when the pollutant loading exceeds the assimilative capacity of water bodies. In the past, in a majority of cases, wastewater was directly discharged into bodies of receiving waters. It was assumed that wastewater was diluted by a factor of hundreds so that **natural purification** was good enough to maintain the quality of water in the receiving waters. However, due to the increase in urbanization and luxurious lifestyles, it is not now feasible. Dilution is not a solution. For obvious reasons, perennial streams and rivers are the best types of bodies for the disposal of sewage. The most stressful period for receiving water bodies is summer, when flows are at their minimum and the BOD reaction rate is accelerated due to a rise in temperature. Organic loading in a wastewater stream is indicated by the BOD loading. To understand natural purification, it is important to understand BOD reaction.

22.1 BOD REACTION

BOD reaction is a first order reaction at a given temperature. The rate at which BOD is exerted (BOD_e) is directly proportional to BOD remaining (BOD_r). On integration, the BOD remaining after duration, t, is given by the following expression:

$$\frac{dBOD_r}{dt} = -KBOD_r \text{ or } BOD_r = BOD_u e^{-Kt} \text{ or } BOD_t = BOD_u\left(1 - e^{-Kt}\right)$$

The value of BOD rate constant k depends on temperature and biodegradability of the wastewater. Typical value of K at 20°C is 0.23/d. Rate constant k for other temperatures can be modified using the following empirical equation.

$$K_T = K_{20}\theta^{(T-20)} = K_{20} \times 1.047^{(T-20)}$$

For temperatures less than 20°C, the value of θ is 1.35. Rate constant of BOD reaction indicates biodegradability of a given wastewater. It can be determined based on BOD test data for a number of days.

Example Problem 22.1

Assuming BOD reaction constant value of 0.23/d, figure out what percentage of ultimate BOD is exerted after a duration of 5.0 d?

Given:

$K = 0.23/d \quad BOD/BOD_u = ?$

DOI: 10.1201/9781003347941-27

Solution:

$$\frac{BOD_t}{BOD_u} = \left(1 - e^{-Kt}\right) = \left(1 - e^{\frac{-0.23}{d} \times 5d}\right) = 0.683 = \underline{68\%}$$

Example Problem 22.2

5-d BOD for a given wastewater is 220 g/m³. Assuming BOD_5 is 70% of the ultimate BOD, find 3-d BOD.

Given:

$K = 0.23/d \quad BOD_5 = 250 \text{ g/m}^3 \quad BOD_u = ?$

Solution:

$$k = -\frac{1}{t}\ln\left(1 - \frac{BOD_t}{BOD_u}\right) = -\frac{\ln(1 - 0.70)}{5d} = 0.2407 = 0.24/d$$

$$BOD_3 = BOD_u\left(1 - e^{-Kt}\right) = \frac{1}{0.70} \times \frac{220 \text{ g}}{m^3} \times \left(1 - e^{-\frac{0.24}{d} \times 3d}\right) = 88.0 = \underline{88 \text{ g/m}^3}$$

Example Problem 22.3

Given that 5-d BOD is 250 g/m³, find the ultimate BOD. Assume BOD reaction constant value of 0.23/d. What would be 5-d BOD at temperature of 25°C?

Given:

$K = 0.23/d \quad BOD_5 = 250 \text{ g/m}^3 \quad BOD_u = ? \quad BOD_5 \, (25°C) = ?$

Solution:

$$BOD_u = \frac{BOD_5}{\left(1 - e^{-5k}\right)} = \frac{250 \text{ g}}{m^3} \times \frac{1}{\left(1 - e^{\frac{-0.23}{d} \times 5d}\right)} = 365 = \underline{370 \text{ g/m}^3}$$

$$K_{25} = K_{20} \times 1.047^{(25-20)} = 0.289 = 0.29/d$$

$$BOD_5 = \frac{365.8 \text{ mg}}{L} \times \left(1 - e^{-\frac{0.289}{d} \times 5d}\right) = 279 = \underline{280 \text{ mg/L}}$$

22.2   NATURAL PROCESS

When wastewater is discharged into a natural stream, organic matter is biodegraded aerobically. As a result of aerobic reaction, ammonia, nitrates, sulphates, carbon dioxide, water, and new cells are produced. The main factors influencing this process are dilution, sedimentation, oxidation, reduction, temperature, and sunlight. High dilution ensures aerobic conditions and completion of biodegradation. During the summer months, stream flows are minimal and biological activity is accelerated due to an increase in water temperature. This represents the worst-case scenario from the point of view of pollution. The dissolved oxygen (DO) content of the stream water is depleted. As the water moves downstream, its DO content goes on depleting until it reaches a minimum level, the so-called **critical point**. Saturation DO values of water at various temperatures are presented in **Table 22.1**.

22.2.1   ZONE OF DEGRADATION

As wastewater is discharged into a water body, it goes through various phases of degradation as shown in **Figure 22.1**. The water in the zone of degradation or

TABLE 22.1
Saturation dissolved oxygen

°C	mg/L	°C	mg/L
0	14.60	23	8.56
1	14.19	24	8.40
2	13.81	25	8.24
3	13.44	26	8.09
4	13.09	27	7.95
5	12.75	28	7.81
6	12.43	29	7.67
7	12.12	30	7.54
8	11.83	31	7.41
9	11.55	32	7.28
10	11.27	33	7.16
11	11.01	34	7.05
12	10.76	35	6.93
13	10.52	36	6.82
14	10.29	37	6.71
15	10.07	38	6.61
16	9.85	39	6.51
17	9.65	40	6.41
18	9.45	41	6.31
19	9.26	42	6.22
20	9.07	43	6.13
21	8.90	44	6.04
22	8.72	45	5.95

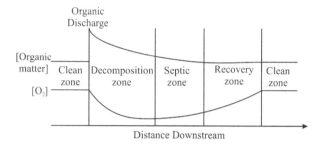

FIGURE 22.1   Zones of natural process

decomposition is turbid and sludge forms at the bottom of the stream. The dissolved oxygen level is suddenly reduced due to mixing and the deoxygenation rate exceeds the reaeration rate. The decomposition of the solid matter takes place in this zone and anaerobic conditions prevail.

22.2.2   ZONE OF ACTIVE DECOMPOSITION

This zone is the zone of most stress due to pollution. As a result, the DO level falls to a minimum and anaerobic conditions set in. Hence it is also called the septic zone. Sludge at the bottom goes through anaerobic biodegradation producing methane and hydrogen sulphide. Towards the end of this zone, the oxygenation rate becomes greater than the deoxygenation rate and DO starts picking up.

22.2.3   ZONE OF RECOVERY

As the name indicates, the process of recovery starts, and stream water starts moving toward its former conditions. Aerobic stabilization of soluble BOD takes place in this zone, thus BOD falls and DO level goes up. Near the end of this zone, microscopic life reappears, fungi decrease, and algae start to become established.

22.2.4   CLEAR WATER ZONE

In this zone, the stream returns to its former conditions, the water becomes clear, and recovery is assumed to be complete. Oxygen level reaches close to saturation.

22.3   OXYGEN SAG CURVE

As waste is discharged into the stream, it lowers the DO level in the stream water. As mixed water moves downstream, oxygen deficit increases as BOD is exerted. The plot of DO versus distance or time from point of mixing downstream is called the **oxygen curve**, as shown in **Figure 22.2**. Immediately after mixing, the **oxygen deficit** is called the **initial deficit**, D_0. As water moves downstream this deficit increases

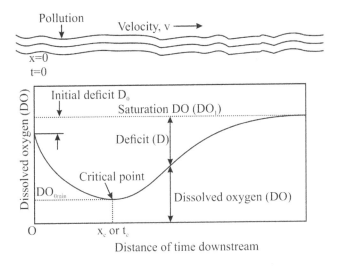

FIGURE 22.2 Oxygen sag curve

to a maximum, D_c, also called the **critical point**. At the critical point, the DO_c level reaches its minimum and oxygenation and deoxygenation rates become equal. Beyond the critical point, the oxygenation rate is greater than the deoxygenation rate, and DO in the stream water starts moving up and the stream starts recovering. The critical point is of greater interest since it allows us to figure out the amount of waste that can be naturally purified while maintaining a minimum DO level.

Reaeration and deoxygenation both are assumed to be first order reactions. Based on this assumption, the oxygen deficit is described by the first order differential equation as suggested by Streeter and Phelps.

$$\frac{dD_t}{dt} = K_1 BOD_r - K_2 D_t$$

Where, D_t = DO deficit at time t BOD_r = BOD remaining at time t k = reaction rate constant Sub 1 = deoxygenation Sub 2 = oxygenation or reaeration

Solving the differential equation, oxygen deficit at any time t after mixing can be predicted. Knowing the deficit, D_t, dissolved oxygen DO can be found.

$$D_t = \frac{K_1 BOD_u}{K_2 - K_1}\left[e^{-K_1 t} - e^{-K_2 t}\right] + D_0 e^{-K_2 t} \quad or \quad DO_t = DO_{sat} - D_t$$

Determination of the **critical point** is of great engineering significance and can be found by equating the first differential to zero. Time and distance are related by the flow velocity, u. Thus the critical point can be expressed in time or distance travelled. Further knowing time to critical point, the maximum oxygen deficit can be found.

Time to reach critical point

$$t_c = \frac{1}{K_2 - K_1} \ln\left[\frac{K_2}{K_1}\left(1 - \frac{D_0(K_2 - K_1)}{K_1 BOD_u}\right)\right] \quad hence \quad D_c = \frac{K_1 BOD_u}{K_2} e^{-K_1 t_c}$$

Example Problem 22.4

A stream with minimum flow is 1.6 m³/s containing on average 5.0 g/m³ of BOD, DO = 8.0 g/m³ and re-aeration constant of 0.60/d. A wastewater plant effluent of 0.10 m³/s with BOD of 20 mg/L and DO = 3.0 g/m³ is discharged into the stream. BOD rate constant is 0.25/d and average flow velocity in the stream is 0.20 m/s. Assuming saturation DO of 9.2 g/m³, calculate the DO in the stream water at locations 25 km and 50 km downstream of discharge point.

Given:

K_1 = 0.25/d K_2 = 0.60/d BOD_w = 20 g/m³ Q_w = 0.10 m³/s DO_w = 3.0 g/m³

DO_{st} = 8.0 g/m³ BOD_{st} = 5.0 mg/L Q_{st} = 1.60 m³/s DO_{sat} = 9.2 g/m³ DO_t =?

Solution:

Mixed stream

$$BOD_m = \frac{BOD_w \times Q_w + BOD_{st} \times Q_{st}}{Q_w + Q_{st}} = \frac{20 \times 0.1 + 5 \times 1.6}{0.1 + 1.6} = 5.88 \, g/m^3$$

$$BOD_u = \frac{BOD_5}{(1 - e^{-5k})} = \frac{5.88 \, g}{m^3} \times \frac{1}{\left(1 - e^{-\frac{0.23}{d} \times 5d}\right)} = 8.24 = 8.2 \, g/m^3$$

$$DO_m = \frac{DO_w \times Q_w + DO_{st} \times Q_{st}}{Q_w + Q_{st}} = \frac{3.0 \times 0.1 + 8.0 \times 1.6}{0.1 + 1.6} = 7.70 = 7.7 \, g/m^3$$

$$D_m = DO_{sat} - DO_m = 9.2 - 7.70 = 1.5 \, g/m^3$$

At a distance of 25 km oxygen deficit

$$t = \frac{d}{v} = 25 \, km \times \frac{s}{0.2 \, m} \times \frac{1000 \, m}{km} \times \frac{h}{3600 \, s} \times \frac{d}{24 \, h} = 1.446 = 1.4 \, d$$

$$D_t = \frac{K_1 BOD_u}{K_2 - K_1}\left[e^{-K_1 t} - e^{-K_2 t}\right] + D_0 e^{-K_2 t}$$

$$= \frac{0.25 \times 8.2 \, g}{(0.60 - 0.25)m^3} \times \left[e^{-0.25 \times 1.4} - e^{-0.6 \times 1.4}\right] + \frac{1.5 \, g}{m^3} e^{-0.6 \times 1.4} = 2.25 \, g/m^3$$

$$DO_t = DO_{sat} - D_t = (9.0 - 2.25) = 5.75 = 5.8 \, g/m^3$$

At a location 50 km (t = 2.9 d) downstream, oxygen deficit

$$D_t = \frac{K_1 BOD_u}{K_2 - K_1}\left[e^{-K_1 t} - e^{-K_2 t}\right] + D_0 e^{-K_2 t}$$

$$= \frac{0.25 \times 8.2\,g}{(0.60 - 0.25)m^3} \times \left[e^{-0.25 \times 2.9} - e^{-0.6 \times 2.9}\right] + \frac{1.5\,g}{m^3}e^{-0.6 \times 2.9} = 2.1\,g/m^3$$

$$DO_t = DO_{sat} - D_t = (9.0 - 2.07) = 6.93 = 6.9\,g/m^3$$

Time to reach critical point

$$t_c = \frac{1}{K_2 - K_1}\ln\left[\frac{K_2}{K_1}\left(1 - \frac{D_0(K_2 - K_1)}{K_1 BOD_u}\right)\right]$$

$$= \frac{d}{0.60 - 0.25}\ln\left(\frac{0.60}{0.25}\right)\left(1 - \frac{1.5(0.60 - 0.25)}{0.25 \times 8.2}\right) = 1.656 = 1.7\,d$$

This further confirms that recovery starts after 1.7 d, hence DO starts picking up as shown by the above calculations.

Example Problem 22.5

A town discharges secondary effluent containing BOD_5 of 25 mg/L at the rate of 5.0 m³/s into a stream with a minimum flow of 65 m³/s. Calculate the critical DO in the stream water downstream. Assume BOD rate constant of 0.22/d and oxygenation rate constant of 0.75/d and saturated DO is 9.0 mg/L. Assume DO of the effluent is 1.0 mg/L.

Given:

$$K_1 = 0.22/d \quad K_2 = 0.75/d \quad BOD_5 = 25\,g/m^3 \quad Q_w = 5.0\,m^3/s \quad Q_{st} = 65\,m^3/s$$

Solution:
After mixing

$$BOD_m = \frac{BOD_w \times Q_w + BOD_{st} \times Q_{st}}{Q_w + Q_s} = \frac{25 \times 5 + 0 \times 5}{5 + 65} = 1.78\,g/m^3$$

$$BOD_u = \frac{BOD_5}{(1 - e^{-5k})} = \frac{1.78\,g}{m^3} \times \frac{1}{\left(1 - e^{-\frac{0.22}{d} \times 5d}\right)} = 2.676 = 2.7\,g/m^3$$

$$DO_m = \frac{DO_w \times Q_w + DO_{st} \times Q_{st}}{Q_w + Q_{st}} = \frac{1 \times 5 + 9 \times 65}{5 + 65} = 8.428 = 8.43\,g/m^3$$

$$D_m = DO_{sat} - DO_m = 9.0 - 8.43 = 0.57\,g/m^3$$

Critical point

$$t_c = \frac{1}{K_2 - K_1} \ln\left[\frac{K_2}{K_1}\left(1 - \frac{D_0\left(K_2 - K_1\right)}{K_1 BOD_u}\right)\right]$$

$$= \frac{d}{0.75 - 0.22} \ln\left[\frac{0.75}{0.22}\left(1 - \frac{0.57\left(0.75 - 0.22\right)}{0.22 \times 2.68}\right)\right] = 0.958 = 0.96\,d$$

$$D_c = \frac{K_1 BOD_u}{K_2} e^{-K_1 t_c} = \frac{0.22}{0.75} \times \frac{2.68\,g}{m^3} \times e^{-\frac{0.22}{d} \times 0.96 d} = 0.636 = 0.64\,g\,/\,m^3$$

$$DO_c = DO_{sat} - D_c = \left(9.0 - 0.64\right) = 8.36 = 8.4\,g\,/\,m^3$$

22.4   DILUTION INTO SEA

Since saturated DO in water decreases with increasing solid content, saturated DO of seawater is about 80% that of fresh water. Since seawater is denser than sewage water, which usually has a higher temperature, sewage discharged into the sea spreads on the surface in a thin layer called **sleek**. Moreover, seawater contains large quantity of solids that react with some of the sewage solids to precipitate, resulting in the formation of **sludge banks**. As more and more waste is dumped, seawater capacity to absorb sewage is reduced and results in undesirable conditions. However, since the sea contains a large volume of water, such problems can be overcome if the sewage is discharged deep into the sea and further away from the coastline. To achieve this, the following point should be kept in mind.

   i. The wastewater should be discharged at least 1 km away from the shoreline.
  ii. The outfall should be designed so as to ensure proper dilution and adequate mixing. This is achieved by providing a multipoint diffuser.
 iii. The minimum depth of water at the outfall point should be 3 to 5 m.
  iv. Wastewater is discharged only during low tides. Provision must be made to hold wastewater during the high tide.
   v. While deciding the position of the outfall, the direction of the wind velocity and direction of ocean currents should be taken into consideration.

22.5   DISPOSAL BY LAND TREATMENT

Land treatment refers to the disposal of treated or raw wastewater spread over the land surface. Some part of the wastewater evaporates and other enters the soil by infiltration. Wastewater percolating through soil pores leaves behind organic solids, which are oxidized aerobically, and adds to the fertility of the soil. For this reason, disposal by land treatment is also called sewage farming.

SEWAGE FARMING

In sewage farming, in addition to the disposal of partly treated wastewater, crops are raised using nutrients and other microelements from wastewater. The most common methods of sewage farming and treatment of wastewater are irrigation, rapid infiltration, and overland runoff.

Irrigation and infiltration are suitable for land with coarse-textured soils with relatively high percolation rates. Raw sewage applied in large amounts can result in plugging the soil surface, thus significantly reducing infiltration rates. Thus, partially treated wastewater can greatly reduce plugging and allow voids to remain open. **Rapid infiltration** is practiced where the percolation rate is in the range of 6–25 mm/min (0.25–1 in/min) **Irrigation** is suitable when rates are 2–6 mm/min, and **overland runoff** is the best choice when the percolation rate is below 2.0 mm/min, or say in soils with low permeability. Recommended loading rates are shown in **Table 22.2**.

Except for overland runoff, straining and biological filtration can reduce the organic content of the wastewater. In sewage farming, as wastewater percolates, the soil holds moisture, nutrients, and organic matter to sustain crops. **Rapid infiltration** is suitable to recharge groundwater. Since wastewater carries pathogens, **sewage farming** must ensure the safety of the workers. In addition, leafy vegetables and root crops like potatoes are not recommended for sewage farming. In sewage farming, wastewater effluents from primary or secondary treatment plants can be applied to the land by **surface irrigation** methods like border or furrow irrigation. Sprinkler irrigation is more common in Western countries.

22.6 COMPARISON OF DISPOSAL METHODS

Dilution methods require large water bodies to assimilate water, whereas the land treatment disposal method requires large area of pervious land. Thus, the following points can be made to compare the disposal methods.

- Since in urban areas land is relatively expensive, the dilution method is more suitable and so is preferred.

TABLE 22.2
Recommended loading rate

Soil Type	Application Rate of mm/d	in/d
sandy	22–25	0.9–1
sandy loam	15–20	0.6–0.8
loam	10–15	0.4–0.6
clayey loam	5–10	0.2–0.4
clay	3–5	0.1–0.2

- Form the point made above, it can also be said that disposal through land application is preferred in rural areas.
- In dilution methods, the assimilative capacity of the receiving water must be considered, and the quality of effluent must be good so as not to cause any pollution and for the water body to remain healthy.
- In the broad irrigation method of wastewater disposal, raw wastewater can be applied. However, care must be taken to avoid **sewage sickness** of soil.
- In disposal by land application, sewage can be applied by flooding the land surface. Topographic features of the land may be such that the pumping of water is required. However, this is not usual in dilution methods.
- In the dilution method, when wastewater effluent is discharged to a relatively small stream, it is very important to check for adverse conditions such as would happen in summer due to low temperatures and low flows.
- In sewage farming, good management is necessary and safeguards must be in place to prevent any health risks to workers and the public.
- In the broad irrigation method, disposal must be managed so as to prevent any risks to groundwater, such as contamination.

Discussion Questions

1. Assuming BOD is a first order reaction, develop the relationship between ultimate BOD and BOD exerted at any time t.
2. Describe the various stages of self-purification after waste is discharged.
3. Dilution is not a solution, comment.
4. What is land treatment? Discuss the conditions under which it is suitable.
5. Discuss the factors that affect the self-purification capacity of a river?
6. Draw a typical oxygen sag curve and label it with various stages of self-purification.
7. Write a note on the disposal of sewage into the sea.
8. Compare the deoxygenation rate constant with oxygenation rate constant.
9. Explain the terms sewage farming and soil sickness.
10. Discuss the critical point in relation to oxygen sag curve.
11. Describe the steps required to carry out a BOD test on a sample of wastewater.
12. Explain why dilution of the wastewater sample is required to carry out a successful BOD test. What is the rule of thumb when deciding about the dilution factor?
13. Define the terms oxygenation and deoxygenation as applied to natural purification in streams. How does it affect the shape of the oxygen sag curve?
14. Explain the difference between the dilution of sewage effluents discharged into a river and those discharged into seawater?
15. In majority of the cases, especially in developing countries, the disposal of sewage by land treatment is better than disposal by dilution. Discuss.

Practice Problems

1. A five-day BOD test was conducted on municipal wastewater using 1:60 dilution. Initial and final DO respectively are: 7.80 mg/L and 4.30 mg/L. Compute 5-d BOD and ultimate BOD assuming a K = 0.23/d. (210 mg/L, 310 mg/L).

2. A sample of sewage is found to have a 5-d BOD of 250 mg/L. Assuming rate constant K of 0.11/d ($\log_{10}$ basis), compute BOD_u, and BOD_8. (350, 300 mg/L).

3. Assuming reaction rate constant of 0.23/d, what is the BOD_5 if the ultimate BOD is known to be 280 mg/L? (190 mg/L).

4. A domestic wastewater sample has BOD_2 and BOD_4 of 110 mg/L and 180 mg/L. What is the rate constant?(0.23/d).

5. Assuming reaction rate constant of 0.23/d, find the ultimate BOD of a wastewater sample that has BOD_5 of 180 mg/L. (260 mg/L).

6. An activated sludge plant effluent of 250 gpm containing BOD of 13 mg/L is discharged in to a stream with a flow of 7.0 ft^3/s(cusecs). Assuming BOD of the stream water is negligible, what is BOD of water after complete mixing? (1.0 mg/L).

7. A secondary wastewater treatment plant effluent of 11 L/s containing BOD of 15 mg/L is discharged in to stream with a flow of 0.22 m^3/s. If the BOD of the stream water is 2.5 mg/L, what is the BOD after complete mixing? (3.1 mg/L).

8. Assuming BOD rate constant of 0.23/d, how many days would it take to reduce the total BOD by 99%? (20 d).

9. A BOD test was done on a wastewater sample by making a 2.0% dilution. Initial and final DO readings in the diluted samples are 7.50 mg/L and 2.15 mg/L, and the same readings for the seeded blanks are 8.15 mg/L and 7.95 mg/L. What is the BOD of the sample? (260 mg/L).

10. Assuming BOD reaction constant K of 0.20/d, what percentage of ultimate BOD is exerted over a 5 d period? (63%).

11. A BOD test is performed on a river sample and 5-d BOD is found to be 35 mg/L. Assuming K of 0.23/d, determine ultimate BOD. (51 mg/L).

12. For making 2.0% dilution, 20 mL of sample aliquot was poured in a graduated cylinder. Dilution water was added to make it to 1000 mL. After mixing, the diluted sample was transferred to three BOD bottles. Initial DO reading was read 8.32 mg/L and average final DO reading of three BOD bottles was observed to be 4.20 mg/L. Compute 5-d BOD and ultimate BOD assuming a K-rate of 0.25/d. (210 mg/L, 290 mg/L).

13. BOD of a sewage after 3 days of incubation at 27°C was found to be 110 mg/L. Determine BOD_5 at 20°C, assuming rate constant K = 0.23/d and temperature coefficient θ = 1.047. (BOD_u = 180 mg/L, BOD_5 = 120 mg/L).

14. In a BOD test, 4.0 mL of a raw wastewater sample was added to 300 mL BOD bottle and filled by adding aerated dilution water. Initial DO of the

diluted sample was observed to be 8.1 mg/L. After incubation for a period of 5 d at 20°C, final DO was read to be 4.5 mg/L. What is BOD of the sample tested? (270 mg/L).

15. Effluent from a wastewater treatment plant enters a stream at a flow rate of 3.5 MGD. The BOD of the discharge is 35 mg/L. How many pounds of BOD is entering the stream per day? (1020 lb/d)

16. Biodegradable matter in a wastewater sample is estimated to be 300 mg/L and rate constant of 0.30/d. How much matter remains after 3 days? (120 mg/L).

17. The 5-d BOD of a certain wastewater has been found to be 600 mg/L. Assuming reaction rate constant of 0.25/d, what fraction of this waste would remain unoxidized after a period of 10 days? (8.2%).

18. A sewage treatment plant discharges treated wastewater at the rate of 1.5 m³/s with BOD of 35 mg/L and DO of 2.0 mg/L. Minimum flow in the river is 6.0 m³/s containing BOD of 5.0 mg/L and DO is at 85% of the saturation level. Temperature of discharge and river water both are at 20°C. Assume an deoxygenation rate constant of 0.20/d and oxygenation rate constant of 0.67/d. Calculate minimum critical level of DO in the river water. (6.6 mg/L)

19. A city discharges untreated sewage @ 130 ft³/s with BOD of 260 mg/L in a river with minimum flow of 1600 ft³/s and DO of 9.0 mg/L fully saturated. Assume BOD rate constant of 0.22/d and oxygenation constant of 0.92/d. Find out the critical DO in the stream. (4.6 mg/L after 2.2 d).

20. A river is flowing at the rate 22 m³/s and receives wastewater discharge of 0.5 m³/s. The initial DO of the river water is 6.3 mg/L and DO content in the wastewater is 0.6 mg/L. Lab tests have indicated that BOD_5 of river water is 3.0 mg/L and the wastewater discharged into river has BOD_5 of 130 mg/L. Consider saturation DO of 8.22 mg/L and deoxygenation and reaeration rate constant values of 0.23/d and 0.69/d respectively. Find critical DO deficit and DO in the river. (D_0 = 2.1 mg/L BOD_u = 8.5 mg/L, t_c = 0.96 d, D_c = 2.3 mg/L, DO_c = 5.9 mg/L).

23 Characteristics of Wastewater

The term **wastewater** refers to water carrying waste. Wastewater from sources such as domestic, commercial, institutional, and infiltration combine in the sewer system and arrive at the sewage treatment plant to be cleaned before being sent back to the environment. The combined flow is called **municipal wastewater**. In many situations, some industrial flows are also part of the municipal wastewater. However, the nature of industrial wastewater has to be compatible with the treatment process at the municipal plant. The purpose of the treatment plant is to remove solids, BOD, nutrients, and pathogens. If contributions from industrial sources is very significant, it may directly affect the quality of municipal wastewater and hence the operation of the wastewater treatment plant.

23.1 TREATMENT FACILITY

In a wastewater treatment plant, the solids, BOD, and pathogens are removed from incoming wastewater as water passes through various unit processes and operations. These processes include physical, chemical, and biological treatments. Based on the level of treatment, a water pollution treatment facility can be primary, secondary, or tertiary, as shown in **Figure 23.1**.

Preliminary treatment includes screening, shredding, and grit removal. These processes remove the coarse material and grit before the water flows to the primary treatment. Irrespective of the stages of treatment, preliminary treatment is always there.

Primary treatment removes solids from incoming wastewater by gravity settling and floatation. Primary treatment typically removes 50% of suspended solids (SS) and 35% of BOD.

Secondary treatment processes usually follow primary treatment and commonly consist of biological processes. Colloidal and dissolved organics not removed by the preceding treatment are converted into biomass, energy, carbon dioxide, and water. The term biomass is to indicate the microorganisms which break down the organics and convert them to more stable forms. A fraction of raw solids is converted to carbon dioxide and water and the remaining fraction to new cells and energy. Typically, the effluent from a secondary treatment plant (secondary effluent) contains less than 15 mg/L each of SS and BOD. In most cases, secondary treatment of municipal wastewater is sufficient to meet effluent standards. However, depending on the use of the receiving waters, further treatment may be required. Secondary treatment has become more or less the norm of the day.

DOI: 10.1201/9781003347941-28

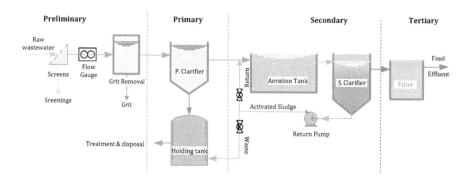

FIGURE 23.1   Flow schematic of a tertiary plant

Tertiary treatment is the third stage of treatment for the further removal of solids and nutrients. Filtration, nitrification–denitrification, and further removal of BOD by polishing lagoons are examples of tertiary treatment. In places where the water downstream is used as a source for drinking water supplies, tertiary treatment in different forms is used to meet more stringent discharge criteria. Tertiary treatment is more expensive, but you have a higher degree of flexibility and control.

Advanced treatment is an add-on process to primary and secondary treatment. Phosphorus removal by chemical precipitation as carried as part of primary or secondary treatment will be called advanced wastewater treatment. However, phosphorus removal following the secondary treatment is called tertiary treatment.

23.2   DOMESTIC WASTEWATER

Domestic wastewater originates in the kitchen, bathroom, and laundry room. Wastewater generated from toilets is also called **black water** and that from kitchen and baths is called **grey water**. Grey water, in some cases, may be recycled. **Sanitary water** is the term applied to domestic water plus water from commercial areas and some industrial water. The term municipal water applies to sanitary water plus **infiltration & inflow**. It may contain large amounts of industrial wastewater. However, any industrial wastewater allowed must be treatable by conventional wastewater treatment processes. A typical municipal wastewater contains only 0.1% solids while the remaining 99.9% is made up of water. The average daily volume of wastewater production in North America is typically 450 L/person/d. In developing countries like India, per capita wastewater production is much less and can be considered in the range of 120–200 L/c·d, with 150 L/c·d as an average value.

23.3   PHYSICAL CHARACTERISTICS

Raw sewage is highly turbid. Normal fresh sewage smells musty but not unpleasant. Obnoxious smells, if present, indicates old septic sewage. Septic or partially decomposed sewage is dark, sometimes black in color with a sulphurous odor due to the

TABLE 23.1
General characteristics of flow streams

Parameter	Raw Sewage	Primary Effluent	Secondary Effluent
temperature	generally warm	lower temperature	lower temperature
turbidity	high in solids	non-settling	no visible solids
color	milky gray to black	grayish to colorless	Clear colorless
odor	musty to sulphurous	musty to sulphurous	fresh

presence of hydrogen sulphide. Hydrogen sulphide is toxic at low levels and causes corrosion. Another gas, methane, produced due to anaerobic conditions is odorless but very explosive. A comparison of the characteristics of three flow streams, that is: raw, primary effluent, and secondary effluent, is shown in **Table 23.1**.

Wastewater **temperature** is typically higher than tap water as water passes through various dwellings. The temperature of wastewater affects the biological reaction rates in the secondary treatment. As a rule of thumb, every 10°C increase in temperature doubles the rate of biological activity. Wastewater becomes less dense at higher temperature, which would affect the settling characteristics of solids. A sudden increase in temperature may be an indication of industrial discharge. On the other hand, intrusion of storm water can cause a sudden drop in wastewater temperature.

Turbidity is related to the presence of particulate matter in a flow stream. It is useful in assessing the quality of secondary effluents. The color of fresh sewage is gray and becomes darker as wastewater gets septic. Any other type of color again may be due to industrial discharges.

23.4   CHEMICAL CHARACTERISTICS

Normal municipal wastewater contains only 0.1% of solids. The **solids** in wastewater are classified as settleable and non-settleable (dissolved, colloidal, floatable). **Settleable** solids can be determined by running a one-hour settling test using an **Imhoff Cone**. This test on raw wastewater can be used to estimate the removal of solids during primary treatment. Typically, the measurements fall in the range of 10–20 mL/L.

23.4.1   Solids

The term **suspended solids** is usually applied to solids retained on a filter paper with opening size of 0.45 μm. They may be settleable or in suspension and can be removed by the process of settling or filtration.

Dissolved solids cannot be seen as they are in solution or in liquid form. The sum of suspended and dissolved solids is termed **total solids**.

Solids can also be categorized as **inorganic** and **organic**. Organic solids can be broken down by biological processes while inorganic solids cannot. Organic solids are measured using the volatile portion of the solids. A **volatile solids test** is where organics are burned off at 550°C, leaving only the inorganic or the inert material.

23.4.2 DISSOLVED GASES

In addition to dissolved oxygen, sewage water may contain other gases such as carbon dioxide, ammonia, and hydrogen sulphide because of decomposition as well as nitrogen dissolved from the atmosphere. The presence of gases like hydrogen sulphide can indicate the septicity of water. This is especially true for large cities with complex wastewater collection systems where sewage takes more time to reach the plant and develop anaerobic conditions.

23.4.3 ALKALINITY AND pH

Whereas alkalinity indicates the capacity of wastewater to neutralize acid, pH indicates the degree of acidity or alkalinity. Both properties define the chemical environment and hence rate of chemical and biological reactions. Raw Wastewater pH typically falls in the range of 6–8. With pH falling below 6, biological activity is almost nonexistent. When wastewater gets septic due to anaerobic conditions, it becomes darker and lowers the pH.

23.4.4 BIOCHEMICAL OXYGEN DEMAND

The strength of sewage is measured by determining the **biochemical oxygen demand (BOD)**. This parameter measures the quantity of oxygen utilized in the decomposition of organic matter in a sample of wastewater, over a specific period, and at a specific temperature. If not indicated, the BOD values refer to 5 days' incubation time at a temperature of 20°C. The most common tests performed for process control and regulatory requirements are the **BOD test** and the **suspended solids test**. BOD is an indirect measure of biodegradable organic matter. Since a BOD test takes a minimum of five days to complete, other tests like **chemical oxygen demand (COD)** are used for process control. Most common regulatory requirements require 85% BOD removal in secondary treatment plants and a monthly average of suspended solids in secondary effluent not exceeding 30 mg/L.

23.4.5 CHEMICAL OXYGEN DEMAND

Chemical oxygen demand is commonly used to describe the strength of industrial wastewater and refers to the total oxygen demand, so is usually greater than the BOD value. For a typical municipal wastewater, BOD is about 60% of COD. In some cases, **total organic carbon (TOC)** is used as a measure of strength of wastewater. As a rough estimate, BOD can be assumed to be 250% of the TOC values.

23.4.6 Nutrients

Nutrients, including nitrogen and phosphorous, are essential for the growth of organisms. A lack of nutrients may hinder their ability to properly decompose organic matter. On the other hand, nutrients can also cause havoc in the receiving streams by fertilizing algae and other aquatic weeds, which can lead to **eutrophication**. The term eutrophication refers to the premature aging of natural water systems. If adequate treatment is not obtained, large amounts of solids may settle on the bottom, covering spawning beds. These solids may also decrease sunlight penetration to plants due to the cloudiness created by suspended material. The organic material can also put a demand on the receiving waters' oxygen supply, which can kill fish through lack of oxygen. The principal nitrogenous compounds in domestic sewage are proteins, amines, amino acids, and urea. Phosphorus is contributed by food residue containing phosphorus and phosphorus-based synthetic detergents. As shown in **Table 23.2**, a secondary treatment plant is able to remove about 25% of the nitrogen and about 15% of the phosphorus content of wastewater.

Nitrogen and phosphorus are essential for the growth of organisms in the biological process, but the generally accepted mass ratio value of biochemical oxygen demand/ nitrogen/phosphorus required for biological treatment is 100/5/1. This indicates that nutrients in municipal wastewater exceed biological needs.

23.4.7 Toxins

Conventional wastewater treatment is not designed to remove heavy metals and pesticides and similar chemicals. On the other hand, these toxins can run havoc in biological treatments. Through sewer use control programs, every effort should be made to prevent discharges containing toxins. An upset of the biological process not only lowers removals, but it also takes a long time to bring it back to normal operating conditions.

23.5 BIOLOGICAL CHARACTERISTICS

Bacteriological testing is performed to determine the efficiency of the disinfection process. This is usually done by testing for indicator organisms like total coliform

TABLE 23.2
Nutrient removal

Flow Stream	N, mg/L				P. mg/L	
	NH$_4$	Organic	Nitrite	Nitrate	Total	Soluble
raw sewage	15–50	25–85	<0.1	<0.5	6–12	4–6
primary effluent	15–50	25–85	<0.1	<0.5	4–8	4–6
secondary effluent	0–1	5–20	<5.0	>10	3–6	2–5

and fecal coliform in a given flow stream. These results are usually reported as number of colonies/100 mL of the sample by performing a membrane filtration test.

In secondary treatment, a microscopic examination of the activated sludge can be used to study the health of the activated sludge process. An important part of this examination is to determine the distribution of various organisms such as protozoa and the presence of filamentous bacteria.

Example Problem 23.1

An influent wastewater stream has a suspended solids concentration of 250 mg/L and average BOD of 350 mg/L. The wastewater flow is 1.9 MGD. Determine SS and BOD loadings on the plant? Given that plant effluent on average contains 30 mg/L of BOD and 15 mg/L of suspended solids, what quantity of pollutants are discharged into the receiving water? Also find the percentage removal of BOD and SS.

Given:

$Q = 1.9$ mil gal/d, $BOD_i = 350$ mg/L, $SS_i = 250$ mg/l

$BOD_e = 30$ mg/L, $SS_e = 15$ mg/L

Solution:
Influent (raw) mass loading

$$M_{SS} = Q \times SS = \frac{1.9\ MG}{d} \times \frac{250\ mg}{L} \times \frac{8.34\ lb/MG}{mg/L} = 3961 = \underline{4000\ lb/d}$$

$$M_{BOD} = Q \times BOD = \frac{1.9\ MG}{d} \times \frac{350\ mg}{L} \times \frac{8.34\ lb/MG}{mg/L} = 5546 = \underline{5500\ lb/d}$$

Effluent mass loading rate

$$M_{SS} = Q \times SS = \frac{1.9\ MG}{d} \times \frac{15\ mg}{L} \times \frac{8.34\ lb/MG}{mg/L} = 237.6 = \underline{240\ lb/d}$$

$$M_{BOD} = Q \times BOD = \frac{1.9\ MG}{d} \times \frac{30\ mg}{L} \times \frac{8.34\ lb/MG}{mg/L} = 475.3 = \underline{480\ lb/d}$$

Percentage removal

$$PR(SS) = \frac{(3961 - 237)\ lb/d}{3961\ lb/d} \times 100\% = 94.01 = \underline{94\%}$$

$$PR(BOD) = \frac{(5546 - 475)\ lb/d}{5546\ lb/d} \times 100\% = 91.4 = \underline{91\%}$$

23.6 PERCENTAGE REMOVAL

The efficiency of a treatment process is its effectiveness in removing various contaminants from wastewater. **Removal efficiency** is defined as the fraction of the constituents removed from the incoming (influent) flow to the treatment process.

Percentage removal

$$PR = \frac{C_r}{C_i} = \frac{(C_i - C_e)}{C_i} \times 100\% \quad or \quad C_e = C_i \left(1 - \frac{PR}{100}\right)$$

Subscripts i and e refer to the influent and effluent flow streams respectively

Example Problem 23.2

A paper mill effluent with an SS content of 450 mg/L is fed to a primary clarifier. If the clarifier's effluent has a SS content of 50 mg/L, what is the SS removal efficiency?

Given:

$$SS_i = 450 \text{ mg/L} \quad SS_e = 50 \text{ mg/L} \quad PR =?$$

Solution:

$$PR = \frac{(SS_i - SS_e)}{SS_i} = \frac{(450 - 50)mg/L}{450 \; mg/L} \times 100\% = 88.8 = \underline{89\%}$$

Example Problem 23.3

Raw municipal wastewater entering a trickling filter plant contains 220 mg/L of BOD. If the average primary BOD removal is 35%, what is the minimum secondary removal required to produce plant effluent with BOD not exceeding 30 mg/L?

Given:

$$SS_i = 220 \text{ mg/L} \quad PR_I = 35\% \quad PR_{II} = ? \quad BOD_e = 30 \text{ mg/L}$$

Solution:

$$BOD_e = BOD_i \left(1 - \frac{PR}{100}\right) = \frac{220 \; mg}{L} \times \left(1 - \frac{35\%}{100\%}\right) = 143.0 = \underline{140 \; mg/L}$$

$$PR = \frac{(BOD_i - BOD_e)}{BOD_i} \times 100\% = \frac{(143 - 30)mg/L}{450 mg/L} \times 100\% = 0.790 = \underline{79\%}$$

23.7 INDUSTRIAL WASTEWATER

The characteristics of industrial waste depend on the industrial processes and raw materials used. Since characteristics vary widely from industry to industry, each industry must customize its pre-treatment system through chemical, biological, or physical processes.

Pre-treatment should only be used after all industrial plant controls like recycling, equipment changes, and process modifications have been applied. Modern industrial plant design dictates the segregation of wastes for individual pre-treatment, controlled mixing, or separate disposal. Municipal sewers use by-laws to control sewer uses or joint treatment agreements between industry and the municipal plants. Such agreements are called extra-strength surcharge (ESS) agreements and these charges are assessed for the BOD and SS over the average municipal limits. An extra-strength surcharge is calculated by determining any extra loading over and above the normal limit. The by-law limits for BOD and SS are typically 250 mg/L and 300 mg/L respectively.

$$ESS = R \times Q \times (C - C_L)$$

R = rate of extra-strength contaminant over the by-law limit;
C = maximum concentration of the pollutant, BOD or SS;
C_L = limit of the pollutant over which surcharge applies, Q − flow rate.

Industrial wastes containing pollutants, which cannot be handled by the municipal plant, are not allowed. As is the practice in developed countries, high-strength industrial waste is allowed provided the industry is ready to pay for the extra strength of the discharge and the plant has the capability to handle that waste.

23.7.1 EQUIVALENT POPULATION

For interpretation purposes, industrial waste strength is related to the number of people using the system. This is termed population equivalence and is often calculated for BOD and flow. The population equivalence is used to compare the strength of industrial waste with municipal waste. In North America, per capita daily flow is 450 L/c.d (120 gal/c.d) containing 200 mg/L of BOD and 240 mg/L of SS. BOD and SS loading in municipal wastewater are as follows:

$$BOD = \frac{450\,L}{c.d} \times \frac{200\,mg}{L} \times \frac{g}{1000\,mg} = 90\,g\,/\,c.d = 0.2\,lb\,/\,c.d$$

$$SS = \frac{450\,L}{c.d} \times \frac{240\,mg}{L} \times \frac{g}{1000\,mg} = 108 = 110\,g\,/\,c.d = 0.24\,lb\,/\,c.d$$

In a developing country like India, the hydraulic equivalent population is much less and is assumed to be 200 L/c. d. A BOD equivalent population is based on the BOD load of 45 g/c.d.

23.7.2 COMPOSITE CONCENTRATION

When industrial contribution is significant, it may seriously affect the plant loading. Since industrial discharges are usually of a high strength, even small flows can cause significant changes in BOD loading. The **composite concentration** of a municipal wastewater is the weighted average of the various discharges contributing to municipal flows.

Composite concentration

$$\bar{C} = \frac{\sum Q_i C_i}{\sum Q_i} = \frac{Q_1 \times C_1 + Q_2 \times C_1 + \dots}{Q_1 + Q_2 + \dots}$$

Example Problem 23.4

Municipal wastewater from a small community of 8200 people consists in addition to sanitary sewage, food processing wastewater flow of 120 m³/d containing 950 mg/L of BOD. A dairy flow of 550 m³/d with BOD concentration of 1200 mg/L also discharges in to the sanitary system. Assuming typical values for sanitary discharge, calculate the municipal wastewater strength and flow.

Solution:

Contributor	Q, m³/d	BOD, g/m³	BOD, kg/d
sanitary	3690	200	738
food processing	120	950	114
dairy	550	1200	660
Total	4360	2350	1512

Sanitary contribution

$$Q = \frac{450 L}{p.d} \times 8200 \, p \times \frac{m^3}{1000 L} = 3690 \, m^3 / d$$

$$M_{BOD} = \frac{3690 \, m^3}{d} \times \frac{200 \, g}{m^3} \times \frac{kg}{1000 \, g} = 738 \, kg / d$$

BOD loading from industrial discharges

$$M_{BOD} = \frac{120 \, m^3}{d} \times \frac{950 \, g}{m^3} \times \frac{kg}{1000 \, g} = 114 \, kg / d$$

$$M_{BOD} = \frac{550 \, m^3}{d} \times \frac{1200 \, g}{m^3} \times \frac{kg}{1000 \, g} = 660 \, kg / d$$

Composite concentration

$$\bar{C} = \frac{\sum Q_i C_i}{\sum Q_i} = \frac{1512 \ kg}{d} \times \frac{d}{4360 \ m^3} \times \frac{1000 \ g}{kg} = 346.7 = \underline{350 \ g / m^3}$$

23.8   INFILTRATION AND INFLOW

Infiltration is defined as runoff water and groundwater leaking into the sewer system through joints, porous walls, manholes, or breaks. Infiltration is a serious problem in old cities and towns due to poor joints and broken pipes. The extraneous flows that enter a sanitary sewer from sources other than infiltration, such as basement drains, roof leaders, and other illegal clean water connections are referred to as **inflow**.

The infiltration flows for a given facility can be estimated by comparing the **dry weather** flow (summer months) and **wet weather** flow hydrograph (spring month). The difference between the two is due to the excess water. Though infiltration is mostly experienced only during a high storm runoff period, the flow due to infiltration can be as much as twice or greater than that of the dry weather flow. The excessive flows create problems including washing out of tanks, poor treatment, high chemical costs, and flooding of streets and basements. Acceptable values of infiltration and inflow are about 10% of the average daily flow. However, in modern wastewater collection systems, due to better pipe joints and more stringent sewer use by-laws, infiltration and inflow are much less than 10%.

23.9   MUNICIPAL WASTEWATER

Municipal wastewater consisting of domestic waste, infiltration and inflow, and industrial discharges enters into the sanitary sewer system at one point or another. Collectors must be designed for specific uses whether the flows are domestic, industrial, infiltration, or a combination of all three. The design must be able to handle peak hourly flows. The population of the collectors as well as types of industrial waste and their characteristics can be used to calculate expected BOD and hydraulic loadings. In new systems, storm flows are carried by separate sewer systems called **storm sewers**. Though a collection system is designed based on the peak flow rates, wastewater treatment plants are designed based on the maximum daily average flow rates.

23.9.1   HYDRAULIC AND ORGANIC LOADING

Wastewater reaching a sewage treatment plant varies from hour to hour in terms of both quality (BOD, SS) and quantity (flow Q). A typical discharge pattern is shown in **Figure 23.2**. As seen in the **flow hydrograph**, the wastewater flows pattern is diurnal in nature.

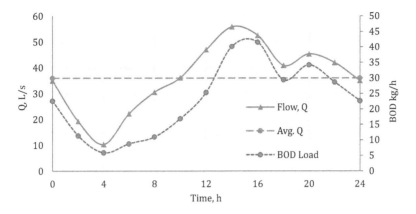

FIGURE 23.2   Flow and BOD pollutograph

23.9.2   MAIN POINTS

The following points should be noted when studying the flow variation at municipal sewage treatment plants.

- Hourly flow rates range from 20% to 250% of the daily average flow.
- In larger communities, peaks tend to level out, therefore hourly flow ranges from a minimum of 50% to a maximum of 200% of the daily average flow.
- Lowest flow occurs in the early morning hours around 3 a.m. to 4 a.m. and maximum flow occurs at midday. In some communities, a smaller second peak flow occurs in the early evening hours.
- Strength of peak wastewater in terms of BOD follows a similar pattern but not exactly the same pattern as maximum flow.
- Summer flows frequently exceed winter flows.
- The flow hydrographs with and without significant infiltration and inflow are respectively called wet weather and dry weather flow hydrographs.
- Large volumes of industrial wastes and or infiltration and inflow can distort the shape of a **flow hydrograph**.
- Knowledge of hydraulic and organic (BOD) loading are essential in evaluating the operation of a treatment plant.
- The difference between the wet weather and dry weather hydrographs represents the infiltration and inflow contribution.
- A flow hydrograph can be converted to a **pollutograph** by knowing the concentration of the pollutant.

23.10   EVALUATION OF WASTEWATER

For an accurate evaluation of the wastewater to be treated, it is important that the sampling techniques are suited to the task at hand. The samples must also provide an accurate representation of the characteristics to be analyzed. The results from sample testing are only as accurate as the techniques used and the quality of the sample. Samples can be collected manually or using **automatic samplers**. Programmable automatic sampling is becoming very common. Manual sampling is usually done for spot-checking and similar uses.

Irrespective of the technique used, a sample must be representative. Sampling locations should be selected where the flow stream is smooth and well mixed. Scoping the sample from the surface or from the bottom would produce a biased sample. There are two basic types of samples: grab and composite.

A **grab sample** is a sample taken at a given place and time. Discrete samples are usually taken for measurements of parameters that are constantly changing or unstable, such as chlorine residuals and dissolved oxygen. Grab samples are also used on parameters that do not change significantly, such as sludge and mixed liquor suspended solids (MLSS).

A **composite sample** is a combination of individual grab samples. A simple composite sample is prepared by combining equal volumes of individual samples.

A **flow propositional sample** is made of individual samples of volume proportional to flow at the time of collection. Composite samples are required for a representative analysis where there is variation in the flow and/or concentration. A simple composite sample is acceptable when variations are less than 15%.

23.10.1   AUTOMATIC COMPOSITING

Flow proportional composite samples are slightly more difficult. A sampling device can be connected to a flow meter, which can increase or decrease the number of samples, or the sample volume size, with an increase or decrease in flow.

23.10.2   MANUAL COMPOSITING

To determine aliquot volume for flow proportional samples, hourly readings of the flow are required. The volume of aliquot required from a given hourly sample is taken in proportion to flow in that period.

Volume of aliquot

$$A_i = Factor \times Q_i \quad and \quad Factor = \frac{V_C}{\Sigma Q_i} = \frac{V_C}{\bar{Q} \times N}$$

V_C = volume of the composite sample   N = number of individual samples
Q = flow in the ith discrete sample ΣQ = total flow

Example Problem 23.5

A 2.0 L composite sample is to be prepared manually. An automatic sampler was programmed to collect individual samples every hour. Flow rate information for each hour is given below.

time t, h	1	2	3	4	5	6	7	8
flow Q, ML /d	11	12	15	22	20	15	12	10

Solution:

$$Factor, F = \frac{V_C}{\sum Q_i} = \frac{2.0\,L}{117\,ML/d} \times \frac{1000\,mL}{L} = 17.0 = \underline{17\,mL/MLD}$$

Computations for the entire period of sampling is presented in **Table 23.3**.

Sample of calculations (noon hour):

$$A_{12} = F \times Q_{12} = \frac{17\,mL}{ML/d} \times 20\,ML/d = 340.0 = \underline{340\,mL}$$

23.10.3 SAMPLE LOCATIONS

Accurate sampling and testing are important in plant operation and control. In addition, regulatory samples are collected for compliance. Most common and standard sampling locations are shown in **Figure 23.3**.

TABLE 23.3
Table of computations

#	Clock hour	Q, ML/d	A$_i$, mL
1	8:00	11	190
2	9:00	12	200
3	10:00	15	260
4	11:00	22	370
5	12:00	20	340
6	13:00	15	260
7	14:00	12	200
8	15:00	10	170
	Σ	117	1990

FIGURE 23.3 Standard sampling locations

Key Points

1. Raw sewage is sometimes taken after the screening if plugging of the sampler is a problem. This composite sample is used to determine the loading to the plant.
2. The second sample location is located after the primary clarifier. This composite sample is used to measure the efficiency of the primary clarifier as well as the loading to the aeration section.
3. The third clarification sampling location measures the mixed liquor solids or MLSS. This grab sample is used in the 30-minute settling test, suspended solids concentration test, oxygen uptake test, and microscopic survey test. These tests, along with visual observations, are used to operate and control the biological treatment process.
4. The fourth sample location is preferably taken before disinfection but after the secondary clarifier. This composite sample is used to determine the efficiency of the plant operation and the load on the receiving stream of water.
5. The fifth location uses a grab sample to monitor the chlorine residual.
6. The sixth sample location is for sampling the return sludge. A grab sample is required to determine the solids concentration in the activated sludge. Return sludge solids is compared with mixed liquor solids and concentration factor is determined.

Discussion Questions

1. What are the common tests carried out in the laboratory of a wastewater treatment facility? Explain their significance.
2. Distinguish between volatile and non-volatile solids. What is the importance of volatile solids in wastewater treatment?
3. Explain the relationship between BOD and COD. Except COD, name another parameter that can be used to estimate BOD.
4. Describe the test used for determining the settleability of solids?
5. Give the characteristics and composition of raw sewage.
6. Name the most important parameters used to characterize sewage and describe their significance.
7. State the general requirements to be observed in ensuring correct sampling.
8. Explain the term "population equivalent". In what cases will BOD population equivalent not be equal to hydraulic population equivalent?
9. What are the advantages of automatic sampling?
10. Distinguish grab, simple composite, and flow proportioned composite types of samples and describe their suitability.
11. As part of wastewater plant operation and control, what are the standard sampling locations in a secondary treatment plant?
12. Sketch a typical wastewater flow hydrograph and explain diurnal variation.
13. Search the internet and find the typical characteristics of raw sewage in India and North America.

14. Compare raw sewage with secondary effluent in terms of temperature, odor, color, and turbidity.
15. Explain why excessive infiltration and inflows are undesirable.
16. What type of industrial wastewaters should be allowed to discharge into a municipal sewer system?
17. In developed countries, industrial dischargers with strong waste are surcharged. What are the typical limits of SS and BOD above which fees are applicable?
18. Define the term nutrients and their role in biological treatment. Does typical municipal wastewater have sufficient or excessive nutrients?

Practice Problems

1. A large dairy discharges 0.50 MGD with BOD content of 850 mg/L. Determine equivalent BOD population @ 0.2 lb/c·d? (18 000 p).
2. The average wastewater flow from a city is 80 ML/d with BOD content of 280 mg/L. what is the BOD equivalent population of @ 80 g/c.d? (280 000 p).
3. Based on the suspended solids removal efficiency of 40% of the primary clarifier, determine the expected concentration of SS in the primary effluent. It is known that suspended solid in raw wastewater is 235 mg/L. What must be the minimum removal if it is desired to achieve a primary effluent concentration not exceeding 150 mg/L? (140 mg/L, 36%).
4. A secondary plant processes an effluent with SS concentration of 10 mg/L. The average concentration of SS in the raw wastewater composite sample is 195 mg/L. What is the SS removal efficiency of the plant? Assuming 85% to be the solids removal efficiency of the secondary unit, determine the removal efficiency of the primary unit? (95%, 66%).
5. A secondary treatment receives domestic flow of 35 ML/d with BOD content of 220 mg/L. Food-processing and dairy industries contribute 8.0 ML/d of wastewater with BOD load of 7800 kg. What is the BOD of the municipal wastewater? (360 mg/L).
6. If the concentration of suspended solids and BOD is 100 g and 110 g per capita per day, estimate the population equivalent of 50 kL of industrial wastewater containing 1000 mg/l of suspended solids and 3300 mg/L of BOD. (550 p, 1500 p).
7. A paper mill plant effluent contains 50 mg/L of SS. If on average the mill discharges 5000 m³/day of water into the river, calculate SS loading into the receiving stream. If the legislative requirement limits the loading to 75 kg/d, what is the maximum content of SS allowed in the effluent? (250 kg/d,15 mg/L).
8. A dairy discharges effluent into the city sewer system at the rate of 6.5 ML/d carrying BOD of 880 mg/L. What is the equivalent population based on BOD contribution of 75 g/c·d? If the daily production of wastewater is

55 ML/d with BOD content of 250 mg/L, what would be the average BOD of the wastewater reaching the municipal plant? (76 000 p, 320 mg/L).

9. For the data of Example Problem 23.5, find the sample aliquot for the 14th and 15th hour to make a 2.5 L composite sample. (250 mL, 210 mL).

10. As shown in **Table 23.4**, hourly flow readings and samples of raw wastewater entering the plant were collected.
   a. What percentage of the average flow is low flow and peak flow? (34%, 170%).
   b. What sample aliquot is required for the low and peak flow hour to make a 3.0 L flow proportioned composite sample? (42 mL, 220 mL).

11. Suspended solids analyses were conducted on a 25 mL sample of mixed liquor in triplicate. The average values of the three trials are as follows:
   filter disc = 0.2165 g filter + dry solids = 0.2695 g
   filter + ignited solids = 0.2285 g
   a. What is the concentration of suspended solids, MLSS? (2120 mg/L).
   b. What percentage of solids is volatile, VF? (77%).

12. For the data of Practice Problem 11, a sample of mixed liquor was obtained from an activated sludge plant that has total aeration tank capacity of 0.35 MG.
   a. What mass of dry solids is kept in the aeration tanks? (6200 lb).
   b. If volatile solids represent biomass, find biomass (4800 lb).
   c. What is the mass of inert solids in the aeration? (1400 lb).

13. Certificate of approval of a wastewater treatment facility limits the phosphorus loading to 16 lb/d. Average daily flow of this plant is 2.2 MGD.
   a. What is the maximum allowable concentration of phosphorus in the plant effluent? (0.87 mg/L).
   b. To be on the safer side, an operating engineer decides a target of 0.50 mg/L. If the raw wastewater contains 7.0 mg/L of phosphorus and primary removal is only 32%, what is the required removal by the secondary treatment? (89%).

TABLE 23.4
Flow hydrograph (Practice Problem 10)

Time	MGD	Time	MGD	Time	MGD
08:00	18.5	16:00	17.5	24:00	13.1
09:00	21.8	17:00	16.0	01:00	10.2
10:00	24.7	18:00	15.8	02:00	8.2
11:00	26.9	19:00	16.7	03:00	7.3
12:00	28.1	20:00	17.5	04:00	5.8
13:00	25.5	21:00	16.7	05:00	5.5
14:00	23.3	22:00	16.0	06:00	7.5
15:00	20.4	23:00	14.6	07:00	11.0

14. It is intended to make a 2.5 L flow proportioned composite sample during the 8 h period of peak loading. Calculate the portions to be used from a noon and 4 p.m. grab sample based on the flow rate data given below. (350 mL, 280 mL).

#	1	2	3	4	5	6	7	8
hour	9	10	11	12	13	14	15	16
flow, L/s	90	104	113	116	112	106	100	95

15. Municipal wastewater from a community of 9500 people is made up of sanitary wastewater and industrial waste from a meat processing and a milk processing plant. Domestic waste is 100 gal/c.d containing BOD of 200 mg/L. Meat processing waste of 55 000 gal contributes 750 lb of BOD and the milk processing plant contributes 25 000 gal of wastewater containing 1000 mg/L of BOD. What is composite BOD of the municipal wastewater? (300 mg/L).

24 Primary Treatment

Depending on the stages of treatment, wastewater treatment plants are classified as primary, secondary, and tertiary. Whereas, primary treatment includes physical and mechanical processes, secondary treatment plants also provide biological treatment. In earlier days, primary treatment, which typically removed 30% to 40% of BOD, was acceptable. However, secondary treatment is the norm these days. Primary treatment plants are not built any more, and existing plants are being upgraded to meet regulatory requirements. In towns and cities where the receiving water body is more sensitive, tertiary treatment in the form of a combination of physical, chemical, and biological treatment is necessary.

Irrespective of the number of sages of treatment, wastewater goes through **preliminary treatment** to make it ready for next level of treatment. Preliminary treatment occurs at the head end of the plant and removes materials that might impair or harm headworks or the operation of downstream processes. These materials usually consist of wood, rags, plastic, and grit. The term **grit** refers to inorganic material like sand that is nonbiodegradable.

24.1 PRELIMINARY TREATMENT

Preliminary treatment devices are designed to remove or reduce large solids, grease, scum, and grit before any further treatment of sewage. The removal of these materials protects pumps and other treatment devices from possible damage. If the preliminary treatment devices do not function as intended, maintenance costs for pump repairs, digesters, and clarifier clean-outs, etc., will increase. In addition, the effective capacity of treatment would be reduced due to the space occupied by the inert material. This would result in lowering the hydraulic efficiency. If grit and other hard and heavy stuff are not removed in the earlier stages of treatment, it can pose serious problems.

  i. Wear out pumps, pipes, and other equipment faster.
 ii. Mix with sewage sludge and interfere with its digestion.
iii. Adulterate the sludge and reduce its manurial and fuel value.
 iv. Occupy large volumes in digestion and air tanks and reduce their effective capacity and possibly block passages in sludge pipes.

To remove pieces of wood, rags, and inorganic material like grit and sand, treatment devices usually associated with preliminary treatment are briefly discussed.

24.1.1 SCREENS

Screens are used to remove debris such as rocks, cans, rags, rubber goods, toys, bits of wood, etc. If this material is not screened, it may damage equipment, interfere with the process, and are aesthetically undesirable in effluent. So this is the first step in the treatment of wastewater. Two basic types available are coarse screens and fine screens.

24.1.2 COARSE SCREENS

Coarse screens, commonly called trash rack or **bar screens**, generally have bars spaced from 2 cm to 15 cm (1 in-6 in). The screens are usually installed at an angle to facilitate manual cleaning, but some units are available that can be mechanically cleaned. Trash racks are normally installed at the pumping stations. In most plants, bar screens with a spacing of 50 mm (2-in) are used.

24.1.3 FINE SCREENS

Fine screens were originally used in place of sedimentation tanks. They are not commonly used in sewage treatment because the mesh will accumulate material and plug very quickly, causing what is called a head loss in the system. There are other operating and economic problems as well.

24.1.4 MECHANICALLY CLEANED SCREENS

A **mechanical rake** cleans vertical or inclined bar screens. The accumulated material on the screen is pulled up the screen and wiped off into a hopper. Screenings are regularly removed from the hopper to prevent nuisance odors and to ensure adequate capacity for incoming screenings.

Cleaning Screens

During dry weather periods, coarse trash racks should be cleaned daily. During storm periods, they should be cleaned two to five times per day to maintain a free flow of sewage through the process. Failure to clean the screens can result in septic action upstream of the sewer, blockage of the sewer upstream, surcharge of the sewers, and shock loads on sewage units when the screens are finally cleaned. When mechanically cleaned, coarse screens reduces labor cost and improve flow conditions.

24.1.5 VOLUME OF SCREENINGS

The quantity of **screenings** will depend on the characteristics of the wastewater and of the screen openings. It is obvious to expect more screenings during storm periods and at the spring runoff. The production of screenings will vary from plant to plant. The volume of material or screenings removed is difficult to estimate accurately. Experience has shown that screens with openings of 25 mm to 50 mm (1–2 in) will collect between 10 L and 100 L of screenings per million liters, (ML) of sewage.

24.1.6 DISPOSAL OF SCREENINGS

The screenings may be disposed of by burial, incineration, grinding, or digestion. Burying and incinerating are the usual methods of disposal because they are the most economical. Most municipalities use one of these methods for disposal. Screenings are removed in covered containers. When burying screening, odor may be prevented by sprinkling powdered lime or other odor-control chemicals on the material. An earth cover of 30–60 cm (1–2 ft) will usually give the best results for bacterial activity. Grinding devices have been used in the past. The ground screenings are redirected to the influent flow for treatment in the process. This method has proven unsatisfactory, however, as it may create digester problems. Screenings received from grinders have caused digester foaming and excessive scum blankets.

24.1.7 Flow through Screens

Excessive head loss will cause back up flow and poor removal. To minimize head loss as wastewater passes through screens, it is recommended to keep the screen flow velocity < 1.0 m/s, typically 0.80 m/s (2.5 ft/s). High head losses will allow the screenings to pass through. Head loss is usually not allowed to exceed 15 cm (6 inches). In modern plants, head loss is monitored and controlled using a programmable logic controller (PLC).

24.2 COMMINUTION OF SEWAGE

Comminutors, barminutors, or rotogrators are trade names used by different manufacturers to identify their shredding devices. By shredding rather than removing the screenings, material is put back in the wastewater stream. It may seem attractive at first sight, but keep in mind that the material has become part of the wastewater and can interfere with other processes following screens. One of the main problems associated with comminutors is high maintenance. This is logical as the main part is the cutting teeth, which will wear out very quickly. In plants today, the trend is more towards staying away from comminutors.

A **barminutor** is a combination of bar screen and a rotating drum with teeth. The rotating drum travels up and down the bar screen. This piece of equipment is used to shred and grind material small enough to pass through the screens of the grinding unit. Shredders should be installed with a by-pass equipped with a bar screen to facilitate removal of settled material and allow inspection of the equipment components such as the cutting edges. **Comminuting** devices are normally operated continuously and are usually located ahead of the grit removal units.

24.3 FLOW MEASUREMENT

Even though flow measurement is not a wastewater treatment process, its importance can be overemphasized. This information is required to adjust chemical feed rates, frequency of sludge withdrawal, recirculation rates, airflow rates, plant operation, plant capacity, and future planning. Some of the common wastewater flow-measuring devices are flumes and weirs.

24.3.1 PARSHALL FLUME

The best equipment for flow measurement in wastewater treatment plants is a **Parshall flume** equipped with an automatic flow recorder and **totalizer**. Flow moving freely through the flume can be calculated by measuring flow depth upstream.

A **stilling** well is usually provided to hold a float, bubbler tube, and other depth sensors like ultrasonic, which is connected to a transmitter and flow recorder. A Parshall flume is basically a venturi meter of open channel flow since it has a converging section, throat, and diverging section. A Parshall flume is preferred since it is self-cleaning and causes minimum head loss.

The size of the Parshall flume is indicated by the width of the throat section, and the flow equation is written in terms of width of throat and head upstream.

24.3.2 PALMER-BOWLUS FLUME

A **Palmer-Bowlus** flume is another type of flume, which is more commonly employed to measure flow in sewer pipes since it can fit into half-sewer sections. Palmer-Bowlus flumes are not generally available in standard sizes.

24.3.3 WEIRS

Weirs or **notches** can also be employed for flow measurement. Their most common shapes are rectangular, triangular (V-notch), and trapezoidal (Cipolletti). Weirs are not preferred for the measurement of raw wastewater since they are prone to get solids built up behind the weir and that affects the flow profile. However, in some cases, weirs can be used to measure plant effluent flows, which are required by regulation in many cases. A triangular or V-notch weir is more accurate for measuring small flows. For standard sharp-crested triangular weirs, the general flow formula is as follows:

$$Q(V\ notch) = \frac{8}{15} \times C_d \times tan\left(\frac{\theta}{2}\right) \times \sqrt{2gH^5}$$

where θ is the notch angle and C_d is the coefficient of discharge. H is the head over the weir and g is acceleration due to gravity. Minimum value of discharge coefficient is 0.58 and typical value is 0.60.

24.4 GRIT-REMOVAL UNITS

Grit such as sand, stones, and gravel may find its way into a sewer system and be carried by the sewage to the treatment plant. Grit material

- is inorganic in nature, inert, or non-biodegradable;
- much denser (SG = 2.3) than organics (SG = 1.1–1.2), hence quick settling; and
- when combined with other material, like grease, can solidify.

Grit-removal units are installed after screening equipment in the process to protect mechanical equipment from abrasion, avoid pipe-clogging, and reduce the
sedimentation load on the primary clarifier and sludge digesters. Grit settling in aeration tanks and digesters occupies dead space and reduces their effective capacity.
Depending on the characteristics of wastewater and its operation, some organic matter will always settle out with grit. In some cases, grit is washed to further remove
the organic matter. Due to the presence of organic matter, grit containers and their
disposal are similar to screenings.

24.4.1 SETTLING VELOCITY

Being inorganic material, sedimentation grit follows discrete settling. Since particles
are usually >0.2 mm, transitional flow conditions prevail and **Stokes' Law** becomes
inapplicable. For particles falling in the range of 0.1–1.0 mm, as in the case of grit
particles, settling velocity is given by Hazen's equation.

Settling velocity of a discrete particle

$$v_s = 418D^2(G_s - 1)\left[\frac{3T + 70}{100}\right] = 61D(G_s - 1)\left[\frac{3T + 70}{100}\right]$$

D = diameter of the particle in m, G_s= SG of grit material, T= temperature in °C

In theory, overflow rate should be set equal to settling velocity. Since there is
always some short-circuiting, actual overflow rate is chosen lower than the settling
velocity. Actual settling velocity will be less than as given by these equations since
the real particles are not spherical in shape. In fact, actual velocity is many times
less.

24.4.2 GRIT CHANNELS

Grit particles will settle faster than organic **putrescible** solids because they are
heavier. Grit channels are usually designed to maintain flow velocity of 0.3 m/s
(1 ft/s) at design flow which is usually sufficient to keep the organic matter in suspension while allowing the heavier particles to settle. Grit channels are usually
rectangular and velocity control is achieved by installing a **proportional weir** at
the effluent end of the channel. In other cases, velocity control can be achieved by
changing the shape of the channel. However, in such cases, a flow control device like
a Parshall flume is installed at the end. There is usually more than one channel to
accommodate peak flows and for cleaning purposes.

24.4.3 AERATED GRIT CHAMBER

Grit chambers using air to separate the lighter materials from the heavier ones
are called aerated grit chambers. Sewage flows into the aerated grit chamber and
the heavier particles settle to the bottom as the sewage rolls in spiral motion from

entrance to exit. The lighter organic particles eventually roll out of the tank. The grit at the bottom of the tank is directed to a grit hopper where it is removed by a clam shell bucket or an air lift unit.

The **aerated** grit chambers are designed to provide 3–5 min of detention time based on peak flow. Aeration keeps the organics in suspension and helps to freshen up the wastewater. The velocity of the rolling action in the chamber dictates the removal efficiency. If too slow, organics will settle out. If the rolling velocity is properly adjusted, the settled material is usually very clean of organics and does not need any washing.

Example Problem 24.1

A grit chamber is to be designed to handle peak flow of 65 ML/d. Assume grit particle size of 0.20 mm and relative density of 2.5. Assume temperature of 20°C.

Given:

$$Q = 65 \text{ ML/d} = 65\,000 \text{ m}^3/\text{d} \quad D = 0.20 \text{ mm} \quad G_s = 2.5 \quad T = 20°C$$

Solution:

$$v_s = 61D(G_s - 1)\left[\frac{3T + 70}{100}\right] = 61 \times 0.0002\,m \times 1.5 \times \left[\frac{60 + 70}{100}\right] = \underline{0.029 \ m/s}$$

Since the grit chamber is not 100% efficient, overflow rate is chosen as 60% of the settling velocity.

$$A_s = \frac{Q}{v_o} = \frac{65\,m^3}{d} \times \frac{s}{0.60 \times 0.029\,m} \times \frac{d}{24\,h} \times \frac{h}{3600\,s} = 43.2 = \underline{43\,m^2}$$

Select two channels for flexibility. Assuming an effective depth of 1.0 m and maximum flow velocity of 0.30 m/s, desired length of grit channel can be found.

$$A_X = \frac{Q}{2 \times v_H} = \frac{65000\,m^3}{d} \times \frac{s}{2 \times 0.30\,m} \times \frac{d}{24 \times 3600\,s} = \underline{1.25\,m^2}$$

$$B = \frac{A_X}{d} = \frac{1.25\,m^2}{1.0\,m} = 1.25 = \underline{1.3\,m}$$

$$L = \frac{A_S}{W} = \frac{43\,m^2}{2 \times 1.3\,m} = 16.53 = \underline{17\,m}$$

$$H = 1.0\,m + 0.25\,m\,(f.b.) + 0.25\,m\,(storage) = \underline{1.5\,m}$$

In addition to grit channels, aerated grit chambers grit-removal devices include detritus tanks, and centrifugal separators. A brief description of these and similar devices follows.

24.4.4 Detritus Tank

Short-period sedimentation in a tank that operates at substantially constant levels produces a mixture of grit and organic solids called **detritus**. A detritus tank is basically a grit clarifier. The lighter organic solids are subsequently removed from or washed out of the mixture. Several manufacturers specializing in sewage disposal equipment have perfected this type of equipment. For example, one such unit not only removes the grit but also washes it.

24.4.5 Cyclone Separators

Grit removal is possible by mechanical means such as a centrifugal unit. Centrifugal units are usually liquid cyclones. The wastewater is introduced tangentially into a cylindrical conical housing. The heavier, larger particles of grit are thrown to the outside wall and collected for disposal. Cyclone separators are very compact and are inexpensive compared to channels and grit chambers. One of the main disadvantages is the flexibility of operation. It is very difficult to regulate the quality of the grit which is removed by the unit.

24.4.6 Grit Disposal

Average figures indicate that from 20 L(gal) to 50 L(gal) of grit can be collected per million liters (MG) of sewage. The grit must be removed before it is carried by the stream flow into the primary clarifier, digester, or chlorine contact chamber. It is good practice to periodically check that the grit is not being carried to the clarifier, or digester, or chlorine contact chamber, where it would still have to be removed but with much more difficulty and expense.

The grit-removal facility can reduce unnecessary maintenance costs more than any other unit. If these facilities malfunction because of problems or improper operation, the result will be plugged lines, abraded impellers, and grit-filled treatment tanks. The disposal of grit is usually done by burial or dumping at the municipal dump or in a landfill site. If the grit is adequately washed (having less than 3% volatile solids remaining as determined by lab tests) it may be used as fill around the plant or may be used to re-sand sludge drying beds.

24.5 PRE-AERATION

Aeration basins may precede or follow screens and grit chambers. In general, **pre-aeration** tanks are designed for detentions of 5 min to 15 min using 0.1–1.0 unit of air per 1000 units of sewage treated. If flocculation of the fine suspended solids in the raw sewage is also attempted, the detention period must usually be extended from at least 15 min to as much as 60 min, the average time being about 30 min. Pre-aeration of the raw wastewater is becoming common, especially in cities and towns where the sewage gets septic before it reaches the plant. Raw sewage is aerated for one or more of the following purposes.

1. To remove gases from the sewage, especially hydrogen sulphide, which creates odor problems and increases the chlorine demand of sewage.
2. To promote floatation of excessive grease, which then can be removed from the raw sewage at an early stage in its treatment.
3. To aid in the coagulation of the colloids (finely divided suspended solids) in the raw sewage for the purpose of obtaining a higher removal by primary settling.
4. To freshen the wastewater and elimination of odors.
5. Pre-aeration is more economical in an activated sludge process since the aeration system is in place for the activated sludge process.
6. Addition of dissolved oxygen makes the water ready for further processing.

24.6   PROCESS CALCULATIONS

In addition to detention time and overflow rate, horizontal flow velocity calculations are made to control the operation of grit removal units especially grit channels. The relationship between settling velocity and flow velocity can be used to determine what size particles will be removed. Referring to **Figure 24.1**, the following relationships will apply.

Hydraulic detention time

$$t_d = \frac{L}{v_H} = \frac{d}{v_o} = \frac{V}{Q} \qquad v_H = \frac{Q}{A_X} = \frac{L}{t_d}$$

v_H = horizontal flow velocity v_o = overflow rate
L = length of the chamber/channel, B = width, d = depth,
H = height = d + f.b. + storage,
A_x = B × d = area of cross section, A_S = L × B = surface area

Example Problem 24.2

A grit chamber is designed to remove grit material with an average particle size of 0.20 mm with settling velocity ranging from 3–4 ft/min. By employing a

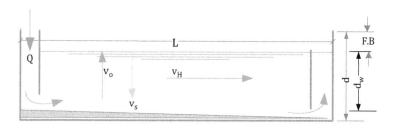

FIGURE 24.1   Flow through a grit channel

proportional weir at the exit end, a flow velocity of 1.0 ft/s will be maintained. Determine the dimensions of the grit channel to carry a peak flow of 2.5 MGD.

Given:

$Q = 2.5$ MG/d $d = 3.5$ ft assumed $v_s = 3.6$ ft/min $= 0.06$ ft/s $v_H = 1.0$ ft/s

Solution:

Chamber section

$$A_x = \frac{Q}{v_H} = \frac{2.5 \times 10^6 \ gal}{d} \times \frac{s}{1.0 \ ft} \times \frac{ft^3}{7.48 \ gal} \times \frac{d}{24 \times 3600 \ s} = 3.868 \ ft^2$$

$$B = \frac{A_x}{d} = \frac{3.868 \ ft^2}{3.5 \ ft} = 1.1 \ ft$$

$$t_d = \frac{H}{v_s} = 3.5 \ ft \times \frac{s}{0.06 \ ft} = 58.3 = \underline{58 \ s}$$

$$L = v_H \times t_d = \frac{1.0 \ ft}{s} \times 58.3 \ s = 58.3 = \underline{60 \ ft}$$

Hence a grit channel 60 ft × 1.1 ft × 3.5 ft will do the job.

24.7   PRIMARY CLARIFICATION

Settling in primary clarifiers is type II settling, as in case sedimentation of coagulated water. The main purpose of **primary clarification** is to separate and remove **settleable** and floatable solids and allow thickening of sludge. The principle of primary settling is to slow the flow of sewage in the clarifier to very close to below 10 mm/s (2 ft/min) to permit solids to settle. Settling under gravity can result in 40% to 60% of the suspended solids and 30% to 40% of BOD being removed. Chemical precipitation with addition of alum or other coagulants can remove much of the colloidal material, bringing the suspended solids removed to 80% or 90%. The basic components of sedimentation tanks include **baffles** to slow down the influent flow while overflow weirs even out the effluent flow. Mechanical skimmers collect scum from the surface and deposit it in scum pits. Collector arms or skimmers then scrape the settled sludge into a hopper.

24.8   CIRCULAR CLARIFIER

The two main designs of primary clarifiers are rectangular and circular. A schematic of a circular clarifier is shown in **Figure 24.2**. Circular tanks can be either center-fed or peripherally fed. In general, sludge collection mechanisms in circular tanks are operated over longer periods than collectors in rectangular tanks. Collectors should be run often enough to prevent a build-up of solids in the tank from causing an undue

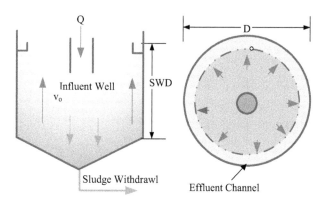

FIGURE 24.2   Circular settling tank

TABLE 24.1
Main components of a circular clarifier

Component	Function
Tank structure	A cylindrical tank with a conical bottom. The bottom slope is about 8%. In the center at the bottom is the hopper.
Influent Line	Usually runs under the tank and rises vertically in the center of the tank to feed the tank. Again, there is always some sort of flow control like gates or a valve.
Influent well	A cylindrical baffle forces the water to flow down and then in the radial direction towards the effluent weir. Its main function is distributing the water evenly and to prevent short circuiting. Small ports near the surface allow floating material to pass through.
Sludge collector	A continuously moving scraper arm at the bottom scrapes the sludge towards the center and into hopper. A motor and reduction gear at the center usually drive the sludge collector.
Access bridge	Bridge over the weir is to provide access to the center for maintenance and operation.
Skimmer arms	Skimmer arms rotating at the top surface move the floating material to the periphery of the tank. Scum is collected in a trough.
Sludge line	A line fitted with a valve is provided to pump the sludge out.
Effluent weir	Weirs in circular clarifiers are along the periphery of the clarifier. Weirs can be adjusted to maintain even flow.
Launder	This is the effluent channel trough into which weirs overflow. Flow from the effluent channel goes to the next process.

load on the mechanism at start-up and damaging the equipment. The main components of a circular clarifier are listed in **Table 24.1**.

24.9   RECTANGULAR CLARIFIERS

Rectangular clarifiers are more commonly used for primary treatment. In the past, a sludge-collection mechanism was a chain and sprocket arrangement, which required

TABLE 24.2

Components of a rectangular clarifier

Component	Function
Tank structure	Usually made of concrete, the floor is horizontal or slopes towards the head end to allow the collection of sludge in the hopper at the head end side.
Influent end	Pipes or channels to distribute water evenly along the shorter side of the clarifier as the flow is along the length. There is always some sort of flow control like gates, valves, or weirs.
Inlet baffle	Baffle plays an important role to distribute the water evenly and prevent short-circuiting. Small ports near the surface allow floating material to pass through.
Sludge collector	Collectors operating in rectangular tanks consist of two endless chains operating on sprocket wheels and supporting wood crossbars or flights. The flights push the sludge to a hopper at the end as they move slowly along the bottom. The system is designed so that flights move across the surface on the return trip and move floating solids to the inlet end. In modern plants, sludge is moved to the hopper by a scraper attached to a moving bridge on rails.
Sludge hopper	Receives sludge scrapped from the floor and is located at the head end. To provide room for sludge, head end is deeper.
Scum trough	Scum is collected in this trough, and it extends along the width of the tank.
Sludge line	A line fitted with a valve is provided to pump the sludge out.
Effluent weir	Effluent from the clarifier flows over the weirs into the effluent channel. The role of the weir is to even out the flow and to reduce flow velocity so as to prevent the carry-over of the solids.
Launder	This is the effluent trough into which the weirs overflow.

high maintenance. These days, a travelling bridge is more common. Sludge is pushed towards the head end and clarified water flows over the weirs at the exit end. The main components of the rectangular clarifier are described in **Table 24.2**.

24.10   SCUM REMOVAL

Foreign matter that rises to the surface forms scum. It should be pumped out of the tank before pumping the sludge, if possible. By doing this, any grease remaining in the pipes will be scoured by the sludge when it is removed. Removal of scum, floating garbage and grease are essential for the efficient operation of settling tanks. A scum barrier or **baffle** is generally provided in the flow path between the center of the tank and the effluent weir. Scum must be removed daily, and, ideally, small amounts should be removed continually rather than a large batch at one time.

24.11   FACTORS AFFECTING SETTLING

There are many factors which will influence settling characteristics in a particular clarifier. The most important ones include temperature, short-circuiting, detention time, surface loading (overflow rate), solids loading, and weir loading. Brief descriptions follow.

24.11.1 TEMPERATURE

In general, settling improves with an increase in temperature as water becomes less dense. Cold water causes poor settling due to the increase in viscosity and the density of water. If possible, provide additional detention time to compensate for a low settling rate. Due to low temperatures in winter, wastewater and sludge could be held for longer without going septic.

24.11.2 SHORT-CIRCUITING

When the flow velocities are non-uniform, short-circuiting occurs and reduces effective detention time. Short-circuiting may easily begin at the inlet end of the settling tank. This is usually prevented using weir plates, baffles, and proper design of the inlet channel. Short-circuiting may also be caused by density currents or stratification due to temperature or density. When warm influent flows across the top of cold water in a clarifier, short-circuiting will occur.

24.11.3 SETTLING CHARACTERISTICS OF SOLIDS

In municipalities where there is significant industrial discharge contributing to municipal wastewater, the concentration and nature of the solids may be quite different. If the industrial effluents contain heavy solids it may help to improve the efficiency of primary clarification. **Chemical precipitation** of phosphorus as part of primary treatment can also enhance the removal of solids. However, this results in increased sludge volume and chemical dosage rates. It would also affect dewatering characteristics of the sludge.

24.11.4 DETENTION TIME

Water should be held in the settling tank long enough to allow for the settling of solid particles. An overloaded hydraulic tank will result in poor-quality effluent due to carryover of the solid particles. However, too much detention time in primary clarification can cause **septicity**. Most engineers design clarifiers to provide a detention time of 1 to 2 hours (**Table 24.3**).

24.11.5 SURFACE SETTLING OR OVERFLOW RATE

Overflow rate indicates the rate at which water will rise in the tank before it exits. Mathematically, it is flow per unit surface area of the tank. For solids to settle out, settling rate or velocity should be greater than overflow rate. Overflow rates vary in the range of 16–32 $m^3/m^2 \cdot d$ with 24 $m^3/m^2 \cdot d$ (600 $gal/ft^2 \cdot d$) being the typical value. For a given clarifier, hydraulic detention time, t_d, and overflow rate, v_O, are related

TABLE 24.3
Design parameters for settling tank

Types of Settling	Overflow Rate m³/m²·d*		Solids Loading kg/m²·d**		Depth, m	Detention time, h
	Average	Peak	Average	Peak		
Primary settling only	25–30	50–60	–	–	2.5–3.5	2.0–2.5
Primary settling + secondary treatment	35–50	60–120	–	–	2.5–3.5	
Primary settling with activated sludge return	25–35	50–60	–	–	3.5–4.5	–
Secondary settling for trickling filters	15–25	40–50	70–120	190	2.5–3.5	1.5–2.0
Secondary settling for activated sludge	15–35	40–50	70–140	210	3.5–4.5	–
Secondary settling for extended aeration	8–15	25–35	25–120	170	3.5–4.5	–

*m³/m²·d = 24.54 gal/ft²·d.
**kg/m²·d = 0.2048 lb/ft²·d.

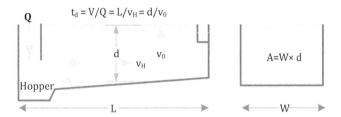

FIGURE 24.3   Flow through a rectangular clarifier

terms. For design purposes, only one is used. Referring to **Figure 24.3**, the relationship between hydraulic detention time and overflow rate is as follows.

Hydraulic detention time and overflow rate

$$v_O = \frac{Q}{A_S} = \frac{d}{HDT} \quad HDT \text{ or } t_d = \frac{V}{Q} = \frac{d}{v_O} = \frac{L}{v_H}$$

V = volume, Q = flow rate, d = side water depth (SWD), v_o = overflow rate

24.11.6   WEIR LOADING

Wastewater leaves the clarifier by flowing over weirs and into an effluent trough or channel. The length of weir over which the wastewater flows should be long enough to prevent short-circuiting and carryover of the solids due to excessive velocities. The weir loading is expressed in terms of flow per linear meter of the weir. Most designers recommend 125 m³/m·d to 250 m³/m·d (10 0000 to 20 000 gal/ft.d). **Secondary clarifiers** need lower weir overflow rates. In some situations, additional weir length is achieved by providing an inboard channel. These will almost double the weir length for a given circular clarifier. The inboard effluent channel is commonly used in secondary clarifiers where low weir loading is desired.

Example Problem 24.3

A primary sedimentation tank is to be designed to treat wastewater of a town that has annual average flow of 300 m³/h and maximum 2 h flow of 550 m³/h. Using the design values given below, find the required diameter and depth of the tank. Assume that the conical bottom part is used for sludge collection and no additional depth is required.

Given:
Average flow: surface overflow rate = 1.5 m/h, detention time = 1.5 h.
Peak flow: surface overflow rate = 3.0 m/h, detention time = 1.0 h.

$$Q_{avg} = 300 \text{ m}^3/\text{h} \quad Q_{peak} = 550 \text{ m}^3/\text{h}$$

Solution:
Surface area of the clarifier

$$Average\ flow : A_s = \frac{Q}{v_o} = \frac{300\,m^3}{h} \times \frac{h}{1.5\,m} = 200\,m^2$$

$$Peak\ flow : A_s = \frac{Q}{v_o} = \frac{550\,m^3}{h} \times \frac{h}{3.0\,m} = 183\,m^2$$

Minimum required surface area is thus 200 m²

$$D = \sqrt{\frac{4A_s}{\pi}} = \sqrt{\frac{4 \times 200\,m^2}{\pi}} = 15.95 = 16\ m\ \ say$$

Water depth of the tank

$$Average\ flow : d = v_o \times t_d = \frac{1.5\,m}{h} \times 1.5\,h = 2.25 = 2.3\,m$$

$$Peak\ flow : d = v_O \times t_d = \frac{3.0\ m}{h} \times 1.0\ h = 3.0\ m$$

Select 3 m as side water depth or 3.5 m as depth of tank

Example Problem 24.4

A wastewater flow of 3500 m³/d enters a plant with two rectangular clarifiers. Each clarifier is 5.0 m wide, 9.0 m long and 3.0 m deep. Calculate the surface settling rate and hydraulic detention time. What is the minimum length of the weir needed not to exceed weir overflow rate of 125 m³/m·d?

Given:

Q = 3500 m³/d (# =2) = 3500/2 = 1750 m³/d, L = 9.0 m, W = 5.0 m, d = 3.0 m

Solution:

$$v_O = \frac{Q}{A_S} = \frac{1750\,m^3}{d} \times \frac{1}{(9.0\,m \times 5.0\,m)} = 38.8 = \underline{39\,m^3/m^2.d}$$

$$t_d = \frac{d}{v_o} = 3.0\,m \times \frac{d}{38.8\,m} \times \frac{24h}{d} = 1.85 = \underline{1.9h}$$

$$L_W = \frac{Q}{Loading} = \frac{1750\,m^3}{d} \times \frac{m.d}{125\,m^3} = 14.0 = \underline{14\,m}$$

Example Problem 24.5

Design a suitable rectangular clarifier for treating wastewater flow of 2.5 MGD to provide detention time of 2.0 h and horizontal flow velocity not to exceed 1.0 ft/min. Make appropriate assumptions, wherever needed.

Given:

Q = 2.5 mil gal/d t_d = 2.0 h v_H = 1.0 ft/min   L, B, d, = ?

Solution:

Rectangular tank

$$V = Q \times t_d = \frac{2.5 \times 10^6\,gal}{d} \times 2.0\,h \times \frac{d}{24\,h} \times \frac{ft^3}{7.48} = 2.785 \times 10^4\,ft^3$$

$$L = v_H \times t_d = \frac{1.0\,ft}{min} \times 2.0\,h \times \frac{60\,min}{h} = 120.0 = \underline{120\,ft}$$

At the end, give an additional space for overflow weirs of about 10 ft. Assume the sidewall depth to be 10 ft. To provide space for sludge accumulation at the bottom of the tank, the overall depth of the tank will be 12 ft.

$$B = \frac{A_s}{L} = \frac{V}{d \times L} = \frac{27852\,ft^3}{10\,ft \times 120\,ft} = 23.2 = 23\,ft$$

Hence a rectangular tank with overall dimensions of 130 ft × 23 ft × 12 ft will meet the requirements.

Example Problem 24.6

Design a primary circular clarifier to treat a flow of 3.0 ML/d such that the overflow rate does not exceed 30 m³/m²·d. Work out the detention time achieved assuming the effective depth of the tank is 3.0 m and the weir loading rate assuming there is an effluent weir along the periphery of the clarifier.

Given:

$$Q = 3.0 \text{ ML/d} = 3000 \text{ m}^3/\text{d} \quad t_d = ? \quad v_o = 30 \text{ m/d} \quad D = ?$$

Solution:

Size of the clarifier

$$A_s = \frac{Q}{v_o} = \frac{3000\,m^3}{d} \times \frac{d}{30\,m} \times \frac{24\,h}{d} = 2400\,m^2$$

$$D = \sqrt{\frac{4A_s}{\pi}} = \sqrt{\frac{4 \times 2400\,m^2}{\pi}} = 55.27 = \underline{55\,m}$$

Detention time and weir loading

$$t_d = \frac{d}{v_o} = 3.0\,m \times \frac{d}{30\,m} \times \frac{24\,h}{d} = 2.40 = \underline{2.4\,h}$$

$$WLR = \frac{Q}{L_W} = \frac{3.0\,ML}{d} \times \frac{1}{\pi \times 55\,m} \times \frac{1000\,m^3}{ML} = 17.1 = \underline{17\,m^3\,/\,m.d}$$

Example Problem 24.7

A secondary clarifier is built with an inboard channel with an effluent weir on both sides of the channel. If the average daily flow to the plant is 7.5 ML/d, what must be the minimum diameter of the inboard channel so that weir loading of 100 m³/m·d is not exceeded?

Given:

$$Q = 7.5 \text{ ML/d} = 7500 \text{ m}^3/\text{d} \quad D = ? \quad WL = 100 \text{ m}^3/\text{m·d}$$

Solution:

$$D = \frac{Q}{2\pi \times WL} = \frac{7500\,m^3}{d} \times \frac{m.d}{2\pi \times 100\,m^3} = 11.93 = \underline{12\,m}$$

24.12 SECONDARY CLARIFIER

In an activated sludge flow process scheme, activated sludge is continuously shifting between the aeration tank and the secondary clarifier. The activated sludge is less dense, hence secondary clarifiers are more sensitive to loading compared to primary clarifiers. A sludge blanket is typically about one-third of the depth of the clarifier. For this reason, engineers are more conservative in the design of secondary clarifiers.

Whereas in primary clarification settling is mainly of the **flocculent** type, in secondary clarification it is primarily **zone settling** due to the fact that the influent to the secondary clarifier is mixed liquor that comprises a high concentration of biological floc. For this reason, the sludge blanket is relatively high and the whole thing settles as one mass unit, hence the name zone settling.

SOLIDS LOADING

A solids loading rate is more commonly used for secondary clarifiers where settling of biological floc is carried out. Much along the same lines as the overflow rate, the **solid loading rate** is expressed as mass of solids per unit surface area. One important difference in the case of a secondary clarifier is the increased solids due to recirculated flows.

Solids loading rate

$$SLR = \left(Q + Q_{RS}\right) \times \frac{MLSS}{A_S}$$

MLSS = mixed liquors suspended solids conc. Q_{RS} = recirculation of return sludge

24.13   SLUDGE HANDLING

Sludge-handling operations are governed by quality in terms of solids concentration or thickness of the sludge and quantity or volume of sludge pumped. Volume of sludge produced indirectly depends on the thickness achieved and the solids removal efficiency of the primary clarifier. Thin sludge with solids concentrations as low as 0.5% is acceptable if being sent to a thickener. However, this may not be acceptable if the downstream process is a digester or a dewatering unit. Under proper operating conditions, an operator should be able to able to obtain a solid concentration of 4% to 6%. The process control requires finding a **sludge blanket** level that will produce sludge of the desired consistency without adversely affecting removal, overloading of collector arm, and causing septicity.

As a general rule of thumb, sludge should be pumped more frequently and for smaller durations. It would allow for higher consistency and prevent septicity. Estimation of sludge production is illustrated in Example Problem 24.8.

Example Problem 24.8

In a wastewater treatment plant, the average daily flow to each primary unit is 1.3 MGD containing 250 mg/L of solids. Assuming a solids removal of 45% by settling, work out the volume of primary sludge produced assuming consistency of raw sludge is 3.5%.

Given:

$$Q = 1.3 \text{ mil gal/d} \quad PR = 45\% \quad SS = 250 \text{ mg/L} \quad SS_{sl} = 3.5\%$$

Solution:

$$M_{SS} = Q \times SS_i \times PR = \frac{1.3 \, mil \, gal}{d} \times \frac{250 \, mg}{L} \times \frac{8.34 \, lb \, / \, mil \, gal}{mg \, / \, L} \times \frac{45\%}{100\%}$$

$$= 1219.7 = 1220 \, lb \, / \, d$$

$$V_{sl} = \frac{M_{SS}}{SS_{sl}} = \frac{1219.7 \, lb}{d} \times \frac{100\%}{3.5\%} \times \frac{gal}{8.34 \, lb} = 4718 = \underline{4200 \, gal \, / \, d}$$

Example Problem 24.9

Design a secondary circular clarifier to treat an average daily flow of 45 ML/d such that overflow rate does not exceed 45 m³/m²·d during peak flow conditions (assume peak = 2.2 × average). Assume MLSS of 2500 mg/L and return sludge ratio of 25% and permissible loading of 100 kg/m²·d.

Given:

$$Q = 45 \text{ ML/d}, \ Q_R = 0.25Q, \ v_o = 24 \text{ m}^3/\text{m}^2\cdot\text{d}, \ SLR = 100 \text{ kg/m}^2\cdot\text{d}, \ D = ?$$

Solution:

$$A_s = \frac{Q}{v_o} = \frac{45 \, ML}{d} \times \frac{1000 \, m^3}{ML} \frac{d}{24 \, m} = 1875 = 1880 \, m^2$$

$$A_s = \frac{(Q + Q_R) \times MLSS}{SLR} = \frac{1.25 \times 45 \, ML}{d} \times \frac{2500 \, kg}{ML} \times \frac{m^2.d}{100 \, kg} = 1406 = 1410 \, m^2$$

Choose the larger value

$$D = \sqrt{\frac{4A_s}{\pi}} = \sqrt{\frac{4 \times 1875 \, m^2}{\pi}} = 48.79 = 49 \, m$$

Two identical clarifiers are suggested

$$D(each) = \frac{D}{\sqrt{2}} = \frac{48.79 \, m}{\sqrt{2}} = 34.5 = \underline{35 \, m}$$

$$WLR = \frac{Q}{L_W} = \frac{0.5 \times 45 \, ML}{d} \times \frac{1}{\pi \times 35 m} \times \frac{1000 \, m^3}{ML} = 204 = \underline{210 \, m^3 \, / \, m.d}$$

Again, on the higher side, so inboard channel with weir on both sides is suggested.

Discussion Questions

1. Differentiate between unit operations and unit processes.
2. Explain why modern treatment plants are at least secondary. What biological treatments are commonly employed in the treatment of municipal wastewater?
3. Which parameter is used to describe surface loading and how it is related to detention time?
4. What is the first unit operation in the treatment of wastewater? Discuss.
5. Discuss the importance of grit removal at the beginning of wastewater treatment.
6. What are major sources of grit in municipal wastewaters? What treatment methods are commonly used to remove grit?
7. Compare aerated grit tanks with pre-aeration units.
8. What do you understand by a velocity control device? What is the most common device used for this purpose?
9. Explain in what situations you would recommend using pre-aeration grit removal chambers.
10. Discuss the importance of flow measurement as part of preliminary treatment.
11. What flow-measuring devices are commonly employed in wastewater treatment plants?
12. The Parshall flume flow formula in USC units is $Q = 4WH^{1.522W^{0.026}}$. Modify this formula for SI units when width of flume is 2.0 ft ($Q = 1.22H^{1.55}$).
13. Modify the triangular weir general equation assuming C_d of 0.60 and angle of weir is 90°.
14. List the advantages of pre-aeration.
15. Explain why it is not necessary to define both detention time and overflow rate in the design of sedimentation tanks.
16. Compare circular tanks versus rectangular tanks for primary treatment.
17. Give the advantages and disadvantages of chemical precipitation.
18. Describe various operating parameters for primary clarifiers and a normal range of values.
19. Describe the dominant types of settling in primary clarifiers versus secondary clarifiers.
20. Describe the additional parameter which is important in the design and operation of secondary clarifiers in the activated sludge process.
21. Compare a primary clarifier with a grit clarifier.
22. Explain why primary treatment should be followed by secondary treatment.
23. Describe the main components of a rectangular clarifier and a circular clarifier.
24. What chemicals are commonly used for chemical precipitation? Chemical precipitation in primary clarification is generally used for the removal of which pollutant?

25. An inboard channel is preferred in secondary clarifiers. Explain
26. For a rectangular sedimentation tank, derive the following relationship.

$$\frac{d}{v_O} = \frac{L}{v_H}$$

Practice Problems

1. A typical grit particle settles at the rate of 4.0 ft/min. In a grit channel flowing with a horizontal velocity of 1.0 ft/s at a depth of 2.5 ft, what must be the minimum length required to capture this grit? (38 ft).
2. A grit chamber is designed to remove 100% of particles with settling velocity of 2.0 cm/s. Flow velocity is maintained at 0.25 m/s by a proportional weir at the exit end. What is the depth of water if the length of the chamber is 15 m? (1.2 m).
3. An 60 ft long grit channel is flowing with a horizontal velocity of 1.0 ft/s at a 3.0 ft depth of flow. Grit particles of what settling velocity (equal or greater than) will all be removed? (3.0 ft/min).
4. A wastewater pollution control plant final effluent is measured by a 90° V-notch. What is the maximum head expected, knowing that peak flow is 3.5 m³/s? (1.4 m).
5. A channel type grit chamber has a flow velocity of 0.30 m/s, length 10 m and depth of 0.90 m. Grit particles of what settling velocity can be completely removed? (0.03 m/s).
6. A grit channel has a flow velocity of 0.28 m/s, length 10 m and depth of 1.0 m. for grit of with specific gravity of 2.5. Determine the largest diameter grit particles that can have 100% removal? (0.28 m/s, 0.24 mm).
7. A channel type grit chamber is to be installed in a wastewater pollution control plant with daily flow of 2.5 MGD to provide minimum detention time of 0.5 min. The flow velocity is controlled by an effluent weir at the rate of 1.0 ft/s. Recommend the channel dimensions for a depth to width ratio of 1:1.5? (30 ft × 2.4 ft × 1.6 ft + f.b).
8. A grit chamber is designed to remove grit particles with an average size of 0.2 mm and specific gravity 2.65 when the average sewage temperature is 20°C. A flow-through velocity of 25 cm/s will be maintained by installing a proportional weir at the exit end. What size channel would you recommend carrying a peak flow of 12 ML/d? Assume design overflow rate to be 60% of settling velocity.
   (v_s = 2.6 cm/s, v_o = 1.6 cm/s, L = 16 m, W = 5.6 m. d = 1.0 m + f.b,).
9. Design a grit chamber to remove grit particles with a settling velocity of 20 mm/s, typical of inorganic particles of size 200 μm and SG of 2.65. Assume that a parabolic flume at the effluent end will maintain a flow velocity of 0.30 m/s. Design the channel to treat a maximum flow of 12 ML/d. Assume the water depth of 1.1 m in the channel. Draw the plan view and cross section of the designed channel. (17 m × 0.42 m × 1.1 m)

10. For a 90° notch, what is the head achieved when a flow of 5.5 ft³/s is passing through it? (1.4 ft).

11. A wastewater flow enters a circular primary clarifier at a peak hourly flow rate of 3500 m³/d. The clarifier has a diameter of 9.0 m and side water depth of 3.0 m. Determine surface loading and detention time. (55 m³/m²·d, 1.3 h).

12. Estimate the detention time in an 85 m³ capacity primary clarifier when treating a flow of 1.0 ML/d? (2.0 h).

13. A clarifier is 3.5 m deep. What will the hydraulic detention time be if the clarifier is designed for an overflow rate of 32 m³/m²·d? (2.6 h).

14. What is the length of a peripheral weir of a 100 ft diameter clarifier? (310 ft).

15. Raw sludge solids concentration on average is 4.0%. If 50 lb of solids are removed in one hour, how many gallons of sludge should be withdrawn from the bottom of a primary clarifier? (150 gal).

16. A wastewater treatment plant has daily average flow of 5.0 MGD and consists of two circular primary clarifiers each with a radius of 40 ft. What is the surface overflow rate? (500 gal/ft²·d).

17. A new clarifier is to be designed to handle an average daily flow of 4.5 ML/d. It is desired to keep the weir loading not to exceed 150 m³/m·d, what diameter of clarifier should be selected? (10 m).

18. In the previous question, if it is desired to limit the overflow rate to 26 m³/m²·d, what should be the minimum diameter of the sedimentation tank? (15 m).

19. A rectangular clarifier size is to be selected such that the length of the clarifier is five times the width. What should be the length of the clarifier if surface overflow rate is not to exceed 600 gal/ft²·d while treating average daily flow of 1.0 MGD? (91 ft).

20. Flow to a primary clarifier is 180 m³/h. Given that SS are removed @ 120 mg/L and primary sludge is thickened to 4.6% solids concentration, how long should the sludge pump be run every hour when pumping @55 L/min? (8.5 min/h).

21. Design a circular clarifier for a town with a population of 35 000. Average demand of water is 150 L/c·d, 75% of which is discharged as wastewater. Maximum allowable surface loading is 24 m³/m²·d. (15 m).

22. Size a rectangular primary clarifier to serve a population of 40 000. Being in a developing country, average demand of water is 45 gal/c·d, 75% of which is discharged as wastewater. Peaking factor is 2.5 and allowable surface loading is 500 gal/ft²·d based on average daily flow. Assume length of the tank is four times the width and effective depth is 9.0 ft. Check for horizontal flow velocity. (100 ft × 26 ft × 9.0 ft, 0.53 ft/min <1.0 ft/min).

23. A city sewage treatment plant treats an average flow of 4500 m³/d with a peaking factor of 2.2. For appropriate removal, it is recommended the overflow rate not exceed 30 m³/m²·d. Check for average detention time assuming depth of 3.1 m. Also check loading during the peak hour and comment on the removal. (14 m, 2.5 h, 66 m³/m²·d.).

24. Determine the recommended size of two new circular clarifiers for an activated sludge system with a design flow of 20 ML/d with a peak hourly flow 32 ML/d. Use 30 m³/m²·d overflow rate for design flow and 50 m³/m²·d for the peak hourly flow. (21 m diameter, 48 < 50 m³/m²·d).

25. For a secondary treatment plant, it is decided to have twin primary clarifiers. What is the minimum diameter required if it is desired not to exceed an overflow rate of 1000 gal/ft²·d for a design flow of 3.0 MGD? (44 ft).

26. Design a pair of circular primary clarifiers to handle raw wastewater flow of 12 ML/d based on surface overflow rate of 24 m³/m²·d.
   a. Find detention time assuming side water depth of 3.0 m. (3.0 h).
   b. What minimum diameter you would need? (18 m).
   c. Estimate the daily production of sludge from each clarifier assuming SS removal @ 100 mg/L and solid concentration of the wet sludge is 3.0 %. (20 m³/d).

27. Design a circular primary clarifier to handle raw wastewater @ 1.8 MGD to provide a minimum detention time of 2.0 h, and weir loading not to exceed 10 000 gal/ft·d. Also calculate the surface flow rate for the designed clarifier. Assume side water depth of 8.5 ft. (57 ft, 710 gal/ft²·d).

28. Design the size of twin primary clarifiers for a wastewater treatment facility in a community of 60 000 people with wastewater contribution of 80 gal/c.d. Maximum allowable overflow rate is 750 gal/ft²·d. Assume side water depth of 8.5 ft.
   a. Design a rectangular tank with length 4 times the width. (110 ft × 28 ft ×11 ft).
   b. What diameter circular tank is needed? (64 ft).
   c. What is the theoretical detention time? (2.0 h).

25 Activated Sludge Process

Primary effluent contains 60% to 70% of its original organic contaminants in terms of biochemical oxygen demand (BOD). These are in the form of very fine (colloidal) or dissolved organic materials not readily removed by normal mechanical or physical methods. If left untreated, these will cause odors and, eventually, pollution in the receiving water bodies. Hence, primary treatment is followed by secondary treatment, which is usually a biological treatment.

25.1 BIOLOGICAL TREATMENT

Naturally occurring bacteria in the presence of oxygen can break down colloidal and soluble BOD left over in the primary effluent. This process is known as **biochemical oxidation**. "Bio" indicates that microorganisms are an essential part of the process. This is usually an **aerobic** process, since microorganisms, primarily bacteria, need molecular oxygen to survive. Biological treatment can be broadly classified into two main categories of suspended growth systems and fixed growth systems.

25.1.1 SUSPENDED GROWTH SYSTEMS

In **suspended growth systems**, the microorganisms are suspended in the wastewater either as single cells or as a cluster of cells called **biological floc**. They are, therefore, surrounded by the wastewater containing food or BOD and molecular or dissolved oxygen for their growth. This suspension of wastewater containing BOD and microorganisms in the form of activated sludge is known as **mixed liquor.** The most common biological process in this category is the activated sludge process, as shown in **Figure 25.1**. In addition to the activated sludge process, stabilization pond or lagoon systems are also suspended growth systems.

25.1.2 FIXED GROWTH SYSTEMS

Fixed growth systems, also called attached culture systems, consist of biomass adhered to inert surfaces with wastewater passing over the microbial layer. Trickling filters and rotating biological contactors (RBCs) are good examples of fixed growth systems. In both types of systems, the biological process is **aerobic**. Whereas in the activated sludge process oxygen is added by pumping air into mixed liquor, in a fixed growth system, wastewater passing over the microbial layer dissolves the necessary oxygen from the atmosphere.

DOI: 10.1201/9781003347941-30

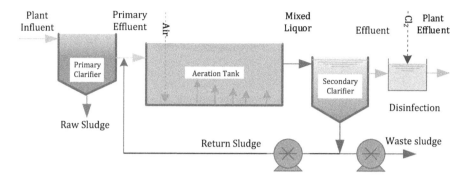

FIGURE 25.1   A conventional activated sludge plant

25.2   PRINCIPLES OF THE ACTIVATED SLUDGE PROCESS

Activated sludge is the most widely used process for the secondary treatment of wastewater. The term **activated sludge** indicates the biomass which causes the breakdown of organics. The soluble and colloidal BOD is converted to new growth, so-called activated sludge, in three steps, namely transfer, conversion, and flocculation.

25.2.1   TRANSFER

Transfer occurs when organic food matter meets microorganisms. During the transfer phase, colloidal organic matter is adsorbed on the cell membrane. Adsorbed organics must be broken into a simple soluble form before they can be absorbed into the cell.

25.2.2   CONVERSION

Conversion of food matter into cell matter occurs after all the food enters the cell. This conversion is called **metabolism**. During conversion, only part of the BOD is converted to new cell growth. More than half of the BOD is converted to carbon dioxide, water, and energy for the microorganisms. In an aerobic process like activated sludge, biodegradable organic matter is oxidized, as shown in the following biochemical reaction.

$$COHN(organics) + O_2 \rightarrow CO_2 + NH_3 + H_2O + C_5H_7NH_2(new\ cells)$$

25.2.3   FLOCCULATION

Biological **flocculation** occurs when cells combine to form clusters called biological floc. The separation of water from sludge is determined by the settlement efficiency of the biological floc. The transfer and conversion steps are completed in the aeration tank and flocculation and separation of solids mainly occur in the secondary clarifier.

25.3 COMPONENTS OF ASP

The principal elements of the activated sludge process are shown in **Figure 25.1**. Three main components are aeration tanks, final clarifier, and recirculation pumps.

25.3.1 AERATION TANKS

The aeration tank is the **bioreactor** in which colloidal and soluble matter contributing to BOD is oxidized under aerobic conditions. The aeration tanks can be square, rectangular, or circular and generally are 3–5 m (10–15 ft) deep. The tanks are relatively deeper to allow proper mixing and oxygen transfer. The tank size depends on the volume of sewage to be treated and its ability to hold the incoming sewage for a period of 4 to 8 hours, called the **aeration period.**

Aeration tanks are generally made of concrete or steel. Oxygen is dissolved into the wastewater in tanks either by **diffused** aeration or **surface** aeration. It is essential that adequate mixing be provided so that the activated sludge is maintained in suspension. To keep the contents aerobic, dissolved oxygen or DO of 1–2 mg/L is maintained. Higher levels of dissolved oxygen do not necessarily help the biological process. However, DO levels below 1.0 mg/L may encourage the growth of **filamentous** organisms that hinders settling. Aeration of mixed liquor requires a lot of energy, so maintaining a DO level significantly higher than 2.0 mg/L is a waste of money and may encourage the growth of filamentous organisms.

Diffused Aeration

In this type of aeration system, air is blown from compressors through various types of diffusers located at the bottom of the aeration tanks, generally on one or both sidewalls of the tank. While oxygen is being dissolved into the liquid, a **rolling action** is generated to ensure thorough mixing and suspension of the activated sludge.

Mechanical Aeration

This technique uses blades of various designs which rotate partially submerged at the surface of the liquid with dissolved oxygen from the atmosphere. These devices splash large volumes of liquid over the surface of the tank entraining and dissolving atmospheric oxygen into the tank contents. This also generates pumping action for the necessary mixing. The amount of oxygen which can be dissolved varies with the speed of the device, its diameter, submergence, and the power of the drive unit. The drive motor ranges from 5 to 100 kW and the device can be as big as 3 m (10 ft) across.

Example Problem 25.1

A water pollution control plant daily average flow is 4.5 MGD. The total aeration capacity is 1.2 MG and is operated by maintaining mixed liquor suspended solids (MLSS) of 2000 mg/L. Determine the aeration period and quantity of mixed liquor solids kept in the aeration tank.

Given:

Q = 4.5 MGD   MLSS = 2000 mg/L V_A = 1.2 MG

Solution:

$$AP = \frac{V_A}{Q} = 1.2\, MG \times \frac{d}{4.5\, MG} \times \frac{24\, h}{d} = 6.40 = 6.4\, h$$

$$m_{MLSS} = V_A \times MLSS = 1.2\, MG \times \frac{2000\, mg}{L} \times \frac{8.34\, lb\,/\,MG}{mg\,/\,L} = 200016 = 20000\, lb$$

25.3.2 FINAL SETTLING TANKS

As seen in **Figure 25.1**, the secondary clarifier, or final settling tank, receives the activated sludge in the form of mixed liquor from the aeration tank. The microorganisms in the form of sludge settle to the bottom of this clarifier where, with the aid of scraper mechanisms, they are collected and returned (recycled) to the aeration tank. The supernatant is clarified wastewater (secondary effluent) that flows over weirs to be disinfected before discharge to the receiving rivers or lakes.

Secondary Clarification

Any solids which escape separation will reduce the quality of the final effluent. Thus it is important that the clarifier is operated so as to remove the maximum possible solids as sludge. Since the activated sludge is relatively lighter, secondary clarifiers are designed to operate at lower overflow rates.

Settling

In the secondary clarifier, solids settle as one mass – a process called **zone settling**. These solids are contained at a larger depth called a **sludge blanket**. For this reason, secondary clarifiers are relatively deeper. In plant operation, the depth of the sludge blanket is monitored to judge the performance of the clarifier. Old sludge, being heavier, may settle too quickly and reduce the particle collision necessary for capturing the fine solids (**pin floc**) in the upper regions of the tank. A sludge depth meter is used by plant operators frequently to monitor the depth of sludge blanket. For continuous monitoring, a sonic depth reader is fitted above the clarifier.

Thickening

Another important function of the clarifier is to allow thickening of the sludge settling at the bottom of the tank. In addition to the hydraulic conditions under which a clarifier is operated, the concentration of solids in the return sludge will depend on the rate at which it is removed.

25.3.3 SLUDGE RECIRCULATION AND WASTING

Variable speed pumps take their suction from the draw-off and return the sludge to the aerator. During plant operation, the **return rate** of sludge is controlled to maintain a healthy biomass in the aeration tank. From the return sludge lines, a certain volume of sludge is occasionally **wasted.** Sludge wastage is necessary, otherwise solids will accumulate in the system daily as a result of new growth. The wastage rate is adjusted so as to maintain steady state conditions.

25.4 FACTORS AFFECTING ASP

Since microorganisms are important parts of the activated sludge system, factors affecting the system are those affecting the bacteria, also called **growth pressures**. As with all life forms, these organisms can only live if conditions remain suitable for their growth. The following are the main considerations in maintaining healthy activated sludge.

- For satisfactory operation, a dissolved oxygen concentration in the aeration tank of at least 1.0 mg/L should always be present. A target value of 2.0 mg/L is most common.
- It is also important that sewage entering the activated sludge system contain no materials toxic to microorganisms.
- In addition to carbon, nutrients including nitrogen (N) and phosphorus (P) are needed for biological growth. In municipal wastewaters there is an abundance of N and P. During treatment of industrial wastewaters, urea and phosphoric acid may be added to bring nitrogen and phosphorus up to the desired level.
- Temperature is an important growth pressure for microorganisms. As a rule, the rate of microbial growth doubles with every 10°C increase in temperature up to a limiting temperature. During winter conditions, biogrowth slows down. Some operators compensate for this by maintaining high mixed liquor suspended solids (MLSS) in the aerator.

25.5 PROCESS LOADING PARAMETERS

The main parameters used in the design and operation of activated sludge plants include aeration period, volumetric BOD loading, F/M ration, and solids retention time.

25.5.1 AERATION PERIOD

Aeration time or period indicates the **hydraulic loading** on the aeration tank. It is essentially the hydraulic detention time in the aeration tank based on the daily average wastewater flow. The recirculation flow is not considered in calculating the aeration period. It is usually expressed in hours and ranges from 4 to 30 h.

25.5.2 VOLUMETRIC BOD LOADING

Volumetric BOD loading on the aerator is usually expressed as mass of BOD entering the aeration tank per unit aeration volume as $g/m^3 \cdot d$ ($lb/1000 \ ft^3 \cdot d$). Depending on the type of activated sludge process, organic loading can vary from 100 to 3000 $g/m^3 \cdot d$. BOD loadings increase with an increase in incoming BOD and decrease with an increase in aeration period.

BOD loading rate

$$BODLR = \frac{M_{BOD}}{V_A} = \frac{Q \times BOD_{PE}}{V_A} = \frac{BOD_{PE}}{AP}$$

For the same quality of water entering the aeration tank, halving the aeration period will double the organic loading. For operational control, the aeration process has the following commonly used variations, including extended aeration, conventional aeration, contact stabilization, high-rate, and high-purity oxygen. Due to the longer aeration period afforded in extended aeration, BOD loading is at minimum.

Example Problem 25.2

In a conventional activated sludge plant, aeration capacity is 5000 m^3. Primary effluent BOD content is 150 g/m^3. Daily average flow is 20 ML/d. What is the BOD loading rate?

Given:

$$Q = 20 \text{ ML/d} = 20\,000 \text{ m}^3/d \quad BOD = 150 \text{ g/m}^3 \quad V_A = 5000 \text{ m}^3$$

Solution:

$$BODLR = \frac{M_{BOD}}{V} = \frac{20000 \ m^3}{d} \times \frac{150 \ g}{m^3} \times \frac{1}{5000 \ m^3} = \underline{600 \ g/m^3 .d}$$

25.5.3 Food to Microorganism (F/M) Ratio

The **F/M ratio** expresses BOD loading in relation to biomass in the system. The microbial mass is indicated by the mass of mixed liquor solids in the aeration tank.

Food to microorganism ratio

$$\frac{F}{M} = \frac{M_{BOD}}{m_{MLSS}} = \frac{Q \times BOD_{PE}}{MLSS \times V_A} = \frac{BOD_{PE}}{MLSS} \times \frac{1}{AP}$$

Because the F/M ratio is of mass rate to mass, the units of F/M are per day. The values typically range from 0.1/d to 1.0/d. Most municipal plants are operated at a F/M ratio of 0.1 to 0.3/d. Some authors express F/M ratio in terms of mass of volatile mixed liquor solids (MVLSS) rather than MLSS for the reason that volatile solids more accurately represent the biomass. The **volatile fraction** is around 75% of the total mixed liquor solids.

F/M ratio is used to maintain the proper balance between food supply and bio-mass. The F/M ratio maintained in the aeration tank defines the operation of the activated sludge. If the F/M ratio is high, there is an abundance of food, and the microorganisms are in a **logarithmic growth** phase. The new growth is young and highly active. Because of a high activity level, the cells do not easily floc together to become heavy enough to settle. A system operated with a high F/M ratio will have DO in the aerator depressed and DO in the final clarifier will be at or near zero. Activated sludge is light brown in color and settles slowly. BOD removal efficiency will be generally poor and rapid increase in MLSS build up.

When the process is operated with a low F/M ratio (old sludge), the growth phase becomes **endogenous,** or starvation occurs. Because of the lack of food during the starving phase, there would continue to be a loss in body weight as more and more cellular material is converted into energy. An activated sludge system operated with a low F/M ratio it is called **extended aeration**. A process operated by maintaining low F/M ratio will have high DO in the aeration tank and sludge will settle quickly. However, some pin floc in supernatant may be present. BOD removal and total oxygen requirements are high. The **conventional activated sludge** process is operated with the F/M ratio in the middle range of 0.2/d to 0.5/d. This creates sludge which is neither old nor young. Hence conventional activated sludge is a good compromise between quality and quantity.

Example Problem 25.3

An aeration tank receives a primary effluent flow of 5.0 MGD with a BOD concentration of 150 g/m³. The mixed liquor suspended solids concentration is 2500 mg/L and the aeration volume is 1.3 MG. Find the F/M ratio.

Given:

$$Q = 5.0 \text{ MGD} \quad BOD = 150 \text{ mg/L} \quad V_A = 1.5 \text{ MG}$$

Solution:

$$\frac{F}{M} = \frac{Q \times BOD_{PF}}{MLSS \times V_A} = \frac{5.0\,MG}{d} \times \frac{1}{1.5\,MG} \times \frac{150\,g/m^3}{2500\,g/m^3} = 0.333 = \underline{0.33/d}$$

25.5.4 Sludge Age

Sludge age, also referred to as mean cell residence time (MCRT), or **solids retention time** (SRT), is an operational parameter related to the F/M ratio. Since biomass as return sludge is recycled from the clarifier back to the aeration tank, the biosolids have more than one pass through the system. Whereas the aeration period varies from 3 h to 30 h, the SRT is much greater and is measured in terms of days.

The **sludge age** is calculated based on the MLSS or MVLSS in the aeration tank related to the total mass of bio solids leaving via the waste sludge stream and final effluent stream. Referring to **Figure 25.2**, the sludge age can be calculated as follows:

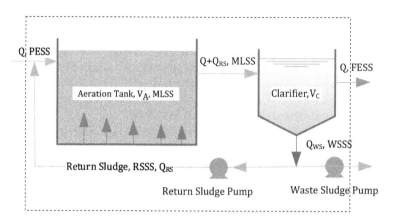

FIGURE 25.2   Solids entering and exiting

Solids retention time

$$SRT = \frac{(V_A + V_C) \times MLSS}{WSSS \times Q_{WS} + FESS \times Q_{FE}}$$

Sometimes a distinction is made when using volatile solids or MVLSS instead of MLSS. Since the volatile component of mixed liquor solids is a better representative of microorganisms, the term MCRT is preferred when MVLSS are used in calculations. Some authors ignore the solids exiting as part of final effluent and volume of mixed liquor in the final clarifier. This is valid when the solids concentration in the final effluent (FESS) is relatively small.

Solids retention time (short form)

$$SRT = \frac{V_A \times MLSS}{WSSS \times Q_{WS}} = \frac{MLSS}{WSSS} \times \frac{V_A}{Q_{WS}}$$

Sludge age indicates the growth rate, hence the metabolic state of the biological process. In an old sludge with longer SRT, the microorganisms are in a starvation phase. In high-rate systems, the SRT is relatively short, which signifies the logarithmic growth phase thus producing young sludge. Sludge age and F/M ratio are inversely related. As the F/M is reduced, it results in longer SRT, which results in higher BOD removal due to improved settlement and longer reaction time. However, due to a higher aeration period, more oxygen is required to remove the same amount of BOD.

Example Problem 25.4

An aerator is 7.0 m deep with a holding capacity of 5000 m³. The MLSS concentration is kept at 2500 g/m³ and the return activated sludge solids concentration

(RSSS) is 5000 mg/L. If 500 m³ of the activated sludge is wasted daily, find the solids retention time.

Given:

$$V_A = 5000 \text{ m}^3, \text{ MLSS} = 2500 \text{ g/m}^3, \text{ WSSS} = \text{RSSS} = 5000 \text{ g/m}^3, Q_{WS} = 500 \text{ m}^3/\text{d}$$

Solution:

$$SRT = \frac{MLSS}{WSSS} \times \frac{V_A}{Q_{WS}} = \frac{5000 \ m^3.d}{500 \ m^3} \times \frac{2500 \ mg/L}{5000 \ mg/L} = 5.00 = \underline{5.0 \ d}$$

25.5.5   SUBSTRATE UTILIZATION RATE

Another parameter commonly used to design aeration units is called the **specific substrate utilization rate, U.** It is very much parallel to the F/M ratio, with the exemption that BOD removed rather than BOD applied is used and the conversion factor of BOD into microbial mass factor is used to translate BOD into new growth.

Specific substrate utilization rate

$$U = \frac{\alpha M_{BODR}}{m_{MLSS}} = \frac{\alpha \times Q \times (BOD_{PE} - BOD_{FE})}{MLSS \times V_A}$$

α = *maximum yield coefficient, g/g* U = *specific substrate utilization rate, /d*

Under steady state conditions, when MLSS is maintained at a given level, mass of solids wasted must be equal to new growth minus constant **endogenous respiration rate**, k_e. Based on this it can be shown that U and SRT are related as follows:

Relationship between SRT and U

$$\frac{1}{SRT} = \alpha \times U - K_e = \frac{\alpha \times Q \times (BOD_{PE} - BOD_{FE})}{MLSS \times V_A} - K_e$$

The values of α and k_e are usually constant for a given municipal wastewater, with typical values of 1.0 and 0.06/d. The value of SRT adopted for the design controls the quality of the secondary effluent as well as the settlement of the mixer liquor in the secondary clarifier. As discussed before, for a selected SRT, the amount of activated sludge to be wasted can be worked out as they are related parameters.

Sludge wastage rate

$$Q_{WS} = \frac{V_A \times MLSS}{WSSS \times SRT}$$

Longer SRT results in better quality effluent. However, oxygen requirements increase exponentially, and operating costs are higher.

Example Problem 25.5

An activated sludge plant operating data is given as V_A = 0.50 MG, WSSS = 5100 mg/L, and MLSS = 2000 mg/L. Based on this data, determine the waste sludge rate to achieve a target SRT of 5.0 d.

Given:

V_A = 2100 m³, WSSS = 5100 mg/L, MLSS = 2000 mg/L

Solution:

$$Q_{WS} = \frac{V_A}{SRT} \times \frac{MLSS}{WSSS} = \frac{0.5 \times 10^6 \; gal}{5.0 \; d} \times \frac{2000 \; mg/L}{5100 \; mg/L} = 39216 = \underline{39000 \; gal/d}$$

Example Problem 25.6

The mixed liquor in a 1600 m³ aeration tank has a MLSS of 2050 mg/L. The waste sludge solid concentration is 4900 mg/L. If the target MLSS is 2000 mg/L, determine the additional volume of waste sludge to be wasted.

Given:

V_A = 1600 m³, WSSS = 4900 mg/L, ΔMLSS = 2050–2000 = 50 mg/L

Solution:

$$\Delta V_{WS} = V_A \times \frac{\Delta MLSS}{WSSS} = 1600 \; m^3 \times \frac{50 \; mg/L}{4900 \; mg/L} = 16.3 = \underline{16 \; m^3}$$

Additional wastage should be spread over days to avoid shock to the system

25.6   FINAL CLARIFICATION

Loading and sludge settling characteristics will affect clarification and removal function of the clarifier.

25.6.1   HYDRAULIC LOADING

Hydraulic loading on the secondary clarifier is expressed in terms of hydraulic detention time, overflow rate, and weir loading. Low weir loading rates are recommended for secondary clarifiers. To achieve this, many clarifiers are designed with an **inboard effluent** channel to provide extra weir length.

25.6.2   SOLIDS LOADING

Solids loading rate is the maximum rate of solids that can be applied to a clarifier. The allowable **solids loading rate** is governed by the volume of the clarifier, the settling characteristics of the sludge. Excessive solids loading results in solids accumulated in the upper layer, which are carried out over the weirs by the overflow velocity. This would cause billowing clouds of particles at the weirs. Typical solids loading rates are in the range of 100–150 kg/m²·d or 20–30 lb/ft²·d.

25.6.3 SLUDGE SETTLEMENT

Under actual operating conditions sludge settling characteristics are known by running a **settlement test**. The **sludge volume index** (SVI) is computed based on the settling test and the recirculation rate is selected accordingly.

Settleometer Test

One of the most common tests for monitoring the operation of an aeration system is the settlement test as shown in **Figure 25.3**. The procedure involves determining the MLSS concentration by running a solids test. The sludge settlement is measured by observing the volume of settled sludge in a graduated cylinder filled with one liter of mixed liquor. The volume of settled solids after 30 min of settling is used to calculate the index. The **sludge volume index** (SVI) is the volume in milliliters occupied by one gram of settled suspended solids.

Sludge volume index and sludge density index

$$SVI = \frac{V_{SSL}}{MLSS} \quad as \quad \frac{mL}{g} \quad SDI(\%) = \frac{1}{SVI} \times 100\%$$

Sludge solids concentration or denseness is related to SVI. A low SVI <100 indicates a dense sludge and SVI >150 indicate bulking sludge. Some authors use **sludge**

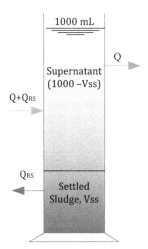

FIGURE 25.3   Settleometer test

density index (SDI) to indicate sludge settlement. Sludge density index represents the maximum concentration of return sludge that could be achieved by maintaining proper recycling rates. Mathematically, SDI is the inverse of SVI. A SVI of 100 mL/g is equivalent to a SDI of 1.0%.

25.6.4 RETURN RATE AND SVI

Referring to **Figure 25.3**, if settling of mixed liquor in a graduated cylinder and clarifier are considered identical, then the ratio that return sludge flow (Q_{RS}) is to settled sludge volume (V_{SS}) as the final effluent flow (Q) is to the volume of supernatant ($1000 - V_{SS}$). This assumes that sludge recirculation is at the required rate. A lesser flow value allows the solids to accumulate in the clarifier and results in eventual loss in the final effluent.

Hypothetical return rate

$$R_{hyp} = \frac{Q_{RS}}{Q} = \frac{V_{SI}}{\left(1000 - V_{SI}\right)}$$

If the sludge is returned as a rate indicated by this hypothetical relationship, the solids concentration in the return sludge, RSSS, equals the **sludge density index** (SDI). In fact, this will indicate the minimum return rate. If the return rate exceeds this value, which is usually the case, the return sludge is diluted.

25.6.5 RETURN RATIO AND SLUDGE THICKNESS

The primary function of the secondary clarifier is clarification and thickening of settled solids. Basically, it concentrates the incoming stream of mixed liquor solids and leaves the clear supernatant to flow over the weirs. Under steady conditions, the mass of solids entering the clarifier must equal the solids drawn in the underflow. For a given return sludge rate, Q_{RS}, or return ratio, R, $= Q_{RS}/Q$, the expected consistency of the return sludge can be predicted by performing the mass balance around the final clarifier.

Applying mass balance relationship

$$\frac{\left(Q + Q_{RS}\right)}{Q} = \frac{RSSS}{MLSS} \quad or \quad \frac{1+R}{R} = \frac{RSSS}{MLSS} \quad or \quad R = \frac{MLSS}{RSSS - MLSS}$$

For example, if the sludge is recycled at a rate of 50% (R = 0.5) of the wastewater flow rate, the return sludge will be three times as concentrated as mixed liquor. If RSSS is observed to be significantly less, it will indicate the solids are not completely drawn from the clarifier. The return rate ratio must be equal to or greater than the hypothetical return ratio. Usually, the return rate is kept at a rate greater than the hypothetical rate so that solids do not become septic.

Example Problem 25.7

The volume of settled solids in a 30-min settling test was read to be 300 mL/L. A solids test on the same sample of mixed liquor yielded concentration of total solids to be 3000 mg/L. Calculate SVI, SDI, and minimum return rate, Q_{RS}.

Given:

$$V_{ssl} = 300 \text{ mL/L} \quad MLSS = 3000 \text{ mg/L} = 3.0 \text{ g/L}$$

Solution:

$$SVI = \frac{V_{SSI}}{MLSS} = \frac{300\,mL}{L} \times \frac{L}{3000\,mg} \times \frac{1000\,mg}{g} = 100.0 = \underline{100\,mL/g}$$

$$SDI = \frac{1}{SVI} = \frac{g}{100\,mL} \times \frac{1.0\,g}{mL} \times 100\% = 1.00 = \underline{1.0\%}$$

$$R_{hyp} = \frac{V_{sl}}{1000 - V_{sl}} = \frac{300\,ml}{L} \times \frac{L}{(1000 - 300)\,mL} \times 100\% = 42.8 = \underline{43\%}$$

Example Problem 25.8

Solids test is performed on a 50 mL sample of mixed liquor, using a dish weighing 0.300g. After drying, the sample weighed 0.450 g. If the return rate is maintained at 40%, calculate the expected solids concentration in the return sludge.

Given:

$$V = 50 \text{ mL} \quad A = 0.300 \text{ g} \quad B = 0.450 \text{ g} \quad R = 40\%$$

Solution:

$$MLSS = \frac{m_{SS}}{V} = \frac{(0.450 - 0.300)\,g}{50.0\,mL} \times \frac{1000\,L}{m^3} \times \frac{1000\,mL}{L} = 3000\,g/m^3$$

$$RSSS = MLSS \left(\frac{1+R}{R}\right) = \frac{3000\,mg}{L} \times \frac{(1+0.40)}{0.40} = 11000\,mg/L = \underline{1.1\%}$$

25.6.6 STATE POINT ANALYSIS

The secondary clarification is one of the most important unit processes and often determines the efficiency of the biological process. Clarifier failure can lead to solids carryover resulting in a reduction in solids retention time (SRT) below that required for meeting process goals. The **state point analysis** (SPA) is a practical tool that enables engineers to examine clarifier behavior under various operating scenarios during the design phase. This can lower design safety factors and avoid overdesign, maintain size clarifiers in conjunction with the biological process, and achieve cost savings. Likewise, when using the SPA approach, operators can predict impending problems early, implement corrective measures in a timely fashion, and adapt to upstream changes in the biological process.

The SPA is based on the solids flux theory, which describes the movement of solids through a clarifier. Type III settling is the predominant solids removal mechanism in secondary clarifiers. It is characterized by flocculated particles settling as a zone or blanket. As they settle, the particles maintain their positions relative to each other.

As shown in **Figure 25.4**, the solids flux curve SFR is developed by plotting solids flux rate G on the y-axis and the corresponding value of SS concentration on the x-axis. The next step is to superimpose the two key operating parameters of a clarifier, the overflow rate (OFR) and the underflow rate (UFR). The slope of the OFR line equals the overflow rate (positive slope) or the upward velocity of water. The UFR represents the downward velocity (negative slope) of the solids due to sludge withdrawal.

The point of intersection of the OFR and UFR lines is the **state point**. The solids concentration (X-axis) corresponding to the point of intersection is MLSS concentration. The state point represents the operating point of a clarifier. Because operating conditions are never constant, the state point is dynamic in nature. A good settling sludge will have a greater area below the solids flux curve relative to a poor settling sludge. The state point curves can be used to assess the behavior of clarifiers that are not limited by hydraulic inefficiencies. State point contained within the flux curve indicates an underloaded condition and a stable clarifier operation. State point located outside the flux curve indicates an overloaded condition. Clarification failure can occur resulting in solids washout resulting in poor quality effluent. A UFR line

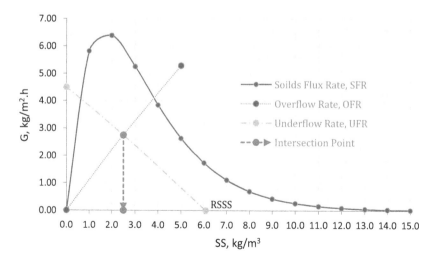

FIGURE 25.4 State point analysis

falling within the flux curve represents underloading conditions, with no appreciable blanket thus thickening.

25.7 VARIATIONS OF ASP

As mentioned earlier, there are many types of aeration systems. Some involve subtle differences, such as rates and points of air or wastewater applications, detention times, reactor shapes and types, and methods of introducing air or oxygen. Others involve more drastic differences such as sorption and settling prior to biological oxidation.

25.7.1 CONVENTIONAL AERATION

This process is similar to the earliest activated sludge systems and is very suitable for treating medium to large flows. The aeration basin is a long rectangular tank with air diffusers along the bottom for oxygenation and mixing. Long aeration basins are generally designed as **plug flow reactors**. In plug flow reactors, the wastewater and return sludge are combined at the head end and the mixed liquor moves along the length of the tank to provide an aeration period of 6 to 8 h. F/M ratio is greatest at the head end and bio-growth is in starvation phase at the exit end.

TAPERED AERATION

In tapered aeration, as shown in **Figure 25.5**, the air supply is tapered along the length of the tank to provide the greatest aeration at the head end where oxygen demand is at the maximum.

STEP AERATION

Air is provided uniformly in step aeration. BOD loading is evened out by introducing wastewater at intervals or steps along the first portion of the tank as shown in **Figure 25.6**

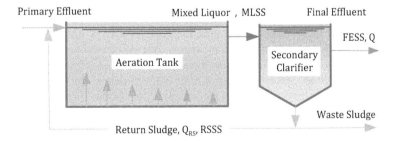

FIGURE 25.5 Tapered aeration ASP process

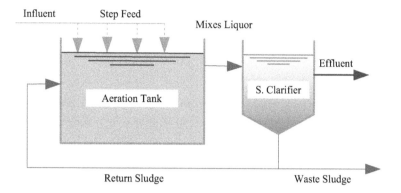

FIGURE 25.6 Step aeration

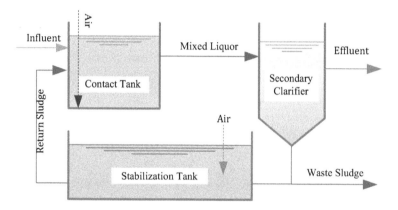

FIGURE 25.7 Contact stabilization

25.7.2 CONTACT STABILIZATION

In this process, the aeration is divided into two portions: the contact zone and stabilization zone. Wastewater flows into the contact or aeration zone whereas return sludge flows into the stabilization as shown in **Figure 25.7**. The effluent from the aeration zone flows into the clarifier. Note that no wastewater flow is introduced into the reaeration zone. The contact zone has an aeration period of 2 to 3 h while reaeration is 4 to 6 h or more. Normal operating sludge circulation is 100 per cent.

In the **contact zone**, the microorganisms quickly adsorb the food. Everything is then settled in the final clarifier before being recycled to a reaeration zone for stabilization of the food. Since the aeration time is much longer, the food adsorbed in the contact zone is stabilized to a greater extent. In the stabilization zone, microorganisms are primarily in the endogenous growth phase. Due to low growth rate, the volume of sludge produced is relatively small. This process is more suited for small communities, schools, resorts, and hospitals. More common are factory-built field-erected plants capable of handling 200–2000 m^3/d.

25.7.3 EXTENDED AERATION

The extended aeration process is characterized by having no primary treatment and a long aeration period of up to 36 hours. These are used for small towns or trailer parks; most units are package plants. The capacity of the process is small due to the tank volumes needed to supply the required retention time and the large oxygen demand. Wasting provisions are not usually provided for small plants. The MLSS can increase for several months, then the air is turned off and the floc allowed to settle. The sludge is then pumped to a digester or hauled away. The MLSS ranges from 1000 to 10 000 mg/L. The SRT is high because the same sludge is used over and over again. This results in a highly treated effluent with 85% BOD removal and a low sludge production. This process often has high concentrations of ash floc due to the high sludge age. The various types of processes are compared in **Table 25.1**.

25.7.4 OXIDATION DITCH

A variation of the extended aeration process is the **oxidation ditch,** as seen in **Figure 25.8**. This has an oval ditch for the aeration tank in which the wastewater is pumped and circulated by mechanical aerators or pumps at a velocity of 0.2 to 0.4 m/s (0.5–1 ft/s). The ditch is usually 1.2 to 1.8 m (4–6 ft) deep.

The ditch configuration is in the form of a racetrack with a surface beater for aerating. This type of plant requires minimum supervision, as the process is mechanical.

TABLE 25.1
Comparison of plant types

Parameters	Conventional	Contact	Extended
aeration period, h	6–8	3–8	18–24
BOD loading (g/m³·d)	500–800	Same	80–250
F/M (1/d)	0.2 0.4	0.2–0.4	<0.1
SRT, d	2–5	2–5	>15
nitrification	> 5 days	> 5 days	>15 days
return rate	> 80%	25% > 100%	>80%
BOD removal	1 mg/L	>1 mg/L	>1 mg/L
nitrification	>2 mg/L	>2 mg/L	>2 mg/L
primary sedimentation	Yes	No	No
grit removal	Yes	Yes	Yes
sludge wasting	Yes	Yes	Yes
sludge reaeration	No	Yes	No
sludge recirculation	Yes	Yes	Yes
mechanical aeration	Yes	Yes	Yes
diffused aeration	Yes	Yes	Yes

g/m³·d = 0.0625 lb/1000 ft³·d.

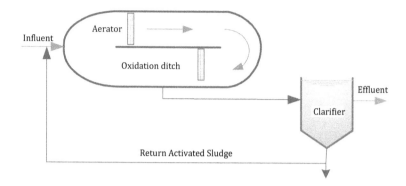

FIGURE 25.8   Schematic of an oxidation ditch

25.7.5   HIGH-RATE AERATION

High-rate aeration systems operate in the logarithmic growth phase thus producing young sludge. Operating at a high F/M ratio, these systems reduce the cost of
construction by providing reduced aeration capacity. The aeration period is of the
order of 3–4 h, whereas the MLSS concentration is maintained at a level as high as
4000 mg/L. Due to rapid growth rate, oxygen requirements are quite high and sludge
settlement is poor.

25.7.6   HIGH PURITY OXYGEN SYSTEM

High purity oxygen systems are employed to treat high-strength wastes and to produce high-quality effluents. The aeration tank is divided into stages by means of
baffles and covered in an enclosure. A slight pressure is maintained in the space
between the cover and the top of the liquid level. Successive aeration chambers are
connected to each other so that liquid flows through the submerged ports and head
gases pass freely from stage to stage with only a slight drop in pressure. In addition
to increased BOD removal efficiency, high-purity oxygen systems have dense sludge
and effective odor control. However, they are costlier.

Example Problem 25.9

An extended aeration plant has an influent BOD of 200 g/m³ and an aeration
period of 36 hours. What size tank would be required to serve a trailer park with
15 hook-ups, averaging 2.0 people per trailer? Also find the operating F/M ratio
if the system is to be operating by maintaining MLSS of 1500 mg/L. Due to less
convenience available at the trailer, assume per capita wastewater production of
200 L/person.d.

Given:

$$BOD = 200 \text{ mg/L} \quad AP = 36 \text{ h} \quad Pop = 15 \times 2 = 30 \text{ p} \quad Q = 200 \text{ L/p·d}$$

Solution:

$$Q = \frac{200\,L}{p.d} \times \frac{2\,p}{trailer} \times 15\,trailers \times \frac{m^3}{1000\,L} = 6.00 = 6.0\,m^3\,/\,d$$

$$V_A = AP \times Q = 36\,h \times \frac{6.0\,m^3}{d} \times \frac{d}{24\,h} = 9.00 = 9.0\,m^3$$

$$\frac{F}{M} = \frac{1}{AP} \times \frac{BOD_{PE}}{MLSS} = \frac{1}{36\,h} \times \frac{200\,g\,/\,m^3}{1500\,g\,/\,m^3} \times \frac{24\,h}{d} = 0.0888 = \underline{0.089\,/\,d}$$

25.8 OXYGEN TRANSFER

Oxygen transfer in to mixed liquor is a two-phase process. In the first phase, oxygen from the air is transferred or dissolved in the mixed liquor. In the second phase, dissolved oxygen is taken up or utilized by the microorganisms in the biochemical oxidation of the waste organic matter.

The rate of **oxygen uptake** depends on several factors, including the temperature of the wastewater and food to microorganism ratio. Increase in temperature and organic loading would increase the uptake rate causing DO in the aeration tank to drop. **Table 25.2** show a typical uptake rate and oxygen requirement for selected types of activated sludge processes. Irrespective of the aeration system used, there is always some way of adjusting its output. This is usually done by increasing the air-flow in diffused aeration systems and by raising the tank level in mechanical aeration systems. The DO level in the aeration rate will increase, decrease, or remain stable depending on how the **oxygen transfer** rate (OTR) compares with the oxygen uptake or utilization rate (OUR).

When the DO level in the aeration tank remains stable, this indicates that the oxygen transfer rate equals the **oxygen uptake rate**. In the early morning hours, OUR runs low, thus DO concentration starts increasing until it reaches an equilibrium value. During the peak organic loading period, an increase in OUR causes a drop in DO level. When toxins are present in wastewater, the DO level will remain high even when BOD loading is relatively high. Biochemical reaction rate drops due to the presence of toxins resulting in increase of DO and drop in OUR.

TABLE 25.2
Uptake rates and air requirement

Activated Sludge Process	Uptake Rate (g/m³·h)	Air Requirement (m³/kg of BOD)
extended aeration	<10	125
conventional	30	95
high rate	100	4

25.8.1 MASS TRANSFER EQUATION

Oxygen transfer rate (OTR) from air bubbles into solution can be expressed by the following mass transfer equation.

$$OTR = KD = K\left(\beta \times DO_{sat} - DO\right)$$

K = mass transfer coefficient, 1/h.
β = oxygen saturation coefficient of wastewater, usually 0.8–0.9.
DO_{sat} = saturation DO for clean water, mg/L, DO = actual DO, mg/L.
D = DO deficit, mg/L OTR = oxygen transfer rate, mg/L.h.

The mass transfer equation indicates that OTR is basically determined by the **transfer coefficient** K and the oxygen deficit maintained. The transfer coefficient factor depends on wastewater characteristics and, more importantly, on the physical features of the aeration system, including fineness of diffuser, liquid depth (shallow or deep), degree of mixing, basin configuration, and characteristics of wastewater. The oxygen transfer coefficient factor will be higher in the case of a fine diffuser installed in deep tanks. However, fine bubble diffusers are more prone to clogging. If some diffusers become clogged, the oxygen transfer rate will drop. The second term in the oxygen transfer equation is the dissolved oxygen deficit. Deficit can be kept high by maintaining low DO levels in the mixed liquor to achieve high transfer efficiency.

25.8.2 SPECIFIC UPTAKE RATE (SUR)

When a comparison needs to be made between plants, uptake rate should not be used until the MLSS is the same. In such cases an uptake rate is usually expressed as specific uptake rate. SUR is the amount of oxygen in mg utilized by one gram of the mixed liquor solids in one hour (mg/g·h).

25.8.3 OXYGEN TRANSFER EFFICIENCY

Oxygen transfer efficiency is the ratio of the mass of oxygen transferred to oxygen supplied. The mass of oxygen supplied can be worked out from the airflow rate and the oxygen content of the air. At standard temperature and pressure, oxygen concentration is 0.279 kg O_2/m³ of air. The aeration transfer efficiency is typically found in the range of 5% to 20%. Diffused aeration systems are usually designed based on air requirements per kg of BOD, as shown in **Table 25.2**. For mechanical aeration systems, the equipment should be capable of transferring at least 1.0 kg of oxygen per kg of BOD applied to the aeration tank.

Example Problem 25.10

An aeration system has a transfer coefficient K of 3.0/h for a wastewater at 20°C with a β of 0.85. What is the rate of oxygen transfer when the DO level in the aeration tank is (a) 2.0 mg/L and (b) 3.0 mg/L?

Given:

K = 3.0/h, DO = 2.0 mg/L, DO$_s$ = 9.2 mg/L at 20°C

Solution:

Oxygen transfer rate

$$OTR = K\left(\beta \times DO_{sat} - DO\right) = \frac{3.0}{h}\left(9.2 \times 0.85 - 2.0\right)\frac{mg}{L} = 17.5 = 18\ mg\,/\,L.h$$

$$OTR = K\left(\beta \times DO_{sat} - DO\right) = \frac{3.0}{h}\left(9.2 \times 0.85 - 3.0\right)\frac{mg}{L} = 15.8 = \underline{16\ mg\,/\,L.h}$$

25.9 OPERATING PROBLEMS

When the activated sludge is operating well, it usually produces effluent containing less than 15 mg/L of solids and BOD each. However, keeping in mind that it is a biological process, it does not take much to upset the process. The key is to detect the problem early and take necessary action before it is too late. Here are some of the operating problems most frequently encountered in the operation of an activated sludge plant.

25.9.1 AERATION TANK APPEARANCE

Under normal operating conditions, the color of mixed liquor is medium brown with an earthy odor. A dark blackish color indicates septic conditions, which may be due to inadequate operation or improper discharge of recycle streams. Another possible reason might be high-strength industrial waste. All these situations may require an increase in air supply.

Turbulence

Much can be said about air distribution by observing turbulence patterns in the aeration tank. One tank may be receiving more air than another, indicating adjustment is necessary. Partially plugged diffusers may create high turbulence at some spots and dead spots at other places. Highly localized turbulence is usually due to a broken or missing diffuser.

Foaming

Another observation that provides clue to the operation of the activated sludge process is foaming in the aeration tank. A small amount of white to light brown-colored foam is an indication of good operating conditions. During initial start-up of the plant or when mixed liquor is relatively dilute, it is normal to expect thick billows of white foam. It is logical to reduce sludge wasting to allow the build-up of mixed liquor solids. Dense, dark-brown foam usually indicates old sludge. As a first step to correct this problem, try increasing sludge wasting. Foam may spill tiny grease particles onto the walking areas and cause unsafe conditions. Such deposits should be cleaned up immediately.

25.9.2 Secondary Clarifier Appearance

The first sign of impending clarification problems is an increase in turbidity of effluent and an increase in the solid content of the effluent. If attention is not paid, it can lead to serious problems.

Bulking is by far the most common problem you might face when operating an activated sludge plant. The first sign of sludge bulking is the rise in sludge blanket. The trend in the sludge volume index will be upwards. This is where the real value of the settling test lies. Operating conditions should be reviewed and adjustments should be made to correct the problem. It has been seen that many of the operators have the tendency to increase return rate to correct bulking. Doing this only makes the problem worse. It is suggested to check the operating data – more importantly the F/M ratio and sludge age to find out what has caused the change, if anything.

A slide of the mixed liquor under the microscope should be checked for **filamentous growth**. If under the microscope you see an abundance of hairlike structures this indicates filamentous growth. Nonfilamentous type bulking is very rare. In this case, the sludge contains a large amount of water trapped in the floc. DO levels significantly higher than 2.0 mg/L are favorable to filamentous growth. If this problem persists, it may be necessary to destroy these organisms by controlled chlorination.

Rising sludge at the top of the final clarifier is usually confused with bulking. The sludge settles well at the bottom of the clarifier. However, after settling, it becomes lighter and clumps of it rise to the surface. The sludge is usually dark gray in color and *rising gas bubbles* are usually associated with it. This problem is caused by **denitrification** and septicity. This usually happens when the sludge age is high and sludge is well oxidized. Again, it might be necessary to adjust the sludge wasting or adjust the loading to the aeration unit.

Pin floc refers to very small floc, usually less than 1 mm in diameter. Some pin floc will always be there. Excess of pin floc is usually caused by overoxidized sludge and or due to unfavorable hydraulic conditions in the aeration tank.

Straggler floc refers to very fluffy, almost transparent and buoyant solids, typically 3–5 mm in diameter. Straggler floc is usually accompanied by clear effluent. In most cases this is due to new growth when the SRT is on the low side.

Deflocculation occurs when sludge breaks up into very small particles that settle poorly, resulting in a turbid effluent. The turbidity is caused by small particles of the broken floc. Deflocculation usually occurs due to the presence of some inhibiting substance – for example, toxins or acid wastes, anaerobic conditions in the aeration tank, organic overloading, nutrient imbalances, and excessive hydraulic loading.

Grease balls of varying sizes can sometimes be found floating in aeration tanks or secondary clarifiers. The joining of the grease particles in wastewater by the gentle rolling action in aeration tank or clarifier forms the balls.

Solids washout refers to flowing out with the effluent even when bulking is not a problem. This problem may be caused by excessive hydraulic loading, uneven weirs, and poor baffling

Toxic substances can cause sudden changes in terms of color, DO, types of dominating organisms, and plant removal. Prevention is the best solution. Any mishap like this should be followed by thorough investigation.

Discussion Questions

1) What parameter is best to describe biomass in an activated sludge process?
2) Compare two activated sludge control parameters, F/M ratio and SRT.
3) Explain with the help of a sketch the components of an activated sludge process plant.
4) Discuss the advantages and disadvantages of various variations of the activated sludge process.
5) Based on the mass balance, derive the relationship for finding expected concentration of return sludge solids, RSSS, in terms of mixed liquor solid concentration, MLSS, and return rate ratio, R.
6) Aeration is an important function in an activated sludge process. Describe various methods of aeration.
7) Describe briefly the main problems encountered in the operation of an activated sludge process.
8) Which parameter is unique to a secondary clarifier and how it is used to control the operation of a secondary clarifier?
9) Assuming a settlement test on mixed liquor exactly simulates settling in a secondary clarifier, how you can decide the rate of sludge return.
10) Explain why the extended rate process does not need primary clarification.
11) Compare pin floc versus straggler floc problems in secondary settling.
12) In the operation of an activated sludge process, DO in the aeration tank is monitored. What typical pattern of DO should you expect on a normal day and how should aeration rate be adjusted?
13) During the peak BOD loading hours of the day, should you expect DO in the aeration tank suppressed? If that is not the case, what inferences do you draw from this and how would you correct it?
14) Discuss how the modifications of step aeration and tapered aeration will improve the efficiency of the activated sludge process.
15) List the factors affecting oxygen transfer efficiency. What role does a plant operator have in achieving higher efficiency?
16) Discuss how state point analysis can be used to study the performance of secondary clarification.

Practice Problems

1) In an extended aeration system with an aeration period of 24 h, mixed liquor solids are maintained at 1500 mg/L. When the incoming BOD is 150 mg/L, at what F/M ratio is the plant operated? (0.10/d).
2) A settling test is performed on a sample of mixed liquor (MLSS = 1500 mg/L) and the volume of settled solids is observed to be 120 mL/L. What is the sludge volume index? (80 mL/g).
3) In an activated sludge plant, the daily average flow is 1.1 MGD and the BOD of the primary effluent is 140 mg/L. What is the F/M ratio, if 4400 lb of solids are kept in the aeration tank? (0.29/d).

4) For the plant indicated in the previous practice problem, it is known that 50% of the incoming BOD becomes new growth. How much of the return sludge with 0.40% solids should you waste to maintain the constant MLSS? (19 000 gal).

5) In a solid test the following weighings were made: crucible = 19.9850 g, crucible + dry solids = 20.0503 g, crucible + ash = 20.0068 g. What fraction are volatile solids? (67%).

6) Mixed liquor solids are estimated at 800 kg. If return sludge concentration is 0.8%, how many m^3 of sludge should be wasted to achieve a SRT of 8 d? (13 m^3/d).

7) Design an activated sludge process for a town with daily average flow of 10 ML/d. Assume primary BOD removal is 40% of the incoming BOD of 250 mg/L. It is desired that the secondary effluent BOD should not exceed 15 mg/L. Based on the pilot studies α is 0.5 and K_e is 0.05/d. Assuming an MLSS of 2200 mg/L, RSSS of 8000 mg/L and SRT of 10 d, find aeration capacity, sludge wasting rate, and sludge return ratio (2.1 ML, 58 m^3/d, 38%).

8) What is the required aeration capacity to treat primary effluent of 1.5 MGD with a BOD of 120 mg/L? It is desired to operate the process at F/M ratio of 0.20/d while maintaining MLSS of 2000 mg/L. What is the required length of each of the twin aeration tanks assuming a width of 20 ft and depth of 16 ft? (0.45 MG, 94 ft).

9) A conventional activated sludge receives wastewater containing BOD of 200 mg/L. Assume 35% BOD removal by the primary unit what is the minimum removal required by the secondary process to produce and final effluent with BOD of 20 mg/L find BOD removal by the secondary process, (85%)

10) For a conventional activated sludge plant, the operational data is as follows: Q = 25 ML/d, BOD_{raw} = 320 mg/L, primary BOD removal = 35%, secondary effluent BOD_{II} = 25 mg/L, aeration capacity, V_A = 9500 m^3, MLSS = 2200 mg/L, WSSS = 0.90%, wastage rate, Q_{WS} = 200 m^3/d. Calculate F/M, SRT, and overall BOD percent removal. (0.25/d, 12d, 92%)

11) A conventional activated sludge plant receives a daily wastewater flow of 9.0 MGD with BOD content of 260 mg/L. Aeration capacity of the plant is 2.8 MG and return sludge solid concentration is 0.98%. Return sludge is wasted at the rate of 58 000 gal/d to maintain mixed liquor solid concentration at 2500 mg/L. Assuming primary BOD removal of 35%, work out aeration period, F/M ratio, and SRT, neglecting the solids leaving in the effluent. (7.5 h, 0.22/d, 12 d).

12) Design an activated sludge process to produce secondary effluent with BOD not exceeding 20 mg/L. Assume BOD of the primary effluent is 150 mg/L and daily wastewater flow is 15 ML/d. Assume yield coefficient α = 0.65, k_e = 0.05/d and SRT = 10 d, recirculation rate = 50% and MLSS = 2000 mg/L (V_A = 4.2 ML, RSSS = 6000 mg/L, Q_W = 140 m^3/d, F/M = 0.27/d).

13) Determine the aeration capacity to treat a wastewater flow of 8 MGD with BOD of 210 mg/L to provide aeration period of 6.5 h. What should be the MLSS to maintain F/M ratio of 0.25/d? (1.1 MG).

14) A conventional aeration tank is to treat a flow of 4.0 ML/d of primary effluent. Raw wastewater BOD is 200 mg/L and 35% removal is expected in the primary treatment. The MLSS concentration is to be maintained at 2000 mg/L and F/M ratio of 0.22/d is specified. Compute the capacity of the aeration tank. If the side water depth is 4.0 m, and length is three times the width, how long should the tank be? (1180 m³, 30 m).

15) A suspended solids test on a sample of mixed liquor from an aeration tank indicates MLSS of 1850 mg/L. In the 30 min settling test, the sludge volume is measured to be 150 mL/L. Compute SVI. Assume settling test exactly simulates settling in secondary clarifier.
   a)   What minimum return ratio would you recommend? (81 mL/g, 18%).
   b)   If actual return ratio is 25%, what is the expected concentration of solids in the return sludge? (0.93%).

16) The mixed liquor in a 0.4 MG capacity aeration tank is operating at MLSS of 2200 mg/L. Return sludge recirculation ratio is maintained at 40%. If the target MLSS is 2000 mg/L, what additional volume of sludge needs to be wasted? (10 000 gal).

17) The flow to a 320 m³ high-rate aeration system is 3.0 ML/d. The BOD concentration of the primary effluent is 350 g/m³. What should be the concentration of MLSS to achieve F/M ratio of 1.0/d? (3300 mg/L).

18) A wastewater pollution control plant treats daily flow of 4500 m³ with average primary effluent BOD of 150 mg/L, MVLSS of 1500 mg/L, and aeration capacity of 900 m³. Calculate the F/M ratio. (0.5/d).

19) The aeration tank is 22 ft deep with a surface area of 6500 ft². The MLSS concentration is 2500 mg/L and the solids concentration in the return activated sludge is 6000 mg/L. If 0.15 MG of activated sludge is wasted per day and final effluent has a SS concentration of 10 mg/L at 1.8 MGD, what SRT is achieved? (2.9 d).

20) In a conventional activated sludge plant, MLSS is determined to be 3200 g/m³ and activated sludge is returned at the same rate as the daily wastewater flow rate of 30 ML/d. Calculate the rate at which activated sludge should be wasted to attain a SRT of 10 d, given that the aeration volume is 4500 m³. (230 m³/d).

26 Stabilization Ponds

Stabilization ponds is the term used for **oxidation ponds** and **lagoons**. They are the most common secondary treatment process in small suburban and rural communities. Their popularity in rural areas is due to the fact that large land areas available at a lower cost and their operation is simple and easy.

Some authors distinguish lagoons from ponds in that oxygen is provided by artificial aeration. The stabilization ponds are shallow, impervious, or watertight basins, formed by excavating the topsoil and building earthen dikes. These basins are then lined with clay to prevent leakage. The adjoining groundwater is monitored for any possible contamination. Non-aerated stabilization ponds can be aerobic, anaerobic, or facultative. **Aerobic ponds** are shallow to maintain aerobic conditions. Such ponds are common for providing tertiary treatment to secondary effluents. In such cases, they are called **polishing lagoons**.

In **facultative ponds**, oxygen requirements are met by the transfer of oxygen at the air–water interface and by **photosynthesis** within algae and wind aeration. The water temperature is also a factor due to the solubility of oxygen. These types of lagoons are operated maintaining water depths of 0.9 to 1.5 m (3–5 ft) to accommodate oxygen requirements. A free board of 90 cm (3.0 ft) is kept above the high-water level. In Northern climates, depths of 1.5 m are usual to prevent freezing of the entire depth. **Anaerobic ponds** are relatively deeper and are usually employed to treat strong industrial wastes. A scum can form at the top and keeps odors under control.

26.1 FACULTATIVE PONDS

The term facultative indicates that both aerobic and anaerobic conditions occur in the pond. As shown in **Figure 26.1**, wastewater enters the facultative pond and heavy solids settle out at the inlet, where anaerobic bacteria break down the complex organics. These organics become organic acids upon which the aerobic bacteria in the surface layer feed and finally convert to gases and nutrients.

Algae consume nutrients and some by-products of the anaerobic reaction to release oxygen into water. This reaction is called **photosynthesis** since sunlight is required for this bio-reaction to take place. The dissolved oxygen level is highest at midday, when photosynthesis is at its peak. In normal operation, dissolved oxygen can reach levels as high as 20 mg/L. Since carbon dioxide is consumed by the algae during photosynthesis, the pH of water rises as photosynthesis progresses. During normal operating conditions, both DO and pH will show diurnal variation.

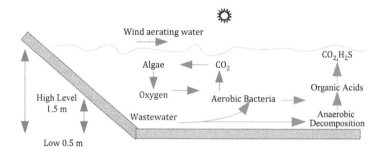

FIGURE 26.1   Facultative pond biology

In Northern climates, photosynthesis may not happen at all due to blocking of sunlight by the frozen layer. In such cases, wastewater is stored until spring when algae are re-established. Where the **facultative lagoons** are organically overloaded, some artificial aeration is used to maintain aerobic conditions in the top layer. However, aeration is limited to allow some settling and keep the bottom layer anaerobic. **Aerated facultative** lagoons can handle as much as ten times more loading. Detention time can be as small as 2–5 days compared to 10–30 days in a normal facultative lagoon.

26.2   LOADING PARAMETERS

Design parameters for lagoons depend on whether they are aerobic or anaerobic. Anaerobic lagoons are dependent on detention time for their treatment; therefore, volumetric loadings rather than surface loading are used. The organic loading in anaerobic lagoons is measured in relation to the volume available. Facultative and aerated lagoons use organic loading based on water surface area. Typically, BOD loading rate in facultative lagoons is 2–5 $g/m^2 \cdot d$ (2–50 lb/acre.d). In warmer climates, where ice coverage does not form, organic loading as high as 3–5 times may be afforded. Hydraulic loadings on lagoons are usually expressed as overflow rate. The retention time may be in months depending on the applied load, depth of wastewater and losses due to seepage and evaporation.

Example Problem 26.1

A facultative pond for a small town consists of a 6-ha primary cell and two smaller cells of 3 ha each. The average daily wastewater flow is 1.0 ML/d containing a BOD of 200 mg/L. Calculate the BOD loading based on the area of primary cell.

Given:

> Q =1.0 ML/d, BOD = 200 mg/L = 200 kg/L
> Area = 12 ha (total), 6.0 ha (primary)

Solution:

$$BODLR = \frac{Q \times BOD}{A_S} = \frac{1.0\ ML}{d} \times \frac{200\ kg}{ML} \times \frac{1000\ g}{kg} \times \frac{1}{6\ ha} \times \frac{ha}{10000\ m^2}$$

$$= 3.33 = 3.3\ g/m^2.d$$

Example Problem 26.2

A facultative pond has an average length of 700 ft with an average width of 450 ft. Daily average flow rate to the pond is 0.6 MGD. Determine the hydraulic detention time when operated maintaining a water depth of 6.0 ft.

Given:

$$Q = 0.60\ mil\ gal/d \quad L = 700\ ft \quad W = 450\ ft \quad d = 6.0\ ft$$

Solution:

$$t_d = \frac{V}{Q} = \frac{L \times W \times d}{Q} = 700\ ft \times 450\ ft \times 6.0\ ft \times \frac{d}{0.6 \times 10^6\ gal} \times \frac{7.48\ gal}{ft^3}$$

$$= 28.5 = 24\ d$$

26.2.1 BOD Removal

As the wastewater flows through an oxidation pond, BOD is removed by biological oxidation. As flow pattern is basically plug flow, BOD removed can be related to detention time by knowing the BOD rate constant, K. Since BOD is a first-order reaction, BOD remaining at any time t is a function of initial BOD. Difference between influent and effluent BOD equals BOD removed in the process. In terms of detention time t, BOD rate equation can be written as:

$$BOD_t = BOD_i \times e^{-Kt} \quad or\ t = \frac{1}{K} \times \ln\left(\frac{BOD_t}{BOD_i}\right)$$

BOD_t = BOD in the effluent leaving after time t
BOD_i = BOD in the influent entering the pond
t = detention time K = BOD rate constant.

The value of K should be modified for the operating temperature. Because of the shape of the reactor, inlet outlet design, wind action flow conditions in ponds are somewhere between plug flow and completely mixed flow.

Example Problem 26.3

Find out the detention time required to remove 90% of BOD in a stabilization pond with minimum operating temperature of 12°C. Assume BOD rate constant of 0.22/d at 20°C. Also determine the pond area needed to treat a flow from 9500 p @ 180 L/c.d. Assume the pond is operated maintaining a water depth of 1.5 m.

Given:

$$K_{20} = 0.22/d \quad BOD_i/BOD_e = 10 \quad T = 12°C \quad Q = 180 \text{ L/p·d}$$

$$Pop. = 9500 \quad d = 1.5 \text{ m}$$

Solution:

$$Q = \frac{180 \ L}{p.d} \times 9500 \ p \times \frac{m^3}{1000 \ L} = 1710 = 1700 \ m^3/d$$

$$K_{12} = K_{20}(1.047)^{12-20} = \frac{0.22}{d \times (1.047)^8} = 0.152 = 0.15/d$$

$$t = \frac{1}{K}\ln\left(\frac{BOD_i}{BOD_i}\right) = \frac{d}{0.152} \times \ln(10) = 15.1 = 15 \ d$$

$$A = \frac{V}{d} = \frac{Q \times t}{d} = \frac{1710 \ m^3}{d} \times \frac{15.1 d}{1.5 \ m} \times \frac{ha}{10000 \ m^2} = 1.726 = 1.7 \ ha$$

26.2.2 WINTER STORAGE

The BOD removal in ponds is very much dependent on climatic conditions. During winter, bacterial activity and algae growth are both severely retarded by cold temperatures. In areas with snow, this is evident from the strong odors in the spring thaw.

In facultative ponds, operating water depths range from 0.5 to 1.5 m (1.5–5 ft). The minimum depth is needed to prevent growth of aquatic weeds. In Northern climates, the water level in the pond is lowered to the minimum level before the winter sets in. Discharge in the winter is minimized or completely stopped and the incoming wastewater is stored until spring. The pond area should be large enough to store the wastewater over the winter months.

Example Problem 26.4

The stabilization pond of 12 ha receives average daily flow of 1200 m³/d. Since the lagoon is operating in colder climates, discharge is stopped in the early winter and the water level is dropped to 0.60 m. Estimate the number of days of winter storage available between 0.60 m and 1.5 m water levels assuming an evaporation and seepage loss of 2.5 mm/d.

Given:

$$d_{sto} = 1.5 \text{ m} - 0.60 \text{ m} = 0.9 \text{ m}, \ A_S = 12 \text{ ha} = 120 \ 000 \text{ m}^2$$
$$Q = 1200 \text{ m}^3/d, \ d_{loss} = 2.5 \text{ mm/d}$$

Solution:

$$Q_{loss} = \frac{2.5\,mm}{d} \times \frac{m}{1000\,mm} \times 120000\,m^2 = 300.0 = 300\,m^3/d$$

$$Q_{net} = 1200 - 300 = 900\,m^3/d$$

$$t_{sto} = \frac{V_{sto}}{Q_{net}} = \frac{A_S \times d_{sto}}{Q_{net}} = 120000\,m^2 \times 0.9\,m \times \frac{d}{900\,m^3} = \underline{120\,d}$$

Example Problem 26.5

A sewage lagoon with discharge control has a total surface area of 90 acres. Average daily flow is 1.1 MGD. Calculate the minimum storage depth required to provide storage time of 90 d. Assume that losses due to evaporation and seepage minus precipitation are 0.04 in/d.

Given:

A_S = 90 acres Q = 1.1 mil gal/d t_{stor} = 90 d d_{sto} = ?

Solution:

$$d_{in} = \frac{Q}{A} = \frac{1.1 \times 10^6\,gal}{d} \times \frac{ft^3}{7.48\,gal} \times \frac{acre}{90\,acre \times 43560\,ft^2} \times \frac{12\,in}{ft} = 0.45\,in/d$$

$$d_{sto} = t_{sto}\left(d_{in} - d_{loss}\right) = 90\,d \times \frac{(0.45 - 0.04)\,in}{d} \times \frac{ft}{12\,in} = 3.07 = \underline{3.1\,ft}$$

26.3   ALGAE

Though growth of algae is important, if discharged with plant effluent it has negative effects including increased turbidity, suspended solids, and biochemical oxygen demand. It is for this reason that algae need to be monitored carefully to ensure that the algae perform properly without affecting the lagoon's overall performance.

Algal blooms refer to rapid mass growth of algae. This usually happens 7 to 12 days after wastes have been introduced into the lagoon. After another week, bacterial decomposition of bottom solids will usually become established, limiting the food produced for the algae.

26.4   BERMS

To prevent erosion, the berm materials must be of a rocky nature or protected by riprap. A gentle slope will erode the least. Also, it is easier to operate equipment to perform routine maintenance. The usual slope on a berm for a lagoon is 1:4. The other important aspect with the berm slope is the correlation with evaporation within the lagoon. A steeper slope will result in less evaporation from the lagoon than a shallow slope. If high winds are expected in the area where the lagoon will be constructed, try to arrange the lagoon so the winds will blow across the short width of the lagoon rather than the length in order to reduce berm erosion caused by waves.

26.5 DAILY MONITORING

One of the most effective types of monitoring in the operation of a lagoon is daily observation. There are no equipment requirements and with some operational experience the operator can determine the condition of the lagoon by the following observations.

26.5.1 VISUAL MONITORING

There are several signs that indicate the level of pH in the lagoon. For example, a deep green sparkling color generally indicates a high pH and satisfactory dissolved oxygen content. A dull green color or lack of color generally indicates a declining pH and lowered dissolved oxygen content.

26.5.2 WATER COLOR

As shown in **Table 26.1**, algae color is directly related to pH and dissolved oxygen and is a good indication of the health of the pond. A change to a less desirable color has a cause and may require correction. Records of color and depth of water should be made daily.

26.5.3 WATER LEVEL

Usually, lagoons are equipped with a post having markings to check water depth. Water depth should be recorded at the same time each month to monitor changes. This provides information for normal operation of the lagoon, for changes and trends.

TABLE 26.1
Color of algae (visual monitoring)

Color	Conditions	Symptoms or Cause
dark sparkling green	good	pH and dissolved oxygen (DO) ideal
dull green to yellow	not so good	DO and pH are dropping; blue-green algae type becoming predominant
gray to black	very bad	lagoon is septic; anaerobic conditions prevail
tan to brown	OK if brown algae	erosion or inflow of surface water

26.6   OPERATIONAL PROBLEMS

Lagoon operation usually does not require very high skills. Common problems that might be encountered during the operation and maintenance of a lagoon are detailed below.

26.6.1   SCUM CONTROL

The accumulation of scum is common in the spring when the water warms and vigorous biological activity resumes. If scum is not broken up it will dry and become crusted, providing a home for blue algae creating odors. Also, scum would block the sunlight, thereby reducing the production of oxygen by algae.

26.6.2   ODOR CONTROL

Most public complaints are related to odors. Most odors are caused by overloading or poor housekeeping practices and can be remedied by taking corrective measures. If a lagoon is overloaded, it is necessary to stop loading and divert the influent to other lagoons, if available, until the odor problem stops. The lagoon should then be gradually loaded again.

Odors occur during the spring warm-up in colder climates because biological activity has been reduced during cold weather. The use of floating aerators and heavy chlorination might help treat odors, but these treatments are usually very expensive. Recirculation from an aerobic lagoon to the inlet of an anaerobic lagoon will reduce or eliminate odors.

Example Problem 26.6

Due to an increase in odor problems in a facultative lagoon, it is decided to apply sodium nitrate @ 55 kg/ha·d. The chemical is to be applied in the wake of a motor boat. How many kg of the sodium nitrate will be needed to treat a lagoon surface measuring 130 m × 200 m?

Given:

application rate = 55 kg/ha·d   surface area = 130 m × 200 m

Solution:

$$Application = \frac{55 \, kg}{ha \cdot d} \times 130 \, m \times 200 \, m \times \frac{ha}{10000 \, m^2} = 143 = \underline{140 \, kg/d}$$

Chemicals that act to mask odors are also used. Make sure to order the chemical before the spring thaw when the odor problem is expected to more severe. Some facilities opt for the use of sodium nitrate as a source of oxygen for microorganisms rather than sulphate compounds. Once the sodium nitrate is mixed into the lagoon it acts very quickly because many common organisms may use the oxygen in nitrate compounds instead of dissolved oxygen.

26.7 LAGOON MAINTENANCE

The maintenance of a lagoon can be done in two ways: artificial or natural. The artificial method tries to regulate every single aspect of the process of the lagoon, surrounding berm, and environment. This process will commonly remove unplanned plant growth and animal life. The disadvantages to this form of maintenance are that it can cause extra work for the operating personnel and one forced change may require a further change to balance the lagoon again. The natural method will allow for natural solutions to some of the problems faced when operating a lagoon. The natural method usually tries to use naturally occurring vegetation and animals to control the operations of a lagoon.

26.7.1 LAGOON WEEDS

One method to control weeds is by conducting daily inspections and immediately removing young plants. The reason that they are removed is because some types of weeds will harbor mosquitos, hinder lagoon circulation, and lead to scum accumulation.

26.7.2 BERM EROSION

Berm slope erosion caused by wave action or surface runoff from precipitation is probably the most serious maintenance problem. If allowed to continue, it can result in a narrowing of the berm crown that will make accessibility with maintenance equipment most difficult. The other long-range solution is to plant grasses and plants that will help anchor the berm slopes against erosion. Portions of the lagoon berm or dike not exposed to wave action should be planted with a low-growing spreading grass to prevent erosion by surface runoff.

26.7.3 MOSQUITOS

One solution to mosquitos is to stock the lagoon with mosquito fish which will eat the mosquito larvae. Another solution is to use duckweed in the pond because the mosquito larvae will not survive when covered in duckweed. A further solution is to encourage mosquito-eating birds to live in the area. One of the more effective ways of doing this is to plant low bushes around the lagoon.

26.7.4 DAPHNIA

Minute shrimp-like animals, called daphnia, may infest the lagoon from time to time during the warmer months of the year. The daphnia will reproduce in great numbers, usually appearing in the lagoon 3 to 7 days after an algae bloom. These predators live on algae and at times will appear in such numbers as to almost clear the lagoon of algae. During the more severe infestations there will be a sharp drop in the dissolved oxygen of the lagoon, accompanied by a lowered pH. However, this is a temporary condition because the predators will die off due to algae overload and will be followed by a rapid growth of algae.

Discussion Questions

1. Explain the role of algae in a facultative pond.
2. Compare oxidation pond with oxidation ditch.
3. Explain the term symbiosis as applied to facultative stabilization ponds. Which microorganisms are responsible for the symbiotic relationship?
4. Make a comparison between an aerated lagoon and a stabilization pond.
5. Explain why stablization ponds are more common in rural settings.
6. Explain the mechanism by which BOD is removed in a facultative pond.
7. How are scum and odor controlled in a lagoon system?
8. Describe the adavantages and disadvantages of natural and artificial maintence of a lagoon.
9. Related to lagoon maintenance, explain mosquito control, weed control, berm erosion, and screenings.
10. In a faculatative lagoon, which of the two parameters shows diurnal variation and why?
11. List the parameters which must be monited by sampling to assist in the operation of a facultative system?
12. How can the colour of algae growth in a facultative pond be used to monitor the health of the operation?
13. Visual monitoring is important in the operation of a lagoon system. In the operation of a facultative pond, what interpretations can you make based on the colour of algae?
14. Explain in brief the priciples of working and application of aerobic, anaerobic, and aerated lagoons.
15. In colder climates, no effluent is dicharged over the winter months. How does it affect the operation of a facultative lagoon?

Practice Problems

1. How many millions of gallons of water are needed to increase the water depth in a 10 acre lagoon by 4 inches? (1.1 MG).
2. In a community of 1100 people, the wastewater treatment system is a lagoon. The average daily wastewater flow is 320 m^3/d. What is the daily per capita flow? (290 L/c.d).
3. A pond measuring 230 m long and 130 m wide receives a wastewater flow of 80 m^3/d. Find the hydraulic loading. (2.7 mm/d).
4. Select the size of a rectangular stabilization pond for treating wastewater from a new development with 5000 people, contributing 140 L/c.d of

wastewater containing BOD of 280 mg/L. Due to warmer climatic conditions, BOD loading of 25 g/m².d can be afforded.(130 m × 65 m).

5. A 1 ha pond lagoon is operated by maintaing a water depth of 1.3 m. What detention time is provided if the pond is treating flow contributed by 6000 people at the rate of 120 L/c.d.(18 d).

6. For the data of Practice Problem 2, select the size of inlet pipe to flow at an average velocity of 1.0 m/s. Assume most of the flow is contributed over a peiod of 10 h. (110 mm).

7. What detention period must be provided to achieve 90% BOD removal in a stabilization pond? Assume BOD reaction rate constant of 0.20/d. (12 d).

8. Design an oxidation pond for serving a community of 10 000 people assuming per capita wastewater contribution of 190 L/c.d containg BOD of 280 g/m³. As this community is in the tropical region, BOD loading of 30 g/m².d can be afforded.

   a. Assume two primary and one secondary pond, all of the same size and with length twice as much as width. (# = 3, each 110 m × 55 m).

   b. Assume per capita sludge production of 75 L/c.a on an annual basis. If bottom 0.40 m depth is for storage of sludge, how long would it last before sludge needed to be removed? (10 years)

9. Based on allowable loading of 25 g/m².d, what pond area is needed to treat a wastewater flow of 1.5 ML/d with BOD content of 260 mg/L. Determine the minimum operating depth for providing retention time of 15 d with net loss of water of 2.0 mm/d. (1.6 ha, 1.4 m).

10. Design a facultative pond to serve a community of 8000 people with per capita wastewater discharge of 160 L/c.d and BOD of 250 mg/L. Water regulation requires effluent BOD not to exceed 30 mg/L. Assume BOD rate constant of 0.16/d at the operating temperature and allowable BOD loading of 20 g/m².d. Determine pond area required, detention period, and depth. (1.6 ha, 13 d, 1.1 m).

11. From the data of Practice Problem 10, it is planned to have a parallel-series system consisting of a total of 6 ponds, with 4 primary and 2 secondary ponds. Each row contains two primary and one secondary pond. Assuming the width of each pond is 2.5 times the width, select the size of each pond. (82 m × 33 m).

12. Design a rectangular stabilization pond for treating wastewater from a subdivision of 5500 people, contributing @ 110 L/c.d containing BOD of 210 mg/L. It is planned to have two primary cells in parallel and a common secondary cell. Assume width of each cell is 1/3rd of length and, due to warmer conditions, BOD loading of 25 g/m².d can be afforded. Select the size of each cell. (# 3, 2 primary, 1 secondary, each 72 m × 24 m).

13. A stabilization pond with total surface area of 16 acres receives a wastewater flow of 0.25 MGD with a BOD concentration of 220 mg/L. Calulate the BOD loading and winter storage available between 2 ft and 5 ft depths. Assume net loss of 0.05 in/d during the winter months. (29 lb/acre.d, 69 d).

14. Based on a BOD loading of 30 lb/acre.d, what primary pond area is needed for an average flow 0.50 MGD containg 170 mg/L of BOD. Assuming a secondary pond with half the area of the primary pond is provided, what should be the operating depth to achive a hydraulic retention time of 50 d? (24 acres, 2.1 ft).

15. A sewage lagoon with discharge control has a total surface area of 22 ha. Average daily flow is 3200 m³/d. Calculate the minimum water depth required for achieving a retention time of 90 d. The difference between evaporation and seepage and precipitation is estimated to be 1.0 mm/d. (1.2 m).

16. The stabilization pond for a town consists of two cells for series operation. The first cell has an area of 2.7 ha and the second is smaller with 1.5 ha. Water loss from the pond averages 2.1 mm/d and the daily average influent wastewater is 300 m³/d. Calculate the minimum storage depth required to ensure zero discharge for 120 d. (0.61 m).

17. Stabilization ponds with a total surface area of 15 acres receive a wastewater flow of 1.4×10^5 gal/d. Work out the winter storage available between 2.0 ft and 5.0 ft depths assuming a water loss 0.02 in/d during the winter months. (110 d).

18. A facultative pond serves a population of 4500 people. The average dimensions of the pond are 200 m × 150 m with an average water depth maintained at 1.0 m. The average daily flow is 1.5 ML/d with BOD concentration of 150 g/m³. Calculate detention time, hydraulic loading, population loading, and BOD loading. (20 d, 50 mm/d, 1500 person/ha, 7.5 g/m²·d).

27 Attached Growth Systems

27.1 TRICKLING FILTERS

Trickling filters, one of the oldest forms of biological treatment, can achieve a good-quality effluent in 3 to 4 hours. Compared to the activated sludge process, a trickling filter falls in the category of **fixed growth** or **attached culture** systems. Trickling filters are essentially a biological contact bed and are also called **biological filters**. A typical trickling filter plant flow diagram is shown in **Figure 27.1**.

The words filtration and filter are misnomers, as there is no straining or filtering action involved. Biological filtration consists of a tank filled with a fixed media on which the biological growth lives, and the primary effluent is sprayed. There is also a thin **anaerobic** section at the bottom.

As wastewater moves through the filter bed, the non-settleable colloidal and dissolved solids convert to readily settleable solids. These largely organic solids are converted to microorganisms and get attached to the biological media. This build-up of solids is continuously unloading in small amounts resulting in the need for a secondary clarifier. Falling of the slime layer attached to the filter media is called **sloughing**. The solids sloughing (breaking) off the media is called **humus**. The wastewater is applied on the top of the filter by a water distribution system like sprinklers. As the water trickles through the filter, organisms attached to the media pick up BOD from the passing water.

The hydraulic loading needs to be very tightly controlled. Too much flow can cause ponding and anaerobic conditions. A flow which is too small may not be able to move the distributor arm of the sprinkler system. In trickling filter systems, one important aspect of recirculation is maintaining the minimum hydraulic loading during the minimum flow period. Although trickling filters are known for their ease of operation and sturdiness, the problem of plugging is a major disadvantage. Plugging occurs due to excessive organic growth, which plugs the air passages. This results from organic overloading which creates anaerobic conditions as well as odor problems. Other disadvantages are the high cost of construction, the need to be covered in colder climates, the large area required for set-up as well as the odor and concomitant fly nuisance. Advantages include ease of operation. Moreover, the system is quite forgiving and quite stable, no aeration equipment is needed, and there is a lower cost of operation.

DOI: 10.1201/9781003347941-32

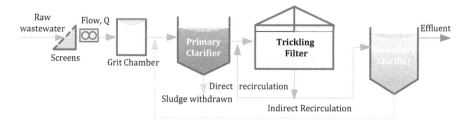

FIGURE 27.1   A typical trickling filter plant

COMBINED SYSTEMS

Although an activated sludge process demands skilled operation, it offers the flexibility of controlling the quality of the effluent. Recently, combined systems are becoming popular to take advantage of the strengths offered by fixed growth and suspended growth systems like the activated sludge process. For example, combining the trickling filter and the activated sludge process has helped to eliminate shockloads to the highly sensitive activated sludge process and provide high-quality effluent. Using a trickling filter alone cannot be expected to yield high removals.

27.2   MAIN COMPONENTS OF THE TRICKLING FILTER

The main components of a biological filter are filter media, filter underdrains, and a wastewater distribution system.

27.2.1   FILTER MEDIA

The filter media supports biological growth. This slime growth, sometimes called a **zoogleal film**, contains the microorganisms. The media may be rock, coal, bricks, or molded plastics. The media should provide sufficient porosity (voids) for air to ventilate the filter. The media depth ranges from 1 m to 2.5 m (3–8 ft) for rock media filters and from 5 m to 10 m (15–30 ft) for synthetic media, also called **biological towers**. The material used for the filter media should be clean, inert, and range in size from 25–75 mm (2–3 in). Large-size media is placed at the bottom above the underdrains.

27.2.2   UNDERDRAINS

The underdrain system carries away the effluent and supplies an air source. The holes in the underdrain are determined by the amount of air required. The underdrain system is generally made of prefabricated blocks of concrete or vitrified clay. The flow from the underdrains is led to a main effluent channel graded to flow partially full to allow ventilation and the passage of air from the bottom to the top of the filter. This main effluent channel may be provided adjoining the central column of the distributor or along the periphery of the filter. The slope of the effluent channel should be sufficient to ensure a flow velocity of 0.90 m/s (3 ft/s). Since the wastewater in the top part of the media is relatively warmer, it creates natural circulation of air from bottom to top. Unless there is a serious plugging problem, this natural circulation is sufficient to maintain aerobic conditions.

27.2.3 WASTEWATER DISTRIBUTION

The **distributor arm** fitted with **orifices** provides a uniform hydraulic loading to the filter. The number of orifices in each section of the arm varies in proportion to the areal coverage. The arm is usually driven by hydraulic action of the wastewater flowing out. The return rate is essential to arm rotation. The fixed-nozzle distribution system is not as common as the rotary type. Rotary-type distributor requires less maintenance and provide uniform application. In fixed rotary distributors, the shape of the filter is usually square or rectangular, but in rotary-type distributors, filter enclosures are always circular. The loading on filters can vary quite a bit depending on the type of filter and the number of stages.

27.2.4 LOADING ON FILTERS

There are three types of filters, standard or low rate, high rate, and super high rate or roughing. The main features, including loading, recirculation, depth, and type of filter media are shown in **Table 27.1**. Organic loading on filters is described as BOD per unit volume of the filter media while hydraulic loading is described as flow rate per unit surface area of the filter top surface.

Standard Filter

A **standard** or low-rate filter has an organic loading of 50–400 g/m^3·d (3–50 lb/1000 ft^3.d) and hydraulic loading of 10–40 m^3/m^2·d. (250–1000 gal/ft^2.d). This type of filter produces quite nitrified effluent, with BOD removal of only 80% to 85% without any recirculation. The low rate trickling rate filters are more suitable for treatment of low- to medium-strength sewage.

High Rate

The basic difference between a high rate and a conventional filter is increased loading**.** The high rate trickling filters, single stage and two stage, are recommended for medium- to high-strength wastewaters. **High-rate** filters achieve BOD removals of only 65% to 80% and there is little nitrification. Organic loading is 500–2000 g/m^3·d and hydraulic loading of 4–40 m^3/m^2·d. due to increased flows. The biofilm is relatively thinner and is more efficient to supply oxygen and nutrients to aerobic bacteria.

Roughing Filters

Super high-rate filters are sometimes used for treatment of high-strength wastes or pre-treatment of domestic wastes. They are characterized by low removal efficiencies and high loading rates. For this reason, they are also called **roughing** filters.

TABLE 27.1
Design features for trickling filters

Design Features	Low Rate	High Rate	Roughing
hydraulic loading, $m^3/m^2 \cdot d^*$	1–4	4–40	20–40
organic loading, $g/m^3 \cdot d$	50–400	500–2000	8000–6000
depth, m	1.8–3.0	0.9–2.5	4.5–12
recirculation ratio	usually absent	0.5–3.0	1–4
filter media	rock, gravel slag	rock, synthetic	plastic media
size of media, mm	25–75	25–60	25–60
sloughing	intermittent	continuous	continuous
BOD removal, %	80–85	65–80	low
quality of effluent	nitrified	partially nitrified	not nitrified
quality of sludge	black, highly oxidized	brown, not fully oxidized	brown, not fully oxidized
flexibility	less flexible	more flexible	–
cost of operation	more	less	
flies	many	varies	few
land requirement	more	less	
supervision/operation	less skilled	more skilled	

$^*g/m^3 \cdot d = 0.0625 \ lb/1000 \ ft^3 \cdot d$ $m^3/m^2 \cdot d = 1.07 \ MG/acre \cdot d.$

Super-rate filters use synthetic media, usually interlocking corrugated sheets of plastic, thus producing a non-clogging filter. Due to better transfer oxygen efficiency, the depth of such filters can be as much as 10 m (33 ft), hence the name **biological towers**.

27.2.5 RECIRCULATION

Though recirculation is not practiced in low-rate filters, in high-rate filters it is a must. **Indirect recirculation** includes the recirculation of underflow from the final clarifier to the head unit during low flows periods. This help to maintain minimum flows required to keep the distributor arm moving and to prevent the filter surface from becoming dry. **Direct recirculation** will represent the flow being returned directly to the tank they just exited. Both types of circulations and staging are shown in **Figure 27.2**. In the two-stage process these recycling flows will be further subdivided into Stage I and Stage II.

The advantages of recirculation are as follows.

- Increases BOD removal efficiency.
- Maintains higher hydraulic loading that helps to reduce plugging and aids in uniform organic load over the filter surface.
- Dampens the variations in strengths and flows of wastewater applied to filters.

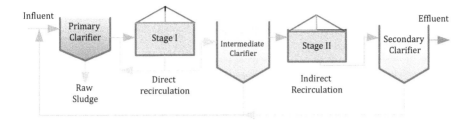

FIGURE 27.2   Recirculation and staging in TF filter plants

- Reduces thickness of the biological film and fly breeding by forced sloughing.
- Keeps distributor arm running during low flow periods in the middle of the night.
- Loads more effectively the lower parts of the filter.
- Freshens the sewage and thus helps to keep odors under check.

Recirculation Ratio

The recirculation ratio is defined as the ratio of recirculated flow, Q_R, to the influent flow, Q. It usually ranges from 0.5 to 3.0, and >3 is used with high-strength wastewaters and super-rate filters. The recirculation ratio is used to determine the required capacity of recirculation pumps and the hydraulic load placed on the filter.

27.2.6   STAGING

High-rate or single-stage trickling filters are commonly used in industrial pre-treatment or in treating strong municipal wastewaters. Two-stage trickling filters are used when a BOD effluent of 30 mg/L is required for strong sewage. The **two-stage** system as shown in **Figure 27.2** consists of two identical trickling filters in series with an optional intermediate settling tank in between. The system is designed with several recirculation points from a single stage to a two-stage trickling filter.

27.3   BOD REMOVAL EFFICIENCY

BOD removal efficiency of biological filtration primarily depends on such factors as depth of bed, kind of media, temperature, recirculation, and of course organic loading. Empirical equations have been developed to predict the BOD removal efficiency based on organic loading and recirculation ratios.

One of the most popular formulations evolved from filter plants at military installations in the USA. The equation given below is applicable to single-stage stone-media filters, followed by a final clarifier and for treating settled domestic wastewater with a temperature of 20°C. For single-stage or for the first stage of a two-stage filter system, BOD removal can be estimated using the following empirical relationship.

BOD removal efficiency empirical formula

$$E_{20} = \frac{100\%}{\left(1 + K\sqrt{(L/F)}\right)} \quad F = \frac{1+R}{\left(1+.1R\right)^2} = \frac{1+R}{1+0.2R}$$

In this equation L is the BOD loading on the filter expressed as g/m^3·d (lb/ft^3.d), unit conversion factor K = 0.014 (SI) and 1.77 (USC). Factor F is based on the recirculation ratio. BOD removal in the second stage is given by the following expression.

BOD removal in second stage

$$E_{II} = \frac{100\%}{\left(1 + \dfrac{0.014}{\left(1-E_I\right)}\sqrt{(L/F)}\right)} \quad E_T = E_{20}\left(1.035\right)^{T-20}$$

BOD removal is strongly dependent on wastewater temperature. Filters in cold climates operate at lower efficiencies for the same loadings. Based on the actual operating data, BOD removal efficiency can be calculated by knowing the BOD concentration of the influent and effluent from a process or treatment plant.

27.4   OPERATING PROBLEMS

The major operating problems with the trickling filters are ponding or clogging of the filter media and fly and odor nuisance.

27.4.1   PONDING

Ponding of water on the filter surface due to plugging of the filter media can happen due to excessive BOD loading, inadequate hydraulic loading, and high fungus growth in the filter media. Due to plugging, ventilation drops sharply and the filter becomes partially anaerobic. This problem can be remedied by raking or forking the filter surface. Washing the filters by applying a high-pressure water stream helps. Chlorination of influent to filter for 2 to 5 hours during low flow periods have been successfully tried.

27.4.2   FLY NUISANCE

Trickling filters, especially those operated at low rates, are prone to fly infestations. Though these kinds of flies do not bite they are troublesome because they bother the operating personnel. Heavy infestation is associated with thick films and high temperatures. They can be destroyed by flooding the filter and chlorinating the influent. Jetting the filter walls with high-pressure hoses and removing excessive growth should be undertaken. High-rate filters are less prone to such infestations because of high hydraulic loading and continuous sloughing of the growth or humus.

27.4.3 ODOR NUISANCE

Odors are usually a result of anaerobic conditions due to partial plugging of filter media. An odor problem is more serious when septic wastewaters are fed to low-rate filters. Odors can be eliminated by providing recirculation of filter effluent to favor aerobic conditions. Aeration or chlorination of plant influent and maintaining a well-ventilated filter will keep odors under control.

27.5 SECONDARY CLARIFICATION

Secondary sludge in trickling filter systems is quite different from the ones in the activated sludge process. The term **humus** is commonly used to describe well-oxi-dized solids in trickling filter systems. Secondary sludge from tricking filter plants is dark brown in color, relatively inoffensive, and flocculent in consistency. Bulking is usually not a problem as is the case in activated sludge plants. Settling in the second-ary clarifier of a trickling filter system is the predominantly discrete type of settling.

Example Problem 27.1

Design a single-stage trickling filter with a media depth of 2.0 m based on hydraulic loading of 20 m/d. The wastewater influent is 10 ML/d and settled wastewater BOD is 150 mg/L. The desired BOD of the filter effluent is not to exceed 30 mg/L. Make suitable assumptions.

Given:

d = 2.0 m   Q = 10 ML/d BOD_i = 150 mg/L BOD_e = 30 mg/L   R = 2 assumed

Solution:

Filter surface area

$$A_s = \frac{(1+R) \times Q}{v_o} = \frac{3 \times 10000\, m^3}{d} \times \frac{d}{20\, m} = 1500\, m^2$$

BOD loading rate

$$L = \frac{M_{BOD}}{V_F} = \frac{10000\, m^3}{d} \times \frac{150\, g}{m^3} \times \frac{1}{1500\, m^2 \times 2.0\, m} = 500\, g/m^3.d$$

Factor F

$$F = \frac{1+R}{1+0.2R} = \frac{1+2}{1+0.2 \times 2} = 2.14$$

Removal efficiency

$$E_{20} = \frac{100\%}{\left(1 + 0.014\sqrt{(L/F)}\right)} = \frac{100\%}{\left(1 + 0.014\sqrt{500/2.1429}\right)} = 82.38 = \underline{82\%}$$

$$E_{reqd} = \frac{(150-30)}{150} \times 100\% = 80\% \; so \; okay$$

Select diameter

$$D(2\#) = \sqrt{\frac{4A}{\pi}} = \sqrt{\frac{4}{\pi} \times \frac{1500 \, m^2}{2}} = 30.9 = 31 \, m \; each$$

Example Problem 27.2

A trickling filter plant has a filter tank with a diameter of 70 ft and media depth of 6.5 ft. The wastewater influent is 0.80 MGD with a BOD of 200 mg/L. The plant is operated with indirect recirculation during low flow equal to 0.40 MGD and constant direct recirculation at 1.0 MGD. Calculate BOD removal efficiency of the plant at a wastewater temperature of 16°C. Assume 35% of BOD is removed by the primary treatment.

Given:

$$D = 70 \, ft \quad d = 6.5 \, ft \quad E_1 = 35\% \quad T = 16°C$$

$$Q = 0.80 \, MGD \quad BOD_{raw} = 200 \, mg/L \quad Q_R = 0.4 + 1.0 = 1.4 \, MGD$$

Solution:

BOD loading

$$L = \frac{Q \times BOD_{PE}}{V_F} = \frac{0.80 \, MG}{d} \times \frac{200 \times 8.34 \, lb}{MG} \times 0.65 \times \frac{4}{\pi(70 \, ft)^2 \times 6.5 \, ft}$$

$$= 3.467 \times 10^{-2} = 0.0355 \, lb/ft^3.d$$

Recirculation ratio and F factor

$$R = \frac{Q_R}{Q} = \frac{1.4}{0.80} = 1.75$$

$$F = \frac{1+R}{1+0.2R} = \frac{1+1.75}{1+0.2 \times 1.75} = 2.03$$

Removal Efficiency

$$E_{20} = \frac{100\%}{\left(1+1.77\sqrt{L/F}\right)} = \frac{100\%}{\left(1+1.77\sqrt{3.467 \times 10^{-2}/2.03}\right)} = 81.1 = 81\%$$

Removal at 16°C operating temperature

$$E_T = E_{20}(1.035)^{T-20} = 81.1\%(1.035)^{(16-20)} = 70.67 = 71\%$$

27.6 ROTATING BIOLOGICAL CONTACTORS

Rotating biological contactors (RBCs) create a biological slime similar to that of the trickling filter, which is grown on plastic discs mounted on a long horizontal rotating shaft. This is the most recent type of fixed culture systems and came about in the early 1970s. RBC systems can be adopted for small and medium towns for the treatment of both domestic and industrial wastewaters. They can be adopted as **package plants**. They differ from the trickling filter in that the media is rotated into the settled wastewater and then into the atmosphere for oxygen. RBCs are placed between the primary and secondary clarifiers as shown in **Figure 27.3**. There are no solids removed by the discs, just the breaking down of the dissolved and suspended solids by bacteria. Stages can be added or removed with the use of baffles. The main disadvantage of RBC is that they must be kept inside a building of some sort to prevent freezing, algae growth, UV radiation, and washing of media due to rain.

Biological discs are available in diameters up to 3.7 m (12 ft) and may be assembled to form a drum of length 7.5 m (25 ft). The spacing between sheets in the media used for BOD removal is 19 mm (3/4 in) and spacing used for nitrification is 12 mm (½ in). A typical 7.5 m long drum with 3.7 m diameter discs will have a total surface area of 10 000 to 15 000 m² (100 000 to 150 000 ft²). The submergence is about 40% and the typical operating speed is 1–10 rpm. The rotation of the disc unit ensures that the media are alternately in air and wastewater resulting in the formation of biofilm. The peripheral speed must be limited to 0.30 m/s (1 ft/s) to avoid stripping of biomass.

The sludge from a secondary clarifier is usually pumped to a primary clarifier for storage. A cover is needed to protect **biofilm** from heavy rain, frost, snow, and

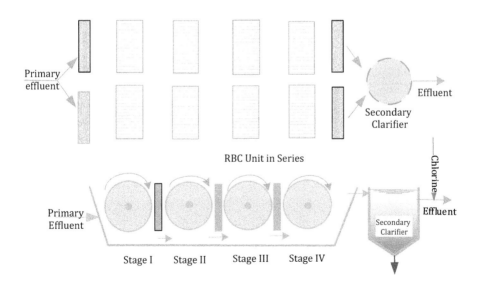

FIGURE 27.3 Rotating biological contactors

for safety. **Recirculation** in RBCs is generally not practiced and underflow from the final clarifier is allowed to settle in the primary clarifier. Biomass is like that of trickling filters.

The advantages of RBC are:

- Ease of operation and high BOD removal efficiency.
- Good solids settling, low head loss, and lower power requirements.
- Ability to lend itself to modular fabrication to suit required effluent quality.
- Low maintenance, with the possibility of nitrification and denitrification.
- Ability to work under shock loading, low noise levels, and no fly nuisance.

The disadvantages of the RBC system are:

- Some commissioning problems may occur.
- Structural damage may be expensive, media must be sheltered.
- Pumps may be necessary to move sludge from the bottom of the secondary clarifier to the primary clarifier.
- It is necessary to avoid grease or oil coating on biofilm.

27.6.1 STAGING

RBC units are arranged in series or in parallel formation. RBC in series results in an effluent of better quality if influent organic loading is not too high. Large plants overcome this problem by placing their RBC shafts perpendicular to the flow. Small plants prefer RBC in series as they can add baffles to create extra stages.

A series of four **stages** is normally installed for BOD removal. Additional stages may be required for introducing nitrification. Each stage acts as a completed reactor and the different stages combined act as a plug flow system. BOD loading decreases exponentially as wastewater moves from stage 1 to stage 4.

27.6.2 OPERATION

Operation inspections are the most important process control tools. The first stage of an RBC should be uniformly brown and be distributed in a thin layer with a dissolved oxygen level of 2.0 mg/L to be healthy. If the biomass is heavy and shaggy with white or gray patches, then there is, or has been, an organic overload.

It is common to get some sloughing of the biomass as it gets washed off and carried to the final clarifiers for settling and ultimate removal. The discs are spaced to allow for sloughing while at the same time preventing plugging. This also allows air in as the wastewater trickles out. The sludge from RBCs can be filamentous, which settles slowly. It is suggested that weir overflow rate in the secondary clarifier should be less than 100 m³/m·d.

27.7   PROCESS CONTROL PARAMETERS

Control parameters are hydraulic detention time, rotation velocity, and arrangement of disc stages. The shafts are air driven or mechanically driven.

BOD loadings are based on mass of BOD per unit surface area expressed in g/m²·d. *Note: in RBC system, loadings are based on per unit surface rather than volume.*

Hydraulic loading rate is calculated as flow divided by the disc surface. The commonly used units to express hydraulic loading are m³/m²·d. In new models, the calculation includes recirculation rates.

Total disc surface on a shaft

$$A_S = \frac{\pi D^2}{4} \times \frac{2\ faces}{disc} \times \frac{L}{\Delta L}$$

L = length of shaft (7.5 m standard) ΔL = disk spacing D = diameter of disk

Example Problem 27.3

A 5 m long RBC shaft is packed with 250 discs of 3.6 m diameter and 20 mm spacing. Work out the total disc surface per shaft length. For a rotating speed of 1.5 rpm, determine the peripheral speed in m/s.

Given:

$$D = 3.6\ m \quad N = 1.5\ rpm \quad \# = 250\ discs/shaft$$

Solution:

Disk surface, A_S and peripheral velocity v_p

$$A_S = \frac{\pi(3.6m)^2}{4} \times \frac{2\ faces}{disc} \times \frac{250\ disc}{shaft} = 5086 = \underline{5100\,m^2}$$

$$v_p = \frac{1.5\ rev}{min} \times \frac{\pi(3.6\ m)}{rev} \times \frac{min}{60\ s} = 0.282 = \underline{0.28\ m/s}$$

Example Problem 27.4

Based on hydraulic loading of 0.11 m³/m²·d, how many ML of flow can be treated by pre standard shaft with a surface area of 9300 m²?

Given:

$$A_s = 9300\ m^2 \quad HLR = 0.11\ m^3/m^2 \cdot d \quad Q = ?$$

Solution:

$$Q_{design} = A_S \times HLR = \frac{011\,m^3}{m^2.d} \times 9300\ m^2 \times \frac{ML}{1000\,m^3} = 0.99 = \underline{1.0\ ML/d}$$

27.7.1 Soluble BOD

Usually, the RBC unit is designed to remove soluble BOD. Total BOD in wastewater has two components: organic solids (**particulate BOD**) plus soluble and colloidal biodegradable matter. Soluble BOD can be thought of as BOD of a filtered sample. Experience has shown that for a given wastewater BOD contributed by SS, the constant of proportionality is typically in the range of 0.5 to 0.7 for municipal wastewater. The K factor needs to be determined for a given operation. This can be done by observing total BOD, soluble BOD, and suspended solids concentration data over a certain period. The K value is the ratio of average particulate BOD to the average suspended solids concentration.

$$Particulate\ BOD_{ptl} = BOD_{tot} - BOD_{sol} = k \times SS$$

Example Problem 27.5

Over a period of two months, average values of total BOD, soluble BOD, and SS in the primary effluent respectively are 175 mg/L, 94 mg/L, and 148 mg/L. What fraction of suspended solids contributes to BOD?

Given:

TBOD = 175 mg/L   SBOD = 94 mg/L   SS = 148 mg/L   k = ?

Solution:

   Particulate BOD

$$BOD_{ptl} = BOD_{tot} - BOD_{sol} = 175\ mg/L - 94\ mg/L = 81.0 = 81\ mg/L$$

$$k = \frac{BOD_{ptl}}{SS} = \frac{81\ mg/L}{148\ mg/L} = 0.547 = \underline{0.55 = 55\%}$$

27.7.2 Organic Loading

Organic loading on RBC units is based on soluble BOD or total BOD. It is expressed as mass load of BOD in the primary effluent per unit disc surface. Some of the key points regarding operation of RBCs are as follows:

- Typical loadings are 7.5 g/m²·d of soluble BOD or 15 g/m²·d (1 lb/1000 ft³·d) of total BOD.
- In larger plants, RBC shafts are placed perpendicular to the direction of flow, thus each shaft acts as one stage of BOD removal.
- The various stages of RBC simulate plug flow, with maximum loading on the first stage. Thus, the loading on the first stage is an important consideration.
- If operated properly, the biological growth on the first stage should be fairly uniform and light brown in color. The following stages should look similar except with an additional gold or reddish tone.
- Overloading conditions are evidenced by the gray or white biomass.

- A loading of 60 g/m²·d of total BOD on the first stage should not be exceeded.
- For operating temperature below 13°C, temperature must be correction for additional disc surface @ 15% for each 3°C below 13°C.
- DO in the first-stage effluent should not be allowed to fall below 0.5 mg/L and DO of the final-stage effluent should preferably be more than 2.0 mg/L.

Example Problem 27.6

A treatment plant processes domestic wastewater by primary sedimentation, RBCs, and final clarification. Each RBC shaft has a length of 5.3 m with a 3.6 m diameter disk for a nominal surface area of 5600 m². The installation has 16 shafts arranged with 4 rows of shafts of 4 stages each. The influent wastewater slow is 9.0 ML/d containing 180 mg/L of BOD. Assuming 30% BOD removal in the primary clarification, calculate the BOD loading based on the total RBC area and for the first stage.

Given:

$$\text{Disk surface} = 5600 \text{ m}^2/\text{shaft} \quad \text{\# of shafts} = 16 \quad \text{\# of stages} = 4$$

$$E_I = 30\% \quad BOD_{raw} = 180 \text{ mg/L}$$

Solution:

$$A_S = \frac{5600\,m^2}{shaft} \times \frac{4\,shafts}{stage} = 22400\,m^2\,/\,stage$$

$$M_{BOD} = Q \times BOD_{PE} = \frac{9.0\,ML}{d} \times \frac{180\,kg}{ML} \times (1-0.30) = 1134\,kg\,/\,d$$

$$BODLR = \frac{M_{BOD}}{A_S} = \frac{1134\,kg}{d} \times \frac{1}{21400\,m^2} \times \frac{1000\,g}{kg} = 50.6 = \underline{51\,g\,/\,m^2.d}$$

Example Problem 27.7

An RBC unit consists of 16 shafts with each shaft having a disc surface of 5600 m². The installation has 16 shafts arranged with 4 rows of shafts of 4 stages each. On average, the primary effluent flow of 8500 m³/d containing 150 mg/L of BOD and 120 mg/L of SS is treated. Assuming k = 0.50, calculate the soluble BOD loading on the RBC process in g/m² d.

Given:

$$\text{Disc surface} = 5600 \text{ m}^2/\text{shaft shafts} \quad \text{\#} = 16 \quad Q = 8500 \text{ m}^3/\text{d}$$

$$BOD = 150 \text{ mg/L} \quad SS = 120 \text{ mg/L} \quad k = 0.50$$

Solution:

$$A_S = \frac{5600\,m^2}{shaft} \times 16\,shafts = 89600\,m^2$$

TABLE 27.2

Operating conditions and slime color

Slime Color	Process Condition
gray shaggy	normal process
reddish brown, golden	nitrification occurring
white chalky	high sulphur content
no slime growth	severe temperature or pH changes

$$BOD_{ptl} = k \times SS = 0.50 \times 120\,mg\,/\,L = 60\,mg\,/\,L$$

$$BOD_{sol} = BOD_{tot} - BOD_{ptl} = 150\,mg\,/\,L - 60\,mg\,/\,L = 90\,mg\,/\,L$$

$$BODLR = \frac{Q \times BOD}{A_S} = \frac{8500\,m^3}{d} \times \frac{90\,g}{m^3} \times \frac{1}{89600\,m^2} = 8.53 = \underline{8.5\,g\,/\,m^2\,.d}$$

27.8   OPERATION OF RBC SYSTEM

Operation inspections are the most important process control tools. During normal operation, the operation needs to keep an eye on RBC movement, slime color, and appearance (see **Table 27.2**).

The first stage of an RBC should be uniformly brown and be distributed in a thin layer with a dissolved oxygen level of 2.0 mg/L to be healthy. Since the first stage receives the maximum load, the DO level will remain suppressed. However, the DO level in the first stage should not be allowed to fall below 0.5 mg/L. If the biomass is heavy and shaggy with white or gray patches, then there is, or has been, an organic overload. In a nutshell, slime or bio-growth indicates the process condition.

It is common to get some sloughing of the biomass as it gets washed off and carried to the final clarifiers for settling and ultimate removal. The discs are spaced to allow for sloughing while at the same time preventing plugging. This also allows air in, as the wastewater trickles out. As part of routine testing and sampling, the operator should observe DO content at various stages, pH, and suspended solids content. These results aid in assessing the performance and adjusting the process.

Discussion Questions

1. Compare trickling filter with rotating biological contactors.
2. Discuss the working principle of a tricking filter.
3. Describe various components of a trickling filter with the help of a sketch.
4. List the advantages and disadvantages of biological filtration.
5. With the help of a flow diagram, explain direct and indirect circulation in tricking filters.
6. What is main purpose of each of type of recirculation, direct and indirect?

7. Differentiate between a standard-rate filter and high-rate filter?
8. The advent of plastic media has led to the creation of super-rate filters. Comment.
9. How does a trickling filter differ from a rapid sand filter?
10. What modification would you need to make in the NRC formula for BOD removal in the second stage?
11. Describe different ways to introduce nitrification in an RBC biological system.
12. List the merits and demerits of a RBC system.
13. How might the color of the slime on the RBC indicate the health of the RBC process?
14. Explain why trickling filter and RBC are put under the category of fixed growth systems. Activated sludge falls into the category of suspended growth systems for biological treatment. Which of these three is the oldest and which one is the most recent biological treatment.
15. Trickling filters are also called biological filters. What do the terms trickling, biological, and filter signify?
16. How is staging introduced in an RBC system of a small plant and a large plant?

Practice Problems

1. Calculate the kg of BOD per day entering the trickling filter given that raw wastewater flow is 6.0 ML/d containing BOD of 150 mg/L and there is 30% reduction in BOD across the primary clarifiers. (630 kg/d).
2. What is the hydraulic loading on a 25 m diameter trickling filter when the daily flow is 9.5 ML/d and recirculation rate is 15% of the daily flow rate? ($22 \text{ m}^3/\text{m}^2 \cdot \text{d}$).
3. What is the plant influent flow rate if BOD load to tricking filter is 1500 lb/d and the average BOD of the plant influent is 210 mg/L? Assume 35% BOD removal by primary clarification. (1.3 MGD).
4. What is the BOD loading on a trickling filter 22 m in diameter and 2.0 m deep? The average flow fed to the filter is 4.1 ML/d containing BOD of 140 mg/L. ($750 \text{ g/m}^3 \cdot \text{d}$).
5. Overall BOD removal efficiency of a tricking filter plant is 85%. If the BOD removal by the primary clarification is 35%, what is BOD removal by the secondary treatment? (77%).
6. A bio-disc with a total surface area of 100 000 m^2 treats a flow of 15 ML/d. If the soluble BOD in the primary effluent is 75 g/m^3, what is the BOD loading? ($11 \text{ g/m}^2 \cdot \text{d}$).
7. A town produces sewage flow of 5.5 ML/d containing BOD of 270 mg/L. After primary clarification it is planned to have trickling filters for secondary treatment. Assuming primary BOD removal of 35%, work out the volume and depth of the filter unit based on the following design parameters:
   BOD loading rate = 150 $\text{g/m}^3 \cdot \text{d}$ Hydraulic Loading = 4.0 $\text{m}^3/\text{m}^2 \cdot \text{d}$

For the selected filter, calculate the BOD removal efficiency based on the NRC formula. (D = 42 m, # =2, d = 2.1 m, R = 1, 88%).

8. Select the diameter of a trickling filter to serve a population of 500 persons contributing wastewater @ 35 gal/c·d with BOD content of 300 ppm. Assume 40% of BOD is removed in the primary clarification and design BOD loading is 30 lb/1000 ft³·d. (13 ft, 7.0 ft deep).

9. How many RBC shafts (100 000 ft²/shaft) are required to treat a flow of 3.0 MGD with BOD content of 240 mg/L? Assume primary removal of BOD is 40%. The design BOD loading rate is 3.0 lb/1000 ft²·d. (12#, 3 × 4).

10. What diameter trickling filter is required to treat a flow of 5 ML/d with BOD content of 240 mg/L. Assume primary BOD removal of 35%, BOD loading of 550 g/m³·d and filter depth of 2.5 m? (27 m).

11. A standard rate trickling filter has a diameter of 26 m and an average media depth of 2.1 m. The daily wastewater flow is 5200 m³/d with an average BOD of 180 mg/L. During periods of low influent flow, 2.5 ML/d of underflow from the final clarifier is returned to the wet well.
    a. Calculate hydraulic loading, return ratio, and BOD loading assuming 35% BOD removal by the primary treatment.
       (15 m/d, 0.48, 550 g/m²·d).
    b. Applying the NRC formula, determine the expected BOD removal assuming operating temperature of 20°C (78%).
    c. If during winter months, the operating temperature drops to 16°C, what BOD removal can be expected? (68%).

12. Over a period of three months, average values of total BOD, soluble BOD, and SS in the primary effluent respectively are 190 mg/L, 95 mg/L and 150 mg/L. What fraction of suspended solids contributes to BOD? (63%).

13. Find maximum allowable BOD loading on a high-rate filter to treat a wastewater flow of 5 ML/d with recirculation rate of 150%. Assume the BOD in the primary effluent is 160 mg/L and desired BOD removal is 80%. (610 g/m³·d).

14. Design a single-stage high-rate filter for treating wastewater from a community of 38 000 people. Per capita wastewater production is 170 L/c·d with BOD of 200 mg/L. Allowable BOD loading and hydraulic loading respectively are 800 g/m³·d and 16 m/d. Assume BOD removal of 30% by the primary treatment and recirculation rate of 100%. Find
    a. volume of filter media based on organic loading. (1130 m³).
    b. surface area based on hydraulic loading. (810 m²).
    c. diameter and depth. (32 m, 1.4 m).
    d. the expected BOD in the plant effluent. (77%).

15. A trickling filter is operated with organic loading 30 lb/1000 ft³·d while maintaining recirculation rate of 120%. What is the expected BOD removal if the operating temperature is 22°C.? (87%).

16. A trickling filter is designed based on BOD loading of 550 g/m³·d. To meet effluent quality standards, the required BOD removal is 80%. What should be the minimum recirculation ratio? (1.1).

17. A 22 m diameter 2.3 m deep high-rate trickling filter is to treat a wastewater flow of 3.5 ML/d with a BOD of 180 mg/L.
    a. Expected BOD removal, assuming recirculation ratio of 1.4. (78%).
    b. BOD removal at an operating temperature of 18 degree C (73%).
18. An RBC unit consists of 12 shafts with three trains each with four stages. Each shaft has a disc surface of 5500 m². On average, the primary effluent flow is 6500 m³/d containing 150 mg/L of BOD and 110 mg/L of SS. Assuming particulate BOD is 50%, determine soluble BOD loading rate. (9.4 g/m²·d).
19. Determine maximum allowable BOD loading on a high-rate trickling filter that is operated with 100% recirculation to remove 80% of the BOD. (33 lb/1000 ft³.d).
20. A trickling filter is designed for BOD loading of 35 lb/1000 ft²·d. What is the minimum recirculation required to remove 80% of the BOD? (120%).

28 Anaerobic Systems

In **anaerobic** systems, the main bio-reaction takes place under anaerobic conditions and hence there is no need to provide molecular oxygen by aeration. One of the main end products is methane gas, which can be used as an energy source in some cases. However, due to the production of foul gases, there is the problem of strong smells and hence such systems need to be well ventilated. The most common anaerobic systems are **septic** tanks, **Imhoff** tanks, and anaerobic **filters** and **reactors**. The main advantages of such systems are energy savings, low capital costs, and ease of operation.

28.1 SEPTIC TANKS

Septic tanks are **onsite** systems and are more common in rural areas and in isolated buildings and institutions, hotels, schools, hospitals, and small residential areas. To prevent contamination, septic systems must be away from sources of drinking water and should not be located in swamp areas or areas prone to flooding. In addition, soil should be porous to absorb effluent from septic tanks.

A typical section of a septic tank is shown in **Figure 28.1**. The **septic tank** is a watertight underground tank usually made of concrete. The capacity of the tank is such as to provide a long retention period of 1–3 days. Solids settle to the bottom where anaerobic sludge digestion takes place. Generally, a septic tank has two chambers separated by a baffle wall. The first chamber allows the removal of grit and the second chamber is primarily for the settling of organic solids and anaerobic digestion. The sludge tank must be provided with a **vent** to eliminate foul gases produced as a result of anaerobic digestion. Digested sludge is pumped out periodically and is not allowed to exceed 3 years.

28.2 DESIGN CONSIDERATIONS

28.2.1 CAPACITY

Capacity of the septic tank should be such to provide adequate detention time for solids to settle out and enough room for the storage of digested sludge until the withdrawal period. The recommended capacity for sewage flow is 90–150 L/c.d (25–40 gal/c.d). Sludge and scum will accumulate in the tank at varying rates depending on the characteristics of the raw sewage. The typical value of sludge accumulation is 30 L/c.a (8 gal.c.a). The minimum capacity of a septic tank is usually more than 2.0 m³ (500 gal).

28.2.2 FREE BOARD

A free board of 0.3 to 0.5 m is provided above the top of the sewage line to provide room for scum to accumulate. The scum layer prevents the spread of foul gases.

DOI: 10.1201/9781003347941-33

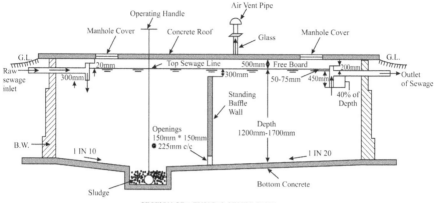

SECTION OF A TYPICAL SEPTIC-TANK

FIGURE 28.1   Section of a typical septic tank

28.2.3   INLET AND OUTLET

The design of the inlet and outlet should be such as to prevent **short-circuiting**. To achieve this, baffles, usually tees, should extend to the top level of scum but below the ceiling of the tank. To avoid short-circuiting, the inlet should extend to a depth of about 30 cm (1.0 ft) below the sewage line, as shown in **Figure 28.1**. The outlet is submerged and is located 5–8 cm (2–3 in) below the inlet.

28.2.4   DETENTION TIME

As previously mentioned, the detention time for septic tanks varies from as little as 12 h to as much as 3 d. However, 24 h (1d) is used for most designs.

28.2.5   SHAPE OF THE TANK

Septic tanks are usually rectangular in shape with length 3–4 times width. A minimum depth of 1.0 m (3 ft) is provided below the water level. The bottom of the tank is sloped for sludge to accumulate in sump form, where it is pumped out periodically.

Example Problem 28.1

Design a rectangular septic tank with length to width ratio of 3:1 to serve a population of 100 people discharging wastewater @ 25 gal/c.d. Assume detention period of 30 h and sludge withdrawal every second year @ 15 gal/c.

Given:

$$Q = 100 \text{ p @ 25 gal/c·d} \quad Q_{sludge} = 15 \text{ gal/c} \quad t_d = 30 \text{ h}$$
$$L = 3 \times W \quad d = 5.0 \text{ ft (assumed)}$$

Solution:

Capacity of the tank

$$V_I\left(\text{Waste water}\right) = \frac{25\,\text{gal}}{\text{p.d}} \times 100\,\text{p} \times 30\,\text{h} \times \frac{d}{24\,\text{h}} \times \frac{\text{ft}^3}{7.48\,\text{gal}} = 417.78\,\text{ft}^3$$

$$V_{II}\left(\text{Sludge}\right) = \frac{15\,\text{gal}}{\text{p}} \times 100\,\text{p} \times \frac{\text{ft}^3}{7.48\,\text{gal}} = 200.5\,\text{ft}^3$$

$$V_T = V_I + V_{II} = 417.7\,\text{ft}^3 + 200.5\,\text{ft}^3 = 618.2\,\text{ft}^3$$

Dimensions of tank

$$W = \sqrt{\frac{V}{3d}} = \sqrt{\frac{618.2\,\text{ft}^3}{3 \times 5.0\,\text{ft}}} = 6.41 = 6.5\,\text{ft}\quad L = 6..41\,\text{ft} \times 3 = 19.2 = 20\,\text{ft}$$

Assuming free board of 1 ft, dimensions of the tank are 20 ft × 6.5 ft × 6.0 ft

Example Problem 28.2

Estimate the size of septic tank to serve a community of 250 people in a developing country. Assume wastewater production is 100 L/c·d and sludge production on a yearly basis is 40 L/c.a. Assume length to width ratio of 2.5:1 and average detention time is 2 d and depth of the tank 2.0 m. If desludging is done when the sludge fills up one-fourth of the tank volume, estimate the frequency of desludging.

Given:

$$Q = 250\,\text{p @ 100 L/c·d}\quad Q_{sludge} = 40\,\text{L/c·d}\quad t_d = 2\,\text{d}$$
$$L = 2.5 \times W\quad d = 2.0\,\text{m (assumed)}$$

Solution:

Capacity of the tank

$$V_I\left(\text{Wastewater}\right) = \frac{100\,\text{L}}{\text{p.d}} \times 250\,\text{p} \times 2\,\text{d} \times \frac{\text{m}^3}{1000\,\text{L}} = 50\,\text{m}^3$$

Sludge occupies 25% of the space when desludged, hence 75% space is for wastewater

$$V_T = \frac{V_I}{0.75} = \frac{50\,\text{m}^3}{0.75} = 66.6\,\text{m}^3$$

$$A_T = \frac{V_T}{d} = \frac{66.6\,\text{m}^3}{2.0\,\text{m}} = 33.3\,\text{m}^2$$

Dimensions of tank

$$W = \sqrt{\frac{A}{2.5}} = \sqrt{\frac{33.3\,\text{m}^2}{2.5}} = 3.65 = 3.7\,\text{m}\quad L = 3.65 \times 2.5 = 9.1 = 9.0\,\text{m say}$$

Assuming free board of 40 cm, dimensions of the tank are 9.0 m × 3.7 m × 2.4 m

Desludging period

$$t = \frac{V_{ll}}{Q} = 0.25 \times 66.6 \text{ m}^3 \times \frac{a}{40 \text{ L} \times 50 \text{ p}} \times \frac{1000 \text{ L}}{\text{m}^3} \times \frac{12 \text{ mo}}{a} = 19.9 = 20 \text{ mo}$$

Sludge should be pumped out after a 20-month period of accumulation

28.2.6 DISPOSAL OF THE TANK EFFLUENT

Effluent from the tank still carries a large amount of organic load thus needs to be taken care of. Since septic tank effluent carries BOD in the range of 100–200 mg/L, disposal practices are such as to provide some sort of biological treatment.

28.3 SOIL ABSORPTION SYSTEM

The **soil absorption system**, also called leach fields, remove the effluent from septic tanks. When effluent is spread on pervious land, microorganisms in the soil reduce BOD by aerobic decomposition and add nutrients to the soil. This method is applicable if the soil is pervious and the ground water is well below the surface. Perviousness of the soil is indicated by the **percolation rate**, which is defined as the time in minutes required for seepage of 1 cm depth of water. That is to say, the percolation rate will be quite low for sandy soils and high for clay soils. Soils suitable for an absorption field or pit should have a percolation rate less than 30 min and must not exceed 60 min.

28.3.1 PERCOLATION TEST

The step by step procedure is as follows.

- Dig a square or a circular hole with side width or diameter respectively 100 to 300 mm (0.3 ft–1.0 ft) and vertical sides. Carefully scratch the bottom and sides of the holes to remove any smeared soil surface.
- Remove all the loose material from the hole and add 5.0 cm (2 in) of coarse sand or fine gravel to protect the bottom from scouring and sediment.
- Pour water into the hole to achieve a minimum depth of 300 mm (1.0 ft) over the gravel. Let the soil swell for 24 h. If the water remains in the test hole after the overnight swelling period, add more water to bring the depth to 150 mm (6 in) over the gravel.
- From a fixed reference point, record the water level drop for a 30 min period.
- If after the swelling period no water remains in the hole, add more water to bring the depth of the water in the hole until it is 150 mm over the gravel. Record the water level drop at 30-minute intervals for 4 hours, refilling 150 mm over the gravel as necessary. The drop that occurs during the final 30-minute period can be used to calculate the percolation rate.

- In sandy soils or other porous soils in which the first 150 mm of water seeps away in less than 30 minutes after the overnight swelling period, the time interval between measurements should be 10 minutes and the test run for one hour. The drop that occurs during the final 10 minutes should be used to calculate the percolation rate.
- Knowing the **percolation rate**, the maximum allowable application rate of the septic tank effluent in L/m²·d can be worked out using the following empirical relationship.

$$\text{Application}, Q = \frac{204}{\sqrt{t}} \text{ for } t = 10 \text{ min} \quad Q = \frac{204}{\sqrt{10}} = 64.5 = 65 \text{ L} / \text{m}^2.\text{d}$$

28.3.2 ABSORPTION FIELD

An absorption field is constructed by laying trenches in a field. Trenches are laid with drainage tiles with a diameter of 75–100 mm (3–4 in) to allow effluent to infiltrate into the soil. Dispersion trenches are 0.5–1.0 m deep and 0.3–1.0 m wide, excavated to a slight gradient. Trenches are provided with 150–250 mm of washed gravel or crushed stones. The length of a dispersion trench should not exceed 30 m (100 ft) and trenches should not be placed closer together than 2.0 m (6.5 ft). Many of the operating problems in septic system are the result of plugging of the absorption field. Poor distribution of the effluent, hydraulic overloading, broken pipes, compaction of the field due to heavy machinery and tree roots are some of the culprits responsible for plugging. Plugging of the field is indicated by odors and back-up of sewage.

28.4   SOAK PIT

As seen in **Figure 28.2**, a soak pit is usually a circular covered pit through which the effluent is allowed to percolate and be absorbed into the surrounding soil. The pit may be empty or filled with stone aggregate. When the pit is empty, the pit is lined with brick or concrete such that joints are open to allow absorption into the surrounding soil. In addition, backing of 75 mm (3-in) thick coarse aggregate supports the pit surface below the inlet. The space above the inlet should be plastered.

Whatever the subsoil dispersion system, it must not be closer than 20 m (65 ft) from any source of drinking water, such as a well, to mitigate the possibility of bacterial pollution of the water supply. In limestone or crevice rock formations, the soil absorption system is not recommended as there may be channels in the formation which may carry contamination over a long distance. In such cases, and generally where suitable conditions do not exist for the adoption of soil absorption systems, the effluent, where feasible, should be treated in a biological filter or by **up-flow** anaerobic filters.

28.5   BIOLOGICAL FILTERS

Biological filters are suitable for the treatment of septic tank effluent where the soil percolation rate exceeds 60 min, in waterlogged areas, or where limited land

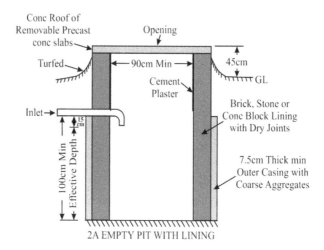

FIGURE 28.2   A soak pit

area is available. In a **biological filter**, the effluent from a septic tank is brought into contact with a suitable medium, the surfaces of which become coated with an organic film. The film assimilates and biochemical oxidation takes place. The biological filter requires ample ventilation and an efficient system of underdrains leading to an outlet.

It is essential that the volume of the filter medium is sufficient, especially in small systems where more variations are expected. For serving fewer than 10 persons, provide a filter volume of 1 m³ (35 ft³) per head. To serve more than 10 persons, use a 0.8 m³/head. For populations larger than 50 people, the recommended volume is 0.6 m³/head. The final effluent is either discharged into the city surface drain or evenly spread over a grass field. In case contamination of the receiving water is suspected, provide disinfection before discharge.

28.6 UPFLOW FILTERS

This method of septic effluent treatment is **anaerobic** and is used for areas with limited land, high water table conditions, or clayey soils with high percolation rates. It is also called a reverse filter since septic tank effluent is introduced at the bottom. Filter stone media remains submerged, thus anaerobic conditions prevail. The capacity of the unit is typically 40–50 L/c·d (10–12 gal/c.d) and high BOD removals can be expected when operated properly.

A single-chamber upflow filter is shown in **Figure 28.3** and the double-chambered type of rectangular upflow filter is common to treat effluent from septic tank units. Filter material usually consists of 20 mm stones resting on a concrete slab with a false bottom. The septic tank effluent enters at the bottom through a 150 mm (6 in) pipe and is fitted with a tee at the bottom. One end of the tee is at the bottom of the filter and other end, placed towards the cleaning chamber, remains plugged during

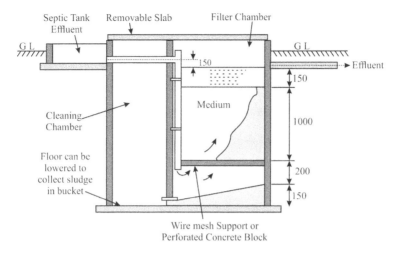

FIGURE 28.3 A single-chamber upflow filter

operation. During cleaning, the plug is removed to empty the filter into a cleaning chamber. During the operation of the filter, treated water exits over a V-notch weir to maintain a water depth 150 mm above the top of the filter media.

In a double-chamber filter, both chambers are filled with filter media. In the first chamber, septic tank effluent enters at the top of the media and flows upward through the media into the second chamber. This allows more contact time and hence better BOD removals can be expected. A galvanized iron pipe at the bottom fitted with a valve leading to an adjacent chamber can be used for cleaning the filter when required.

Example Problem 28.3

Design a rectangular septic tank to serve a population of 85 people discharging wastewater @ 90 L/c.d. Assume sludge is withdrawn once in a year and a length to width ratio of 2:1.
   a. Size the soak pit based on percolation rate of 1200 L/m³·d
   b. If the tank effluent is to be discharged into an absorption tank with a percolation rate of 5.0 min, select the size of the trenches.

Given:

   Q = 85 p @ 90 L/c·d   Percolation = 1200 L/m³·d   L:W: 2:1   d = 2.0 m assumed

Solution:
   Capacity of tank

$$Q = \frac{90 L}{p.d} \times 85\, p = 7650\, L / d$$

$$V_I = \frac{7650\,L}{d} \times 24\,h \times \frac{d}{24\,h} \times \frac{m^3}{1000\,L} = 7.65\,m^3$$

$$V_{II} = \frac{30\,L}{p.a} \times 85p \times 1a \times \frac{m^3}{1000L} = 2.55\,m^3$$

$$V_T = V_I + V_{II} = 7.65\,m^3 + 2.55\,m^3 = 10.1\,m^3$$

Dimensions, assuming water depth of 1.5 m

$$W = \sqrt{\frac{V}{2\,d}} = \sqrt{\frac{10.1\,m^3}{2\times1.5\,m}} = 1.83 = 1.8\,m \quad L = 2\times1.83 = 3.66 = 3.7\,m$$

Providing a f.b. of 0.4 m, the dimensions of the tank are 3.7 m × 1.8 m × 1.9 m

Capacity of soak pit, assuming percolation rate of 1200 L/m³·d is

$$V_{pit} = \frac{7650\,L}{d} \times \frac{m^3.d}{1200\,L} = 6.37 = 6.4\,m^3$$

$$D = \sqrt{\frac{1.27\,V}{d}} = \sqrt{\frac{1.27\times6.37\,m^3}{2.0\,m}} = 2.01 = 2.0\,m$$

Note 4/π = 1.27.
   The dispersion trench is based on maximum percolation allowed, which is found from the empirical equation.

$$q = \frac{204}{\sqrt{t}} = \frac{204}{\sqrt{5}} = 91.23 = 91\ L/m^2.d$$

$$A_{trench} = \frac{7650\,L}{d} \times \frac{m^2.d}{91L} = 85\,m^2$$

Assuming each trench is 1 m wide, with 4 trenches

$$L = \frac{85\,m^2}{4\times1.0\,m} = 21.25 = 21\,m$$

Four trenches each 21 m × 1.0 m will do the job of the absorption of septic effluent.

28.7   UPFLOW ANAEROBIC SLUDGE BLANKET

The upflow anaerobic sludge blanket (UASB) reactor was developed in 1970 in the Netherlands. This reactor has become very popular in places like India to treat domestic and industrial wastes. This type of units is one in which no special media is required since the sludge granules themselves act as the media and stay in suspension. A typical UASB reactor is shown in **Fig. 28.4**.

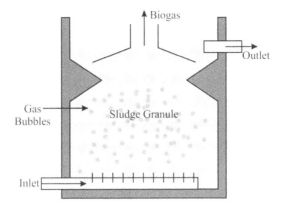

FIGURE 28.4 A typical UASB reactor

The anaerobic unit does not need to be filled with stones or any other media; the up flowing sewage itself forms millions of small granules or particles of sludge, which are held in suspension and provide a large surface area on which organic matter can attach and undergo biodegradation. A high solid retention time (SRT) of 30–50 d or more occurs within the unit. No mixers or aerators are required. The gas produced can be collected and used if desired. Anaerobic systems function satisfactorily when temperatures inside the reactor are above 18–20°C. Excess sludge is removed from time to time through a separate pipe and sent to a simple sand bed for drying.

28.7.1 ZONES AND COMPONENTS

A single module of UASB can handle wastewater flows of 10–15 ML/d (2.5–4 MGD). For large flows, a number of modules could be provided. There are four major zones in a UASB reactor.

Inlet Zone

In the inlet zone incoming wastewater is introduced at the bottom and distributed uniformly. Water coming out of the orifices moves upward through the suspended sludge blanket.

Sludge Blanket Zone

This is the reaction zone in which the anaerobic reaction takes place. The micro-organisms in the sludge blanket (granular biomass) convert volatile solids to new cells and gases like methane. Flow velocity should be controlled to allow reasonable reaction time.

Clarifier Zone

Clarifier zone provides opportunity for solids to settle down. The treated wastewater moving upward is a mixture of gases and some biomass. Solids are retained by the filtration effect of sludge blanket and settle to the bottom.

Gas-Liquid–Solid Separator

The upward motion of gas bubbles allows mixing with any mechanical agitation. In this zone solids are separated from the liquid. This is the zone where the three phases are separated. Solids are pushed downwards to the settling zone and gases are collected and effluent flow over the weirs at the top.

Advantages

- High reduction of BOD.
- Can withstand high organic and hydraulic loading rate.
- Low sludge production.
- No aeration required.
- Biogas produced can be used for energy.
- Though scrubbing is required.
- Effluent rich in nutrients, can be used for irrigation.
- Can be constructed underground with local available materials.

Disadvantages

- Not suitable for colder regions.
- May be unstable for fluctuating organic and hydraulic loading rates.
- Proper design and construction is required.
- Skilled operation is required.
- Long start time to reach full capacity.
- Effluent and sludge need further treatment.

28.7.2 Design Approach

Generally, UASBs are considered where temperature in the reactors will be above 20°C (68°F). At equilibrium condition, sludge withdrawn must be equal to sludge produced daily. The sludge produced daily depends on the characteristics of the raw wastewater since it is the sum of the following:

(i) The new VSS produced because of BOD removal, the yield coefficient being assumed as 0.1 g VSS/g BOD removed.
(ii) The non-degradable residue of the VSS coming in the inflow assuming 40% of the VSS are degraded and residue is 60%.
(iii) Ash received in the inflow, or non-volatile fraction, TSS-VSS mg/L.

At steady state conditions, the reactor volume must be so chosen that the desired SRT value be achieved. This is done by solving for HRT from SRT equation assuming depth of reactor, the effective depth of the sludge blanket, and the average concentration of solids in the sludge blanket (7.0%).

The full depth of the reactor for treating low BOD municipal sewage is often 4.5–5.0 m (15–17 ft) of which the sludge blanket itself may be 2.0–2.5 m (6–8 ft) depth. For high strength wastes, the depth of both the sludge blanket and the reactor may have to be increased so that the organic loading on solids may be kept within the prescribed range. Once the size of the reactor is fixed, the upflow velocity can be determined.

Using average flow rate, one gets the average HRT while the peak flow rate gives the minimum HRT at which minimum exposure to treatment occurs. To always retain any flocculent sludge in reactor, the upflow velocity should not exceed 0.5 m/h (1.5 ft/h) at average flow and not more than 1.2 m/h (4 ft/h) at peak flow. At higher velocities, carry over of solids might occur and effluent quality may be deteriorated. The feed inlet system is next designed so that the required length and width of the UASB reactor are determined.

The sloping hoods for gas collection form the settling compartment. The depth of the compartment is 2.0–2.5 m and the surface overflow rate kept at 20–28 m^3/m^2.d (500–700 gal/ft^2.d) at peak flow. The flow velocity through the aperture connecting the reaction zone with the settling compartment is limited to not more than 5 m/h (15 ft/h) at peak flow. It is important to ensure proper working of the Gas-Liquid-Solid-Separator (GLSS), the gas collection hood, the incoming flow distribution to get spatial uniformity and the outflowing effluent.

Process Design Parameters

A few process design parameters for UASBs are listed in **Table 28.1**. The values furnished in this are specific to municipal sewage with BOD about 200–300 mg/L and temperatures above 20°C.

TABLE 28.1
Design Parameters of UASB Reactor

HRT	8–10 h at average flow (minimum 4 h at peak flow)
SRT	30–50 days or more
Sludge blanket concentration (average)	15–30 kg VSS/m^3. About 7.0% TSS
Organic loading on sludge blanket	0.3–1.0 kg COD/kg VSS day (even up to 10 kg COD/kg VSS/day for agro-industrial wastes)
Volumetric organic loading	1–3 kg COD/m^3 day for domestic sewage (10–15 kg COD/m^3/d for agro-industrial wastes)
BOD/COD removal efficiency	Sewage 75–85% for BOD. 74–78% for COD
Inlet points	Minimum 1 point per 3.7–4.0 m^2 floor area
Flow regime	Either constant rate for pumped inflows or typically fluctuating flows for gravity systems
Upflow velocity	About 0.5 m/h at average flow, or 1.2 m/h at peak flow, whichever is low
Sludge production	0.15–0.25 kg TSS per m^3 sewage treated
Sludge drying time	Seven days (in warmer climates)
Gas production	Theoretical 0.38 m^3/kg COD removed. Actual 0.1–0.3 m^3 per kg COD removed
Gas utilization	Method of use is optional. 1 m^3 biogas with 75% methane content is equivalent to 1.4 kWh electricity
Nutrients (nitrogen, and phosphorus) removal	5–10% only

Example Problem 28.4

Design an upflow anaerobic sludge blanket reactor for an average daily flow of 5 ML/d containing COD of 450 mg/L. Make following assumptions:

   i. Design hydraulic residence time = 8 h
   ii. Design COD loading 1–3 kg /m³·d
   iii. Rise rate in the reactor = 0.5 m/h
   iv. Overflow rate in the settling chamber = < 30 m³/m²·d
   v. Flow area covered by each inlet =1–3 m²

Given:

$$Q = 5.0 \text{ ML/d} = 5000 \text{ m}^3/\text{d} \quad COD = 450 \text{ mg/L} \quad HRT = 8.0 \text{ h}$$

$$v_{up} = 0.50 \text{ m/h} \quad A = 2.0 \text{ m}^2/\text{inlet}$$

Solution:

$$V = Q \times t = \frac{5000 \text{ m}^3}{d} \times 8.0 \text{ h} \times \frac{d}{24 \text{ h}} = 1666.6 = 1700 \text{ m}^3$$

$$\text{Loading} = \frac{5.0 \text{ ML}}{d} \times \frac{450 \text{ kg}}{\text{ML}} \times \frac{1}{1700 \text{ m}^3} = 1.32 = 1.3 \frac{\text{kg/m}^3 \cdot d}{\text{kg/m}^3 \cdot d} \quad \text{so okay}$$

$$\text{Reactor}, H = v_{up} \times t = \frac{0.5 \text{ m}}{h} \times 8.0 \text{ h} = 4.0 \text{ m} \quad A = \frac{V}{H} = \frac{1700 \text{ m}^3}{4.0 \text{ m}} = 425 \text{ m}^2$$

Assuming total height of the reactor = 5.5 m, and a square section and # = 2

$$\text{Reactor}, L = \sqrt{\frac{425 \text{ m}^2}{2}} = 14.57 = 15 \text{ m}$$

Hence each reactor is 15 m × 15 m × 5.5 m for allowable overflow rate of 40 m/d at average flow.

Discussion Questions

1. Briefly describe various methods of disposing of and treating septic tank effluent.
2. Discuss advantages and disadvantages of UASB reactor.
3. Describe the type of wastes and the conditions for which you would recommend high rate anaerobic reactors.
4. What is the main concept on which UASB works? Give a brief account of design approach.
5. What is typically BOD of septic tank effluent. Is it good enough to discharge without any further treatment? If not, which treatment would you recommend to make it acceptable from an environmental protection point of view.

6. How would you determine the suitability of a given method for treating septic tank effluents?
7. One of the biggest problems related to anaerobic systems for treating domestic wastewater is odors. What things which can worsen the problem?
8. Scum is allowed to build up in the operation of septic tanks. Explain.
9. Write down the steps to perform a percolation rate test.
10. In the design of septic units, two chambers are preferred over a single chamber. Explain.
11. What considerations would you make in the design of the inlet and outlet of a septic tank?

Practice Problems

1. Design a septic tank with a soak pit to absorb the tank effluent. This system is to serve a resort community of 500 people with daily wastewater production of 150 L/c.d. Assume annual sludge accumulation @ 15 L/c·d.
   a. Allow minimum detention time of 12 h in the septic tank and length to width ratio of 1:3 with effective depth of 2.0 m. (# = 2, 6 m × 2 m × 2.5 m).
   b. Assume absorption value of soil is 1500 L/m³.
      (# = 2, 4.0 m diameter, 2.0 m deep).
2. Design a septic tank based on the following data: population served = 400, daily wastewater production = 100 L/c.d.
   a. What is the flow capacity to provide a detention period of 24 h and storage of sludge @30 L/c.a? (52 m³).
   b. If soil absorption is used as effluent disposal, find the length of each of the ten 1.0 m wide trenches in soil with a percolation rate of 8 min (# = 10, 55 m × 1 m).
3. Design a septic tank to serve a population from a small community of 100 people.
   a. Detention time = 30 h, water supply = 130 L/p·d.
   b. Sludge removal every year; sludge production of 30 L/p·d.
   c. Length to width ratio 1:4, depth 1.5 m (# = 2, 5.0 m × 1.3 m × 2.0 m).
4. Design a UASB reactor for an average flow of 3500 m³/d containing COD of 600 mg/L. Make the following assumptions:
   a. Design residence time = 10 h
   b. Design COD loading rate 1–3 kg/m³.d
   c. Rise rate in the reactor = 0.5 m/h
      (#2 each 13 m × 13 m × 6.0 m)
5. Select the number, diameter, and height of upflow anerobic sludge blanket reactors required to treat an average daily flow of 1 MGD containing BOD of 300 mg/L. Make the following assumptions:
   a. Design hydraulic detention time = 12 h
   b. Design BOD loading rate = 40–120 lb/1000 ft³.d
   c. Rise rate in the reactor = 1.5 ft/h
      (#2 each D = 50 ft, H = 20 ft)

29 Bio-Solids

During the processing of wastewater, solids removed as slurry is called sludge, or more appropriately, **biosolids**. The processing of sludge is a costly and difficult operation. In a secondary wastewater treatment plant, the solids slurry collected at the bottom of the primary clarifier is termed **primary or raw sludge**. Similarly, the sludge pumped from the bottom of a secondary clarifier is called **secondary sludge** (**Figure 29.1**). In some plants, secondary sludge is pumped back into the primary clarifier. In such cases, sludge pumped out of the primary clarifier will be a combination of primary and secondary sludge solids.

29.1 PRIMARY SLUDGE

Primary sludge is raw since it is produced because of the settling of solids under gravity. Primary sludge is septic or anaerobic, dark in color and offensive. Solids are concentrated without undergoing any biological or chemical treatment; hence sludge is unstabilized. The solids content of this sludge will depend on the characteristics of the raw wastewater, solids removal by the primary treatment, and any chemical addition upstream, for example alum addition for phosphorus removal. Due to chemical coagulation, the removal of solids will be enhanced and thus results in increased raw sludge production. Since primary sludge lacks molecular oxygen, sludge left for long in the clarifier may get gasified and become buoyant. The concentration of raw sludges typically falls in the range of 1%–2%.

29.2 SECONDARY SLUDGE

Secondary sludge is not offensive as it comes out of an aerobic process. In the case of activated sludge plants, secondary sludge is golden brown, light, and has an earthy odor. However, sludge from a trickling filter is dark brown and bulky. The solids in secondary sludge are partially oxidized compared to solids in raw sludge. To stabilize this sludge, it needs to go through further biodegradation known as **digestion**. Digested sludge with minimal metal content can be used as a **soil conditioner** on agricultural land.

The quantity of secondary sludge produced in a sewage treatment plant depends on the fraction of BOD converted to new cells (biomass). The fraction of BOD that is converted is represented by a coefficient K that depends on the operating F/M ratio, which ranges from 0.25 to 0.50. The lowest value is for extended aeration and for conventional aeration a typical value is 0.35. Sludge consistency of secondary sludges is usually less than 1%.

DOI: 10.1201/9781003347941-34

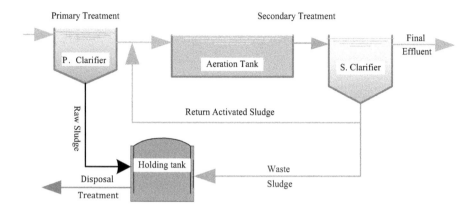

FIGURE 29.1   Primary and secondary sludge flow streams

Blending of Sludge

When sludges of different solid content are mixed, the resulting sludge has a solid content somewhere in between the sludge contents of the original sludge. The resulting value will be close to the solid content of the sludge, but with more volume. It is assumed that the mass density of the sludge is the same irrespective of the solid content. This is valid for the majority of situations. When dry solid concentration of sludges is less than 10%, the density of sludge can be safely assumed to be equal to that of water (that is 1000 kg/m³ (8.34 lb/gal)).

Example Problem 29.1

Calculate the solids concentration in the sludge fed to a digester which consists of the mixing of a 4% primary sludge flowing at 6900 gal/d and 6% thickened secondary sludge flowing at 5000 gal/d.

Given:

Parameter	Primary Sludge, 1	Secondary Sludge, 2
Q, gal/d	6900	5000
SS, %	4.0	6.0

Solution:

$$SS_{mix} = \frac{\Sigma \left(SS_i \times Q_i \right)}{\Sigma Q_i} = \frac{6900 \times 4.0\% + 5000 \times 6.0\%}{6900 + 5000} = 4.84 = \underline{4.8\%}$$

Example Problem 29.2

A flow of 1.5 MGD with SS = 240 mg/L and BOD of 220 enters a trickling filter plant. Removal efficiency in the primary clarifier is 50% for SS and 30% for BOD. Compute mass of solids generated as primary sludge and secondary sludge assuming 25% of incoming BOD is converted to microbial mass. Primary and secondary sludges are combined in a holding tank. Assuming blended sludge has solids concentration of 4.0%, what is the total volume of sludge generated per day?

Given:

$Q = 1.5$ MGD   BOD $= 220$ mg/L   SS $= 240$ mg/L $K = 0.25$ $SS_{sl} = 4.0\%$

Solution:

Mass of solids in primary sludge

$$M_{ss}(I) = Q \times SS_{rem} = \frac{1.5\,MG}{d} \times \frac{240 \times 8.34\,lb}{MG} \times \frac{50\%}{100\%} = 1501\,lb/d$$

Mass of solids in secondary sludge

$$M_{ss}(II) = Q \times BOD \times K = \frac{1.5\,MG}{d} \times \frac{220 \times 0.70 \times 8.34\,lb}{MG} \times \frac{25\%}{100\%} = 481.6\,lb/d$$

Volume of blended sludge

$$V_{sl} = \frac{M_{SS}}{SS_{sl}} = \frac{(1501 + 481.6)\,lb}{d} \times \frac{100\%}{4.0\%} \times \frac{gal}{8.34\,lb} = 5943 = 5900\,gal/d$$

29.3   PROCESSING OF SLUDGES

As mentioned above, sludge processing is a costly and difficult operation. Before disposal, the various treatments used to process sludge can be put in two groups:

First stage: conditioning, thickening, dewatering, and stabilization.

Second stage: digestion, composting, thermal drying, incineration, pyrolysis, and wet air oxidation.

The objective of the first stage of treatment is to reduce the sludge volume as well as stabilize it. This involves the use of chemical, physical, and biological processes. Water held in the pores of sludges can be free capillary water or bound (absorbed) with the sludge particles. To remove bound water, the addition of polymers is required. The addition of polyelectrolytes causes changes in the chemical charge and hence helps the release of intercellular water. Some of the common processes are discussed in the following sections.

29.4   SLUDGE THICKENING

Sludge thickening reduces the volume of sludge by removing water. For example, in a sludge with 3% of solids (97% water), when thickened to 6% solids, the volume of sludge is halved. Thickening of sludge makes further processing or disposal easier and more efficient. Hence, sludge is usually thickened before dewatering and sludge digestion. Depending upon the characteristics of the sludge and the size of the plant, sludge can be thickened using gravity or floatation thickeners, gravity belt thickeners, rotary drum thickener, and centrifuges.

29.4.1 GRAVITY THICKENER

A gravity thickener is designed to further concentrate sludge before sending it to additional sludge handling and treatment processes, such as digestion, conditioning, and dewatering, as shown in **Figure 29.2**. A gravity thickener works on the same principle as a clarifier. The calculation of hydraulic loading is important in determining whether the process is underloaded or overloaded. Hydraulic loading for a gravity thickener is expressed as flow per unit surface area. Sludge detention time refers to the length of time the solids remain in the gravity thickener, which depends on the volume of the sludge blanket and the pumping rate of the sludge from the bottom of the thickener. The efficiency of a gravity thickener is a measure of the effectiveness in capturing the suspended solids from the influent sludge into the thickened sludge (underflow).

29.4.2 CONCENTRATION FACTOR

Concentration factor, CF, is another way of determining the effectiveness of the gravity thickener. This parameter indicates the factor by which sludge has been thickened. A concentration factor of three will mean that the thickened sludge is three times as concentrated as the influent sludge or the volume is reduced to one-third.

29.4.3 FLOATATION THICKENER

Dissolved air floatation is achieved by releasing fine air bubbles that attach to sludge solids to make them buoyant and rise to the surface. This type of thickening is more common in activated sludge plants since aeration equipment is already there.

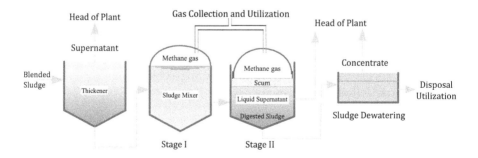

FIGURE 29.2 Sludge process scheme

Air under pressure is released at the bottom of the tank. The process underflow is returned to the wastewater treatment, and the overflow discharged by a mechanical skimming device is the thickened sludge. Floatation thickeners are able to thicken the sludge to 4% silds concentration (**Figure 29.3**).

29.4.4 GRAVITY BELT THICKENER

A gravity belt thickener consists of a continuously moving permeable fabric belt that passes over a horizontal belt. Free water drains out and thickened sludge is retained on the moving belt. Polymers are usually used to condition the sludge to release more water. Rows of pickets assist in releasing additional water by opening more drainage passages. Dewatered sludge falls into a small hopper at the end of belt. Rotary drum thickeners function similarly to the gravity belt, where free water from a flocculated sludge drains through a wire stainless steel screen.

29.4.5 CENTRIFUGE THICKENING

As the name indicates, centrifugal forces drain the free water out of sludge. Centrifugal thickening machines similar to dewatering centrifuges are discussed in a later part of this chapter.

Example Problem 29.3

Blended sludge in Example Problem 29.1 is fed to a gravity thickener. The effluent from the thickener contains 1200 mg/L solids. What is the solids removal efficiency of the thickener? Also, determine the volume of thickened sludge containing 7.4% solids.

Given:

$$Q = 11\ 900\ \text{gal/d} \quad SS_i = 4.8\% \quad SS_e = 1200\ \text{mg/L} = 0.12\ \% \quad SS_{thick} = 7.4\%$$

Solution:

$$PR = \frac{(SS_i - SS_e)}{SS_i} \times 100\% = \frac{(4.8 - 0.12)\%}{4.8\%} \times 100\% = 97.5 = 98\%$$

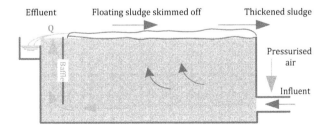

FIGURE 29.3 Schematic of a floatation thickener

$$V_{sl} = \frac{M_{SS}}{SS_{sl}} = \frac{11900\,gal}{d} \times \frac{4.8\%}{7.4\%} \times 0.975 = 7525.9 = 7500\,gal\,/\,d$$

$$CF = \frac{V_{in}}{V_{out}} = \frac{11900\,gal}{d} \times \frac{d}{7500\,gal} = 1.586 = 1.6$$

Volume of sludge is reduced by a factor of 1.6.

Example Problem 29.4

Given the data below, determine the change in sludge blanket solids.

Parameter	Feed Sludge	Effluent	Thickened Sludge
Q, m³/d	630	240	290
SS, %	3.5	0.01	8.0
kg/m³	35	0.1	80

Solution:

$$M_{feed} = Q \times SS = \frac{630\,m^3}{d} \times \frac{35\,kg}{m^3} = 22050\,kg\,/\,d$$

$$M_{eff} = Q \times SS = \frac{240\,m^3}{d} \times \frac{0.10\,kg}{m^3} = 24\,kg\,/\,d$$

$$M_{thick} = Q \times SS = \frac{290\,m^3}{d} \times \frac{80\,kg}{m^3} = 23200\,kg\,/\,d$$

Change, $\Delta M = 22050 - 24 - 23200 = -1174 = 1170\ kg/d\ drop$

29.5 MASS VOLUME RELATIONSHIP

Solids concentration of sludges would vary depending on the point of production, the removal of solids, and the specific gravity (SG) of the sludge. As mentioned earlier, in sludge with solid concentration less than 10%, it is safe to assume that the SG of the sludge is close to that of water. Solids in sludge consist of volatile and non-volatile or fixed solids. Non-volatile solids being inorganic in nature are heavier. The specific gravity of total solids in sludge can be found as follows:

Specific gravity of TSS

$$\frac{1}{SG_{TSS}} = \frac{VSS_f}{SG_{VSS}} + \frac{FSS_f}{SG_{FSS}}$$

Sub f = decimal fraction TSS = total suspended solids
VSS = volatile suspended solids FSS = fixed suspended solids

Wet sludge or slurry consists of dry solids and water. Knowing the specific gravity of solids in sludge, the SG of sludge slurry can be found.

Specific gravity of sludge

$$\frac{1}{SG_{SL}} = \frac{SS_f}{SG_{SS}} + \frac{WC_f}{1} \quad or \quad SG_{SL} = \frac{SG_{SS}}{\left(SS_f + SG_{SS} \times WC_f\right)}$$

Sub f = decimal fraction   SL = sludge slurry
SS = dry solids   WC = water content

Example Problem 29.5

Determine the volume of wet sludge before and after digestion produced for every 1000 lb for dry solids as feed sludge having the characteristics shown in the table below.

Parameter	Feed Sludge, 1	Digested Sludge, 2
SS, %	5.5	11
VF, %	65	60% destroyed
SG_{FS}	2.5	2.5
SG_{VS}	1.0	1.0

Given:

$$SS_1 = 5.5\% \quad SS_2 = 11\% \quad VF_1 = 65\%$$

Solution:

Feed Sludge (1)

$$\frac{1}{SG_{TSS}} = \frac{VSS_f}{SG_{VSS}} + \frac{FSS_f}{SG_{FSS}} = \frac{0.65}{1} + \frac{0.35}{2.5} = 0.79 \quad or \quad SG_{TSS} = 1.265$$

$$SG_1 = \frac{SG_{SS}}{\left(SS_f + SG_{SS} \times WC_f\right)} = \frac{1.265}{\left(0.055 + 1.265 \times 0.945\right)} = 1.0116$$

$$V_1 = \frac{M_{SS}}{SS} = 1000\,lb \times \frac{gal}{8.34\,lb \times 0.055 \times 1.012} = 2154 = 2150\,gal$$

Digested sludge (2)

$$M_{SS} = fixed + volatile = \frac{1000\,lb}{d} \times \left(0.35 + 0.65 \times 0.60\right) = 740\,lb$$

$$VF = \frac{M_{VSS}}{M_{TSS}} = \frac{0.65 \times 0.6 \times 1000\,lb}{740\,lb} = 0.527 = 53\%$$

$$\frac{1}{SG_{TSS}} = \frac{VSS_f}{SG_{VSS}} + \frac{FSS_f}{SG_{FSS}} = \frac{0.53}{1} + \frac{0.47}{2.5} = 0.718 \ \ or \ SG_{TSS} = 1.39$$

$$SG_2 = \frac{SG_{SS}}{(SS_f + SG_{SS} \times WC_f)} = \frac{1.39}{(0.11 + 1.39 \times 0.89)} = 1.0318$$

$$V_2 = \frac{M_{SS}}{SS} = 740 \ lb \times \frac{gal}{0.11 \times 1.032 \times 8.34 \ lb} = 781.6 = 780 \ gal$$

SLUDGE VOLUME PUMPED

The quantity of sludge pumped per day, in terms both of mass and volume, is an important variable that operating personnel need to determine. In smaller plants using positive displacement pumps, the volume displaced by the pump during each revolution determines the volume of raw sludge. Positive displacement pumps are equipped with a counter at the end of the shaft and are rarely operated faster than 200 L/min. Actual volume will be less due to slow or incomplete valve closures and slippage. The pump may be calibrated by recording the number of strokes required to fill an empty tank of known volume.

In the case of centrifugal pumps, it would be necessary to determine the volume pumped within the system. A pump can be calibrated by determining how long it took to pump a given volume of sludge by observing the pumping time to raise a fixed depth of sludge in the digester.

29.6   SLUDGE STABILIZATION

In the second stage of sludge treatment, sludge is stabilized. The term **stabilization** means to make the sludge innocuous. This will need to further reduce the sludge in BOD and volatile solids content. In addition to sludge digestion, other common methods of sludge stabilization include wet air oxidation, chemical oxidation, lime treatment, and composting.

29.7   SLUDGE DIGESTION

Sludges produced during wastewater treatment are not fully stabilized and must be stabilized before final disposal. For non-stabilized sludge, the most common disposal method is landfill, whereas stabilized sludge can be used for conditioning agricultural soils or other similar uses. Sludge can be stabilized using an aerobic process or an aerobic process.

29.7.1   ANAEROBIC SLUDGE DIGESTION

Anaerobic digestion is more common and will be discussed in more detail. Anaerobic sludge digestion takes place in two distinct stages. In the **first stage** of digestion **acid-forming** bacteria break down large organic compounds to organic acids. In the second stage, **gasification** occurs, and **methane-forming** bacteria convert organic acids to carbon dioxide and methane gas. Whereas acid-forming bacteria are sturdier, methane formers are very sensitive to pH, temperature, and loading changes. Any rapid change will inhibit the methane formers and result in souring or **pickling** of the digestion process. End products of anaerobic digestion include: methane, carbon dioxide, some hydrogen sulphide, and organic acids. A single tank digester, where both stages take place in the same unit is common for small plants. For larger plants, separate tanks are used for each stage as shown in **Figure 29.4**.

As part of daily operation, certain tests must be done, including: volatile acid concentration, temperature, pH, alkalinity, gas composition, volatile fraction of solids, scum blanket depth, and suspended solids in the supernatant. In addition, records of the quantities of sludge and gases plus the mixer schedule should be maintained.

29.7.2   AEROBIC SLUDGE DIGESTION

Aerobic sludge digestion is commonly used for digestion of the activated sludge. Since this is an aerobic process, all the contents of the sludge are kept aerated. Aerobic digestion can be thought of as the aeration process for the sludge to stabilize by further reducing BOD and volatile solids. A comparison of anaerobic and aerobic sludge digestion is shown in **Table 29.1**. Disposal methods of digested sludge include spreading on farmland, lagooning, or drying on beds. The **elation** period is typically 10 to 20 days. A common problem experienced is poor settling of the digested sludge and so chemical conditioning might be necessary.

29.7.3   ANAEROBIC DIGESTER CAPACITY

The determination of digester tank volume is a critical step in the design of an anaerobic system. The digester volume must be sufficient to prevent the process from failing under all accepted conditions. Process failure is defined as accumulation of volatile acids that results in a decrease in pH, what is usually called pickling of the sludge. When the **volatile acids/alkalinity** ratio becomes greater than 0.5,

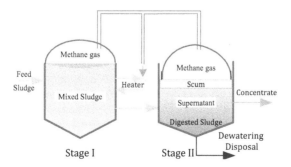

FIGURE 29.4   Schematic of a two-tank anaerobic sludge digester

TABLE 29.1

Aerobic versus anaerobic digestion

Characteristic	Aerobic	Anaerobic
Source of sludge	waste activated sludge	primary + secondary
End products	CO_2 + cellular protoplasm	CH_4 + unused organics + small portion of cellular protoplasm
Growth	limited by availability of carbon	limited by lack of H ion acceptor
Completion	close to complete	complete degradation not possible

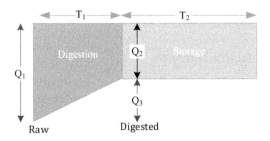

FIGURE 29.5　Sludge digestion and storage capacity

acid-forming bacteria take over and the second stage virtually stops – and no methane is produced. Once the digester turns sour, it usually takes several days to return to normal operation after corrective actions are taken.

The digester capacity must also be large enough to ensure that raw sludge is adequately stabilized and there is room for storage, if needed. The relationship between the volume of sludge reduction and detention time is shown in **Figure 29.5**. During digestion, the reduction in volume is assumed to be linear. Hence, in a single-stage digester, the volume of the digestor equals the area of a trapezoid and the area of the rectangle is as seen in the figure.

Single tank digester capacity

$$V = \frac{(Q_1 + Q_2)}{2} \times T_1 + Q_2 \times T_2$$

Example Problem 29.6

Work out the digester capacity needed for a low-rate anaerobic digester required to serve a population of 50 000 people. Assume per capita production of sludge is 90 g/c·d, with a solid content of 4.0% and volatile fraction of 70%. Assume a digestion period of 25 d and storage period of 20 d and volatile reduction of 50% during digestion, and solid content of digested sludge is 8.0%.

Given:

$$SS_1 = 4.0\% \quad SS_2 = 8.0\% \quad T_1 = 25 \text{ d} \quad T_2 = 20 \text{ d}$$
$$VF_1 = 70\% \quad VF_2 = 50\% \times 70\% = 35\%$$

Solution:

Feed sludge (1)

$$M_{SS} = \frac{90\,g}{p.d} \times 50000\,p \times \frac{kg}{1000\,g} = 4500\,kg\,/\,d$$

$$Q_1 = \frac{M_{SS}}{SS} = \frac{4500\,kg}{d} \times \frac{m^3}{40\,kg} = 112.5 = 110\,m^3\,/\,d$$

Digested sludge (2)

$$M_{SS} = \frac{4500\,kg}{d} \times \left(0.30 + 0.70 \times 0.50\right) = 2925\,kg\,/\,d$$

$$Q_2 = \frac{M_{SS}}{SS} = \frac{2925\,kg}{d} \times \frac{m^3}{80\,kg} = 36.56 = 37\,m^3\,/\,d$$

Digester capacity

$$V = \frac{(112.5 + 36.56)m^3}{2\,d} \times 25\,d + \frac{36.56\,m^3}{d} \times 20\,d = 2594 = \underline{2600\,m^3}$$

29.7.4   TWO-STAGE DIGESTION

In a high-rate digester, two tanks are used, one for each stage, thus separating the mixing and storage. This produces better conversion, and control is better too. The capacity of each tank can be worked out as follows.

$$\text{Stage I}, V_I = Q_1 \times T_I$$

$$\text{Stage II}, V_{II} = 0.5\left(Q_1 + Q_2\right) \times T_{II} + Q_2 \times T_2$$

Time, T_I is the solids retention time in stage I, typically 10–15 d. Time T_{II} is the digestion period in stage II, and T_2 is the storage period required during a winter or monsoon period. Two-stage process digestion happens more efficiently and quickly and the quality of the supernatant is much better. With the second stage, there is no mixing, so it provides the freedom to build up the scum depth, as desired.

Example Problem 29.7

Work out the digester capacity required for a high-rate anaerobic digester required to serve a population of 50 000 people. Assume per capita production of sludge is 90 g/c·d with a solid content of 4.0%, with volatile fraction of 70%. Sludge is kept for a period of 10 d stage I before it is transferred to stage II. In the second stage,

assume a digestion period of 15 d and storage period of 20 d, volatile reduction of
50% during digestion, and solid content of digested sludge is 8.0%.

Given:

$$SS_1 = 4\% \quad SS_2 = 8\% \quad T_1 = 10 \text{ d} \quad T_{||} = 15 \text{ d} \quad T_2 = 20 \text{ d}$$

$$VF_1 = 70\% \quad VF_2 = 50\% \times 70\% = 35\%$$

Solution:

Feed sludge

$$M_{SS} = \frac{90 \text{ g}}{p.d} \times 50000 \, p \times \frac{kg}{1000 \text{ g}} = 4500 \text{ kg} / d$$

$$Q_1 = \frac{M_{SS}}{SS} = \frac{4500 \text{ kg}}{d} \times \frac{m^3}{40 \text{ kg}} = 112.5 = 110 \, m^3 / d$$

Digested sludge

$$M_{SS} = \frac{4500 \text{ kg}}{d} \times (0.30 + 0.70 \times 0.50) = 2925 \text{ kg} / d$$

$$Q_2 = \frac{M_{SS}}{SS} = \frac{2925 \text{ kg}}{d} \times \frac{m^3}{80 \text{ kg}} = 36.56 = 37 \, m^3 / d$$

Digester capacity

$$V_1 = Q_1 \times T_1 = \frac{112.5 \, m^3}{d} \times 10 d = 1125 = \underline{1130 \, m^3}$$

$$V_{||} = \frac{(112.5 + 36.56) m^3}{d} \times \frac{15 d}{2} + \frac{36.56 \, m^3}{d} \times 20 d = 1849 = \underline{1850 \, m^3}$$

29.7.5   VOLATILE SOLIDS REDUCTION IN DIGESTION

During digestion, volatile solids are oxidized. Thus, the total amount of volatile solids is reduced. However, the concentration of solids in the sludge increases due to
thickening. Due to these factors, the volume of digested sludge is significantly less
than feed sludge. During the digestion process, a large quantity of volatile solids is
destroyed, as shown by the following expression.

Volatile reduction

$$VR = \frac{(VF_1 - VF_2)}{VF_1 - VF_1 \times VF_2}$$

$VF_1 =$ *Volatile fraction of feed sludge,* $VF_2 =$ Volatile fraction of digested sludge

Example Problem 29.8

What is the percentage reduction in volatile matter in a primary digester if the volatile content of the raw sludge is 69% and the volatile content of the digested sludge is 51%?

Given:

$$VF_1 = 69\% = 0.69 \quad VF_2 = 51\% = 0.51$$

Solution:

Volatile solids reduction

$$VR = \frac{(VF_1 - VF_2)}{VF_1 - VF_1 \times VF_2} = \frac{0.69 - 0.51}{0.69 - 0.69 \times 0.51} = 0.532 = \underline{53\%}$$

29.7.6 GAS COMPOSITION

Sludge gases produced during anaerobic digestion are normally composed 60% to 70% of methane and 25% to 35% of carbon dioxide by volume. Smaller quantities of other gases like hydrogen sulphide, hydrogen, nitrogen, and oxygen are produced. The combustible constituent is primarily methane. Hydrogen sulphide is not only corrosive but it also causes problems during burning of the gas. In terms of volatile acids destroyed during digestion, average gas production is 0.9 m^3/kg of volatile acids destroyed at a normal operating pressure of 1.5 to 2.0 kPa. Sludge gases are collected under positive pressure to prevent mixing with air and causing explosions. The gas may be collected directly from under a floating cover or from a fixed cover by maintaining a constant water level.

29.7.7 DIGESTER SOLID MASS BALANCE

The principle of mass balance dictates what comes to process, or a combination of processes must go out. If 90% of this solid material is leaving in the form of sludge, liquid (effluent), or gas (digester gas), then the digestion process is in control. This accounting process provides a good check on your metering devices, sampling procedures, and analytical techniques.

29.8 DEWATERING OF SLUDGES

Digested sludge or blended sludge usually contains more than 90% water. It would be expensive to carry this watery sludge for disposal. Hence sludge is dewatered before disposal. Depending on the method of **dewatering**, sludge water content can be brought as low as 20%-30% thus reducing the volume many times. In warmer climates, the most common practice is to spread the sludge on **drying** beds and let it dry in the open. When the space required for dry beds is not available, sludge is conditioned followed by mechanical dewatering. Mechanical dewatering methods include vacuum filters, filter presses, or centrifugation.

29.8.1 SLUDGE DRYING BEDS

This method can be used in all places where adequate land is available and dried sludge can be used for soil conditioning. The sludge drying beds are affected by weather, sludge characteristics, system design that includes depth of sludge layer, and frequency of scraping the beds after drying.

Area of Beds

The area needed for dewatering and drying the sludge is dependent on the volume of sludge. The cycle time between two successive dryings depends on the characteristics of sludge including water content, drainage ability and rate of evaporation, and acceptable water content in the dried sludge.

Drying Beds

A sludge drying bed usually consists of a bottom layer of gravel of uniform size, which is laid over a bed of clean sand. Graded gravel is placed in the underdrain in layers up to 30 cm(I ft), with a minimum of 15 cm(6 in) above the top of the underdrains. The top layer should be at least 3.0 cm thick and consist of gravel 3–6 mm in size. Gravel should be covered with a 20–30 cm thick layer of clean sand of effective size 0.50–0.75 mm and uniformity coefficient less than 4.0. The finished surface should be level. Open jointed underdrain pipes of diameter 100–150 mm are laid with spacing of 6.0 m or less. The pipes are laid on a gradient of 1%. The beds are about 15 m × 30 m in plan and are surrounded by about 1.0 m high brick walls above the sand surface.

Example Problem 29.9

Design sludge drying beds for digested sludge as discussed in Example Problem 29.7. The population served is 50 000 and the average production of digested sludge is 37 m³/d. Assume 25 cm of digested sludge is spread every couple of weeks. Though drying in summer is done over 10 day, a drying cycle of 14 days is assumed to compensate for wet weather.

Given:

Q_{dsl} = 37 m³/d, t = 14 d, d = 25 cm/cycle, each bed = 30 m × 15 m

Solution:

$$Area, A = \frac{37\,m^3}{d} \times \frac{cycle}{0.25\,m} \times \frac{14d}{cycle} = 2072\,m^2$$

$$\# = 2072\,m^2 \times \frac{bed}{30\,m \times 15\,m} = 4.6 = 5\,beds\,say$$

Making 100% allowance for space for storage, repairs, and resting of beds, it is suggested to have 10 beds, each of area 30 m × 15 m.

29.8.2 MECHANICAL METHODS OF DEWATERING SLUDGE

Mechanical methods of dewatering include filter press, vacuum filtration, and centrifuges.

Vacuum Filtration

Rotary **vacuum filtration** basically consists of a cylindrical drum covered with a filter media which rotates partially submerged in a vat of sludge. The physical mechanisms which take place during vacuum filtration may be divided into three phases. The first phase, which refers to the cake pick-up or form phase, occurs when a segment of the drum rotates into the sludge. A vacuum is applied to that segment so filtrate is drawn through the media and discharged. Concurrently, sludge solids are deposited on the media to form a partially dewatered cake. As the sludge cake increases in thickness, its resistance to the passage of filtrate increases.

The second phase, cake drying, occurs during that time the drum segment leaves the sludge and before the cake is removed. As the drum leaves the sludge, the cake is still under vacuum and additional moisture within the cake is drawn out.

The third phase, cake discharge, occurs after an acceptable cake dryness has been achieved and without vacuum. All operations are continuous in nature such that all three phases occur simultaneously on different portions of the drum.

Precoat vacuum filtration is like conventional filtration with the exception of the application of a precoat prior to filtration. The precoat is normally diatomite – the siliceous skeletal remains of single cell aquatic plant life called diatoms. These diatoms form a permeable coating on the filter allowing filtrate to pass through easily while trapping sludge solids. The use of a precoat produces filtrate of a very high quality.

Belt Filter Press

The continuous **belt filter press** was originally developed and in subsequent years modified and improved in West Germany. Installation of the latest and best models in the USA have only recently experienced popularity. These systems were developed to overcome the sludge pick-up problem occasionally experienced with rotary vacuum filtration. A combination of sludge conditioning, gravity dewatering, and pressure dewatering is utilized to increase the solids content of the sludge.

With all units, the infeed sludge is mixed with **polymer** (or other chemicals) and placed onto a moving porous belt or screen. Dewatering occurs as the sludge moves through a series of rollers which squeeze the sludge to the belt or squeeze the sludge between two belts much like an old washing machine wringer. The **sludge cake** formed is then discharged from the belt by a scraper mechanism. There are basically three processing zones which occur along the length of the unit.

   i. Drainage zone, which is analogous to the action of a drying bed.
   ii. Pressure zone, which involves the application of pressure.
   iii. Shear zone, in which shear is applied to the partially dewatered cake.

Shearing action is accomplished by positioning the support rollers of the filter belt and the pressure rollers of the pressure belt in such a way that the belts and the sludge between them describe an S-shape curve.

Pressure Filtration

Of the several types of pressure filters available, the most widely used consists of a series of vertical plates held rigidly in a frame which are pressed together between a fixed and moving end. Mounted on the face of each individual plate is a filter cloth to support and contain the cake produced. **Pressure filters** do not produce a cake by pressing and squeezing. Instead, sludge is fed into the press "batch mode" through feed holes in trays along the length of the press. Pressure up to 1600 kPa (250 psi) are applied to the sludge causing water to pass through the cloth while the solids are retained forming a cake on the surface of the filter cloth. Sludge feed is stopped when the chambers between the trays are filled. Drainage ports are provided at the bottom of each chamber where the filtrate is collected, taken to the end of the press, and discharged.

The dewatering phase is complete when the flow of filtrate through the filter cloth nears zero. At this point, the sludge feed pump is stopped and any back pressure in the piping released. Each plate is then turned over the gap between the plates and the moving end to allow for cake removal. Filter cake usually drops below onto a conveyor for further removal.

The main advantage of using a filter press for sludge dewatering is in the reduced sludge disposal costs associated with producing a drier cake. However, a detailed cost analysis should be performed to determine if these savings are sufficient to offset its high capital cost.

Centrifugation

A centrifuge is essentially a sedimentation device in which the solids from liquid separation are enhanced using centrifugal force. This is accomplished by rotating the liquid at high speeds to subject the sludge to increased gravitational forces. There are basically three types of centrifuges available for sludge dewatering.

CONTINUOUS SOLID BOWL CENTRIFUGE

This centrifuge consists of two principal elements: a rotating bowl which is the settling vessel and a conveyor which discharges the settled solids. The rotating bowl is supported between two sets of bearings and includes a conical section at one end. This section, which is not submerged, forms the dewatering beach or drainage deck. Sludge enters the rotating bowl through a stationary feed pipe extending into the hollow shaft of the rotating screw conveyor and is distributed through ports into a pool within the bowl. As the bowl rotates, centrifugal force causes the slurry to form an annular pool, the depth of which is determined by the effluent weirs.

Basket Centrifuge

The basket, or imperforated bowl-knife discharge unit, is a batch dewatering unit introduced primarily for use as a partial dewatering device for small operations.

Disc Centrifuge

The disc centrifuge is a continuous flow variation of the previously described basket centrifuge. The incoming sludge is distributed between a multitude of narrow channels formed by stacked conical discs.

29.8.3 Sludge Conditioning

Whatever the methods of dewatering, sludge needs **conditioning** to facilitate the extraction of moisture held by the sludge solids. Coagulants and coagulant aids like alum, lime, and polymers are commonly used. Digested sludge, because of its high alkalinity, needs very high doses of coagulant if not elutriated. **Elutriation** of sludge is done to reduce chemical dosage and is carried out by washing the sludge followed by decantation.

Polymers are less affected by the pH and work better in the case of finely dispersed solids. It is difficult to specify any formula suitable for a polyelectrolyte. Even though there are quite a few such polyelectrolytes in the market, it is best to carry out an actual laboratory scale testing before launching into the procurement.

29.9 DISPOSAL OF SLUDGE

In Indian conditions, digested sludge is usually disposed of on land as manure to soil, or as a soil conditioner. In developed countries, unstabilized sludges are sent to sanitary landfill, incinerated, or barged into the sea. The most common method is to utilize digested sludge as a fertilizer. Ash from incinerated sludge is used as a landfill. In some cases, wet sludge, raw or digested, as well as supernatant from a digester can be constructed as lagoons as a temporary measure, but such practice may create problems like odor nuisance, groundwater pollution, and other public health hazards. Wet or digested sludge can be used as sanitary landfill or for mechanized composting with city refuse. Burial is generally resorted to for small quantities of putrid sludge. In Ontario, Canada, sludge is treated as hazardous waste and its disposal (handling and management) is undertaken according to the rules of the Ministry of Environment and Forestry (MOEF), as shown in **Table 29.2**.

In general, digested sludge is indelicate, but has definite value as a source of slowly available nitrogen and some phosphate. It is comparable to farmyard manure

TABLE 29.2
Ceiling concentration of heavy metals

Chemical	Concentration, mg/kg	Chemical	Concentration, mg/kg
As	75	Cr	500
Cd	85	Se	100
Cu	4300	Zn	7500
Hg	57	Mb	75
Ni	420	Pb	840

except for its deficiency in potash. It also contains elements essential to plant life, and minor nutrients in the form of trace metals.

The sludge humus also increases the water holding capacity of the soil and reduces soil erosion, making it an excellent soil conditioner, especially in arid regions by making available needed humus content which results in greater fertility. Dewatered cake typically is stored before additional treatment (e.g., heat drying) or being hauled off-site for use or disposal.

29.9.1 INCINERATION

The purpose of **incineration** is to burn out the organic material at high temperatures, the residual ash being generally used as landfill. During the process, all the gases released from the sludge are burnt off and all the organisms are destroyed. Dewatered or digested sludge is subjected to temperatures between 650°C and 750°C. Cyclone or multiple-hearth and flash type furnaces are used with proper heating arrangements, with temperature control and drying mechanisms. Dust, fly ash, and soot are collected for use as landfill.

It has the advantages of freedom from odors and a great reduction in volume and weight of materials to be disposed of finally, but the process requires high capital and recurring costs, the installation of machinery, and skilled operation. Controlled drying and partial incineration have also been employed for dewatering of sludge, before being put on drying beds.

29.9.2 SANITARY LAND FILL

When organic solids are placed in a landfill, decomposition may result in odor if sufficient cover is not available. Surface-water contamination and leaching of sludge components to the groundwater must be considered. Decomposition may result in soil settlement resulting in a surface water pond above the fill. Typical depths of soil cover over the fill area are 20 cm after each daily deposit and 60 cm over an area that has been filled completely.

Surface topography should be finished to allow rainfall to drain away and not allow it to infiltrate into the solid landfill. Landfill **leachate** requires long-term monitoring and should satisfy the relevant water pollution control standards for land applications. Vegetation must be established quickly on completed areas to provide for erosion control. It is general practice not to crop the landfill area for several years after completion.

29.9.3 DISPOSAL IN WATER OR SEA

This is not a common method of disposal because it is contingent on the availability of a large body of water adequate to permit dilution at some seacoast sites. The sludge, either raw or digested, may be barged to sea far enough to make available the required dilution and dispersion. The method requires careful consideration of all factors including flora and fauna for proper design and siting of outfall to prevent any coastal pollution or interference with navigation.

29.9.4 SLUDGE COMPOSTING

Sludge **composting** is a method by which microorganisms decompose the degradable organic matter in sludge under aerobic conditions and create stable material that is easy to handle, store, and use for farmland. Sludge compost is humus-like material without detectable levels of pathogens that can be applied as a soil conditioner and fertilizer to gardens, feed crops, and farmland. Sludge compost provides large quantities of organic matter and some nutrients such as nitrogen and potassium. It improves the soil texture and structure and greatly improves the water holding capacity of soil. It elevates the soil cation exchange capacity (an indication of the soil's ability to hold nutrients). Sludge compost is safe to use and generally has a high degree of acceptability by the public.

Sludge composting involves mixing dewatered sludge with a bulking agent to provide carbon and to increase porosity. The resulting mixture is placed in a vessel where microbial activity causes the temperature to rise during the first-phase active composting period. For efficient stabilization and pasteurization, the temperature in a compost pipe should rise to 40°C (104°F) for a couple of weeks, but not above 80°C (176°F). The specific temperatures that must be achieved and maintained for successful composting vary based on the method and use of the end product. After the first phase of active composting, the material is cured, and in the second phase it becomes compost and can be distributed. Sludge-composting methods include aerated static pile, windrow, and in-vessel.

Aerated Static Pile

Dewatered sludge cake is mechanically mixed with a bulking agent and stacked into long piles over a bed of pipes through which air is transferred to the composting material. In static pipe composting, sludge temperatures of 55°C (130°F) are maintained for 3 days to further reduce pathogens. The bulking agent is often reused in this composting method and may be screened before or after curing so that it can be reused.

Windrow

Dewatered sludge cake is mixed with a bulking agent and piled in long rows. Because there is no piping to supply air to the piles, they are mechanically turned to increase the amount of oxygen. This periodic mixing is essential to move outer surfaces of material inward so they are subjected to the higher temperatures deeper in the pile. A number of turning devices are available. As with aerated static pile composting, the material is moved into the second phase of composting after the first-phase composting. This is active composting. Several rows may be placed into a larger pile for curing.

In-Vessel

There are two types of in-vessel composting reactor – a vertical type and a horizontal type. A mixture of dewatered sludge cake and bulking agent is fed into a silo, tunnel, channel, or vessel. Augers, conveyors, rams, or other devices are used to aerate, mix, and move the product through the vessel to the discharge point. Air is generally blown into the mixture. After the first-phase composting (i.e., active composting), the finished product is usually stored in a pile for the second-phase composting (i.e., curing prior to distribution). All three composting methods require the use of bulking agents. Wood chips, sawdust, and shredded tires are commonly used, but many other materials are suitable.

Discussion Questions

1) Show that for sludge with solids concentration less than 10%, it is safe to assume that the density of the sludge is close to that of water.
2) What is the purpose of thickening the sludge? Describe various methods of thickening.
3) Write a short note on sludge composting.
4) Briefly describe various methods of the disposal of sludge.
5) Most plant personnel are of the opinion that sludge processing is a most difficult and expensive process. Justify.
6) What is dewatering of sludge? What processes can be used to dewater the sludge?
7) In a country like India, what is the most common method of sludge dewatering? Discuss its merits and demerits.
8) Compare the following:
   a) primary sludge versus secondary sludge;
   b) activated sludge plant sludge versus trickling filter plant sludge;
   c) aerobic versus anaerobic sludge digestion.
9) What is the purpose of conditioning sludge? What role does elutriation play in sludge conditioning?
10) What methods can be used to dispose of undigested sludge?

Practice Problems

1. A dewatered sludge with 25% solids has SG of 1.1. What volume of sludge would contain 100 lb of dry solids? (44 gal).
2) What is the percentage reduction in volatile matter in a primary anaerobic digester if the volatile content of the raw sludge is 71% and the volatile content of the digested sludge is 49%? (61%).
3) Size a low-rate anaerobic digester required to serve a population of 35 000 people. Assume per capita production of sludge is 100 g/c·d with a total solid content of 3.5%, of which 65% are volatile. Assume a digestion period of 25 days and a storage period of 30 days Half of the volatile solids

are destroyed during digestion and digested sludge solid content is 7.0%. (2700 m³).

4) Calculate the capacity needed for a high-rate anaerobic digester required to serve a population of 75 000 people. Assume per capita production of sludge is 95 g/c·d with a solid content of 3.5% and a volatile fraction of 70%. Sludge is retained in stage I for a period of 10 days and then transferred to stage II. During stage II, assume a digestion period of 15 days and a storage period of 20 days During digestion, half the volatile solids are destroyed, and sludge is thickened to 7.5% solids. (2000 m³, 3200 m³).

5) Show that sludge concentration of 1.0% is the same as mass concentration of 10 kg/m³ or 10 g/L?

6) Design an anaerobic digester to treat a blend of primary and secondary sludge. The feed sludge daily production is 250 m³/d. The solid content of the feed sludge and digested sludge respectively are 5.0% and 9.5%. Based on the prevailing temperatures, a digestion period of 25 days is considered and a minimum storage period of 40 days is required. Make other assumptions as appropriate. (70% volatile, 60% reduction, 7100 m³).

7) How many times the volume of sludge is reduced when the moisture content is reduced from 95% to 90%? (2).

8) A wastewater pollution control plant produces on average 2970 lb of dry solids as sludge with solid concentration of 5.0%. The solids are 70% volatile, with SG of 1.05 and 30% fixed with a SG of 2.5. Find the SG of dry solids and hence the volume of sludge produced at this plant before digestion. (7050 gal).

9) Calculate the solids concentration in the sludge fed to a digester which consists of the mixing of a 3.5% primary sludge flowing at 3100 gal/d and 5.0% thickened secondary sludge flowing at 2100 gal /d. (4.1%).

10) Design sludge drying beds for digested sludge. The population served is 65 000 and average production of digested sludge is 0.2 gal/c.d. Assume 1.0 ft of digested sludge is spread every couple of weeks. Though drying in summer is done over 12 days, a drying cycle of 14 days is assumed to compensate for wet weather. (8 beds, each 100 ft × 30 ft).

11) At a given facility, during digestion, 53% of volatile matter is destroyed. Determine the expected volatile content in the digested sludge when sludge with volatile content of 69% is fed? (51%).

12) In a sewage treatment plant, the average primary SS removal is 45%. Estimate the primary sludge production for treating wastewater flow of 3.0 MGD with SS content of 250 ppm. Assume moisture content of primary sludge is 96%. (10 000 gal/d).

13) Blended sludge containing 750 kg of dry solids is fed to a digester for stabilization. The characteristics of feed sludge and digested sludge are shown below. Determine the volume of wet sludge before and after digestion. (15 m³, 4.6 m³).

Parameter	Feed Sludge, 1	Digested Sludge, 2
SS, %	5.0	10
VF, %	60	60% destroyed
SG_{FS}	2.5	2.5
SG_{VS}	1.0	1.0

14) In an activated sludge plant, on average 150 mg/L of BOD is removed by the secondary treatment. Estimate the secondary sludge production for treating a wastewater flow of 3.0 MGD assuming each unit of BOD removed produces 0.5 unit of SS and the water content of the secondary sludge is 99%. (23 000 gal/d).

15) Sludge containing 1600 lb of dry solids is fed to a digester for stabilization. Characteristics of feed sludge and digested sludge are shown below. Determine the volume of wet sludge before and after digestion. (3200 gal, 880 gal).

Parameter	Feed Sludge, 1	Digested Sludge, 2
SS, %	6.0	12
VF, %	65	65% destroyed
SG_{FS}	2.5	2.5
SG_{VS}	1.0	1.0

30 Advanced Wastewater Treatment

Conventional wastewater treatment consists of preliminary treatment and primary treatment, followed by secondary treatment, which is usually a biological process. Early treatment objectives were concerned with the removal of suspended and floatable material, the treatment of biodegradable organics, and the elimination of pathogenic organics. **Table 30.1** illustrates residual constituents in treated wastewater.

From the information listed in **Table 30.1**, it is evident that limits on wastewater plant effluents are becoming more stringent as unconventional pollutants are found in municipal wastewaters. This requires more-effective and more-widespread treatment of wastewater. The required degree of treatment is significantly increased, and additional treatment objectives and goals have been added. Recent treatment objectives are the removal of nutrients, SS and BOD, and toxic substances. This phase of wastewater treatment is known **advanced wastewater treatment**. Advanced treatment can be part of primary or secondary treatment or a third stage of treatment, called **tertiary treatment**.

Thus, **advanced wastewater treatment** refers to the processes and methods that remove more contaminants from the wastewater than are usually taken out by conventional treatments and techniques. Several of the pollutants or contaminants not taken out by secondary biological methods can adversely affect aquatic life in the receiving waters, accelerate eutrophication in lakes and hinder the reuse of surface water for domestic needs. The reclamation and reuse of wastewaters are becoming more important as water needs cannot be met by the fixed natural supply currently available. Severe droughts and floods due to global warming have posed new challenges in water conservation. The reuse and reclamation of wastewater effluents is more or less is becoming the norm, so advanced and tertiary treatments are being added to serve this objective.

30.1 SUSPENDED SOLIDS REMOVAL

SS removal is one of the most important and common applications of advanced wastewater treatment. The effluent from a secondary treatment plant may contain SS concentration ranging from 20 to 40 mg/L, depending upon the type of treatment method. SS in treated wastewater are in part colloidal and in part discrete, ranging from 10 to 100 microns. Because of their size, these solids are not easily settleable. For further removal of suspended solids, some of the commonly used advanced treatment are here briefly discussed.

DOI: 10.1201/9781003347941-35

TABLE 30.1
Residual constituents in treated wastewater

Constituents	Effect	Critical conc. mg/L
Suspended solids	may cause sludge deposits or interfere with receiving water quality	variable
Biodegradable organics	may deplete oxygen resources	variable
Volatile organic compounds	toxic to humans; carcinogenic; forms photochemical oxidants	varies by individual constituent
Nutrients		
Ammonia	increases chlorine demand; can be converted to nitrates and, in the process, can deplete oxygen resources	any amount
Nitrate	stimulates algal and aquatic growth and can cause methemoglobinemia in infants	0.3 (for quiescent lakes)
Phosphorus	stimulates algal and aquatic growth	45 (as NO_3) 0.015 (for quiescent lakes) 0.2–0.4
Other inorganics		
Calcium and magnesium	interferes with coagulation; interferes with lime-soda softening; increases hardness and total dissolved solids; imparts salty taste	250
Chloride	interferes with agricultural and industrial processes; cathartic action	75–200
Sulfate	causes foaming and may interfere with coagulation	600–1000
Other organics surfactants		1.0–3.0

30.1.1 MICROSCREENING

Microscreening utilizes a special woven metallic or plastic filter fabric which is mounted on the periphery of a revolving drum provided with continuous backwashing. The drum rotates about a horizontal axis at a variable low speed (up to 4 rpm). The size of the openings of the filter fabric may range from 23 to 60 μm. Wastewater enters through the open upstream end of the drum and flows radially outward through the microfabric leaving behind the SS. The solids which are retained on the fabric are washed through a trough, which recycle the solids to the sedimentation tank. Water for back flushing is drawn from the filtered water effluent.

30.1.2 ULTRAFILTRATION

Ultrafiltration is a system similar in operation to reverse osmosis and requires the membrane to be far coarser and the pressure lower. Ultrafiltration membranes are thin film cast from organic polymer solutions, the film thickness being 0.1 mm to 0.3 mm. The film has an extremely thin separation layer on a relatively porous structure, consisting of pores of closely controlled size ranging from 0.3 to 10 nm. Ultrafiltration membranes may be packed either as a plate device or as a tube device.

As shown in **Figure 30.1**, the ultrafiltration technique is used in conjunction with biological oxidation processes such as the activated sludge process. When the effluent from the activated sludge process is passed through an ultrafiltration system, the filter membranes filter out the biological cells while allowing the passage of treated effluent. The retained solids are returned to the activated sludge reactor. This gives rise to a high concentration of biological solids in the activated sludge reactor, resulting in the quick degradation of organics, the prevention of fouling of the membrane surface, and a reduction in the size of the activated sludge reactor.

30.1.3 GRANULAR MEDIA FILTRATION

Although granular media filtration is one of the principal unit operations in the treatment of potable water, the filtration of effluents from wastewater treatment processes is a relatively recent practice. It is used to achieve the supplemental removal of SS, including particulate BOD from wastewater effluents of biological and chemical treatment processes, and the removal of chemical-precipitated phosphorous.

Two configurations of wastewater filtration are shown in **Figure 30.2**. The design of gravity filtration systems must consider the higher SS content and the fluctuating rate of wastewater flow not common in water treatment. Another difference is that solids in wastewater effluents are mostly organic in nature and hence long filter runs cannot be afforded. Multimedia filters are recommended for in-depth-filtration and greater solids holding capacity, resulting in higher filter runs. Efficient backwashing requires an auxiliary scour. The filters may be either gravity or pressure units, depending on the size of the treatment plant. Chlorination prior to filtration prevents growth within the filter.

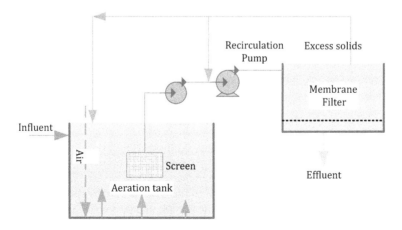

FIGURE 30.1 Ultrafiltration and biodegradation combined

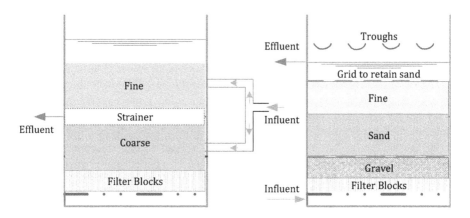

FIGURE 30.2   Configurations for wastewater filtration

30.2   CONTROL OF NUTRIENTS

Discharges containing nitrogen and phosphorus may accelerate the eutrophication of lakes and reservoirs and may stimulate the growth of algae and rooted aquatic plants in shallow streams. Other effects of excessive nitrogen are the depletion of dissolved oxygen and toxicity toward aquatic life. Nitrogen in the form of ammonia is toxic to fish at levels as low as 0.2 mg/L. Algae growth prevents sunlight reaching other microorganisms, thus affecting aquatic life. At the end of its life cycle, algae die off and settle at the bottom, thus creating oxygen demand.

30.3   PHOSPHORUS REMOVAL

The principal sources of phosphorus in wastewater are from domestic sewage, agricultural return water, and land runoff. Phosphorus, like nitrogen, contributes to the eutrophication of surface water. It is often cited as the culprit responsible for the stimulation of aquatic plants, since the concentration of phosphorus necessary to support algal blooms is as low as 10 µg/L. Phosphorus can be removed biologically or chemically. Sometimes, chemicals may be added to biological reactors. In other cases, phosphorus may be chemically precipitated, usually in primary clarification.

30.3.1   Biological Phosphorous Removal

The presence of phosphate in raw waste in a greater proportion to that required for bacterial growth (BOD: N:P = 100:5:1) may result in an effluent phosphate concentration sufficient to cause eutrophication in the receiving stream. Taking phosphates out of solution by photosynthesis has led to the concept of removing nutrients from wastewater by growing algae in stabilization ponds and then separating the cells from suspension by physical or chemical means. The activated algae process and bacterial assimilation processes have the potential of removing additional phosphorus, but if stoichiometric ratios are maintained it would be necessary to add both carbon and nitrogen to the latter and the resultant sludge would be perhaps five times that produced in ordinary plants.

Some microbes uptake phosphorus compounds as an energy source. Such organisms are called phosphorus-accumulating organisms or PAOs. These microorganisms release phosphorus in an anaerobic environment. Thus, cycling mixed liquor through two aerobic and anaerobic zones a number of times and wasting will remove phosphorus. However, this process is not recommended when effluents with phosphorus levels significantly below 1.0 mg/L are desired. When effluents with a phosphorus level 0.05 or below are desired, chemical precipitation is used.

30.3.2 Chemical Phosphorus Removal

Phosphorus in wastewater may exist as organic phosphate, polyphosphate, or orthophosphate, the last consisting of four different ionic forms. For simplicity, phosphorus is considered to be present as a phosphate ion (PO_4^{3+}). Phosphate can be removed by chemical precipitation, and the principal chemicals used for this purpose are lime, alum, and ferric chloride.

$$5Ca^{2+} + 3HPO_4^{2-} + 4OH^- = Ca_5(PO_4)_3 \downarrow +3\ H_2O$$

$$Al_2(SO_4)_3.14.3H_2O + 2PO_4^{3-} = 2AlPO_4 \downarrow +3SO_4^{2-} + 14.3H_2O$$

$$FeCl_3 + PO_4^{3-} = FePO_4 \downarrow +3Cl^-$$

For the efficient removal of phosphorus, an excess of alum is required. Unused alum reacts with alkalinity similar to the removal of turbidity in water treatment. This would increase the solids in the sludge.

$$Al(SO_4)_3.14.3H_2O + 6HCO_3^- = 2Al(OH)_3 \downarrow +3SO_4^{2-} + 6CO_2 + 14.3H_2O$$

Since polyphosphates and organic phosphorus are less easily removed than orthophosphorus, adding aluminum or iron salts after secondary treatment (where organic phosphorus and polyphosphoric acid are transformed into orthophosphorus) gives best removal. In **Figure 30.3**, alternative points for the addition of coagulants are shown. The precipitation of phosphorus in a primary clarifier can be done easily, though a higher volume of sludge is produced. In addition, the characteristics of sludge are changed from totally organic to inorganic, which is difficult to dewater. The addition to the primary clarifier enhances both SS and BOD removal. However, due to the addition of sulphate ions, water becomes corrosive under anaerobic conditions, which may corrode thickeners and digesters. So higher doses of coagulant are required.

A second alternative is to add coagulant to the mixed liquor entering the secondary clarifier. Adding coagulant before secondary clarification allows phosphorus removal by biological uptake. Thus, the dose of chemical required is less. Tertiary treatment, though expensive, requires the minimum dosage of the chemical and allows for better

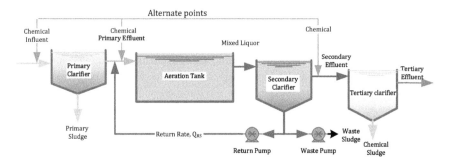

FIGURE 30.3   Alternative points of chemical addition

process control for the removal of phosphorus. Chemical methods are expensive due to the added costs of chemical-feeding equipment, chemicals, and handling of the additional volume of sludge produced. However, phosphorus removal less than 1.0 mg/L can only be achieved by chemical precipitation or a combination of method\s.

30.3.3   Biological–Chemical Phosphate Removal

Biological–chemical phosphate removal utilizes an activated sludge system and an anaerobic cell in which phosphorus taken up in the aeration tank is released to the liquid phase, producing a concentrated phosphates stream and a phosphate defi-cient sludge. This is returned to the aeration tank where the uptake of phosphorus is repeated. Phosphorus is precipitated by dosing the concentrated phosphate stream with lime.

30.4   NITROGEN REMOVAL

Nitrogen in wastewater is found as organic nitrogen, ammonia nitrogen, nitrate nitro-gen, and very small minor amounts of nitrite. All these forms of nitrogen in wastewa-ter effluents are potentially harmful. When present as organic nitrogen or ammonia, nitrogen exerts oxygen demand in accordance with the following equation:

$$NH_4 + 2O_2 \rightarrow NO_3^- + 2H^+ + H_2O$$

Nitrogen enters the aquatic environment from both natural and manmade sources. Natural sources include precipitation and biological fixation. Manmade sources that contribute nitrogen in domestic wastewater are feces, urine, and food-processing dis-charges. The largest single source of nitrogen is urea, which, together with ammonia, comprises approximately 85% of the nitrogen from human excreta.

30.4.1   Biological Nitrification–Denitrification

Nitrogen can be removed from wastewater by progressive biological oxidation of nitrogen compounds to nitrites and nitrates followed by conversion into nitrogen gas.

In **nitrification**, ammonia is oxidized to nitrites and then to nitrates by aerobic nitri-fying autotrophic bacteria. In **denitrification**, nitrates are reduced to nitrogen gas by either autotrophic or heterotrophic anaerobic bacteria. The nitrogen forms of interest in advanced wastewater treatment are organic, inorganic, and gaseous nitrogen. As a first step, bacterial decomposition releases ammonia by **deamination** of nitrogenous organic compounds:

$$Organic\ N\ \overrightarrow{Deamination}\ NH_4^+$$

Nitrosomonas in the first stage and **nitrobacter** in the second stage of nitrification use energy derived from the reaction for cell growth and maintenance. As shown by the following chemical reactions, nitrification is an aerobic process which consumes molecular oxygen and alkalinity. Stoichiometric requirements for nitrification are 4.6 units of oxygen and 7.2 units of alkalinity for each unit of ammonium nitrogen.

$$NH_4^+ + 1.5\,O_2\ \overrightarrow{Nitrosomonas}\ NO_2^- + 2H^+ + H_2O$$

$$NO_2^- + 0.5\,O_2\ \overrightarrow{Nitrobacter}\ NO_3^-$$

Whereas nitrification is an aerobic process, denitrification takes place under anoxic conditions. An anaerobic process is called **anoxic** when nitrates are present. Heterotrophic bacteria using a carbon source break up nitrates to release nitrogen as gas. If carbonaceous BOD is lacking, an artificial source like methanol may be used.

$$NO_3^- + CBOD \overrightarrow{Denitrification}\ N_2 \uparrow + CO_2 + H_2O$$

Example Problem 30.1

In an 12 MGD activated sludge plant, nitrification is introduced to maintain lon-ger solids retention time (SRT) and dissolved oxygen (DO) levels above 2.0 mg/L. Assume that the primary effluent has ammonia nitrogen of 25 mg/L and alkalinity of 110 mg/L. A residual alkalinity of 50 mg/L is recommended. If 85% CaO is used, how many lb should be added?

Given:

$$Q = 12\ MGD \quad NH_4^-N = 25\ mg/L \quad residual\ alkalinity,\ (alk) = 50\ mg/L$$

Solution:

$$Alk\ Required = 7.2 \times NH_4\_N = 7.2 \times 25\ mg/L = 180\ mg/L$$

$$Alk\ Added = 180 - 110 + 50 = 120\ mg/L$$

$$Dosage = \frac{120\,mg}{L} \times \frac{meq}{50\,mg} \times \frac{28\,mg\ of\ CaO}{meq} \times \frac{1}{0.85} = 79.05 = 79\,mg/L$$

$$Dosage\ rate = \frac{79.05\ mg}{L} \times \frac{12\ MG}{d} \times \frac{8.34\ lb}{MG} \times \frac{L}{mg} = 7912.2 = 7900\ lb/d$$

Suspended-Growth Process

In the suspended-growth process, nitrification is achieved in an activated sludge process along with the usual carbonaceous oxidation, with some modification in the design. Nitrifying bacteria are **autotrophic** and very sensitive to pH (pH varying from 7.8 to 8.9, with optimum value of 8.4) and have a very slow rate of growth compared to heterotrophic bacteria. The nitrification rate drops rapidly in temperatures below 10°C. thus nitrification is difficult to achieve in colder climates. It has been shown that to complete nitrification in the conventional process, higher levels of mixed liquor volatile suspended solids (MLVSS) and longer SRT should be maintained. In addition, DO level should be kept above 2.0 mg/L. and residual alkalinity of more than 50 mg/L.

Biological Denitrification

Biological denitrification is achieved under anaerobic conditions by **heterotrophic** microorganisms that utilize nitrate as a hydrogen acceptor. In this process, the nitrate present in the waste process essentially requires a fully nitrified influent, an anaerobic condition, and a supply of proper substrate. In the process of biological denitrification, a balanced amount of substrate is supplied to reduce the nitrate in the process of stabilization of the supplied substrate under anaerobic conditions. In the denitrification process, the heterotrophic bacteria found in activated sludge reduce NO_2^- and NO_3^- to nitrogen gas. When a carbon source like methanol is added, denitrifying microorganisms metabolize methanol according to the following reaction:

$$6NO_3^- + 5CH_3OH \rightarrow 3N_2 \uparrow + 5CO_2 + 7H_2O + 6OH^-$$

As per the above equation, 5 moles of methanol are required to reduce completely 6 moles of nitrate to molecular nitrogen. The overall reaction requires methanol for the reduction of nitrite, nitrate, and DO in the nitrified wastewater. In some modifications of the process, BOD in the primary effluent is used as the carbon source and the reaction takes place in an **anoxic zone**.

30.4.2   THREE-STAGE NITRIFICATION–DENITRIFICATION

In the three-stage process, nitrification and denitrification are carried out separately, as shown in **Figure 30.4**. The recommended MLVSS is 1500–2000 mg/L and the SRT longer than 7 days. Typical loading on a nitrification basin is 15–200 g/m³·d for operating temperatures greater than 15°C (60°F). Correction should be applied for the peaking factor, which is 1.2–1.6 for the average nitrogen load. DO of 3.0 mg/L is desirable and must not fall below 1.0 mg/L.

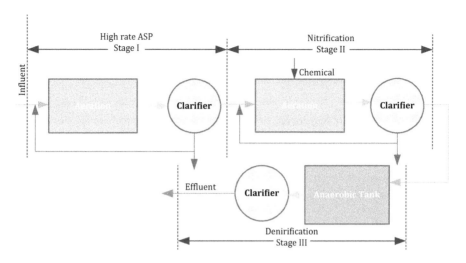

FIGURE 30.4   Three-stage nitrification–denitrification

Stage II (Separate-Stage Denitrification)

The carbon oxidation nitrification/denitrification occurs in a separate reactor and sludge is collected in a separate reactor, thus its name is the "separate-sludge" system. The design of a suspended growth denitrification system is similar to the design of an activated sludge system. A complete mix and plug flow can be used. The N_2 released during the process can attach to biological solids, thus a nitrogen release step is required using aeration. Factors affecting the denitrification process include nitrate concentration, carbon, temperature, and pH.

30.5   TREATMENT METHODS FOR THE REMOVAL OF TOXINS

Refractory organics are compounds resistant to microbial degradation in a conventional biological treatment process. The presence of such substances in wastewater may inhibit the biological process and hence BOD removals.

30.5.1   CARBON ADSORPTION

Carbon adsorption is used to remove nitrogen, sulphides, and heavy metals. Both granular and powdered carbon is used in this process. After treatment, the effluent BOD ranges from 2 to 7 mg/L and COD from 10 to 20 mg/L. Types of carbon contactors are upflow column, downflow column, fixed bed, and expanded bed

The economical application of activated carbon depends on an efficient means of regenerating the carbon after its adsorptive capacity has been reached. Granular carbon can be regenerated by heating in a multiple hearth furnace (at 800°C) wherein the adsorbed organics are volatilized and released in gaseous form from the carbon surface. With proper control, granular carbon can be restored to a near virgin adsorptive capacity with only 5% to 10% weight loss. It combines the use of powered activated carbon with an activated sludge process. It facilitates system stability during shock loads, reduction of refractory priority pollutants, color and ammonia removal, and improved sludge settleability.

30.5.2 CHEMICAL OXIDATION

Chemical oxidation is used to remove ammonia, to reduce the concentration of residual organics, and to reduce the bacterial and viral content of wastewater. Typical chemical dosages for both chlorine and ozone for the oxidation of organics in wastewater are used.

30.6 WASTEWATER DISINFECTION

The methodologies for disinfection of wastewaters are similar to those discussed previously in regard to potable water. As potable water is free from BOD and SS, disinfection by chlorine or other oxidant is usually successful. In the case of wastewater, however, oxidants will first oxidize BOD before being effective as disinfectant. As such, to disinfect wastewater effluent, it is usual to carry out, to begin with, the advanced treatment of wastewater to reduce BOD and SS content.

Dosages of chlorine for disinfection purposes vary widely depending on the type of wastewater. Typically, municipal wastewater requires about 10–12 mg/L. Dosages may be significantly higher if wastewater is septic and or has large industrial components.

30.7 IMPROVED TREATMENT TECHNOLOGIES

To meet more stringent requirement for wastewater effluent quality, old sewage treatments need to be upgraded. Engineers have come up with innovations to meet this challenge. Some of these technologies include the sequencing batch reactor (SBR), membrane bioreactor process (MBP), ballasted flow reactor (BFR), biological aerated filter (BAF), and the integrated fixed-film activated sludge (IFAS) processes.

30.7.1 SEQUENCING BATCH REACTOR (SBR)

As the name implies, SBR is a non-steady process in which the reactor is filled with raw wastewater for a discrete time and operated as a batch mode. Two or more SBRs are used so that when one is filling, the other one is operating. There is no primary treatment, and the aeration tank performs functions of bio-oxidation, settling, and clarification. After a given period of aeration, the aeration is turned off, and solids settle, and clarified water is decanted off. No sludge recirculation is needed as sludge remains in the aeration tank. Since it is a batch process, operating parameters can be tightly controlled to meet the desired quality. With the advent of better control systems, SBR systems can remove as much as 95% of BOD and takes less area than conventional activated sludge plants.

30.7.2 MEMBRANE BIOREACTOR PROCESS (MRP)

In the MRP process, aeration, secondary clarification, and filtration occur in a single reactor. Filtration is achieved by microfiltration membrane modules submerged in the bioreactor. Vacuum pumps pull the effluent through the membranes and leave solids behind in the tank.

30.7.3 BALLASTED FLOC REACTOR (BFR)

This process is like ballasted flocculation in potable water treatment. A coagulant is added to the influent to initiate coagulation and flocculation. Sand and polymers are then added to make heavier floc that readily settles. The addition of polymers makes the organics stick on the surface of the sand particles and thus can easily settle out. Settled sludge is pumped to a **hydrocyclone**, where the sand is separated and recycled back to the BFR.

30.7.4 BIOLOGICAL AERATED FILTERS (BAFs)

In a BAF reactor, basin media is submerged that serves both as a contact surface for biological activity and as a filter to separate solids. Primary effluent flows up through the basin and fine-air bubble-diffusion is used for aeration. Routine backwashing is used to remove captured solids from the media. Two reactors can be used if both nitrification and BOD removal are intended. Organic loading can be as much as six times that of a conventional activated sludge process and the footprint is much smaller. BAFs are more common in Europe.

30.7.5 INTEGRATED FIXED-FILM ACTIVATED SLUDGE (IFAS)

This technology is used particularly for plants being upgraded to introduce nitrification, especially when there is no space for expansion. In IFAS, small plastic sponges or rings, called carriers, are suspended in the aeration tank. The carriers increase the total biomass undergoing nitrification and allow the system to operate at higher BOD loading.

30.8 WATER RECYCLE AND REUSE

As the demand for fresh water is increasing due to rises in population and changes in lifestyles, the need for water conservation and the reuse of water cannot be overstressed. Global warming is further adding to the challenge of water needs for survival. Terms like reclamation, recycling, and reuse are commonly used when referring to the development and expansion of natural water sources. Though the terms are used interchangeably by some engineers, a distinction should be made between reclamation and reuse. Water **reclamation** is defined as the treatment of water to meet predefined water quality criteria whereas **water reuse** refers to the use of treated water for beneficial uses, like urban irrigation needs, including parks and agricultural irrigation. In environmental terms it makes sense to recycle water.

30.8.1 WATER CONSERVATION

In the case of the municipal water supply, water conservation is a way to create a greater supply. Water conservation is applicable to all water users. Broadly, water conservation measures can be categorized as soft measures, like public education, and hard measures, like by-laws and regulations. Soft measures include public education about the use of low-flow toilets and appliances and encouraging native vegetation for landscaping.

When soft measures do not work, measures including restrictions on car washing and watering lawns are introduced, and increased fees are implemented. Some of these measures do result in the conservation of wastewater. However, water reuse and reclamation are more appropriate for wastewater producers. Municipal wastewater reuse is common, especially for arid regions where there is limited rainfall and evaporation loss is high. Environmental factors like global warming, water scarcity, and cost are expected to increase the reuse of wastewater.

30.8.2 REUSE OF PROCESSED WASTEWATER

One of the major uses of processed wastewater in the USA is in agricultural irrigation and in industry for cooling and processing. Since the public is not generally exposed to this reused water, high-quality biological treatment with or without disinfection is often considered satisfactory.

Other uses of processed wastewater water is for urban landscaping, irrigation, and recharging of aquifers. As mentioned before, such uses are gaining popularity in developing countries too. For these applications, however, tertiary treatment with disinfection is necessary.

30.9 WATER QUALITY AND REUSE

Protection of public health is the primary concern in establishing water quality standards for water reuse. Important environmental considerations include the protection of groundwater, soils, and crops. In the selection of quality standards for water reuse, cost must also be considered. For example, if the reuse is for growing fodder crops, conventional treatment may be considered satisfactory. In comparison, for the recharging of aquifers, the requirements for augmenting groundwater have to be much stricter.

The most common applications of wastewater recycling are for non-potable purposes. After some level of treatment, wastewater effluents can be used for watering golf courses and recreational fields, lawn, and for landscape irrigation, crop irrigation, and other agricultural purposes. They can augment, enhance, and sustain natural water bodies, aquatic environments, and ecosystems. They can also be used in industry for cooling and for use in boilers, and for other purposes.

30.9.1 URBAN LANDSCAPE

For urban irrigation use, treated wastewater must meet some quality standards. The suggested water quality for urban reuse is biological treatment, filtration, and disinfection, with turbidity levels not exceeding 2 NTU, and no detectable fecal coliforms. This criterion is usually applicable to unrestricted public areas, such as parks and other recreational areas.

30.9.2 RECLAIMED WASTEWATER

Reclaimed wastewater refers to wastewater treated to its original quality. There are two kinds of planned wastewater recycling protocols for drinking water use. **Indirect potable reuse** (IPR) for intentional and unintended groundwater recharge by seepage of treated wastewater spread or water impounded on the ground surface. The recommended treatment is specific for each site, based on soils, percolation rate, thickness of unsaturated soil profile, natural groundwater quality, and dilution. In general, reclaimed water should meet drinking-water standards and contain no measurable levels of pathogens after percolation through the vadose (unsaturated) zone.

Direct groundwater recharge is the injection of reclaimed water into potable aquifers. The recommended processing is secondary treatment, filtration, disinfection, and advanced wastewater treatment such as chemical precipitation, carbon adsorption, and reverse osmosis. There are many facilities around the world that make indirect use of reclaimed wastewater. In the USA, there is the Fred Hervey Water Reclamation Plant in El Paso, Texas, and the Orange County Plant in California. Similar IPR facilities are in Toowoomba, Australia, and Singapore.

Direct potable reuse (DPR) can be defined as the intentional introduction of reclaimed wastewater directly into the source water intake of a potable water plant. There are no criteria or standards yet for DPR implementation, although research is geared to take DPR to the next level. Another issue is the public acceptance of DPR. As the need for sustainable water supply in the world is growing, DPR seems to be the most economical and viable solution.

30.10 INDUSTRIAL WASTEWATER TREATMENT

Water is generally used by all types of industry, and it is discharged in various water resources. Untreated or partially treated industrial wastewater can pollute the receiving waters to the extent that they are unfit for domestic, recreational, and commercial purposes. Rapid growth in developing countries like India has created multifarious forms of the disposal of industrial wastes. Sites of untreated wastewater discharges on land producing strong odors and contaminating groundwater are common. In some cases, damage to water supply sources can result in chronic diseases and make

TABLE 30.2
Industrial waste characteristics

Undesirable Characteristics	Effect
Soluble organics	depletion of oxygen in river water
Suspended solids	disturbs aquatic life; forms benthic deposits
Phenol and trace organics	bad taste and odor
Color and turbidity	unaesthetic condition; retards photosynthesis
Nitrogen and phosphorus	undesirable algal growth; increases eutrophication
Oil and floating matter	retards reaeration

TABLE 30.3
Selected industrial wastewaters

Industry	BOD (mg/L)	COD (mg/L)	pH	TSS (mg/L)	N (mg/L)	Cl (mg/L)	Grease (mg/L)
milk processing	1000	1900	8	1000	50	–	–
meat packing	1400	2100	7	3300	150	–	500
synthetic textile	1500	3300	5	8000	30	–	–
chlorophenol manufacture	4300	5400	7	53 000	–	27 000	–

them unfit for drinking purposes. In **Table 30.2**, the most common effects of industrial pollutants are shown.

30.11 INDUSTRIAL WASTEWATER DISCHARGES

From the view of treatment, industrial wastewater dischargers can be broadly categorized as direct or indirect. **Direct dischargers** include those industrial effluents which are directly discharged into the receiving waters. Industries like steel making, electroplating, pulp and paper, and tanneries fall into this category. The onus is on the industry to meet standards as stipulated by their regulating body. **Indirect dischargers** include those industries whose wastewater contains conventional pollutants which are directly discharged into municipal sewers to be treated along with domestic sewage, such as the dairy, brewing, and food processing industries. Typical characteristics of some industrial wastes are shown in **Table 30.3**.

Here are some key points related to industrial discharges.

- Industrial discharges are usually stronger than domestic wastes.
- Total solids are usually greater but vary in character from colloidal to dissolved organics.
- Discharges from chemical and material industries are usually deficient in nutrients.

- Industries like metal plating require pre-treatment of their wastewater.
- Industries must consider the segregation of wastes, flow equalization, and reduction of waste strength.
- Process changes, equipment modifications, by-product recovery, and in-plant reuse of wastewater can result in cost savings for the industry.

PRE-TREATMENT OF INDUSTRIAL WASTE

Pre-treatment of industrial waste can result in savings for the industry and make the waste more suitable for further treatment. The main purpose of pre-treatment is:

1. the reduction of waste strength and volume, and
2. equalization and neutralization.

Strong and weak waste may be separately collected, treated, and disposed of, as mixing both wastes creates large volume and many difficulties. So, the reduction of waste, strength, and volume is necessary. Wastewater may be conserved by reusing or recycling. Many substances like chromium, potash, and silver can be recovered easily for economic gains. Fluctuations or variations in the input quality can also be controlled.

30.12   INDUSTRIAL WASTEWATER TREATMENT

The quality of wastewater varies from industry to industry; therefore, each industry or group of industries requires thorough study to choose the best treatment method for its effluent. **Figure 30.5** lists process schemes for treating industrial wastewaters. Industrial wastewater needs a process to remove chromium, phenol, mercury, and nitrogen. Adsorption is used for removing color, phenol, etc. Activated carbon is most commonly used.

30.12.1   REMOVAL OF CHROMIUM

Hexavalent chromium is toxic (0.1mg/l threshold limit) and is present in tanning, electroplating, fertilizers, and other industries. Procedure of removing chromium is as follows:

1. Cr_6 is reduced to Cr_3 by the addition of sulphuric acid (to reduce pH to 2 or 3) and ferrous sulphate. Then it can be precipitated in settling tanks after neutralizing with sodium hydroxide or calcium hydroxide.
2. Cationic resins can be used for the recovery of chromium in the form of sodium chromate or chromic acid. The regeneration of resin is achieved with sodium chloride and sulphuric acid.
3. Lime coagulation and adsorption with activated carbon.
4. Reverse osmosis process can also be used.

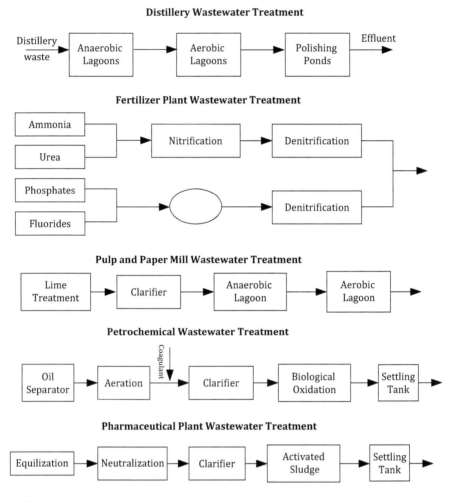

FIGURE 30.5　Flow schemes for industrial wastewaters

30.12.2　REMOVAL OF PHENOL

Phenol is released by the chemical, petroleum, pharmaceutical, plastic, metallurgical, printing, and textile industries. Procedure of removing chromium follows.

1. Oxidizing agents such as hydrogen peroxide, potassium permanganate, and sulphur dioxide are effective. Aeration and biological action are also used.
2. Steam stripping, adsorption, ion exchange, and solvent extraction are adopted.

30.12.3 REMOVAL OF MERCURY

Chlor-alkali plants, the mining industry, paper and pulp mills, and pesticide units are the main source of mercury. Steps for removal of mercury are given below:

1. By adding H_2S or Na_2S mercury, which can be precipitated.
2. Ferric chloride may be added.
3. Activated carbon, clays, and silica gel are used for adsorption.
4. Mercury salt in solution may be reduced by the addition of active metals like iron.
5. Natural or synthetic resins may be used in the ion exchange process.

30.13 COMMON EFFLUENT TREATMENT PLANTS

When there are significant indirect industrial dischargers in a community, municipal wastewater includes industrial wastewater and domestic sewage. This arrangement allows small and medium-scale industries to dispose of their effluents. Otherwise, it may not be economical for these industries to treat their wastewaters or there may be space constraints. Such an arrangement is a win–win for both the community and industry. In addition, it helps to prevent pollution by the discharge of untreated or partially treated wastewater into the receiving waters.

Some of these industries may require pre-treatment at the facility for reducing the strength of wastewater or to correct pH or for the removal of a specific pollutant before discharges into sewers. Combined wastewater from domestic and industrial sources are usually expected to have high strength and a flow hydrograph may show a different pattern compared to a typical municipal plant. A typical process scheme for combined treatment effluent treatment system is shown in **Figure 30.6**.

As seen in the flow diagram, the equalization of wastewater is an integral part of the treatment scheme. If not done, the wastewater strength would show tremendous

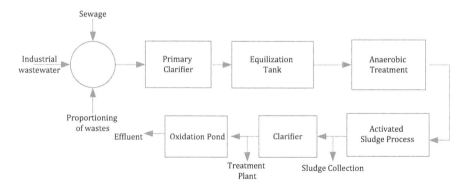

FIGURE 30.6 CETP flow diagram

variation in strength and flows. **Equalization** may help to freshen wastewater and afford uniform and balanced loading. This would allow the designer to size the tanks and devices more economically. The common effluent treatment plant or CETP is designed based on:

- Quality and flow rate of wastewater.
- Effluent standard required by CETP.
- Possibility of recycling and reuse of treated wastewater.
- Availability of land, manpower, energy, and expertise in specific treatment methods.
- Willingness of the industries located in the industrial estate to contribute towards the capital and operating expenses of CETP.

Discussion Questions

1. In the chemical precipitation of phosphorus, coagulant can be added as part of primary, secondary, or tertiary treatment. Discuss the advantages and disadvantages of each.
2. List the common characteristics of secondary effluent from a conventional activated sludge plant. What contaminants are most likely to cause pollution if the plant effluent is discharged with limited dilutional flow? Which contaminants adversely affect a lake or a reservoir receiving plant effluent?
3. What are the potential benefits of flow equalization?
4. What kind of pollutants are removed by granular carbon columns?
5. By appropriate calculations verify that the mass ratio of alum to phosphorus is 9.7 and alum to alkalinity as $CaCO_3$ is 2.
6. What are the necessary conditions to perform biological nitrification with BOD removal in an aeration tank of activated sludge process?
7. Why is methanol or another carbon source needed in biological denitrification?
8. What is the difference between water reuse and water reclamation?
9. Why is water conservation gaining popularity in modern times?
10. What are the various beneficial uses of processed wastewater?
11. What are the different types of membrane filtration, and which one is more suitable for processing secondary effluents? What are advantages and disadvantages of membrane filtration?
12. What is IPR and DPR. Do you think DPR is the way for the future? Discuss.
13. List five alternative technologies to provide tertiary treatment when space is limited and more removals are required.
14. Search the internet and find the scheme of processes used to reclaim wastewater at the Orange County plant in the USA.
15. List the harmful effects due to the disposal of industrial wastes without adequate treatment.
16. Discuss the merits and demerits for combining industrial and domestic waste before treatment and disposal.

17. Describe the use of aerobic and anaerobic lagoons in the treatment of industrial waste.
18. Why is there a need for special processes for industrial effluent and what are these processes?
19. Write a short note on the pre-treatment of industrial waste.
20. What is the concept of a common effluent treatment plant (CETP)? On what basis it is designed? Draw a flow diagram of CETP.
21. In what sense may disinfection of wastewater differ from that of potable water?

Practice Problems

1. Calculate the chemical sludge production as aluminum phosphate for phosphorus removal by dosing alum at the rate of 80 mg/L into the aeration tank. Raw wastewater phosphorus concentration is 7.0 mg/L and assume primary removal of 1.0 mg/L and secondary biological removal is 1.5 mg/L to produce secondary effluent with P not exceeding 1.0 mg/L. (14 mg/L).
2. Based on allowable loading of ammonia nitrogen of 150 $g/m^3 \cdot d$, calculate the required aeration volume for biological nitrification (stage I) following conventional secondary treatment. The wastewater characteristics are as follows: $Q = 40$ ML/d, $NH_3^-N = 20$ mg/L, MLVSS 1600 mg/L, peak load = $1.5 \times$ avg (8000 m^3).
3. In chemical precipitation of phosphorus, the excess dosage of liquid alum @ 40 mg/L is applied. This excess alum reacts with alkalinity. Determine how much of the alkalinity in consumed and inorganic solids produced as aluminum hydroxide? (20 mg/L, 10 mg/L).
4. Calculate the surface area of gravity filters to remove SS from a wastewater effluent with an average daily flow of 15 ML/d and peak 4 h discharge of 33 ML/d. The nominal filtration rate should exceed neither 180 $m^3/m^2 \cdot d$ with all filters operating nor 350 $m^3/m^2 \cdot d$ during peak flow hours with one out of four filters out of service. (130 m^2).
5. A secondary wastewater treatment plant is being upgraded to a tertiary plant to meet more stringent requirement of 10 mg/L of BOD in the plant effluent. Average BOD removals at this plant are 35% in primary and 85% in secondary units. What should be the minimum BOD removal of the tertiary unit, knowing plant influent contains 240 g/L of BOD? Also determine the BOD content of primary effluent and secondary effluent. (57%, 160 mg/L, 23 mg/L).

Appendices

A. PROPERTIES OF WATER

Temp.	Specific gravity	Mass Density	Weight Density	Surface Tension	Dynamic Viscosity	Kinematic Viscosity	Saturated DO	Vapour Pressure	
°C		kg/m³	kN/m³	N/m	Pa.s	m²/s	mg/L	kPa	bar
0	1.000	999.9	9.809	7.56E-02	1.79E-03	1.79E-06	14.60	1.946E+00	6.23E+00
5	1.000	1000.0	9.810	7.49E-02	1.52E-03	1.52E-06	12.80	1.706E+00	8.89E+00
10	1.000	999.7	9.807	7.42E-02	1.31E-03	1.31E-06	11.30	1.506E+00	1.25E+01
15	0.999	999.1	9.801	7.35E-02	1.14E-03	1.14E-06	10.10	1.346E+00	1.74E+01
20	0.998	998.2	9.793	7.28E-02	1.00E-03	1.00E-06	9.10	1.213E+00	2.38E+01
25	0.997	997.1	9.781	7.20E-02	8.90E-04	8.93E-07	8.30	1.106E+00	3.23E+01
30	0.996	995.7	9.768	7.12E-02	7.98E-04	8.01E-07	7.54	1.005E+00	4.33E+01
35	0.994	994.1	9.752	7.04E-02	7.19E-04	7.23E-07	6.93	9.237E-01	5.73E+01
40	0.992	992.3	9.734	6.96E-02	6.53E-04	6.58E-07	6.41	8.544E-01	7.52E+01
50	0.988	988.1	9.693	6.79E-02	5.47E-04	5.54E-07	5.90	7.864E-01	1.26E+02
60	0.993	993.2	9.744	6.62E-02	4.66E-04	4.69E-07		0.000E+00	2.03E+02
70	0.978	977.8	9.592	6.44E-02	4.04E-04	4.13E-07		0.000E+00	3.18E+02
80	0.972	971.8	9.534	6.26E-02	3.55E-04	3.65E-07		0.000E+00	4.83E+02
90	0.965	965.3	9.470	6.08E-02	3.15E-04	3.26E-07		0.000E+00	7.15E+02
100	0.958	958.4	9.402	5.89E-02	2.82E-04	2.94E-07	0.00	0.000E+00	1.03E+03

B. CONVERSION FACTORS (SI TO USC)

Symbol	When you know	Multiply by	To find	Symbol
mm	millimeter	0.039	inch	in
m	meter	3.28	feet	ft
m	meter	1.09	yard	yd
km	kilometer	0.621	mile	mi
mm²	square millimeter	0.0016	square inch	in²
m²	square meter	10.764	square feet	ft²
m²	square meter	1.196	square yard	yd²
ha	hectare	2.47	acre	ac

(Continued)

Symbol	When you know	Multiply by	To find	Symbol
km²	square kilometer	0.386	square mile	mi²
mL	millilitre	0.034	fluid ounce	fl oz
L	litre	0.264	Gallon	gal
m³	cubic meter	35.314	cubic feet	ft³
m³	cubic meter	1.307	cubic yards	yd³
g	gram	0.035	ounces	oz
kg	kilogram	2.202	pound	lb
Mg	megagram	1.103	short tons	T
t	metric ton		2000 pound	
°C	Celsius	1.8C +32	Fahrenheit	°F
N	newton	0.225	pound force	lbf
kPa	kilopascal	0.145	pound per square inch	lbf/in²

C.   CONVERSION FACTORS (USC TO SI)

Symbol	When you know	Multiply by	To find	Symbol
In	Inch	25.4	millimeter	mm
ft	Feet	0.305	meter	m
yd	Yard	0.914	meter	m
mi	Mile	1.61	kilometer	km
in²	square inch	645.2	square millimeter	mm²
ft²	square feet	0.093	square meter	m²
yd²	square yard	0.836	square meter	m²
ac	acre	0.405	hectare	ha
mi²	square mile	2.59	square kilometer	km²
fl oz	fluid ounce	29.57	millilitre	mL
gal	gallon (US)	3.785	litre	L
ft³	cubic feet	0.028	cubic meter	m³
yd³	cubic yard	0.765	cubic meter	m³
oz	ounce	28.35	grams	g
lb	pound	0.454	kilogram	kg
T	short ton (2000 lb)	0.907	megagram metric ton	Mg t
°F	Fahrenheit	5(F-32)/9	Celsius	°C
lbf	pound force	4.45	newton	N
lb/in²	pound per square inch	6.89	kilopascal	kPa

D.   CT VALUES

D.CT Values in (mg/L).min for inactivation of Giardia lambia cysts

Free Chlorine		Water Temperature in °C			
Res., mg/L	pH	0.5	5	10	20
<0.4	6.5	163	117	88	44
	7.0	195	139	104	52
	7.5	237	166	125	62
	8.0	277	198	149	74
1.0	6.5	176	125	94	47
	7.0	210	149	112	56
	7.5	253	179	134	67
	8.0	304	216	162	81
2.0	6.5	197	138	104	52
	7.0	236	165	124	62
	7.5	286	200	150	75
	8.0	346	243	182	91
3.0	6.5	217	151	113	57
	7.0	261	182	137	68
	7.5	316	221	166	83
	8.0	382	268	201	101

CT Values in (mg/L).min for log-0.5 and log-1 inactivation of Giardia lambia cysts

Disinfectant	Log inactivation	pH	Water Temperature in °C				
			0.5	5	10	15	20
Free Chlorine	0.5	6.0	25	18	13	9	7
	1.0	6.0	49	35	26	18	13
	0.5	7.0	35	25	19	13	9
	1.0	7.0	70	50	37	25	18
	0.5	8.0	51	36	27	18	14
	1.0	8.0	101	72	54	36	27
Combined	0.5	6-9	640	370	310	250	190
	1.0	6-9	1300	740	620	500	370
Chlorine	0.5	6-9	10	4.3	4.0	3.2	2.5
Dioxide	1.0	6-9	21	8.7	7.7	6.3	5.0
Ozone	0.5	6-9	0.48	0.32	0.23	0.16	0.12
	1.0	6-9	0.97	0.63	0.48	0.32	0.24

CT Values in (mg/L).min for inactivation of Viruses at pH 6-9

Disinfectant	Log inactivation	Water Temperature in °C				
		0.5	5	10	15	20
Free Chlorine	2.0	6	4	3	2	1
	3.0	9	6	4	3	2
	4.0	12	8	6	4	3
Combined	2.0	1200	860	640	430	320
	3.0	2100	1400	1100	710	530
Chlorine Dioxide	2.0	8.4	5.6	4.2	2.8	2.1
	3.0	25.6	17.1	12.8	8.6	6.4
Ozone	2.0	0.9	0.6	0.5	0.3	0.2
	3.0	1.4	0.9	0.8	0.5	0.4

Source: EPA-USA

Index

A

Absorption, 55, 209, 218, 480, 483
Absorption field, 480, 481
Acidity Tests, 132
Actiflow, 143
Activated carbon, 55, 155, 179, 218, 220, 521
Activated silica, 123, 209
Activated sludge process
    contact stabilization, 434
    conventional, 436
    extended, 437
    high rate, 438
    step aeration, 435
    tapered aeration, 435
Adsorption, 55, 120, 153, 179, 217, 220, 521
Adverse water quality, 108
Aerated grit chamber, 403
Aeration
    diffused, 423
    period, 425
    surface, 423
    tanks, 423
Aerobic digestion, 499
Affinity laws, 274, 276
Air binding, 165
Algae, 62–64, 111, 447
Algal blooms, 62, 451, 481, 516
Alignment, 244, 280, 309, 350
Aliquot, 65, 68, 390
Alkaline, 47, 171, 209
Alum Floc, 122, 125, 148
Anaerobic digester
    capacity, 499
    digestion, 498
Anchoring of pipes, 245
Arithmetical increase, 98, 99
Arsenic removal, 213, 214
Artesian
    aquifer, 86, 89
    well, 84, 85
Artificial
    aeration, 445
    reservoir, 78
Asbestos-cement, 239, 241, 348, 354
Atomic mass, 43, 44

Automatic
    compositing, 392
    residual control, 186
    sampler, 312
Autotrophic, 62, 519

B

Backfilling, 357
Backflow, 231, 249, 291
Back pressure, 248, 506
Back siphoning, 232, 249
Backwashing, 153, 156, 166, 214, 514, 523
Ballast, 143
Ballasted flocculation, 523
Ball Test, 357
Barminutors, 401
Bar screen, 315, 401
Basket Centrifuge, 506, 507
Bedding, 242, 305, 351, 366
Belt Filter Press, 505
Bentonite, 123
Berm Erosion, 452, 454
Berms, 451
Bimodal, 300
Biodegradable, 384, 402, 424, 470, 513
Biological
    aerated filters, 522
    characteristics, 385
    contactors, 421, 456, 467
    denitrification, 520
    filters, 459
    floc, 415, 421
    nitrification, 518
    phosphorus removal, 516
    towers, 460
    treatment, 61, 385, 399, 421, 521
Bleach, 46, 50, 56, 174
BOD
    particulate, 470
    reaction, 369
    test, 68, 384
Borehole, 267
Breakpoint
    chlorination, 179
    curve, 175

Broad irrigation, 376
Building sewer, 297, 303, 306, 311

C

Calcium hypochlorite, 55, 173
Calcium phosphate, 209
Canal intakes, 79
Carbon
    adsorption, 521, 525
    monoxide, 55, 311
Carbonate hardness, 51, 114, 123, 191, 193
Carcinogenic, 8, 171, 180, 217, 514
Cast iron, 223, 242, 310, 330, 346
Cast iron pipe, 240, 349
Cavitation, 277, 317
Cement concrete, 240, 301, 332, 348
Centrifugation, 503, 506
CETP, 529
Check valve, 240, 249
Chemical
    characteristics, 81, 92, 223, 383
    dosing, 110, 218
    feeding, 52, 207, 518
    oxidation, 212, 219, 498, 522
    oxygen demand, 67, 384
    phosphorus removal, 517
    precipitation, 382, 407, 517, 525
    reactions, 51, 110, 114, 124, 193
Chemistry of
    chlorination, 174
    coagulation, 124
    corrosion, 223
    softening, 195
Chloramination, 178
Chlorides, 191
Chlorination equipment, 184, 188
Chlorinator, 184
Chlorine
    compounds, 172
    concentration, 181
    demand, 176, 178, 186, 406, 514
    dioxide, 169, 174, 217, 219
    practices, 178
    safety, 173
Circular
    clarifier, 141, 407
    pipes, 38, 302, 329
Class I systems, 255
Class II systems, 256
Class III systems, 260
Clean Out, 306
Clear Well, 7, 156, 232
Closed-circuit television, 307
Coagulant aids, 123, 142, 507
Coagulants, 56, 121, 131

Coarse screens, 400
Coliforms, 64–66, 525
Collector wells, 82
Colour
    apparent, 110
    true, 110
    units, 110
Combined sewers, 297
Comminutors, 315, 401
Completely mixed flow, 449
Complex pipe networks, 260
Composite concentration, 389
Composting, 493, 498, 507, 509
Concrete Pipe (CP), 301, 344, 349
Contact stabilization, 426, 436
Contact time, 56, 179, 181, 483
Continuity equation, 24
Continuity of flow, 24, 253, 261
Continuity of pressure, 263
Continuous solid bowl, 506
Conventional aeration, 426, 435, 491
Conventional filter, 221, 461
Copperas, 121–123
Corporation stop, 250
Corrosion
    control, 226
    inhibitors, 226
Cross-contamination, 248, 285
Crown corrosion, 305, 348
CT Factor, 56, 181, 183
Curb stop, 250
Current meter, 312
Cyclone separator, 405

D

Daily variation, 106
Daphnia, 452
Darcy–Weisbach flow equation, 33, 343
Dead loads, 358, 362
Deamination, 519
Dechlorination, 179
Declining rate control, 159
Deflocculation, 442
Defluoridation, 208
Degree of mixing, 130, 440
Depressed sewer, 307
Desalination, 78, 222
Design of
    inverted siphon, 308
    sewers, 327
Design period, 93, 97
Detritus tank, 405
Dewatering, 351, 410, 493, 503
Diatomaceous, 155
Diffused aeration, 423, 437

Dilutions, 66, 68
Dimensions, 11, 12
Direct
    dischargers, 526
    potable reuse, 525
Disinfectants, 169, 180
Disinfection by-products, 169
Dispersion, 226, 481
Disposal of
    screenings, 400
    sludge, 507
    tank effluent, 480
    wastewater, 4
Distribution
    grid iron, 237
    radial, 238
    ring, 238
Domestic wastewater, 382, 463, 518
Double deck, 141
Drain hole, 250
Drop manholes, 310
Dry barrel, 250
Drying beds, 405, 503
Dry weather flow, 300, 390
Dry well lift stations, 314
Dual water systems, 251
Ductile iron, 240, 242, 304
Dye test, 306, 307

E

E. coli, 63, 65
Effective size, 132, 155, 504
Effective water depth, 144
Elation, 499
Electrical conductivity, 111
Elevated storage, 233, 236
Elutriation, 507
Embankment conditions, 358
Empirical relationship, 360
Equalizing demand, 233
Equalizing storage, 233, 234
Equation
    Bernoulli's, 28
    chemical, 47, 51
    continuity, 24
    Darcy Weisbach, 33
    energy, 3, 255
    equilibrium, 85
    flow, 33, 272, 282
    Hazen, 403
    Hazen-Williams, 36, 286
    Manning, 38, 302, 328
    mass transfer, 440
    non-equilibrium, 89
    well unconfined, 87

Equivalent
    CaCO$_3$ 52, 192
    hydraulic elements, 334
    length, 33, 273
    mass, 46, 49
    pipe, 253
    population, 388
Eutrophication, 62, 64, 513
Excavation of trenches, 244
Excess lime treatment, 196
Exercising of valves, 248
Exfiltration, 299, 307
Expansion joint, 241, 243
Extended aeration, 411, 426, 437, 438

F

Factors affecting
    ASP, 423
    chlorine dosage, 180
    flocculation, 130
    per capita, 104
    settling, 409
Facultative
    bacteria, 62
    lagoons, 448
    ponds, 447
Failure of wells, 90
Feed
    control, 185
    equipment, 184
    pump, 53, 126, 213, 226
    rate, 53, 207
    sludge, 497, 512
    solution, 53, 126
    system, 184, 211
Fiberglass pipe, 347
Fibreglass, 222, 312
Fibreglass tube, 222
Field supporting strength, 362
Filamentous, 386, 423
Filter
    biological, 460, 481
    box, 156
    breakthrough, 165
    media, 132, 156, 209, 460
    ripening, 160
    roughing, 461
    run, 133, 155, 159
    standard, 461
    underdrain, 157
    up flow, 482
Filtering, 11, 158, 221
Filtering to waste, 158, 161
Filtration
    in depth, 155

mechanisms, 153
membrane, 221
micro, 221
nano, 222
operation, 123, 164, 166
rate, 153, 155, 161
Final settling tanks, 424
Fine screens, 400
Firefighting, 101, 231, 236, 249, 285
Fixed growth systems, 421, 459
Flanged joint, 241, 242
Flash mixing, 121, 128, 131
Flexible joint, 241, 243
Floc
    biological, 415, 421, 422
    pin, 424, 427, 442
    straggler, 442
Flocculation
    ballasted, 523
    phenomenon, 127
    tanks, 128
Floor maintenance, 316
Flow
    classification, 327
    measurement, 312, 401
    through screens, 401
    velocity, 23–26
Flow equation
    Darcy Weisbach, 33
    Hazen Williams, 36
    Manning, 38, 302, 308
Fluorex, 209
Fluoridation, 205, 207
Fluoride chemicals, 205
Flushing velocity, 290
Foaming, 312, 401, 441
Food to microorganism ratio, 426, 439
Force mains, 297, 304, 318, 327
Forecasting population, 97
Free board, 447, 477

G

Gas chlorination, 172, 184
Gasification, 499
Gate valve, 246
Geometrical increase, 98
G-Factor, 128
Granular activated carbon, 155, 221
Granular media filtration, 515
Gravel mounding, 166
Gravel pack, 83
Gravity
    settling, 137, 381
    sewer mains, 303, 304

thickener, 494
Grease, 312, 355, 402, 447, 526
Grey water, 382
Grid-iron system, 238
Grit
    channels, 401, 402
    disposal, 405
    removal units, 400, 404
Ground-level storage, 236
GUDI, 92, 119, 153

H

Hardness
    carbonate, 191, 194
    non-carbonate, 191
    permanent, 191
    temporary, 191
    total, 192
Hardy Cross method, 261
Hatched boxes, 308
Hazen–Williams, 36
Head
    elevation, 26
    energy, 27
    hydraulic, 25
    kinetic, 25
    loss, 33, 155, 253, 260
    shut off, 268
    total, 27
    velocity, 25
Heterotrophic, 62, 519, 520
High-purity oxygen, 426, 438
Homologous pumps, 275
Hourly variation, 107
HTH, 51, 55, 173
Humus, 459, 464, 465, 508
Hydrant
    discharge, 287
    fire, 316
    flow, 287
    flowing, 287
    nozzle, 293
    testing, 287
Hydraulic
    gradient, 26, 244, 325
    head, 25
    loadings, 390, 448
    radius, 21, 38, 302, 329
    slope, 327, 328
Hydrogen sulphide, 55, 62, 92, 116, 305, 348,
        384, 499, 503
Hydrologic cycle, 5
Hypochlorination, 173, 186
Hypochlorinators, 186

I

Imhoff
  cone, 70, 381
  tanks, 477
Incineration, 398, 400, 508
Incremental increase, 99
Index
  aggressive, 225
  corrosivity, 224
  Langelier, 224
  sludge density, 431
  sludge volume, 430
  stability, 224
Indicator organisms, 64, 385
Indirect
  dischargers, 526
  IPR, 525
  recirculation, 460, 462
Industrial
  treatment, 525, 527
  wastewater, 382, 388
Infiltration and inflow, 390, 391
Infiltration galleries, 82, 92
Injector, 184, 186
Intake
  lake, 81
  pipe, 79
  reservoir, 79
  river, 80
  well, 79, 80
  works, 79
Inverted Siphon, 307, 308
Ion exchange
  arsenic, 214
  fluoride, 209
  mercury, 529
  phenol, 528
  softening, 198
  with zeolites, 212
Iron and manganese, 43, 114, 119, 174, 179, 198,
    209

J

Jar testing, 131
Joint
  Bell and Spigot, 354
  expansion, 241, 243
  flanged, 242
  flexible, 243
  mechanical, 242
  simplex, 241
  tongue and groove, 354
Jointing
  ACP, 241

  plastic pipe, 355
  VCP, 353

L

Lagoon
  aerated, 446
  aerobic, 453, 528
  anaerobic, 447, 528
  depth, 447
  facultative, 62, 448
  maintenance, 454
  operation, 453
  polishing, 447
  weeds, 454
Lake intake, 81
Laminar, 88
Langelier index, 224
Lasers, 357
Lateral, 297, 357
Leachate, 508
Level setting, 317
Lift station
  controls, 316
  dry well, 314
  maintenance, 316
  pumps, 315
  wet well, 315
Lime reactions, 193
Linear valves, 246
Load factor, 352, 362
Log removal, 181

M

Manhole
  cover, 309
  dead end, 310
  drop, 310
  inspection, 311
  ordinary, 309
  safety, 311
Manning's equation, 38, 328
Manual compositing, 392
Mass
  atomic, 43, 44
  equivalent, 46
  molecular, 45, 123, 206
  transfer equation, 440
  volume relationship, 496
Maximum velocity, 321, 332
Mean flow velocity, 23, 145
Mechanical
  aeration, 421
  dewatering, 503
  equipment, 137

joint, 241, 242
mixers, 127
rakes, 161
Media boils, 166
Media breakthrough, 166
Membrane bioreactor, 523
Membrane filtration, 65, 170, 221
Membrane technology, 222
Methemoglobinemia, 64, 115, 514
Microbiological tests, 64
Microfiltration, 221
Microorganisms
    aerobic, 61
    anaerobic, 61
    autotrophic, 62
    facultative, 62
    heterotrophic, 62
    pathogens, 64, 109, 133
Microscreening, 514
Minimum velocity, 331
Mirror test, 357
Molarity, 49
Moody diagram, 34
MPN, 66
Mud balls, 165, 179
Multiple pumps, 276
Municipal wastewater, 379, 381, 388, 390, 429

N

Nanofiltration, 222
Natural purification, 39, 61, 369
Net positive suction head, 278
Neutrons, 43
Nitrification, 461, 467, 518, 519
Nitrobacter, 519
Nitrosomonas, 519
Non-carbonate hardness, 194
Normality, 49
Nucleus, 43
Number
    atomic, 43, 45
    mass, 43
    Reynolds, 33, 34
Nutrients, 8, 62, 385, 516, 526

O

Operating point, 269, 272–277
Operating problems, 165, 441, 464
Optimum
    dosage, 131, 212
    filter operation, 166
    level, 208
Ordinary manhole, 309
Organics, 175, 217–221

Orifice, 158, 184, 247
Overflow rate, 137, 144, 410
Oxidation with
    aeration, 211
    chlorine, 211
    permanganate, 212
Oxygen
    dissolved, 67, 115, 210, 421
    sag curve, 372
    transfer, 439
    uptake, 439
Oxygen demand
    biochemical, 4, 67, 116, 385, 451
    chemical, 67, 384
    theoretical, 67
Ozonation, 171

P

Package plants, 437, 467
Palmer-Bowlus, 312, 402
Parshall Flume, 402
Partially full, 297, 460
Particle counters, 133
Particulate
    BOD, 70, 133, 470, 515
    matter, 133, 137, 153, 383
Pathogens, 64, 109, 120, 133
Penstocks, 80
Percolation, 81, 375, 480, 525
Perforated, 82, 185
Performance curves, 268, 318
Periodic table, 44
Permissible suction lift, 279
Pharmaceutical, 528
Phosphate
    removal, 518
    treatment, 210
Photosynthesis, 62
Physical characteristics, 382
Pickling, 499
Pipe
    anchoring, 244
    joints, 241, 356
    laying, 242
    material, 34, 286, 304
    size, 301
    slope, 244
Pipeline
    layout, 237
    systems, 253, 262
Plug flow reactor, 144
Points of chlorination, 179
Polyelectrolytes, 124, 507
Polymers
    anionic, 123

cationic, 123, 124
neutral, 124
Ponding, 459, 464
Ponds
aerobic, 62
anaerobic, 62
facultative, 62
Population growth, 97
Post-chlorination, 179, 180
Potassium permanganate, 218, 528
Powder activated carbon, 220
Pre-aeration, 405
Precision
absolute, 16
numerical, 16
relative, 16
Preliminary treatment, 120, 381, 399, 513
Preservation of samples, 67
Pressure filtration, 506
Primary
disinfection, 169
sludge, 489
treatment, 399
Primary coagulants, 122, 124
Principle of ASP, 422
Process control, 68, 468, 518
Protons, 43
Pulsator clarifier, 143
Pumping stations, 231, 236, 241, 400
Pump operating sequence, 317
Pumps
centrifugal, 267, 498
jet, 268
positive displacement, 267
submersible, 268
turbine, 267

R

Radial system, 239
Radius of influence, 85–88
Rainfall intensity, 336, 338
Rapid sand filters, 4, 155
Rate constant, 369, 373, 449
Rational method, 336, 339
Recarbonation, 195, 209
Recharge, 8, 119, 377, 525
Re-chlorination, 170
Recirculation
direct, 460–462
indirect, 460–462
ratio, 462–463
Reclaimed wastewater, 525
Rectangular basins, 141
Rectangular clarifier, 409, 411
Removal capacity, 198

Reservoir intakes, 79, 80
Residual
chlorine, 111, 169–172
combined, 170, 178
free, 113, 174, 179, 211
Return rate, 424, 432, 437
Return ratio, 432
Reverse osmosis, 221, 222, 514
Ring system, 138, 239
Rotary valves, 247
Rotating biological contactor, 421, 467
Roughing filters, 61, 302, 328
Roughness factor, 320, 326
Run-off coefficient, 336, 339

S

Sample
collection, 66, 312
composite, 71, 312
flow proportioned, 392
grab, 392
SCBA, 56, 311
Scouring velocity, 302, 332
Screening baskets, 315, 318
Seasonal variation, 106, 311
Secondary
clarifier, 394, 414, 422, 432
disinfection, 169, 179, 290
sludge, 465, 491
Sedimentation
basins, 141
plain, 109, 137, 141
primary, 412, 437, 471
Seepage, 448, 450, 525
Septic tanks, 477–478
Sequencing Batch Reactor (SBR), 522
Series and parallel, 253, 277
Service connections, 250
Settleometer, 431
Settling velocity, 110, 403
Sewage farming, 376, 377
Sewage sickness, 378
Sewer
combined, 297
main, 297, 303
rod, 304
separate, 299, 390
storm, 299, 348
ventilation, 311
Short-circuiting, 403, 408–410, 478
Shut off head, 268
Significant figures, 15–18
Simplex joint, 241, 243
Slime, 459, 472
Sloughing, 459, 472

Slow sand filters, 154, 162
Sludge
  age, 427
  blanket, 142, 415, 424, 484
  bulking, 442
  composting, 509
  conditioning, 507
  density index, 432
  digestion, 477, 498
  disposal, 506
  drying beds, 504
  handling, 415
  pickling, 499
  primary, 491
  recirculation, 424, 522
  rising, 442
  secondary, 491
  stabilization, 498
  thickening, 493
  volume index, 431
Smoke test, 357
Soak pit, 481
Socket and spigot joint, 241
Soda Ash equations, 194
Sodium hypochlorite, 46, 50, 56
Softening, 51, 114
Soil adsorption, 481
Solids
  contact units, 142
  dissolved, 70
  organic, 70
  retention time, 425, 427
  settleable, 459
  suspended, 70
  volatile, 70
  washout, 434
Soluble BOD, 470
Specific
  capacity, 85
  speed, 275
  uptake, 440
Split flow control, 159
Split treatment, 197
Springs, 82
Stability index, 224
Stabilization ponds, 447
Staging, 462, 468
Standard
  filter, 461
  solution, 49
Standpipe, 235
State point, 434
State point analysis, 433
Step aeration, 435
Stoichiometry, 51
Storm drainage, 335

Storm sewers, 299
Straining, 377, 460
Streaming current, 133
Submersible pumps, 268, 314
Substrate utilization Rate, 429
Sulphur dioxide, 55, 179, 528
Sump pump, 317
Super-chlorination, 219
Superimposed loads, 358
Surface
  overflow rate, 137, 144
  wash system, 156, 158
Suspended growth systems, 421, 460
System classification, 255
System head, 272–274
System head curve, 273
System of units, 11

T

Tanneries, 526
Tapered aeration, 435
Taste and odour, 111, 171, 217
Thickening, 407, 424, 493
THM, 171, 218, 226
Thrust, 244, 251
Thrust block, 245, 251
Timbering, 350, 351
Time of concentration, 336–342
Ton container, 184
Total demand, 97, 101, 104
Total solids, 111, 123, 381
Toxins, 385, 439, 521
Treatment
  facility, 71, 120, 234, 381
  processes, 47, 112, 381
  stages, 110, 379
Trench conditions, 358–364
Trickling filters, 459–464
Trunk sewer, 297
Tubercles, 224
Tuberculation, 223, 240, 286
Tube settlers, 142
Tunnel conditions, 358
Turbidity tests, 132
Turbine pumps, 267
Turbulence, 31, 141, 441
Twin tower, 80, 81
Two-Stage
  digestion, 501
  excess lime, 196
  trickling filter, 462

U

Ultrafiltration, 221, 514, 515
Underdrain system, 157–161, 460

Uniformity coefficient, 155, 157, 460
Unit filter run volume, 163
Upflow filters, 483
Urban catchments, 335, 341
UV Light, 171

V

Vacuum filtration, 505
Valve
    air relief, 246
    altitude, 247
    butterfly, 247
    check, 247, 249
    diaphragm, 246
    exercising, 248
    foot, 247
    gate, 246
    globe, 246
    plug, 247
    PRV, 248
    PSV, 247
    sluice, 246
    vacuum relief, 247
Variation in demand, 106, 232
Variations of ASP, 435
Velocity pumps, 267
Viscosity, 112
Visual monitoring, 452
Vitrified Clay Pipe (VCP), 347
Volatile solids
    digestion, 498
    mixed liquor, 426
    reduction, 502
    sludge, 496
    test, 70, 384
Volume of screenings, 398
Volume of sludge, 125, 413
Volumetric BOD loading, 423

W

Wash-water troughs, 156, 158
Wastewater
    characteristics, 440
    collection, 297
    effluent, 69, 378, 522
    flows, 300, 390, 411
    reclamation, 78

Water
    analysis, 203
    audit, 291
    black, 382
    color, 452
    conservation, 513, 524
    demand, 97
    distribution, 231
    glass, 227
    grey, 382
    meters, 232, 250
    quality, 290
    reuse, 522
    softening, 191
    stabilization, 223
    standards, 251, 524
    transmission, 81
    treatment capacity, 198
Weighing scale, 184
Weighting agents, 123
Weir loading, 146
Weirs, 402
Well
    abandonment, 90
    artesian, 84
    dry, 313, 314
    efficiency, 85
    hydraulics, 84
    intake, 85
    types, 83
    water table, 87, 88
    wet, 315
Wet barrel, 250
Wet well lift stations, 314
Windrow, 509
Winter storage, 450

Z

Zeolite, 212
Zeta potential, 133
Zone of
    active decomposition, 372
    degradation, 371
    recovery, 372
Zoogleal film, 460

For Product Safety Concerns and Information please contact our EU
representative GPSR@taylorandfrancis.com
Taylor & Francis Verlag GmbH, Kaufingerstraße 24, 80331 München, Germany

www.ingramcontent.com/pod-product-compliance
Ingram Content Group UK Ltd.
Pitfield, Milton Keynes, MK11 3LW, UK
UKHW021114180425
457613UK00005B/85